W9-AVE-738

Annual Review of

INFORMATION SCIENCE AND TECHNOLOGY

Annual Review of
INFORMATION SCIENCE AND TECHNOLOGY

Volume 40 • 2006
Blaise Cronin, Editor

Published on behalf of the
American Society for Information Science and Technology
by Information Today, Inc.

Medford, New Jersey

ISBN: 1-57387-242-3
ISSN: 0066-4200
CODEN: ARISBC
LC No. 66-25096

Published and distributed by
Information Today, Inc.
143 Old Marlton Pike
Medford, NJ 08055-8750

On behalf of

The American Society for Information Science and Technology
1320 Fenwick Lane, Suite 510
Silver Spring, MD 20910-3602, U.S.A.

Information Today, Inc. Staff
President and CEO: Thomas H. Hogan, Sr.
Editor-in-Chief and Publisher: John B. Bryans
Managing Editor: Amy M. Reeve
Proofreader: Pat Hadley-Miller
VP Graphics and Production:
  M. Heide Dengler
Cover Designer: Victoria Stover
Book Designer: Kara Mia Jalkowski

ARIST Staff
Editor: Blaise Cronin
Associate Editor: Debora Shaw
Copy Editors: Dorothy Pike,
  Thomas Dousa
Indexer: Amy Novick

Contents

SECTION I
Information and Society

SECTION II
Technologies and Systems

SECTION III
Information Needs and Use

SECTION IV
Theoretical Perspectives

Introduction

Blaise Cronin

ARIST is 40, a suitable moment to pause and reflect. Volume 1 appeared in 1966 and was published by John Wiley & Sons in conjunction with the American Documentation Institute (ADI), the forerunner of the American Society for Information Science and Technology (ASIST). The inaugural volume contained 389 pages and cost $12.50. Its founding editor, Carlos A. Cuadra, was a System Development Corporation (SDC) employee and graduate of the University of California at Berkeley with a Ph.D. in clinical psychology. The idea for a series of comprehensive reviews in information science and technology came to him in the early 1960s and found enthusiastic support from ADI notables such as Pauline Atherton Cochrane, Charles Bourne, Robert Hayes, and Donald Swanson

Cuadra launched *ARIST* with the financial support of both SDC and the National Science Foundation (NSF) but not without misgivings within the ranks of the ADI itself (see Bjørner, 2003 for some historical background). The NSF, however, had no doubts: in her foreword to Volume 1, Helen Brownson details the organization's commitment to both the editor and the idea of an annual review for the nascent field of information science and technology. Their early optimism was not misplaced. Forty volumes, several publishers, and three editors later, *ARIST* sails on, fulfilling its original remit and continuing to attract plaudits. Since this is an anniversary volume, a little blowing of one's own trumpet may be permitted: According to the most recent ISI data I have seen, *ARIST* ranks first in terms of its impact factor when compared with more than 50 other serials in its subject group, including its stable mate, the *Journal of the American Society for Information Science and Technology*. Dr. Cuadra and his redoubtable successor, Professor Martha Williams, have much of which to be proud; reputations are not built overnight.

Cuadra's expansive introduction makes for interesting reading. Parts of it could have been written yesterday; indeed, his scene-setting remarks brought to mind the title of Bruno Latour's (1987) book, *We Have Never Been Modern*. There is, as Cuadra (1966, p. 1) notes, "very little agreement about the boundaries of Information Science and Technology, or about its parentage, its essential nature, or its future." This, after all, is a field that "draws on fragments and fringes of a number of sciences, technologies, disciplines, arts, and practices." Still, Cuadra (1966, p. 2) is a pragmatist at heart and rather than fret about definitions and essentialism, he focuses on what can be done to encourage "intercommunication in the field," specifically the role *ARIST* can play in creating an interdiscipline (although he doesn't use that term) and promoting what Gernot Wersig (1992, p. 214) later called "interconceptual work." Some of the chapter titles in the first volume (e.g., *Automated Language Processing, Selected Hardware Developments, Content Analysis, Representation and Control, National Information Issues and Trends*) could have been stripped out of the current *ARIST,* suggesting not only that Dr. Cuadra was a prescient and knowledgeable editor but also that some themes and topics are ever-present in the information science canon, even if, on occasion, the language of the first *Review* (e.g., *Man-Machine Communication*) signals that we are in pre-PC (political correctness) times.

Conversely, some of the chapters making up Volume 40 could have featured in Volume 1 (e.g., *Information Seeking, Semantic Relationships, Information History*). Even the chapter on social epistemology could, conceivably, have seen the light of day in 1966, as Jesse Shera (1961) had published on that very subject five years earlier. But some things do change, and Volume 40 addresses a number of topics that even the most farsighted founding editor could not have anticipated (e.g., *The Geographies of the Internet, Open Access*). Constancy and change: Such is *ARIST*.

One of the questions posed by Cuadra is why anyone would willingly undertake to write a comprehensive annual review chapter, given the enormous amount of work involved. This, as you may imagine, is a question often asked by the present editor, but, *mirabile dictu*, there are still enough individuals motivated by enlightened altruism to keep *ARIST* on course. Now, whether all of these selfless souls meet the requirements for the ideal author as specified by Cuadra is a question for you, the reader, to answer. Permit me to quote in full Cuadra's (1966, p. 8) five desiderata: "(1) he must have a strong grasp of the basic issues in his field and must be able to understand and express them in their historical perspective; (2) he must have an established habit of keeping informed by reading reports and published literature and by making effective use of his contacts in the 'invisible college'; (3) he must be able to write lucid, incisive prose and must be willing and able to make objective value judgments—in public—about the merit and implications of given lines of reported work, research, and experience; (4) he must have, in addition to this technical and literary talent, sufficient prestige in the field to invite the reader's respectful attention to his contribution; and (5) he must be willing to do an immense amount of sifting, reading, and evaluation on an extremely tight schedule."

I certainly could not have put it better myself—this is a near perfect pen portrait of the ideal *ARIST* author. I hope that 40 years on we still come respectably close to the benchmark set by Carlos Cuadra. Given the increasing pressure on authors to publish frequently and communicate widely in order to maintain their professional salience (Cronin, 2005), we are mightily grateful to those who commit the time and energy needed to craft a good *ARIST* chapter. At the risk of sounding like a fogey, there is something to be said for deliberative writing and deferred gratification. Paradoxically, in an economy of attention (Simon, 1971), less may well be more. Perhaps what academia needs now is a Slow Writing movement akin to the Slow Food, Slow Cities movement (see http://www.slowfood.com).

References

Bjørner, S. (2003). *Online before the Internet, Part 3. Early pioneers tell their stories: Carlos Cuadra*. Retrieved March 1, 2005, from http://www.infotoday.com/searcher/oct03/CuadraWeb.shtml

Cronin, B. (2005). *The hand of science: Academic writing and its rewards*. Lanham, MD: Scarecrow Press.

Cuadra, C. A. (1966). Introduction to the ADI Annual Review. *Annual Review of Information Science and Technology, 1*, 1–14.

Latour, B. (1993). *We have never been modern* (C. Porter, Trans.). Cambridge, MA: Harvard University Press.

Shera, J. (1961). Social epistemology, general semantics and librarianship. *Wilson Library Bulletin, 35*, 767–770.

Simon, H. (1971). Designing organizations for an information-rich world. In M. Greenberger (Ed.), *Computers, communications and the public interest* (pp. 37–72). Baltimore, MD: Johns Hopkins University Press.

Wersig, G. (1992). Information science and theory: A weaver bird's perspective. In P. Vakkari & B. Cronin (Eds.), *Conceptions of library and information science: Historical, empirical and theoretical perspectives* (pp. 201–217). London: Taylor Graham.

Acknowledgments

Many individuals are involved in the production of *ARIST*. I should like to acknowledge the contributions of both our Advisory Board members and the outside reviewers. Their names are listed in the pages that follow. The great bulk of the copyediting and bibliographic checking was carried out by Thomas Dousa and Dorothy Pike, for which I am most grateful. Amy Novick produced the thorough index to this volume. As always, Debora Shaw did what a good associate editor is supposed to do.

Acknowledgments

ARIST Advisory Board

Judit Bar-Ilan
Hebrew University of Jerusalem, Israel

Micheline Beaulieu
University of Sheffield, UK

Nicholas J. Belkin
Rutgers University, New Brunswick, USA

David C. Blair
University of Michigan, Ann Arbor, USA

Christine L. Borgman
University of California at Los Angeles, USA

Terrence A. Brooks
University of Washington, Seattle, USA

Elisabeth Davenport
Napier University, Edinburgh, UK

Susan Dumais
Microsoft Research, Redmond, USA

Abby Goodrum
Syracuse University, USA

E. Glynn Harmon
University of Texas at Austin, USA

Leah A. Lievrouw
University of California at Los Angeles, USA

Katherine W. McCain
Drexel University, Philadelphia, USA

Charles Oppenheim
Loughborough University, UK

Chapter Reviewers

Judit Bar-Ilan
Nicholas Belkin
Christine Borgman
Terence Brooks
Michael Buckland
Donald Case
Carol Choksy
Ian Cornelius
Christina Courtright
Elisabeth Davenport
Ron Day
Lorcan Dempsey
Susan Dumais
David Ellis
Karen Fisher
Nigel Ford
Jonathan Furner
David Goodman
Abby Goodrum
Noriko Hara

Glynn Harmon
Birger Hjørland
Gert-Jan Hospers
Elin Jacob
Kathryn La Barre
Brian Lavoie
Leah Lievrouw
Charles Oppenheim
Christopher Peebles
Uta Priss
Edie Rasmussen
Boyd Rayward
Yvonne Rogers
Pnina Shachaf
Dagobert Soergel
Lee Strickland
Pertti Vakkari
Kiduk Yang

Contributors

Alistair Black is Professor of Library and Information History in the School of Information Management, Leeds Metropolitan University. He is author of *A New History of the English Public Library: Social and Intellectual Contexts 1850–1914* (Leicester University Press, 1996), *Understanding Community Librarianship* (Avebury, 1997), and *The Public Library in Britain 1914–2000* (The British Library, 2000). He is editor of Volume 3 of the forthcoming *Cambridge History of Libraries in Britain and Ireland*. Since 2003 he has served as chair of the International Federation of Library Associations and Institutions (IFLA) Section on Library History.

Sandra Braman is a Professor in the Department of Communication, University of Wisconsin-Milwaukee. She has been studying the macrolevel effects of the use of new information technologies and their policy implications since the mid-1980s. Her current work includes *Change of State: An Introduction to Information Policy* (in press, MIT Press) and the edited volumes *Communication Researchers and Policy-makers* (MIT Press, 2003), *The Emergent Global Information Policy Regime* (Palgrave Macmillan, 2004) and *The Meta-technologies of Information: Biotechnology and Communication* (Lawrence Erlbaum Associates, 2004). With Ford Foundation and Rockefeller Foundation support, Braman has been working on problems associated with the effort to bring the research and communication policy communities more closely together. She has published over four-dozen scholarly journal articles, book chapters, and books. Braman earned her PhD from the University of Minnesota in 1988 and previously served as Reese Phifer Professor at the University of Alabama, Henry Rutgers Research Fellow at Rutgers University, Research Assistant Professor at the University of Illinois at Urbana-Champaign, and the Silha Fellow of Media Law and Ethics at the University of Minnesota.

Donald O. Case holds a PhD in Communication Research from Stanford University (1984) and MLS from Syracuse University (1977). He has held the post of Professor in the University of Kentucky College of Communication and Information Studies since 1994; between 1983 and 1994 Case was a faculty member at the University of California, Los Angeles. During 1989 he received a Fulbright Fellowship to lecture at the Universidade Nova de Lisboa, Portugal. His research interests include information behavior, social informatics, and information policy. Case's book, *Looking for Information: A Survey of Research on Information Seeking, Needs, and Behavior* (2002) was given the "Best Book of the Year" Award by the American Society for Information Science and Technology.

Hsinchun Chen is McClelland Professor of Management Information Systems at the University of Arizona. He is author of nine books and more than 200 articles covering intelligence analysis, data/text/Web mining, digital libraries, knowledge management, medical informatics, and Web computing. He serves on the editorial board of *ACM Transactions on Information Systems*; *IEEE Transactions on Intelligent Transportation Systems*; *IEEE Transactions on System, Man, and Cybernetics*; *Journal of the American Society for Information Science and Technology*; and *Decision Support Systems*. Dr. Chen is a Scientific Counselor/Advisor to the National Library of Medicine (USA), Academia Sinica (Taiwan), and National Library of China. He is founding director of the Artificial Intelligence and Hoffman E-Commerce Labs. Dr. Chen is conference chair of the ACM/IEEE Joint Conference on Digital Libraries (JCDL). He is also a pioneer in intelligence and security informatics research. He has served as conference co-chair of the IEEE International Conferences on Intelligence and Security Informatics (ISI) in 2003, 2004, and 2005.

Chun Wei Choo is Professor in the Faculty of Information Studies at the University of Toronto and Visiting Professor at the Department of Business Studies, Faculty of Economics and Econometrics, University of Amsterdam. His recent books include: *The Knowing Organization* (Oxford University Press, 2nd ed., 2005); *The Strategic Management of Intellectual Capital and Organizational Knowledge* (co-edited with Nick Bontis, Oxford University Press, 2002); *Information Management for the Intelligent Organization* (Information Today, 3rd ed., 2002); and *Web Work: Information Seeking and Knowledge Work on the WWW* (co-authored with Brian Detlor and Don Turnbull, Kluwer, 2002).

Mark E. Dawes is a nontraditional graduate student at the University of Cincinnati, studying ethnomethodology and conversation analysis. He is also a retired executive with more than 20 years experience in managing entrepreneurial organizations. Mark's research interest is in applying a workplace studies perspective to the examination of corporate culture and sustained product innovation in organizations. He is currently working on a long-term analysis of corporate cultural influences on sustained product innovation in a manufacturing company.

M. Carl Drott is an Associate Professor in the College of Information Science and Technology at Drexel University. He holds a PhD in Industrial and Operations Engineering from the University of Michigan. His areas of special competence include information systems, computer programming, computer applications, and business information systems. His work has been published in *Information Processing & Management, Journal of the American Society for Information Science,* and *Journal of Documentation.*

Don Fallis is an Associate Professor of Information Resources and Library Science at the University of Arizona. The main focus of his research is social epistemology and its applications to library and information science. He received a PhD in Philosophy from the University of California, Irvine. His articles have appeared in such journals as the *American Mathematical Monthly, British Journal for the Philosophy of Science, Journal of the American Society for Information Science and Technology, Journal of Philosophy, Library Quarterly,* and *Philosophical Studies.* He also edited an issue of *Social Epistemology* on "Social Epistemology and Information Science."

Jonathan Foster is a Lecturer in Information Management at the Department of Information Studies, University of Sheffield, U.K. He has higher degrees in information systems and in education. He is currently Sub-Dean for Undergraduate Affairs, Faculty of Social Sciences, University of Sheffield.

Angela Cora Garcia is an Associate Professor of Sociology at the University of Cincinnati. Her research is based on conversation analysis, ethnomethodology, workplace studies, and other qualitative approaches. She has conducted studies of mediation hearings, computer-mediated communication, emergency phone calls to the police, moral reasoning, and gender differences in language and communication.

Stephan F. Groschwitz is a graduate student of sociology at the University of Cincinnati. He was born and grew up in East Germany and went to college at the Technical University of Dresden where he majored in political science with a minor in sociology and economics. He transferred to the Humboldt University in Berlin to major in an integrated social sciences program, before moving to Cincinnati. His interests lie in the sociology of knowledge, especially at the intersection of politics and culture. Currently he is working on his master's thesis on the representation of racialized masculinity in mass media accounts of capital punishment.

Donna K. Harman graduated from Cornell University with a degree in electrical engineering and has been involved with research in new search engine techniques for many years. She retired from the National

Institute of Standards and Technology (NIST) in 2005 after leading a group working in the area of natural language access to full text, both in search and browsing modes. In 1992 she started the Text REtrieval Conference (TREC), an ongoing forum that brings together researchers from industry and academia to test their search engines against common corpora of more than one million documents, with appropriate topics and relevance judgments. She received the 1999 Strix Award from the U.K. Institute of Information Scientists for this effort. Since 2000 she has worked with Paul Over (also at NIST) and the text summarization community to form a new effort, the Document Understanding Conference (DUC), to evaluate text summarization.

Christopher S. G. Khoo is an Associate Professor in the School of Communication & Information, Nanyang Technological University (NTU), where he is the program director for the MSc in Information Studies program. He obtained his PhD from Syracuse University in 1996 and his MS from the University of Illinois in 1987. At NTU, he teaches data mining, knowledge classification, information architecture, Web-based information systems, and research methods. Prior to joining NTU, he worked for several years at the National University of Singapore Library. His main research interests are in text mining, information extraction, information retrieval, and intelligent interfaces for information systems. He was the editor of the *Singapore Journal of Library & Information Management* from 1997 to 2002.

Mary Lou Kohne is a doctoral candidate at the University of Cincinnati's College of Business, Department of Marketing. Her research interests include innovation, ethics, and consumer behavior. Prior to pursuing her doctorate, she received her MBA with concentrations in Marketing and Quantitative Analysis. She has taught master's level and undergraduate business courses. She has served on top management teams within both high technology and financial services firms. She has also worked at major consumer products firms, and served as a consultant to global companies and organizations on marketing strategy, research, innovation, leadership, training, new product development, and growth through acquisitions or strategic alliances. Mary Lou is a member of the American Marketing Association, Product Development Management Association, and the Society for Consumer Psychology.

Felicia M. Miller is a PhD candidate in the Department of Marketing at the University of Cincinnati. Her research interests include brand-consumer relationships and marketing communication.

Anu MacIntosh-Murray is a Postdoctoral Fellow (funded by the Canadian Health Services Research Foundation) with the Knowledge Translation Program, Department of Health Policy, Management, and Evaluation, at the University of Toronto. She obtained her PhD from the

Faculty of Information Studies, University of Toronto. Her research interests include the use of information to make improvements in patient safety in healthcare organizations and genres of knowledge exchange used in academia and by clinical improvement teams.

Jin-Cheon Na is an Assistant Professor in the School of Communication and Information at Nanyang Technological University, Singapore. Previously, he worked for more than six years at a Korean military research institute, the Agency for Defense Development, as a senior researcher. He obtained his PhD in 2001 from the Department of Computer Science at Texas A&M University. As part of his dissertation, he developed caT (context-aware Trellis), a context-aware hypertext model with associated tools. He is currently investigating effective ways of information processing using a variety of techniques from machine learning, data mining, information retrieval, and natural language processing.

Uta Priss is a Lecturer in the School of Computing, Napier University, Edinburgh, Scotland. Previously she was an Assistant Professor at the School of Library and Information Science, Indiana University. She holds a doctoral degree from the University of Darmstadt, Germany. Her research focuses on mathematical theories of conceptual and semiotic structures. She has been interested in Formal Concept Analysis (FCA) ever since conducting her doctoral research on Relational Concept Analysis under the supervision of the founder of FCA, Dr. Rudolf Wille, 10 years ago. She has published numerous papers on FCA, mostly focusing on its applications to lexical databases, thesauri, and classification systems. She chaired the International Conference on Conceptual Structures in 2002.

Ellen M. Voorhees is a Group Leader in the Information Access Division of the U.S. National Institute of Standards and Technology (NIST). Her primary responsibility is management of the Text REtrieval Conference (TREC) project. She received a BSc in computer science from the Pennsylvania State University and MSc and PhD degrees in computer science from Cornell University. Her research interests include information retrieval and natural language processing, especially the development of appropriate evaluation schemes to measure system effectiveness.

Jennifer Xu is a doctoral candidate in Management Information Systems at the University of Arizona, where she is a member of the Artificial Intelligence Lab. Her research interests include knowledge management, social network analysis, information retrieval, human–computer interaction, and information visualization. She received her MS in Computer Science and MA in Economics from the University of Mississippi in 1999 and 2000, respectively.

Matthew Zook is an Assistant Professor of Economic Geography at the University of Kentucky. His research focuses on the geography of cyberspace and the role of technology, finance, and knowledge in regional development. His most recent book, *The Geography of the Internet Industry: Venture Capital, Dot-coms and Local Knowledge* (Blackwell, 2005) examines the localized factors behind the dot-com boom and bust. Future work (funded by the NSF) will focus on the forms, processes, and geographies of e-commerce adoption in the U.S. He holds graduate degrees from the University of California, Berkeley and Cornell University.

About the Editor

Blaise Cronin is the Rudy Professor of Information Science at Indiana University, Bloomington, where he has been Dean of the School of Library and Information Science for 14 years. From 1985 to 1991 he held the Chair of Information Science and was Head of the Department of Information Science at the University of Strathclyde Business School in Glasgow. He has also held visiting professorships at the Manchester Metropolitan University and Napier University, Edinburgh. Professor Cronin is the author of numerous research articles, monographs, technical reports, conference papers, and other publications. Much of his research focuses on collaboration in science, scholarly communication, citation analysis, the academic reward system, and cybermetrics—the intersection of information science and social studies of science. He has also published extensively on topics such as information warfare, information and knowledge management, competitive analysis, and strategic intelligence. Professor Cronin sits on a number of editorial boards, including *Journal of the American Society for Information Science and Technology*, *Scientometrics*, *Cybermetrics*, and *International Journal of Information Management*.

Professor Cronin has extensive international experience, having taught, conducted research, or consulted in more than 30 countries: Clients have included the World Bank, NATO, Asian Development Bank, UNESCO, U.S. Department of Justice, Brazilian Ministry of Science and Technology, European Commission, British Council, Her Majesty's Treasury, Hewlett-Packard Ltd., British Library, Commonwealth Agricultural Bureaux, Chemical Abstracts Service, and Association for Information Management. He has been a keynote or invited speaker at scores of conferences, nationally and internationally. Professor Cronin was a founding director of Crossaig, an electronic publishing start-up in Scotland, which was acquired in 1992 by ISI (Institute for Scientific Information) in Philadelphia. He was educated at Trinity College Dublin (MA) and the Queen's University of Belfast (PhD, DSSc). In 1997, he was awarded the degree Doctor of Letters (D.Litt., *honoris causa*) by Queen Margaret University College, Edinburgh, for his scholarly contributions to information science.

About the Associate Editor

Debora Shaw is a Professor at Indiana University, Bloomington and also Executive Associate Dean of the School of Library and Information Science. Her research focuses on information organization and information seeking and use. Her work has been published in the *Journal of the American Society for Information Science and Technology*, the *Journal of Documentation*, *Online Review*, and *First Monday*, among others. She serves on the editorial board of the *Journal of Educational Resources in Computing*.

Dr. Shaw served as President of the American Society for Information Science & Technology (1997), and has also served on the Society's Board of Directors. She has been affiliated with *ARIST* as both a chapter author and as indexer over the past 18 years. Dr. Shaw received bachelor's and master's degrees from the University of Michigan and her PhD from Indiana University. She was on the faculty at the University of Illinois before joining Indiana University.

Information
and Society

CHAPTER 1

The Micro- and Macroeconomics of Information

Sandra Braman
University of Wisconsin–Milwaukee

Introduction

Economic products and processes have always involved information, but technological innovation has changed society in such a way that information is now at the center of economic thinking and practice. At the microeconomic level, several strands of work dealing with problems that range from decision making under uncertainty to the nature of risk are now collectively referred to as the "economics of information," with the phrase "the information economy" often used to refer to informational issues at the macroeconomic level. The two interact: As consciousness of the evolution of the information economy has grown, the amount of pertinent microeconomic work has increased; and as new ways of thinking about microeconomic informational issues appear, they in turn inspire new macroeconomic questions. Both micro- and macroeconomic thought deal with the same problem—how to understand information creation, processing, flows, and use from an economic perspective. Accordingly, this chapter uses the phrase "the economics of information" as an umbrella term to refer to issues raised by information at every level of analysis.

The question of how to define information itself has also been problematic. As Porat (1977) noted, there is not even a single definition that embraces all aspects of the information industries as a primary sector. Babe (2004) has suggested that it may be this very definitional difficulty that has led to so much confusion for economists working in this area. Information changes over time, depending on context and relationships with other information, just as it has multiple identities at any given point in time, depending upon the perspective of the entity accessing and/or using it. When information is understood as a commodity, it is a commodity that is heterogeneous (Allen, 1990; Porat, 1977). Alternatively, however, as Lamberton (1998a) has suggested, a taxonomy of definitions of information can be used for purposes of economic analysis. The approach used in this chapter does just that by defining

3

information from a theoretically pluralist perspective. As fully articulated elsewhere (Braman, 1989b), this approach acknowledges the richness of the concept of information in both meaning and use and suggests a fourfold typology of definitions—information as a resource, as a commodity, as perception of pattern, and as a constitutive force in society—each of which has validity and utility for particular purposes. All of these ways of defining information appear in the works cited here and it would be interesting, although beyond the scope of this chapter, to organize economic thought about information according to this definitional typology.

In addition to linking together micro- and macroeconomic analysis within a single frame, the approach to the economics of information taken here brings together the poles of a second axis. Certainly the greatest amount of work falling into this area has been undertaken by economists whose work, in the aggregate, yields a picture of the economics of information describable as "narrowly defined." But neither the concept of information nor its empirical manifestations in economic and other dimensions of social life form the bailiwick of any one academic discipline. Issues raised by the economic aspects of information creation, processing, flows, and use have also been the subject of discussion among political scientists, historians, organizational sociologists, information scientists, psychologists, and communication scholars explicitly since at least the late 1940s and implicitly for much longer than that. Indeed, one of the most profound consequences of the effort to grapple with informational issues from an economic perspective has been an intensification of interest in ways in which economic processes are intimately intertwined with other cultural and social processes. The economics of information as understood by noneconomists as well as economists can be said to constitute the field "broadly defined" and it is that definition of the field that is used here.

Despite the fact that the economics of information takes up only a few pages in general histories of economic theory (e.g., Blaug, 1997; Landreth & Colander, 1994), the literature on it is now vast. Even single topics within the domain, such as the role of information in auctions (Klemperer, 1999), have received such extensive treatment that they are the subject of full literature reviews in their own right (Riley, 2001). A search in Econlit in July of 2004 found 237 entries for "economics of information," 245 entries for "information economy," and 57,539 entries for the word "information" in article titles in this database devoted exclusively to work in economics. There have been two efforts to bring together all pertinent work on a topic into a single database, one by Hal Varian at the University of California-Berkeley on the economics of information per se (http://www.sims.berkeley.edu/resources/infoecon) and another by Jeffrey MacKie-Mason at the University of Michigan on the economics of information as it applies to telecommunications policy (http://china.si.umich.edu/telecom/telecom-info.html). However, despite all the knowledge, skills, and resources these two leading scholars brought to their respective projects, both had to give up after just a few years because of the explosive growth of the field: Varian started work in 1994 and stopped adding to his Web site in 1998; MacKie-Mason also started in 1994 and lasted until 2003 before giving up. MacKie-Mason reports that, at its peak, his Web site included links to over 8,000 resources dealing with the topic that were available on the Internet alone.

The *Annual Review of Information Science and Technology (ARIST)* has regularly covered the subject. In what looks in retrospect like a publishing coup, A. Michael Spence, who won a Nobel prize for his work in the economics of information, published a piece for *ARIST* on the subject in 1974. Hindle and Raper (1976) and Lamberton (1984a) subsequently provided additional surveys on the same topic. Numerous other *ARIST* chapters have dealt with specific issues in the economics of information, including the value of information for managers (Bergeron, 1996; Bergeron & Hiller, 2002; Katzer & Fletcher, 1992; Koenig, 1990; Lytle, 1986; Mac Morrow, 2001; Mick, 1979), pricing of information products (Arnold, 1990; Griffiths, 1982; Spigai, 1991; Webber, 1998), information as property or capital (Lipinski, 1998; Snyder & Pierce, 2002), economic issues facing the library sector in particular (Tucci, 1988), and the nature of the information economy in specific societies (Boon, Britz, & Harmse, 1994). Most recently, Davenport and Snyder (2004) examined the ways in which information and communication technologies can be used to develop and manage a particular form of capital, social capital.

In numerous pieces published across several decades and cited throughout this chapter, Australian economist Donald M. Lamberton has documented the emergence of the economics of information in its narrow definition. Canadian economist and communications scholar Robert E. Babe, in a single-volume masterwork (Babe, 1995) and an edited collection (Babe, 1993), has produced the most thorough critique of neoclassical treatments of information creation, processing, flows, and use. He introduced the field as broadly defined through pointers to other strands of social thought that could, and should, valuably enrich how we think about information from an economics perspective. This chapter takes a third step in this conceptual history, complementary to the work of these authors, by contextualizing the subject in its narrower definition vis-à-vis the field as more broadly defined.

To do so, Section I briefly reviews the history of economic thinking about informational matters at both the micro- and macrolevels. Section II looks at ways in which taking information into account affects how we think about the economy itself. An introduction to the multiple problems information presents to neoclassical microeconomics and analysis of the three very different ways of conceptualizing the information economy that have appeared in recent decades are available elsewhere (Braman, 1996). Section III examines various ways of thinking about information as a specific sector of economic activity. Section IV explores the impact of incorporating information into economic theory on a number of key economic concepts. The concluding section considers some of the implications of these developments on policy and on our research agendas.

Because the literature is so large, these tasks are accomplished through reliance on seminal and/or exemplary items that in many cases mark the initiation of, or particular turning points in, important streams of work.

Historical Overview

The story of the economics of information is the history of struggles over how to understand the increasing information intensity of our social and economic lives. Neoclassical economic thought originally avoided informational problems; however, the resulting inconsistencies, paradoxes, and failures led to a new stage of theoretical development. The transformation in economics once it began to deal directly with informational issues has been dramatic. The topic began the 20th century "in a slum dwelling in the town of economics" (Stigler, 1961, p. 171), but by the beginning of the 21st it had been the subject of research awarded a number of Nobel prizes: Of the 34 Nobels given to economists since a prize in that discipline was inaugurated in 1969, seven have been awarded to individuals whose work focused on problems in the economics of information and another three to thinkers whose work has been key to this subject (see Table 1.1).

There was a period during which the history of economic thought largely disappeared from university curricula (Blaug, 1978), but the role of information in the economy is, of course, ancient. Indeed, the very earliest written records document economic activity (Nissen, Damerow, & Englund, 1994). Historians of economic thought typically place the origins of modern economic ideas in the 18th century, with Adam Smith (Blaug, 1996; Landreth & Colander, 1994). Basic ideas about the nature of the economy developed during this period, establishing the framework within which information has come to be understood. By the 19th century, many thinkers identified problems we now understand as information-based even though other language was used to describe those issues and the implications of their insights were not always recognized. Over the course of the 20th century, economists operating from a neoclassical perspective began to acknowledge that differences in the kinds and amounts of information available, and in who has access to that information, are so important that they could not be ignored. This change in attitude, however, was slow. As Babe (1996) notes, despite the appearance of key publications on the topic as early as the 1920s by thinkers such as Coase, Knight, and others whose work is discussed later, full attention was not accorded to the economics of information until the 1960s. Not coincidentally, this was the period during which we also began to use the phrase "the information society" to describe the ways in which society was being transformed as a result of new information technologies (Braman, 1993). Indeed, each development in the history of the economics of information has followed, always with a time lag, a stage of social change stimulated by technological innovation.

Table 1.1 Nobel Prizes for Work Central to the Economics of Information*

Year	Economist	Prize-Winning Achievement
1972	Kenneth Arrow**	pioneering contributions to general economic equilibrium theory and welfare theory
1974	Gunnar Myrdal Friedrich von Hayek	penetrating analysis of the interdependence of economic, social, and institutional phenomena
1978	Herbert Simon	pioneering research into the decision-making process within economic organizations
1982	George Stigler	seminal studies of industrial structures, functioning of markets, and causes and effects of public regulation
1987	Robert Solow	contributions to theory of economic growth
1991	Ronald Coase	discovery and clarification of the significance of transaction costs and property rights for the institutional structure and functioning of the economy
1992	Gary Becker	extended the domain of microeconomic analysis to a wide range of human behavior and interaction, including nonmarket behavior
1996	James A. Mirrlees William Vickrey	fundamental contributions to the economic theory of incentives under asymmetric information
2001	George A. Akerlof A. Michael Spence Joseph E. Stiglitz	analyses of markets with asymmetric information
2002	Daniel Kahneman	integrated insights from psychological research into economic science, especially concerning human judgment and decision making under uncertainty

* The work of all of these individuals is discussed in this chapter. In some cases the work pertinent to the economics of information is at the center of the research for which the award was given; in other cases the pertinent work involved a subset or consequence of the work highlighted by the Nobel committee. Seminal contributions to the economics of information by individuals such as Kenneth Arrow and Ronald Coase appeared several decades before the Nobel prize was awarded.
** Arrow shared this Nobel prize with John R. Hicks, but Hicks' work has been much less important to the economics of information and so he was not included in this chart.

The Nineteenth Century

Nineteenth-century developments of importance to information economics in the 21st century included both attention to the way in which technological development influences economic relations and interest in the economic effects of the content of information flows. Karl Marx introduced the idea that communication itself was critical to the functioning of the economy because it affected the ways in which people thought about their roles within society. His emphasis on social power and the alienating effects of commoditization provided a basis for thinking in terms of "techno-social" (Webster, 1995) or "techno-economic" (Winter, 1988) paradigms as a frame for understanding why innovation in information and communication technologies so influences the economy.

Although some historians of the information society see themselves as post-Marxist (Bell, 1973; Castells, 1996; Webster, 1995; Webster & Robins, 1986), others suggest that Marx himself would be focusing on information, rather than capital, as the basic transformational mechanism of the economy, were he writing today (e.g., Stevenson, 1995).

Between the opening and the close of the 19th century, economic thinking became more abstract and mathematical, pursuing eternal economic truths. An important consequence of this was that the impact of power, human will, and change were neglected because they were not understood in terms that could be described mathematically. This process culminated in the thinking of Alfred Marshall, whose *Principles of Economics* (1890) became the foundation of the neoclassical economics that dominated theory, policy, and practice throughout the 20th century. Marshall idealized a market in which information issues do not arise because everyone has the same information and this information is perfect. In the world portrayed by Marshall, neither the persuasive effects of the content of communication flows nor the relationships such flows build are relevant unless they are explicitly incorporated into individual preferences, for individuals base their decisions according to what will maximize economic "utility." Excluding problems raised by imperfect knowledge—issues we now refer to as problems involving information—also helped to define disciplinary boundaries as the modern university structure, with its departmental divisions, was established. Information and its flows were banished from economics, but under the label of "communication" became the core concept for the field of sociology (Peters, 1988) and the new fields of communication and information science. (The same thing happened with the concept of power, which was abandoned by those studying mainstream economics as too difficult, only to be taken up by the then-new discipline of political science.)

The Twentieth Century

By the opening of the 20th century, however, "information capitalism" (Ewen & Ewen, 1982) had come to dominate the economy. This was a form of capitalism based upon "the active attempts of coalitions within organizations to organize corporate production in such a way as to take advantages of changes in society and information technology" (Kling & Allen, 1993, p. 3). Other factors also encouraged economists to think about information, including the new availability of longitudinal economic data, shifts in production and distribution practices, the growth of large corporations, the appearance of new information industries, and professionalization of many forms of information work. A number of new information industries had appeared, including marketing associations, trade journals, statistical bureaus, advertising agencies, and consulting organizations (Lamberton, 1994). The question of whether pieces of information in the form of "facts" have economic value became a legal issue: In *INS v AP* (1918), the U.S. Supreme Court held that the news does have economic value created through the processes of its production, stimulating the further development of businesses based upon the growth and distribution of information. Knight's (1921) work on informational uncertainty and Coase's (1937) seminal insight into the way in which organizations reduce the "transaction costs" generated by the need to seek information are examples of important work in the economics of information from this period.

During World War II, communication and computing technologies began to converge, and operations research drew increased attention to the economic functions of the management of information. By the close of World War II, as Walter Wriston (1992), ex-chairman of Citicorp put it, the "information standard" had replaced the gold standard. In the 1960s, self-consciousness about the transformation from an industrial to an informational economy became part of discussion about the information society. Bell (1973) is often cited as the thinker who popularized the idea of the transition from an industrial to an information economy, but the notion appeared first in Japan (Ito, 1991). The concept quickly spawned its own sizable literature (see, e.g., Castells, 1996, 1997a, 1997b; Cronin, 1986; Engelbrecht, 1986; Gershuny & Miles, 1983). And it was clear that the phenomenon was global: Jussawalla and Cheah (1983), for example, provided an early report on the ways in which Singapore was using new information technologies to create a niche for itself in the global economy, Karunaratne (1984) looked at the evolution of the information economy in Australia, and a number of researchers examined factors stimulating growth of the information sector in developing countries (Katz, 1986; Kaynak, 1986; McKee, 1988; Saunders, 1983).

In 1966 the American Accounting Association's "Statement of Basic Accounting Theory" announced for the first time that accounting systems must be understood as an application of general theories of information to the problem of efficient economic operations (Crandall, 1969). Lamberton (1995) reports that the first bibliography of information economics, compiled by information scientist Harold Olsen, was offered by the University of Maryland in 1971. In the same year, Lamberton (1971) himself published the first anthology in information economics, identifying as key figures thinkers who are almost all still considered central to this history today: Knight, Hayek, Boulding, Marschak, Shackle, Machlup, Simon, and Arrow. The field grew quickly. By 1973, Hirshleifer (1973) was moved to offer a retrospective. In 1974 Spence announced that information had acquired a secure place in economic analysis. He noted several recurrent themes: the increase in the value of information when it is privately held, the impact of information monopolies, the role of incentives in stimulating or preventing information transmission, and the importance of self-validation to the effectiveness of signals. Although concern over the sheer complexity of dealing with the economics of information remained, in Spence's (1974, p. 60) view, "a large part of recent economic theory is either directly about or related to information problems."

In 1976 the American Economic Association (AEA) officially recognized information as an economic topic in its classification of ideas central to the discipline (Lamberton, 1984b). By the early 1980s, the first annotated bibliography had appeared (Middleton & Jussawalla, 1981), and economists were quarreling over just which topics to include in the category: Hepworth (1989), for example, identified four subfields; Machlup (1983) identified 17. In 1984 Lamberton listed eight literature reviews and nine conferences devoted to the economics of information and highlighted the fact that by that point, questions in every category of the AEA classification system were being addressed from an informational perspective. In that decade, the National Science Foundation began to fund research in the economics of information and the subject began to appear in policy debate in arenas as diverse as the Organisation for Economic Co-operation and Development (OECD) (in committees studying scientific and technical information) and the United Nations Economic, Scientific and Cultural Organization (UNESCO) (in debate over the New World Information Order). In 1985 Stiglitz, another Nobel Prize winner, announced his view that informational considerations had become central to the foundations of economic analysis.

Economics in Transition

The fact that the economics of information was receiving considerable attention by the 1980s did not mean that all economists agreed on the approach to be taken. Poles in the battle over how to think about information from an economic perspective were marked on one hand by Stigler (1983), whose 1982 Nobel lecture took the position that the economics of information had long been accommodated by standard economic theory, and on the other by Arrow (1974a), who insisted that the very foundations of traditional economics had to be challenged in order to adequately think through informational problems. Conceptualizations of the information economy also began to diverge (Braman, 1997/1999). Even among those convinced of the centrality of information to economic thought, there is concern over a lack of clarity in fundamental concepts. Spence's (1974) summary statement that the best that could be said about the economics of information was that its development was proceeding and Repo's (1987) complaint that research in the area was conceptually vague were repeated in Boyle's (1996, p. 45) description of pertinent theory as "so indeterminate that it frequently functions as a Rorschach blot for dominant social beliefs and the prejudices of the analyst."

Nor does the size of the literature mean that all problems have been solved. As Arrow (1996) put it, economists are nowhere near having even a "sort-of good theory" about how to balance the social need to use information and the rights of intellectual property holders. Often information is treated simultaneously as something that is free, complete, instantaneous, and universally available and as a commodity that is costly, partial, and deliberately restricted in its availability. Even where there is theory, it is often difficult to operationalize its insights in particular settings (Repo, 1987). Nevertheless, some relatively early works now appear to have attained the status of classics; not so long ago Lamberton (1998b) reported that he still teaches Arrow's (1974b) work of over two decades earlier on the role of information within organizations.

By now, the field of the economics of information is so well established that those responsible for hiring within universities complain that there are too many job candidates working in this area and that their research agendas all look too much alike (Sandler, 2001). The economics of information now appears not only in specialized texts but also in undergraduate textbooks on economics (e.g., Molho, 1997; Salvatore, 2003) and treatments of economic concepts and theories for the layman (e.g., Sandler, 2001; Wheelan, 2002). Classes in the economics of information are now commonly found not only in economics departments but elsewhere within the university; a study by Weech (1995) found that 43 percent of graduate information science programs offered courses in this area and that a number more were intending to add such courses. Even privacy (Brown & Gordon, 1980; Posner, 1984) and the First Amendment (Posner, 1986) have been analyzed in economic terms, albeit not in great depth (Brennan, 1990, 1994). In some cases, ideas now understood to be key to the economics of information were not necessarily received eagerly when first introduced. For example, as Riley (2001) notes, the seminal ideas of Akerlof (e.g., 1970) were perceived to be so unusual that two leading journals rejected his manuscript before it was published. Other ideas, however, were so simple and widely applicable that they were accepted relatively quickly and—as Sandler (2001) argues is the case with the concept of asymmetric information—even overdone.

The struggle to incorporate the concept and realities of information into economic theory is of course not the only trend in recent economic thought. Other related developments include an increase in interdisciplinarity, a blurring of the distinction between microeconomic and macroeconomic thought (Allen, 2000),[1] and growing sensitivity to economic processes from a cultural perspective and, thus, to significant differences in the nature of the economy from one society to another (Bird-David, 1997; Polanyi, 1957).

The Economy

Thinking about information has affected the way we understand the economy itself. The word "economy" derives from the Greek word *oikos*; in classical thought, it referred to the management of a household. Over the course of the 18th century, however—as part of the broad changes in the nature of European society, including the development of the modern nation-state—the household was unbundled. The individuals of which it had been comprised were transformed into independent economic agents. On the other hand, the society within which the household was embedded became the "economy." With the equation of society and the nation-state, it became possible to formulate new concepts, such as what we today call the "gross national product" and "national income accounts" recording how much of something—such as the amount of money paid for information or the number of pieces of information, however quantified—flows into and out of a country. Perceiving the economy in such terms also made the economy the subject of national policy, as it has been since the first stirrings of what we now describe as "policy science"—cameralism, or *Staatswissenschaften*, in early-19th-century Germany (Wagner, Weiss, Wittrock, & Wollman, 1991). In a world in which the economy is society-wide and each individual operates as an economic agent on his or her own, the focus of economic theory is not on relationships but on transactions involving goods.

As economists struggled to think about how to incorporate information into their theories, basic questions about the nature of the economy were raised, each the subject of a subsection below:

- What is the relationship between information and the factors of production?

- How does information contribute to economic stability and change?

- How does information influence the effects of economic activity?

Information and the Factors of Production

Not everything is deemed by economists to be of economic value. Eighteenth-century economic thought introduced the concept of "factors of production" to refer to the types of resources that have economic importance. Early theorists identified three factors of production: land, labor, and capital. Neither information, knowledge, nor tools or technologies of any kind were originally included in this taxonomy, although industrial technologies were already coming into use. As a result, there was no obvious place for information within this conceptual framework for thinking about economics. Boulding (1984, p. viii) describes the traditional factors of production—even including Alfred Marshall's (1890) additional category of "organization"—as equivalent to the medieval concepts of earth, air, fire, and water in terms of their explanatory utility and validity:

> [W]hat might be called a cookbook theory of production—
> that production comes from mixing together land, labor, and
> capital and out comes potatoes or automobiles—is a prepos-
> terous oversimplification. Production functions are almost as
> worthless as the alchemists' attempts to transmute elements.
> Production ... always begins with know-how.

Information is, of course, essential to each of the traditional factors of production, but only since the 1960s have knowledge and tools been explicitly considered to be forms of capital (see, for example, Rubin & Huber, 1984). Capital is popularly equated with money, but to econo-mists, it constitutes one of the most difficult and contentious theoretical problems in their discipline, as is manifest from ongoing disagreements about just what it is, how it should be measured, and even whether a single measure can be adequate (Hulten, 1990). The concept of capital remains ambiguous and controversial (Nitzan, 1997). Distinctions have been drawn between natural vs. produced, old vs. new capital, informa-tion vs. production capital, and durable vs. nondurable, fixed vs. circu-lating, sunk vs. non-sunk, and appropriable vs. nonappropriable forms of capital.

The received view that return on capital is a function of its produc-tivity has been criticized because of its circularity. The inclusion of infor-mation in economics supports this turn away from approaches to capital as "congealed" labor and deferred consumption toward those that place emphasis on its transformative functions. Money is often used as a mea-sure of the amount of capital, but it cannot represent either the nature of the stock or directions of change. Considered as a relationship, capi-tal has distributional value that is manifested only in part through own-ership. More broadly, capital understood in this way is, in essence, the ability to make things happen; thus the entire realm of social power has potential value.

The effort to account for informational forms of capital began with thinking about the value of individuals in the workplace—what we now refer to as "human capital"—in the form of "human asset accounting." By the early 1980s, human capital was one of the basic categories of the information sector as understood by Machlup. Gary Becker received the 1992 Nobel Prize for his contributions including the development of a theory of human capital. This work is most fully explicated in his 1992 book *Human Capital: A Theoretical and Empirical Analysis, with Special Reference to Education.* Becker explored what it meant to describe the current economy as one based on human capital, a concept that can be defined colloquially as

> the sum total of skills embodied within an individual: edu-
> cation, intelligence, charisma, creativity, work experience,
> entrepreneurial vigor, even the ability to throw a baseball
> fast. It is what you would be left with if someone stripped

away all your assets—your job, your money, your home, your possessions. (Wheelan, 2002, p. 99)

The phrase "intellectual capital" was coined in 1969 by John Kenneth Galbraith in a personal letter to another economist (Snyder & Pierce, 2002). By the second half of the 1990s, corporations were actively engaged in finding ways of quantifying intellectual capital for accounting purposes (Stewart, 1998; Sveiby, 1997), and today's "new growth theory" includes both rivalrous human capital and nonrivalrous information among the factors of production (Hyde, 2003). It is estimated that, in the U.S., the stock of intangible capital began to outweigh that of tangible capital around 1973 (David & Foray, 2002).

There are other forms of intangible capital involving information. Bourdieu (1986) introduced the notion of "cultural capital," referring to the mastery of cultural elements that affect one's position in society, or proficiency in the consumption of and discourse about prestigious cultural goods (DiMaggio, 1991). "Linguistic capital" can include asserting property rights in words and symbols (Lury, 1993) or aggressive translation of economically valuable knowledge (Coulmas, 1992; A. Mattelart & Mattelart, 1992). "Social capital" involves networks of communication and communication-based institutions and their rules, norms of social practice, and relationships of trust (Davenport & Snyder, 2005; Putnam, 2000). The concept of "organizational capital" encapsulates the ability of firms to learn (Lamberton, 1994). In the network economy, networks themselves are considered to represent a stock of knowledge that has value as a factor of production (Kogut, 1993), a perspective that treats the social structure itself as part of its knowledge capital.

Information and Economic Stability and Change

Neoclassical economic theory starts from the assumption that the nature of the economy never changes. Beginning with the work of Veblen (1899), however, several 20th-century developments suggest that the economy itself does change, either simply from one state to another or in ways that are evolutionary (Parker, 1993). Information plays a role in theories about both types of change.

Economic Equilibrium

The world as pictured by neoclassical economics was static. The economy was understood to be structured and to function in the same way across time and cultures. This is not to say that Marshall and others did not understand that demand goes up and down, access to resources may not always be the same, and disruptions such as war or drought will inevitably affect economic processes. But they believed that when such things took place, the economy would, so to speak, "right itself," with compensating developments in other parts of the economy so that a new partial (sector or firm-specific) or general (economy-wide) equilibrium would again be achieved.[2]

Economic Disequilibrium

Many economists, however, view this type of equilibrium as a special case. Stiglitz (1985) has argued that it has only limited validity because it is susceptible to the slightest alterations in informational assumptions. Knowledge production, the accretion of knowledge into technologies and other structures, the role of information in global competition, and events that change the extent to which information is available all influence the nature and structure of the economy in ways that can make it disequilibrious.

The effects of knowledge production and technological innovation, however, became visible only once it became possible to compare systematic data about the economy over long periods of time, in the 1920s. Looking at a century of data, Russian economist Nicholas Kondratieff (1984) recognized "long waves" of rising and falling economic activity now known as Kondratieff Waves in his honor. This discovery was important to the economics of information because Joseph Schumpeter (1939) quickly realized that each wave of economic growth was launched by suites of technological innovations. Information and communication played important roles in earlier Kondratieff Waves and are considered to be primarily responsible for the launch of another in the last decades of the 20th century (Dodgson, 1993; Freeman & Perez, 1988; Hall & Preston, 1988; Hepworth, 1989). Many economists group Kondratieff Waves with other economic and business cycles that reveal upswings and downswings in what is still understood to be an equilibrious economy—change within existing rules of the game. Others, however, argue that Kondratieff Waves are different from other types of economic cycles because they throw the economy out of equilibrium by shifting the very parameters within which it operates—in other words, changing the rules by which the game is played (De Greene, 1993).

The innovation process itself, of course, is always an informational endeavor because it embeds knowledge in processes or material forms. Thus technologies are described as "information-intense" and the more sophisticated a technology the more information-intense it is said to be. Work on "innovation economics" (e.g., Mansfield, 1988; Winter, 1988) relies on correlations between technological development and economic growth with results that support Schumpeter's claims about the economic importance of research-and-development-based innovation. Solow (1957) demonstrated that half of the economic growth in the first half of the 20th century could be explained neither by increases in capital nor by increases in labor. The "Solow residual," he argued, could only be understood as the result of technological innovation. Romer's (1994) new growth theory model starts from the position that technological change accounts for most economic growth (Hyde, 2003). Even so, many innovation theorists continue to feel that economists do not sufficiently account for the role of technological innovation—an important category of applied uses of information—in the economy (Dosi, Freemna, Nelson, Silverberg, & Soete, 1988; Monk, 1992).

Appreciation of the economic role of innovation, however, has led to interest in the economic value of research and development (R&D) (Goldin, 2001), a problem that led to Machlup's (1962) ultimate identification of R&D and other information industries as a specific economic sector. Rostow (1975, 1978) combined Kondratieff Waves, a Schumpeterian approach to innovation, and the work of Bell (1973) and others on the information society to produce a model of successive leading industry complexes today popularly described as "techno-economic paradigms." One example of such a paradigm that has been used in analyses of information industries in recent years is the unfortunately named "Wintelism," meaning the combination of technical standards developed by firms such as Intel and embedded in products such as the Windows operating system and the legal and economic framework within which these technologies are so successful (Hart, 2002).

Evolutionary Economics

Even when Kondratieff Waves are understood to affect the very nature of the economy itself, the changes wrought may be cyclical. Theories of "evolutionary economics," on the other hand, start from the assumption that the economy changes in a progressive way over time, with technological innovation, preexisting knowledge, and learning all playing roles in evolutionary processes (Ellger, 1995). The question of how far the metaphor of biological evolution should be taken is still open: Boulding (1981) introduced the notion in abstract form, Nelson and Winter (1982) tried to rewrite the theory of the firm using evolutionary models, and De Bresson (1987) examined—and then rejected—a strict analogy between biological and economic evolution (Mokyr, 1990). In its extreme form, proponents of the biological metaphor view firms as analogous to species. From this perspective, information architectures and information processing routines are equivalent to DNA and technological change is parallel to mutation and evolutionary change. The widely accepted distinction between "disruptive" technologies that radically transform processes and practices and "incremental" technologies that introduce relatively minor adjustments to systems whose operations remain essentially unaltered provides evidence of a commonsense understanding that technological innovation can, at the very least, significantly change how economic processes unfold and introduce qualitatively new types of products, firms, and behaviors.

Evolutionary economic theory shares much with complex adaptive systems theory (Katsenelinboigen, 1992)—a stream of work that has been of interest to researchers in the economics of information at least since Marschak participated in a famous interdisciplinary group exploring the use of systems theory across the social sciences at the University of Chicago at a time when he was deeply involved in thinking about informational issues (Easton, 1979). An economic manifestation of the famous "butterfly effect," for example, is evident in the fact that isolated pieces of information, even distributed to only a few actors, can have a profound impact on the economy. This suggests an "economics of increasing returns." Marshall had pointed out the first mover effect—the likelihood that the first entity to enter a market will be the most successful—but this insight was largely ignored by theoretical economics until the 1990s. Arthur (1990) has suggested that this was so, in part, because the mathematics and computing capacity necessary to analyze this type of trend were not easily available until then.

Information and the Effects of Economic Activity

All economic activity has consequences—often called externalities—that can be analyzed from an informational perspective; economic activities involving flows of symbolic content have additional cultural effects.

Externalities

In the ideal world pictured by Marshallian neoclassical economics, cause and effect are clear: buyers rationally evaluate the products and prices available in the marketplace and make choices that maximize their utility; sellers produce goods in response to buyers' desires. Even Marshall, however, had to acknowledge that those who are not involved in particular transactions can still be affected by economic activity. He used the term "external economies" to refer to benefits to third parties that result from production or consumption activities without any mediation by the price system. The concept was broadened by Pigou (1920) to include third-party costs involved with external diseconomies (e.g., smoke from a factory). This made it possible to distinguish between private and social net products (and, at the time, to advise governments to narrow the gap between the two). The term "externalities" is now used to refer to these phenomena.

Externalities are intertwined with the economics of information in several ways. First, as Babe (1996) notes, externalities result from, and may even be defined in terms of, inaccurate or incomplete information, as prices are claimed to be sufficient to ensure optimum decision making both for economic agents and for the economy overall with respect to "normal" commodities or activities assumed to have no third-party effects. Second, for networked economic activities—of growing importance in what many now describe as a "network" rather than an "information" economy—externalities are likely to be as important as, or more important than, transactions themselves. In a telephone network, for example, the addition of new customers to the network is of value to the service provider not only because customers engage in additional transactions but also because their participation in the network increases the value of the network to all of its customers. Metcalfe's Law, which defines the value of a network as the square of the number of users, is one approach to quantifying such externalities (Metcalfe, 1995). (For this reason, telephone companies long subsidized the delivery of services to classes of customers—such as those in rural areas—who would otherwise find it too expensive to pay for.) It is for the same reason that in an economy dominated by networked relations among firms, cooperation and coordination become as important as competition for long-term economic success (Antonelli, 1992; Grabher, 1993; Guerin-Calvert & Wildman, 1993). Third, the types of social, cultural, and political consequences of economic activity defined as externalities in neoclassical economics are precisely the kinds of effects and influences now acknowledged to be essential to a full understanding of the political, social, and cultural effects of economic activity involving information.

Cultural Effects

Three arguments have been made regarding the ways in which the symbolic content of information flows affects the economy. The first sees this kind of effect as a subset of the overarching impact of symbolic communication on society, the second identifies a specific role for the media in sustaining a particular set of economic relations, and the third points to ways in which the nature of the economy changes as a result of learning from the content of information flows.

There is a long-standing battle over whether the symbolic content of information makes any difference. There are those who believe that, although the quantities and direction of information flows matter from economic and engineering perspectives, the content of those flows is irrelevant, a perspective summarized in the dictum "a bit is a bit is a bit." The opposite position is taken by those who are interested in information from a cultural, social, and political perspective, often informing analyses undertaken by researchers in scholarly disciplines such as communication, sociology, and political science. From this perspective, the content of information flows is extremely important because it shapes perceptions, understanding, and, in turn, society—and the economy—in fundamental ways. The difference between these two perspectives has run throughout global debate over whether to include information and its flows under international trade agreements. This debate is discussed in more detail below.

Marx's 1860s analysis of the ways in which laborers were psychologically subordinated to the means of production suggested an important role for the content of communication in sustaining the operations of the economy, but it was Lenin who translated this insight into a theory of the press and a set of media practices designed to shape systematically the views of the population. Even so, Smythe (1979) famously insisted that this subject was a "blind spot" for Western Marxism as it ultimately developed. Indeed, Baudrillard (1983) later rejected the basic Marxist distinction between infrastructure (the material world) and superstructure (the symbolic world) precisely because, he argued, it was that distinction that had permitted the intelligentsia of the left to refrain from engaging sufficiently with the role of the media (M. Mattelart & Mattelart, 1990).

Whereas the Marxian approach assumed that the media would sustain the economic system in a particular form, Boulding (1963) claimed instead that knowledge and communication generated by information flows could change the nature of the economic system. His argument was based on the idea that, where knowledge is an essential part of the system, knowledge about the system changes the system itself. Boulding's ideas were viewed as original and unusual when first expressed but became quite common a few decades later; by the 1990s, there was a great deal of attention to the "learning organization" (see, e.g., Garvin, 1993) and related matters that further articulated the economic consequences of self-reflexivity fed by information flows.

Information as an Economic Sector

The first conceptualization of the information economy to appear, in the 1960s, relied upon the notion of an information sector comprising information industries. Those who take this approach claim that an information economy was developing because this sector was proportionately more important vis-à-vis other economic sectors than it had been in the past.[3] This is still the dominant conceptualization of the information economy because it is easy to comprehend and manage methodologically. (Examples of the use of this approach, operationalized for research purposes, include Nass [1987] and Schmoranz [1981].) This sector-based approach is not, however, entirely valid and exclusive reliance upon it leads to a variety of operational problems. A look at just what is meant by an information sector helps explain why this conceptualization is inadequate. Because until very recently industries were delineated by the commodities they produced, this section opens by thinking through issues raised by trying to conceptualize information as a commodity. It then moves to the notion of the "production chain" through which information commodities are produced before looking directly at the sectoral question.

Information as a Commodity

A commodity is a good or service possessing a stable identity that is completely specified physically, temporally, and spatially (Debreu, 1959). Detailed treatment of the problems raised by the effort to treat information as a commodity can be found elsewhere (Braman, 1997/1999); here, it suffices to note that information as such need not have a physical form: although typically embedded in materials, it does not require physical tangibility to be information. Even when information is made tangible, the value of the materials and that of the information are two different things. Further, the spatial location of the materials in which information is embedded does not spatially bound the information itself.

These abstract matters are often concretely problematic. They arise when it is necessary to quantify and/or place a value on information (Hunter, 1984; Jonscher, 1986). They also affect the ability to make policy decisions in areas such as taxation, trade, and investment when the activity being regulated involves transactions across internal or external borders. Such difficulties in treating information as a commodity are among the reasons why some economists claim that markets in information have unique characteristics (Braunstein, 1999).

These problems have not, of course, prevented information from being treated as a commodity. Indeed, one way of reading the history of the 20th century is through the lens of the progressive commodification of ever more types of information and informational activities (Jussawalla, 1984; Mosco, 1999; Murdock & Golding, 1990; Schiller, 1991). Signs of such commodification include the appearance of specialized traders, the development of spatially and temporally extended marketing networks and institutions, an increased distance between producers and end-users, and the potential for development of exchange-based specialization and monopolies of knowledge (Parker, 1993).

Because demand and desire play roles in transforming things into commodities (Appadurai, 1990), culture contributes to the creation of economic value (Bird-David, 1997; Iyer & Shapiro, 1999). In the economics of information, this becomes apparent upon consideration of such matters as the costs of knowledge transfer across cultural as well as organizational boundaries, differences in organizational form across cultures (Kogut et al., 1992), and the "cultural discount" that must be considered when delivering content across borders (Roeh & Cohen, 1992).

The Information Production Chain

For half a century, at least some economists, led by Marschak (1954, 1959, 1968), have pointed out that it is analytically important to distinguish between the costs of information at different stages of an information production chain. Myriad models of such a chain have been proposed. The majority of conceptualizations of an information production chain are oriented around various formulations of information creation, processing, flows, and use. For Marschak, it was important to distinguish between inquiry, communication, and decision making. The United Nations Centre on Transnational Corporations (UNCTC) put forth a model that included the stages of acquisition, production, assembly, storage, monitoring, interpretation, and exchange (Antonelli, 1989). Monk (1992) would treat differently information that is stored through embodiment in a physical medium, the transformation of existing information according to a set of rules, the processing of information that is not rule-driven, access to information, primary uses of information resources, secondary uses of information, and the roles of information in economic transactions. In the business world, the general notion is summarized in the concepts of the "value chain" (Rayport & Sviokla, 1995) and the "value-added" effect of information (Hayes & Erickson, 1982; Hopwood & Miller, 1994; Taylor, 1982, 1986).

Models of an information production chain can even be found embedded in the law. The Office of Management and Budget (OMB) circular that has had such an impact on federal agency practices through OMB evaluation of the costs of regulatory information collection, processing, and access to that information—OMB A-130—describes a life cycle that includes information creation, collection, processing, and dissemination. This author has used a model that synthesizes versions offered by Machlup (1980) and Boulding (1966) as a heuristic to bound the domain of information policy (Braman, 2004a). In studies of the law as diverse as those of international trade (Braman, 1989a), defense (Braman, 1991), and constitutional law (Braman, 1989b), distinctions between information creation (*de novo*, via collection, and via the generation of information as a manifestation of other social processes), processing (cognitive and algorithmic), transportation, distribution, storage, destruction, and seeking are used implicitly, if not explicitly, to justify different approaches to the interpretation and application of the law across cases and situations.

There are several advantages to using the concept of an information production chain as a way of thinking about information from an economic perspective. Doing so makes it easier to incorporate into one's analysis sensitivity to important differences in what is being done to and with information. It draws attention to the important role of the persons who create information, thus serving as a corrective to the common tendency to undercompensate creators because of the invisibility of their function outside of intellectual property rights law. It draws attention to intermediary stages of information processing (Owen & Wildman, 1991), and to the vulnerabilities introduced by moving from one stage to another (Wiley & Leibowitz, 1987). The information production chain provides a context within which possible new information industries become visible. Finally and most importantly, it brings the wide range of types of informational activities of economic importance under a common theoretical umbrella, making it easier to think in terms of a distinct economic sector involving information (Voge, 1983) and to bridge microeconomic and macroeconomic issues (Spence, 1974).

The Information Industries as a Sector

Economic sectors are comprised of industries. An "industry," as traditionally understood within neoclassical economics, produces single products and aligns with a single market.

One striking theme in the history of 20th-century statistics, however, is how problematic it has been to distinguish information industries from each other. The history of antitrust law, frequently applied to firms in the information and communication industries from the late 19th century on, provides vivid evidence of the enduring difficulties of distinguishing activities, products, and types of transactions in the information sector. By the 1980s, work on the information economy led to many quite diverse efforts to "map" the information industries. One edited collection actually included two chapters with completely different approaches (Compaine, 1981/1984; McLaughlin & Birinyi, 1981) and it was clear that the problem was more conceptual than empirical (Bruce, Cunard, & Director, 1986). Here, we look at three aspects of the information sector: its original conceptualization, the impact of efforts to incorporate information flows into international trade law on ways in which we understand the information sector, and the recent recategorization of industries stimulated in large part by efforts to find a better way of thinking about the information sector. This set of interactions between industry classification systems, the law, and economic realities provides a particularly interesting example of the structural influence of classification systems themselves (Bowker & Star, 1999) as well as showing how the struggle to deal with information in economic thought has affected very basic ideas about the economy itself. The section concludes with a brief look at other types of issues raised by a sectoral approach.

The Idea of an Information Sector

When, in the wake of work on the patent system, Fritz Machlup (1962) arrived at the concept of what he called the "knowledge industries," Boulding (1963) immediately noted that the concept of a knowledge industry would have explosive effects on traditional economics. In his seminal book, *The Production and Distribution of Knowledge in the United States*, Machlup (1962) identified over 50 industries in what we now refer to as the information sector. His conceptual approach to defining information industries was quickly followed by an alternative suggested by Porat (1977) that relied upon identifying information industries from among classifications in the existing Standard Industrial Classification (SIC) system codes. It was Porat's approach that was taken up by the U.S. government in the 1970s and, soon after, by other governments and international organizations in the form of the International Standard Industrial Classification (ISIC) codes.

Criticisms of the SIC-based approach, however, appeared almost immediately. As Spence (1974, p. 65) noted,

> Economically useful information is invested in and generated by virtually every industry in the economy. To confine oneself to what are thought of as the information industries is to ignore a major part of the process of investment in information in the economy.

To take but one example of particular import in light of the popular idea that there has been a linear progression from an agricultural to an industrial to an information economy: given the percentage of investment devoted to information creation, processing, and distribution, agriculture could today be defined as an information industry (Flor, 1993)—even if one leaves out of account the additional informatization of the agriculture-food-chemical-pharmaceutical complex that has come about through the convergence of biotechnology and digital information technologies (Braman, 2004b). Starting from existing industrial classifications was also problematic because many of the most economically important information activities never produce goods in tangible form and therefore were not included in the SIC system. And it was hard to justify many of the choices regarding just which industries were "in" the information sector and which were not. It seems arbitrary, for example, to treat medicine as an information industry because it requires knowledge for its practice and not to treat radio repair as such an industry, as it also requires specialized knowledge.

Other problems appeared. All final informational goods are simultaneously secondary goods. Although Jonscher (1982) and others have tried to distinguish between material and informational production sectors, all information remains embedded in the material world. Different types of information markets interpenetrate (King & Kraemer, 1995). Neither recognition of content as a source of meaning nor equity as a value to be incorporated into analysis of information flows appeared within treatment of the information industries as a distinct economic sector (Babe, 1995). There were also significant operational differences across governments; while the U.S. and other OECD countries measured activity in the information sector in terms of employees and the value of the output, for example, the Japanese were doing so by counting bits transmitted (Preston, 1989).

The Impact of International Trade

The first major shift in economic and legal treatment of the information industries as a sector was incorporation of "trade in services" within the global economic arrangements established after World War II. In the 1980s, the U.S. began to push for inclusion of trade in services within the General Agreements on Tariffs and Trade (GATT) (Braman, 1989b). Initially, there was much resistance, beginning with confusion over just what was meant by "services." (The most popular definition then, and still, is that offered by the British news magazine *The Economist* in 1985: "Anything that can be bought and sold that cannot be dropped on your foot" [A GAAT for services, 1985].) The definitional problem was resolved through a series of conceptual exercises and negotiations, but a political problem remained: Countries in the developing world, which in most cases had agriculture-based economies and were already concerned about their inability to compete successfully in the global economy, feared that the gap would widen with more favorable international treatment of a sector of the economy in which they were barely involved. Meanwhile, many leading countries in the developed world, including

Canada and France, resisted the change because they felt it would facilitate U.S. cultural hegemony and serve primarily to make it easier for U.S.-based firms in areas like accounting to extend their activities around the world. These difficulties slowed down the Uruguay Round of negotiations over the GATT for a number of years, but by the early 1990s the U.S. had achieved its goal—the World Trade Organization (WTO) was formed as an umbrella organization to manage not only the goods-oriented GATT but also the new General Agreement on Trade in Services (GATS) and the treaty on Trade-Related Intellectual Property Rights (TRIPS).

The process of examining information flows within the international trade context did several things for our understanding of information as a distinct sector. First, it made dramatically clear that there was no obvious or consensual way of identifying which activities were in and which were outside of the sector. Even the SIC- and ISIC-based approach turned out to be inadequate for decision making about international trade. Second, it highlighted the fact that the way in which various economic activities were categorized had real-world consequences. Third, it greatly expanded the range and numbers of individuals, groups, and governments with a stake in resolving the question of how to account for information-related activities for purposes of economic analysis. Together, these factors fueled growing interest in rethinking the economic classification system altogether.

Revising the Industrial Classification System

The U.S. Department of Commerce had begun to tinker with the treatment of information industries within the SIC codes by the early 1980s (Leeson, 1981), but the developments in international law discussed earlier made increasing the validity of accounting frameworks for use in the information economy even more important. Responding explicitly to the need to harmonize classification systems across the countries involved in the North American Free Trade Agreement (NAFTA, which includes the U.S., Canada, and Mexico) but also clearly motivated by the range of accounting problems raised by informational forms of capital, work began on creation of an alternative to the SIC system in the 1990s. As the Bureau of Economic Analysis (BEA), Economic Classification Policy Committee (1994) noted at the launch of this effort, regardless of whether the neoclassical concept of an industry had ever been valid, it was certainly not so by the 1990s. Worse, assuming that an industry involves only a single product produced in a single manufacturing process and that the market and industry coincide impedes efforts at achieving statistical clarity.

After several years of effort, the BEA presented the new North American Industrial Classification System (NAICS) in 1997; for a number of years companies are free to use either the SIC or the NAICS for accounting purposes. Table 1.2 presents two examples of how this

Table 1.2 Examples of NAICS Hierarchy

NAICS Level	NAICS Code	Description
Sector	31-33	Manufacturing
Subsector	334	Computer & electronic product manufacturing
Industry Group	3346	Manufacturing & reproduction of magnetic & optical media
Industry	33461	Manufacturing & reproduction of magnetic & optical media
US Industry	334611	Reproduction of software
Sector	51	Information
Subsector	513	Broadcasting & telecommunications
Industry Group	5133	Telecommunications
Industry	51332	Wireless telecommunications carriers, except satellite
US Industry	513321	Paging

Source: U.S. Census Bureau, www.census.gov/epcd/www/naicsod.htm

classification system distinguishes industries in the information sector from each other.

The job was not concluded at that point, however. The most "difficult" issues raised in the process of updating the classification system to map more accurately onto the contemporary economy were left for last and turned out to involve activities in the information sector. Some of the conceptual problems involving businesses that deal with various aspects of information, communication, and culture remain unresolved.

Other Sectoral Issues

A great deal of applied economics as applied to information creation, processing, flows, and use naturally deals with the dynamics of specific industries within the information sector. This subject is beyond the scope of the present chapter, but a few examples of this level of analysis of the information sector include work on the economics of the media (e.g., Albarran & Chan-Olmsted, 1998; Gomery, 1989), telecommunications (McNamara, 1991), and research and development (Shi, 2001). Another pertinent research path looks at ways in which information industries may operate differently from those in other sectors. For example, whereas those sectors that are resource-based are in large part[4] still subject to diminishing returns, those that are knowledge-based may be characterized by increasing returns (Arthur, 1990);[5] even the role of patents differs by industry (Mansfield, 1988).

Information and Key Economic Concepts

In addition to influencing how we think about the economy as a whole and about the sectors that it comprises, the incorporation of information into economic theory has influenced how we understand other basic economic concepts. Here we will look at three: agents, the market, and the firm.

Economic Agents

As idealized by Marshallian neoclassical thought, economic agents act independently of one another as they seek to maximize their economic position ("utility") by rationally acting upon perfect information. The assumptions of rationality, perfect information, individualistic independence, and economic-only utility have been questioned by those working with informational issues.

Information and Rationality

The concepts of utility and rationality are not only bound up with each other, but both rely upon some expectation about just what the future will be. Beginning in the 1920s, a number of economists began to explore how a sense of the future is formed because they wanted to understand why many people did *not* seek to optimize in the way that neoclassical economics assumed they would. The concept of bounded rationality refers to the limits to the amount of information that individuals can process (Simon, 1957). Moreover, the information available to an individual is often uncertain, so that additional "limiting" or "strategic" factors must be taken into account in evaluating the quality of economic decision making (March & Simon, 1963). Muth (1961) grouped such ideas together under the label of "rational expectations" problems and that is how they are commonly referred to today. Lamberton (1994), however, prefers the phrase "economics of limited foresight" because all these issues derive from the assumption that rational actors work with all the information available. By now we know there are also limits to the ability of individuals to receive, process, and communicate information (Ellig & Lin, 2001). Relationships among pieces of information may be so complex that individuals cannot process all of the implications. Psychological and cognitive factors influence the way information is filtered through selective perception, attention, and use (Pau, 2002); even the sequencing of information searches can influence decisions (Scazzieri, 1993).

The technological and social contexts within which information is received can also affect whether, and how, it is used. Technologies may limit the amount of knowledge available (Jensen & Meckling, 1992) and knowledge in tacit form is not generally diffused (Cowan, David, & Foray, 2000). Although economic agents are idealized as if they were completely autonomous, economic decision making—like all other social processes—is, of course, deeply embedded within relationships, networks, and communities (Grabher, 1993). Cultural factors play a role in determining how successful searches for information will be, as well as the criteria by which the success of such searches is evaluated (Folt, 1975; Grunig, 1971). There are asymmetries in the ability to *access* information as well as the ability to *use* information (Choi, 1993; Cremer & Khalil, 1992), often (but not only) deriving from and reproducing socio-economic class in ways referred to by sociologists since the 1970s as the knowledge gap (Vishwanath & Finnegan, 1996) and, more recently, the digital divide (Cruise O'Brien, 1983; U.S. National Telecommunications and Information Administration, 2000).

Some take the position that modeling offers a useful response to these limits to rationality, but others believe that models based on linear causal relations are insufficient and have turned, instead, to various versions of complex adaptive systems theory that are also capable of dealing with nonlinear causal relations (Hayward & Preston, 1999; Morgan, 1986). Alternative responses to the limits to rationality emphasize the roles of morality (Brennan, 1993), discretion (March & Simon, 1963), and organizational structure (Haas, 1990; Overman, 1989).

Information and Imperfect Information

Although Marshallian neoclassical economic theory assumes there is perfect knowledge, information is rarely, if ever, perfect, and certainly never so when it comes to the future. In 1921 Frank Knight introduced the concept of uncertainty into economic analysis, inspiring a massive literature that has provided a central thread to information economics. Stigler looked at just what uncertainty means for commodity prices (Stigler, 1961) and the labor market (Stigler, 1962). These seminal pieces spawned further work on various consequences of uncertainty (see, e.g., Diamond, 1978; Nelson, 1970; Rothschild, 1974; Telser, 1973). Others, such as Debreu (1959/1989), Radner (1989), and Williams and Radner (1968), looked at the effects of uncertainty on equilibrium, finding that the creation of information is a disequilibrium-causing process, whereas its dissemination is disequilibrium-repairing (Lamberton, 1984a). We now recognize two different types of uncertainty, each of which requires a different response. We try to *overcome* uncertainty about which conditions or states are possible, whereas we try to *adapt* to uncertainty regarding which among those possible conditions or states will actually occur (Hirshleifer & Riley, 1979).

Of course, where there is uncertainty about the future there is also anticipation and, in a series of articles, Marschak (1954, 1959, 1968) examined the impact of anticipated information on economic behavior. When economic agents are uncertain, they will try to influence perceptions of the products or services they offer, known as "signaling" in the economic literature. Four pioneer papers launched a stream of research on signaling and screening—how signals are received—that continues into the 21st century. Vickrey (1961) examined issues raised by the provision of incentives when agents have private information, an approach that has become the basis of modern auction theory. Mirrlees (1971) used the case of income taxation to look at trade-offs between the signaling of incentives and the actual redistribution of resources. Akerlof (1970) demonstrated how a market can collapse if critical pieces of information are missing. And Spence (1973) examined the relationship between signaling and screening, understanding the latter as a costly but effective search strategy on the part of buyers. Other work in this area looks at the role of reputation as a signal (Backus & Driffill, 1985; Barro & Gordon, 1983), the economic value of the ability to monitor performance (Hechter, 1984), and the impact of advertising (Bagwell & Ramey, 1988; Kihlstrom & Riordan, 1984; Milgrom & Roberts, 1986; Nelson, 1974). Through the lenses of game theory, matters such as how and when parties choose to reveal or hide information, or to provide misinformation, have become susceptible to analysis (Allen, 2000). Both signaling and screening are meant to improve the efficiency of the market.

Information and Utility

In neoclassical thought, an economic agent makes decisions in order to maximize his or her utility—the satisfaction or benefits gained from consuming a particular good or service—according to a hierarchical ranking of preferences. Three types of critiques of this approach have appeared that involve information and communication: recognition of the importance of economic decision making for expressive purposes rather than utility maximization, difficulties generated by the assumption of a stable hierarchy of preferences, and interdependence among the utility functions of different people.

Veblen (1899) was one of the first economists to write in opposition to Marshallian economics because he understood economic activity—like all other social processes—to be fundamentally comprised of communications among individuals. Taking the position that even in their economic lives people are fundamentally engaged in expressive activity driven by a variety of desires and habits in addition to utility, Veblen became famous for his concept of "conspicuous consumption." People acquire and use luxury goods or goods in great quantities, he argued, simply to demonstrate that they have the resources to do so even when the practice is otherwise dysfunctional and may even ensure that they shall *not* have sufficient resources in future. Because commodities are semiotic, symbolic, and informational, Veblen reasoned, economic analysis must incorporate the roles of information and knowledge in society.

Looking beyond utility for other motives and effects of economic activity opens up another way of thinking about industries in the information sector—as cultural forces. Critiques of the activities of the "cultural industries" began with Walter Benjamin and other thinkers of the Frankfurt School of critical thought (Adorno & Horkheimer, 1977; Held, 1989; Negt, 1978) and its offspring (e.g., Enzensberger, 1974; Giersing, 1984; Miege, 1979). Based in Marxist thought but going beyond it to focus specifically on the cultural industries, this line of inquiry emphasized the social effects of the commoditization of information. There is still a great deal of work to be done exploring economic activities from behavioral, cognitive, and cultural perspectives (Sengupta, 1987). Doing so, of course, makes economic analysis far more complex; Timothy Brennan's (1993) exquisitely titled "The Futility of Multiple Utility" summarizes some of the difficulties present in trying to take additional motives into account.

Parker (1993) has provided a concise history of thinking about the interdependence of utility, the fact that the actions of one economic agent affect and are affected by the actions of others. Acknowledgment of such interdependence transforms perception of the individual actor into someone inextricably embedded in relationships, a reality that, as described earlier, constitutes one of the limits to rationality. Arrow's (1979, 1985) recognition that human relations need not, and should not, consist solely of commodity exchange relations was an important turning point in stimulating attention to this question (Babe, 1996).

The Market

The concept of the market is one of the original "big ideas" in economic thought, but it has been subjected to both empirical and theoretical critiques. Thinking about information and its flows from an economic perspective has increased interest in alternative ways of conceptualizing the structures within which economic activity takes place, such as the field, the system, and the network.

Early Conceptualizations of the Market

Stiglitz (1982) opened one of the first conferences on the economics of information, held in 1977, by noting that acknowledgment of the importance of imperfect information challenges the conventional notion of a market and forces economists to reconsider central concepts and principles. The inherited concept to which he referred was originally formulated by Adam Smith (1776/1937), whose highly influential *Wealth of Nations*—considered by many to be the foundational work of modern economic thought—presented the argument for a "free market." In Smith's view, the market was an "invisible hand," coordinating the efforts of individuals in such a way that society as a whole benefits. In this view, each individual acts independently to serve only his or her own interests, a perspective that laid the groundwork for the neoclassical view of utility, as has been discussed.

Smith (1776/1937) did not specifically address information as an economic good or service, even though he wrote during the period in which an explicit market in information was first developing. (Referred to as the "encyclopedia era," the 18th century witnessed the initial stirrings of recognition of the economic value both of literary and journalistic texts and of the very act of organizing "useful" information [Stephen, 1907].) Nor did Smith directly confront the question of the influence of differential access to information on the market. He did, however, talk about the difficulties that lenders face in seeking to establish rates that maximize profits without alienating borrowers, a problem today described as an information issue involving "adverse selection."

From a 20th-century perspective framed by neoclassical economic thought, it is evident that Smith's conceptualization of the market also resonated with 18th-century ideas about the nature of political power. This has suggested the metaphor of the "marketplace of ideas" as a principle upon which to base legal analysis of speech-related matters. Peters (2004) notes the inaccuracy of the popular belief that this very phrase was in use during the 18th century; nonetheless, the seeds of the notion of a marketplace of ideas were clearly planted in the work of authors such as Milton and Mill. Economists of the period, meanwhile, believed that commerce was valuable, among other reasons, because it forced people from different places, societies, and cultures to communicate and learn about each other (Boyle, 1996).

Neoclassical Conceptualizations of the Market

In neoclassical economics, it is assumed everyone has the same information and that this universally held information is perfect. Marshall was quite aware of imperfections in the availability of economically useful information but argued that the incorporation of such imperfections into economic analysis would make the analytical task impossibly complex, on the grounds that there were no reliable and valid techniques available for quantifying information or incorporating its flows into economic calculations. (This problem remains today.) To justify this position, later economists claimed that the imperfections made such little difference that analysis need not take them into account. As a consequence, neither the sources nor the consequences of imperfect information were explored.

Hayek (1945), however, pointed out that the market is actually an informational mechanism, achieving its efficiency by utilizing knowledge that is dispersed throughout the population. Prices may not tell potential buyers everything about a product, but they are "sufficient" as quantitative indicators of the significance of any particular scarce resource relative to all others. In Arrow's (1979) further articulation of this approach, prices constitute "congealed" information and markets are decision structures in which communication takes place through the medium of prices. Searches provide information about the probability distribution of prices in addition to the price charged by individual sellers (Rothschild, 1974; Rothschild & Stiglitz, 1976); the more mature a market, the greater the effective amount of search (Stigler, 1962). Economic agents enter markets with certain types of nonprice knowledge, including resources already held, preferences, and alternative ways of translating inputs into outputs. Research of this type does not, however, explain just how the economy processes information. Indeed, as Stiglitz (2000) notes, such models did not appear until several decades later (e.g., Grossman, 1977, 1981; Grossman & Stiglitz, 1976, 1980; Lucas, 1975), when the economic system had to adjust to information in many new forms.

Just as Smith's earlier concept of the market had political implications, so the neoclassical approach to the market was taken up by political theorists of the 20th century. The rise of the modern corporation was tied to the appearance of political liberalism, "free" social relations, and an increase in openness within society (Bourdieu & Coleman, 1991). Hayek (1973) himself drew links between the nature of the market and the possibilities for democratic practice. His insight that there is competition between forms of social organization as well as between specific economic entities became essential to work on the strategic value of different ways of managing information by political scientists during the closing decades of the 20th century.

Alternatives to the Market

Three alternatives to the market as ways of thinking about the social structure or structures within which economic activity takes place have been suggested: the field, the system, and the network. Whether one chooses to use any of the alternative conceptualizations of the overarching economic structure or not, there is a great deal of statistical evidence for the significant shrinkage of the proportion of economic activity that takes place in anything that may be described as a market—a trend due to the fact that so many economic processes now take place within organizations, networks of organizations, and other types of contractual relationships (Castells, 2000).

Bourdieu's (1991) concept of the "field" refers to the environment of possible actions. The field is rich and multifaceted, including cultural predispositions and habits as well as relationships, knowledge, and discourse. Lash (1993) uses the metaphor of the skeleton to describe how to understand the economic market from the perspective of a Bourdieuian field: the market is outlined by just a few of the types of processes and relationships involved, but it could not live on its own without the many other facets of social life with which it is intimately involved. In recent years, sociologists have begun to analyze the ways in which economic activity changes its nature vis-à-vis the field of possibilities in response to factors such as technological and political change, often focusing on shifts in the informational and communicative processes involved (see, e.g., Dezalay & Garth, 1996).

In the neoclassical ideal of the market, firms, like individuals, are self-enclosed entities that operate autonomously (see next section). Acknowledging the multiplicity of ways in which firms are intertwined with each other suggests a systems alternative to the market as the overarching structure for economic activity. One example of such an approach can be found in Babe's (1995) work on flows of resources, a lens through which the economy is actually best understood as a communication system.

Systems perspectives lead rather naturally to conceptualization in terms of networks of relationships, an approach found in the work of those committed to network economics (e.g., Grabher, 1993; Guerin-Calvert & Wildman, 1993; Sabel, 1991). Analysts operating within this theoretical framework often use the project, rather than either the firm or the industry, as the unit of analysis. Coordination and cooperation are considered to be as important as competition for long-term economic success (Antonelli, 1992) and even competition must be understood through analysis of relationships (Burt, 1992). European economists and policymakers go even further, using the concept of the *filière électronique* to refer to the combination of the information infrastructure and those economic entities that exist within and rely upon the network for their existence (e.g., Dyson & Humphreys, 1986).

The Firm

Almost all information-based approaches to economies focus on organizations rather than markets, with the result that ideas about the firm—and, thus, institutional economics—have also been important to the economics of information (Ellger, 1995). It was in the 1920s that the role of information in corporate life first became the subject of scholarly analysis (Pemberton, 1995). By that time, the Taylorist movement and the "methods office" were widespread, the organization of information into tables and the use of forms were considered the height of technological sophistication (Temin, 1991; Yates, 1989, 1994), and records management was becoming so problematic that it began to draw professional attention (Debons, King, Mansfield, & Shirley, 1981; Hatchuel, Le Masson, & Weil, 2002). Managerial capitalism was coming into its own, replacing owner-managers, who made decisions based on personal relationships and idiosyncratic and unsystematic access to information, with professional managers, who made decisions based on information about corporate and market activity (Chandler, 1977). These shifts brought the fundamental economic roles of information creation, processing, flows, and use within the firm into view.

Coase (1937) provided a theoretical way of understanding these changes in corporate form and business processes in his seminal work on the firm as a means of reducing the costs—known as "transaction costs"—of searching for the information needed in order to engage in economic activity. Coase asked why, given the alleged perfections of the market, the firm had come into existence at all and answered this question with the insight that organizations were economically useful because they so significantly reduced transaction costs through trust, habit, and the development of reliable procedures for handling information. Activities for which transaction costs are high are likely to be managed within a firm to the greatest possible extent; activities for which the transaction costs are low will be left to the marketplace. (Digital technologies have had such an impact on the nature of organizations precisely because they shift transaction costs, encouraging corporations to reconsider which functions they would prefer to manage within the firm and which might be better handled outside the firm.) Richardson (1959), however, pointed out that the assumption that firms can buy information in increments of any size and will do so when its benefit outweighs its cost is appealing but may not reflect the unfolding dynamics of information acquisition and its use within firms.

Douglass C. North (1981) played a leading role in directing attention to the importance of differences in the structural characteristics of "hierarchies," or firms, as opposed to "markets" as means of organizing economic activity. By the late 1970s, economists and historians began to study institutions and institutional change involving treatment of information (e.g., Chandler, 1977). Information economics unpacks the black box of the firm, making it, in Arrow's (1974b, p. 147) words, "an incompletely connected network of information flows" instead of "a point." The implications of looking at the firm in this way include the need to trace information flows and communication patterns and to evaluate the costs associated with those flows and patterns. Organizations may be efficient at managing information, but that does not mean they will be equally efficient at organizing economic activity nor does it entail that they will produce distributive justice.

The binary distinction between markets and firms, or "hierarchies" (Williamson, 1975), remained in place until the 1990s, when it had become clear that a third form of social organization—the network—managed information in yet another way that had growing economic importance (Antonelli, 1992). In addition to exploring relationships between firms and networks, current informational questions of interest involving the firm include the role of the treatment of information in productivity (Cronin & Gudim, 1986; Griliches & Mairesse, 1993), competitive intelligence (Cronin, Overfelt, Fouchereaux, Manzvanzvike, Cha, & Sona, 1994; Porter & Millar, 1985; Shapiro & Varian, 1999), knowledge management (Davenport & Cronin, 2000; Hatchuel et al., 2002; MacMorrow, 2001), informational aspects of contracts as a means of reducing transaction costs (Macho-Stadler & Perez-Castrillo, 1997), and the impact of the Web on the amount of information available in a market (Markillie, 2004).

The Impact of the Economics of Information

Many believe that the economics of information has fundamentally changed the field of economics itself. The view that information would turn out to be like other costs has been turned on its head and we now know that even small information costs can have large consequences (Stiglitz, 2000). Information shapes how economists view the world (Persico, 1996) and provides the basis of "predictable predictions" about the future of economic thought (Allen, 1990). There are three additional impacts of the evolution of the economics of information at both the macro- and microlevels: the unfinished work in the field identifies a research agenda, there are influences on other disciplines, and there are policy implications.

Unfinished Work

The effort to understand information from an economic perspective has still not reached its conclusion, despite the assertion by Locksley (1988) that we had reached the limits of our ability to exploit information technologies economically two decades ago and in spite of the repetitiveness and diminishing returns of much current research on topics such as asymmetrical information (Sandler, 2001). Work remains in the reconsideration of earlier concepts, the filling in of theoretical lacunae, the resolution of methodological problems, and the application of new theories to economic issues involving information.

Rethinking Concepts

There is still a great deal of work to be done rethinking concepts that have long been important to economics. Indeed, as Stiglitz (1985) noted, the concept of information itself largely remains a term applied to a residual regarding matters unexplainable by economists in any other way. In pursuing this work, economic debates of the past are being reopened, such as the 1930s disputes regarding the relative advantages of different types of economic organization (Zappia, 1999). Claims of economies of scale and scope are frequently asserted—particularly among those discussing the network economy—but the meaning of these terms is rarely examined in detail (Charles, Monk, & Sciberras, 1989). And fundamental concepts such as uncertainty (Ewald, 1991) continue to need theoretical attention. Reevaluation of the past through the lenses of the economics of information brings additional issues to light. Cochran (1960), for example, looked at the cultural factors that affect economic growth and there has been reconsideration of the motivations driving colonial political activity (Morgan & Morgan, 1986; Parker, 1966).

Lacunae

Lacunae in economic theory that might be usefully addressed by those working on the economics of information include identifying the positive economic value of terminating information systems (Lamberton, 1998b), on one hand, and accounting for the structural abundance of information, on the other (Goldfinger, 2000). The importance of context to analysis of economic processes is being increasingly recognized, but this notion has not yet been thoroughly incorporated into the economics of information nor has it been explicitly related to the concept of externalities. There are interactions between standardization and codification processes and the facilitation of information flows with economic value (Kahin, 2004) that have not yet been sufficiently explored. One of the most important macroeconomic subjects demanding attention is the ability to actually create new forms of property rights in the information environment (Garcia, 2004; Hazlett, 1990; Meyer, 1984). The question of how to evaluate knowledge production economically is currently receiving a great deal of attention (Foray, 2004; Hall & Preston, 1988).

Methodological Issues

The methodological problems that played such a major role in persuading neoclassical economists to essentially ignore information have not disappeared even with the acceptance of its economic importance (Brinberg, 1989). Issues that remain include rapid changes in the nature of the informational processes about which data are being collected (Bureau of Economic Analysis. Economic Classification Policy Committee, 1994). Many kinds of data may not be available or may appear only in unorganized anecdotal form. There is a need to link the individual and social levels of analysis in dimensions that range from social psychology to semiotics (Parker, 1993). In order to develop policy tools that are adequate for the 21st-century information environment, many economists are calling for experimentation with techniques such as statistical pricing for telecommunications rather than pricing by connection, packets, or time (Pau, 2002).

Quantification of informational phenomena and processes makes it possible to use mathematically based analytical techniques, but, despite their power, such approaches also have limitations. (Adam Smith himself didn't find it necessary to use numbers when he put forward his basic theoretical principles.) Some have said that economists use the most sophisticated techniques to arrive at the most irrelevant conclusions. And as Stigler (1985, p. 5), in his Nobel acceptance speech, said, "Any preoccupation with fairness and justice is uncongenial to a science in which these concepts have no established meaning." With increasing acknowledgment of the economic role of context and greater awareness that cultural differences affect the formation and use of capital (Marcus & Fischer, 1986), research methods are also of growing interest. We should also expect to see an increased use of systems theory and methodologically collectivist modes of analysis to explore the information, or network, economy (Cherry, 2004).

Impact on Other Fields

Although this chapter has focused on how the incorporation of the concept of information into economic theory has affected basic economic concepts, the range of applications of the economics of information is constantly expanding. The field of political science offers a number of vivid examples. Problems—or opportunities—raised by limited, or imperfect, information have been the subject of analysis by political scientists at least since Kissinger (1971) pointed out that politics is often a tradeoff between knowledge and scope for action. The long-dominant view within international relations that nation-states operate like individual economic agents—autonomously and as rational actors—has been critiqued for the same reasons the belief that economic agents have equal and perfect information has been treated with skepticism. As Rosenau and Durfee (1995) note, uncertainty increases with growing interdependence among nation-states and turbulence in international relations. Ideas about asymmetric information are finding a place in conflict management and peace analysis (Chatterji, 1992), as well as national security from both the military (Clarke, 1993) and economic (Smith, 1993) perspectives. The economics of R&D and innovation are particularly important to analyses of arms races (Evangelista, 1988), international competitiveness with security implications (Branscomb, 1993), weapons development, and new types of potential threats (Pearton, 1984). Informational aspects of game theory are often applied to the formation of international agreements (e.g., Stein, 1982) and increasing returns economics suggests that countries should use their industrial policies to promote success in technological innovation and an emphasis on high technology in international trade (Arthur, 1990). Problems of incomplete information are important to domestic politics as well in such areas as electoral competition (Banks, 1990a; Rogoff & Siebert, 1988), agenda-setting (Banks, 1990b), budget cycles (Rogoff, 1990), congressional oversight of federal agency activities (Cameron & Rosendorff, 1993), and even evaluation of the credibility of policy reform itself (Rodrik, 1998).

Policy Implications

We continue to struggle with identifying the most important features of the transition from a manufacturing to an information-based economy with regard not only to theory but also to the implications of this transformation for those who make decisions in the public and private sectors.

Some of the questions raised by economic analysis of information are very specific. Should we be taxing Internet-based commerce and, if so, how (Goolsbee, 2000)? If the information infrastructure is so fundamental to all other types of economic activity, should industries in this sector be granted special tax breaks (Greguras, 1980) or permitted to use new types of investment tools (Alleman & Noam, 2003)? Because capital has become conflated with digital information flows, do we need to reconsider monetary policy (Woodford, 2001)? When is it appropriate to use new policy tools that have become available as a result of the economics of information, as in our deepened understanding of auction theory (Persico, 1996)?

Other policy issues are more overarching. With increased appreciation of the economic value of R&D, do we need to reconsider the nature of the patent system? Given the importance of human capital to economic success, should we be defining the development of human skills as a fundamental right (Scazzieri, 1993)? Given that the earliest copyrights were applied to information architectures (maps, primers, calendars, and law books) (Ginsburg, 1990), do we need to develop new forms of intellectual property rights for emergent information architectures of the 21st century? How far should we go in permitting "precompetitive" R&D collaborations that in essence void antitrust law of effect?

At a yet higher level of generalization, thinking in terms of the economics of information draws our attention to numerous linkages among social processes typically analyzed separately. When we view the economy as essentially a communication system, acknowledge the role of information flows in shaping economic power and capacity, and admit that significant asymmetries in access to information have socioeconomic consequences, the importance of continuing to think through the theoretical, conceptual, empirical, and methodological issues raised by information at the macro- and microeconomic levels is clear.

Endnotes

1. Micro- and macroeconomics are distinguished by the level of analysis at which each operates. Microeconomics focuses on economic processes at the individual or firm level; macroeconomics focuses on economic processes at the level of society as a whole.

2. This is an extreme oversimplification of equilibrium analysis in economics, a theory and methodological approach that involves examining multiple economic relationships simultaneously and, in order to achieve this, depends upon methodological individualism and an analytical insistence that many variables remain static.

3. As explained in more detail in Braman (1996), this conceptualization was followed in the 1970s and 1980s by views of the information economy that focused upon the expansion of the economy due to the commodification of information and then, in the 1990s, by a focus on the information economy as one in which the very nature of economic relations has changed.

4. Innovation at the genetic or molecular level, now possible with biotechnology and nanotechnology, can change the extent to which diminishing returns characterize particular industries.

5. Confidence in the inevitability of increasing returns has of course declined since the dotcom bust.

References

Adorno, T. W., & Horkheimer, M. (1977). The culture industry: Enlightenment as mass deception. In J. Curran, M. Gurevitch, & J. Woollacott (Eds.), *Mass communication and society* (pp. 349–383). London: Hodder Arnold.

Akerlof, G. A. (1970). The market for "lemons:" Qualitative uncertainty and the market mechanism, *Quarterly Journal of Economics, 84*(3), 488–500.

Albarran, A. B., & Chan-Olmsted, S. M. (1998). *Global media economics: Commercialization, concentration and integration of world media markets.* Ames: Iowa State University Press.

Alleman, J., & Noam, E. (Eds.). (1999). *The new investment theory of real options and its implications for telecommunications economics.* Boston: Kluwer.

Allen, B. (1990). Information as an economic commodity. *American Economic Review, 80*(2), 268–273.

Allen, B. (2000). The future of microeconomic theory. *The Journal of Economic Perspectives, 14*(1), 143–150.

Antonelli, C. (1981). *Transborder data flows and international business: A pilot study.* Paris: OECD Directorate for Science, Technology & Industry, DSTI/ICCP/81.16. [UNCTC]

Antonelli, C. (Ed.). (1992). *The economics of information networks.* Amsterdam: North-Holland.

Appadurai, A. (1990). Disjuncture and difference in the global cultural economy. *Theory, Culture & Society, 7*, 295–310.

Arnold, S. E. (1990). Marketing electronic information: Theory, practice and challenges. *Annual Review of Information Science and Technology, 25*, 87–144.

Arrow, K. (1974a). Limited knowledge and economics analysis. *American Economic Review, 64*, 1–10.

Arrow, K. (1974b). *The limits of organization.* New York: Norton.

Arrow, K. (1979). *The limits of organization.* Paris: University Press of France.

Arrow, K. J. (1985). Informational structure of the firm. *American Economic Review, 75*(2), 303–307.

Arrow, K. (1996). *Elements of the economics of information: Information and increasing returns.* Nankang, Taipei, Taiwan: Institute of Economics, Academica Sinica.

Arthur, W. B. (1990, February). Positive feedbacks in the economy. *Scientific American,* 92–99.

Babe, R. E. (Ed.). (1993). *Information and communication in economics.* Norwell, MA: Kluwer.

Babe, R. E. (1995). *Communication and the transformation of economics.* Boulder, CO: Westview.

Babe, R. E. (1996). Economics and information: Toward a new (and more sustainable) world view. *Canadian Journal of Communication, 21*, 161–178.

Backus, D., & Driffill, J. (1985). Inflation and reputation. *American Economic Review, 75*(3), 530–538.

Bagwell, K., & Ramey, G. (1988). Advertising and limit pricing. *Rand Journal of Economics, 19*(1), 59–71.

Banks, J. S. (1990a). A model of electoral competition with incomplete information. *Journal of Economic Theory, 50*(2), 309–325.

Banks, J. S. (1990b). Monopoly agenda control and asymmetric information. *Quarterly Journal of Economics, 104*(2), 445–464.

Barro, R. J., & Gordon, D. B. (1983). Rules, discretion and reputation in a model of monetary policy. *Journal of Monetary Economics, 12*(1), 101–121.

Baudrillard, J. (1983). *Simulations.* New York: Semiotext(e).

Bell, D. (1973). *The coming of post-industrial society: A venture in social forecasting.* New York: Basic Books.

Bergeron, P. (1996). Information resources management. *Annual Review of Information Science and Technology, 31,* 263–300.

Bergeron, P., & Hiller, C. A. (2002). Competitive intelligence. *Annual Review of Information Science and Technology, 36,* 353–390.

Bird-David, N. (1997). Economies: A cultural-economic perspective. *International Social Science Journal, 154,* 463–475.

Blaug, M. (1978). *Economic theory in retrospect* (3rd ed.). Cambridge, UK: Cambridge University Press.

Blaug, M. (1996). *The methodology of economics: Or, how economists explain.* Cambridge, UK: Cambridge University Press.

Blaug, M. (1997). *Economic theory in retrospect* (5th ed.). Cambridge, UK: Cambridge University Press.

Boon, J. A., Britz, J. J., & Harmse, C. (1994). The information economy of South Africa: Definition and measurement. *Journal of Information Science, 20*(5), 334–347.

Boulding, K. E. (1956). *The image: Knowledge in life and society.* Ann Arbor: University of Michigan Press.

Boulding, K. E. (1966). The economics of knowledge and the knowledge of economics. *American Economic Review, 56*(2), 1–13.

Boulding, K. E. (1969). Economics as a moral science. *American Economic Review, 59*(3), 1–12.

Boulding, K. E. (1981). *Evolutionary economics.* Beverly Hills, CA: Sage Publications.

Boulding, K. E. (1984). Foreword: A note on information, knowledge, and production. In M. Jussawalla & H. Ebenfield (Eds.), *Communication and information economics: New perspectives* (pp. vii–ix). New York: North-Holland.

Bourdieu, P. (1986). The production of belief: Contribution to an economy of symbolic goods. In R. Collins, J. Curran, N. Garnham, P. Scannell, P. Schlesinger, & C. Sparks (Eds.), *Media, culture and society: A critical reader* (pp. 131–163). London: Sage.

Bourdieu, P. (1991). *Language and symbolic power.* Cambridge, MA: Harvard University Press.

Bourdieu, P., & Coleman, J. S. (Eds.). (1991). *Social theory for a changing society.* Boulder, CO: Westview Press.

Bowker, G. C., & Star, S. L. (1999). *Sorting things out: Classification and its consequences.* Cambridge, MA: MIT Press.

Boyle, J. (1996). *Shamans, software, and spleens: Law and the construction of the information society.* Cambridge, MA: Harvard University Press.

Braman, S. (1989a). Defining information: An approach for policy-makers. *Telecommunications Policy, 13*(3), 233–242.

Braman, S. (1989b). Information and socioeconomics class in U.S. constitutional law. *Journal of Communication*, 39(3), 163–179.

Braman, S. (1990). Trade and information policy. *Media, Culture & Society*, 12, 361–385.

Braman, S. (1991). Contradictions in brilliant eyes. *Gazette: The International Journal of Communication Studies*, 47(3), 177–194.

Braman, S. (1993). Harmonization of systems: The third stage of the information society. *Journal of Communication*, 43(3), 133–140.

Braman, S. (1997/1999). L'économie de l'information: Une evolution des approches. In A. Mayère (Ed.), *La société de l'information: Enjeux sociaux et approches économiques* (pp. 87–113). Paris: Éditions l'Harmattan. Republished as The information economy: An evolution of approaches. In S. Macdonald & J. Nightingale (Eds.), *Information and organisation* (pp. 109–125). Amsterdam: Elsevier.

Braman, S. (2004a). The meta-technologies of information. In S. Braman (Ed.), *Biotechnology and communication: The meta-technologies of information* (pp. 3–36). Mahwah, NJ: Erlbaum.

Braman, S. (2004b). Where has media policy gone: Defining the field in the twenty-first century. *Communication Law and Policy*, 9(2), 153–182.

Branscomb, L. (1993). *Empowering technology: Implementing a US policy*. Cambridge, MA: MIT Press.

Braunstein, Y. M. (1981). The functioning of information markets. In J. H. Yurow & H. A. Shaw (Eds.), *Issues in information policy* (pp. 57–74). Washington, DC: National Telecommunications and Information Administration, Department of Commerce.

Brennan, T. J. (1990). Vertical integration, monopoly, and the First Amendment. *Journal of Media Economics*, 3, 57–76.

Brennan, T. J. (1993). The futility of multiple utility. *Economics and Philosophy*, 9, 155–164.

Brennan, T. J. (1994). Copyright, property, and the right to deny. *Chicago-Kent Law Review*, 68, 675–714.

Brinberg, H. R. (1989). Information economics: Valuing information. *Information Management Review*, 4(3), 59–63.

Brown, J. A., Jr., & Gordon, K. (1980). *Economics and telecommunications privacy: A framework for analysis*. Washington, DC: Office of Plans and Policy, Federal Communications Commission.

Bruce, R. R., Cunard, J. P., & Director, M. D. (1986). *From telecommunications to electronic services: A global spectrum of definitions, boundary lines, and structures*. London: Butterworths.

Bureau of Economic Analysis. Economic Classification Policy Committee. (1994). *Report No. 1: Economic concepts incorporated in the Standard Industrial Classification industries of the United States*. Washington, DC: The Bureau.

Burt, R. S. (1992). *Structural holes: The social structure of competition*. Cambridge, MA: Harvard University Press.

Cameron, C. M., & Rosendorff, P. B. (1993). A signaling theory of congressional oversight. *Games and Economic Behavior*, 51(1), 44–70.

Castells, M. (1996). *The rise of the network society*. Malden, MA: Blackwell.

Castells, M. (1997a). *End of millennium*. Malden, MA: Blackwell.

Castells, M. (1997b). *The power of identity*. Malden, MA: Blackwell.

Castells, M. (2000). *The rise of the network society* (2nd ed.). Malden, MA: Blackwell.

Chandler, A. D., Jr. (1977). *The visible hand: The managerial revolution in American business*. Cambridge, MA: Belknap Press.

Chatterji, M. (1992). Use of management and peace science techniques for conflict management. In M. Chatterji & L. R. Force (Eds.), *Disarmament, economic conversion, and management of peace* (pp. 15–26). New York: Praeger.

Charles, D., Monk, P., & Sciberras, E. (1989). *Technology and competition in the international telecommunications industry*. New York: Pinter Publishers.

Cherry, B. A. (2004, October). *The telecommunications economy and regulation as coevolving complex adaptive systems: Implications for federalism.* Paper presented to the International Telecommunications Society (ITS), Berlin.

Choi, Y. B. (1993). *Paradigms and conventions: Uncertainty, decision making, and entrepreneurship*. Ann Arbor: University of Michigan Press.

Clarke, M. (1993). Politics as government and politics as security. In M. Clarke (Ed.), *New perspectives on security* (pp. 42–60). London: Brassey's.

Coase, R. (1937). The nature of the firm. *Economica, 4,* 396–405.

Cochran, T. C. (1960). Cultural factors in economic growth. *Journal of Economic History, 20*(4), 515–530.

Compaine, B. M. (1981/1984). Shifting boundaries in the information marketplace. *Journal of Communication, 31*(1), 132–142. Republished in B. M. Compaine (Ed.), *Understanding new media: Trends and issues in electronic distribution of information* (pp. 97–120). Cambridge, MA: Ballinger.

Coulmas, F. (1992). *Language and economy*. Cambridge, UK: Blackwell.

Cowan, R., David, P. A., & Foray, D. (2000). The explicit economics of knowledge codification and tacitness. *Industrial and Corporate Change, 9*(2), 211–253.

Crandall, R. H. (1969). Information economics and its implications for the further development of accounting theory. *The Accounting Review, 44*(3), 457–466.

Cremer, J., & Khalil, F. (1992). Gathering information before signing a contract. *American Economic Review, 82*(3), 566–578.

Cronin, B. (1986). Towards information-based economies. *Journal of Information Science, 12*(3), 129–137.

Cronin, B., & Gudim, M. (1986). Information and productivity: A review of research. *International Journal of Information Management, 6*(2), 85–101.

Cronin, B., Overfelt, K., Fouchereaux, K., Manzvanzvike, T., Cha, M., & Sona, E. (1994). The Internet and competitive intelligence: A survey of current practice. *International Journal of Information Management, 14,* 204–222.

Cruise O'Brien, R. (Ed.). (1983). *Information, economics and power: The North-South dimension*. London: Hodder and Stoughton.

Davenport, E., & Cronin, B. (2000). Knowledge management: Semantic drift or conceptual shift? *Journal of Education for Library and Information Science, 41,* 294–306.

Davenport, E., & Snyder, H. W. (2005). Managing social capital. *Annual Review of Information Science and Technology, 39,* 517–550.

David, P. A., & Foray, D. (2002). An introduction to the economy of the knowledge society. *International Social Science Journal, 54*(171), 9–24.

Debons, A., King, D. W., Mansfield, U., & Shirley, D. L. (1981). *The information profession: Survey of an emerging field.* New York: Marcel Dekker.

De Bresson, C. (1987). The evolutionary paradigm and the economics of technological change. *Journal of Economic Issues, 21,* 751–762.

Debreu, G. (1959). *The theory of value.* New York: Wiley.

De Greene, K. B. (1991). *A systems-based approach to policy-making.* Boston: Kluwer.

Dezalay, Y., & Garth, B. G. (1996). *Dealing in virtue: International commercial arbitration and the construction of a transnational legal order.* Chicago: University of Chicago Press.

Diamond, P. A. (1978). Welfare analysis of imperfect information equilibrium. *Bell Journal of Economics, 9*(1), 82–105.

DiMaggio, P. (1991). Social structure, institutions, and cultural goods: The case of the United States. In P. Bourdieu & J. S. Coleman (Eds.), *Social theory for a changing society* (pp. 133–155). Boulder, CO: Westview Press.

Dodgson, M. (1993). *Technological collaboration in industry: Strategy, policy and internationalization in innovation.* New York: Routledge.

Dosi, G., Freeman, C., Nelson, R., Silverberg, G., & Soete, L. (Eds.). (1988). *Technical change and economic theory.* London: Pinter Publishers.

Dyson, K. H. F., & Humphreys, P. (1986). *The politics of the communication revolution in western Europe.* London: Frank Cass.

Easton, D. (1979). *A framework for political analysis.* Chicago: University of Chicago Press.

Ellger, C. (1995). Beyond the economic? Cultural dimensions of services: The RESER Survey of Service Research Literature in Europe in 2000. *The Services Industries Journal, 1,* 33–45.

Ellig, J., & Lin, D. (2001). A taxonomy of dynamic competition theories. In J. Ellig (Ed.), *Dynamic competition and public policy: Technology, innovation, and antitrust issues* (pp. 16–44). Cambridge, UK: Cambridge University Press.

Engelbrecht, H.-J. (1986). The Japanese information economy: Its quantification and analysis in a macroeconomic framework (with comparisons to the U.S.). *Information Economics and Policy, 2*(6), 277–306.

Enzensberger, H. M. (1974). *The consciousness industry: On literature, politics and the media.* New York: Seabury Press.

Evangelista, M. (1988). *Innovation and the arms race: How the United States and the Soviet Union develop new military technologies.* Ithaca, NY: Cornell University Press.

Ewald, F. (1991) Insurance and risk. In G. Burchell, C. Gordon, & P. Miller (Eds.) *The Foucault effect: Studies in governmentality* (pp. 197–210). Chicago: University of Chicago Press.

Ewen, S., & Ewen, E. (1982). *Channels of desire: Mass images and the shaping of American consciousness.* New York: McGraw-Hill.

Flor, A. G. (1993). The informatization of agriculture. *Asian Journal of Communication, 3*(2), 94–103.

Folt, J. H. (1975). Situational factors and peasants' search for market information. *Journalism Quarterly, 52,* 429–435.

Foray, D. (2004). *Economics of knowledge.* Cambridge, MA: MIT Press.

Foray, D., & Lundvall, B.-A. (1996). The knowledge-based economy: From the economics of knowledge to the learning economy. In D. Foray & B.-A. Lundvall (Eds.), *Employment and growth in the knowledge-based economy* (pp. 11–32). Paris: OECD.

Freeman, C., & Perez, C. (1988). Structural crises of adjustment, business cycles and investment behavior. In G. Dosi, C. Freeman, R. Nelson, G. Silverberg, & L. Soete (Eds.), *Technical change and economic theory* (pp. 38–66). London: Pinter Publishers.

A GAAT for services. (1985, October 12). *Economist*, p. 20.

Garcia, D. L. (2004). Networks and the evolution of property rights in the global, knowledge-based economy. In S. Braman (Ed.), *The emergent global information policy regime* (pp. 130–153). Houndsmills, UK: Palgrave Macmillan.

Garvin, D. A. (1993). Building a learning organization. *Harvard Business Review, 71*(4), 78–91.

Gershuny, J. J., & Miles, I. (1983). *The new services economy*. London: Frances Pinter.

Giersing, M. (1984). The commercialization of culture: Perspectives and precautions. In J. Wasko & V. Mosco (Eds.), *The critical communication review* (pp. 245–264). Norwood, NJ: Ablex.

Ginsburg, J. C. (1990). Creation and commercial value: Copyright protection of works of information. *Columbia Law Review, 90*, 1865–1938.

Goldfinger, C. (2000). Intangible economy and financial markets. *Communication and strategies, 40*, 59–89.

Goldin, C. (2001). The human capital century and American leadership: Virtues of the past. *Journal of Economic History, 61*, 263–291.

Gomery, D. (1989). Media economics: Terms of analysis. *Critical Studies in Mass Communication, 6*, 43–60.

Goolsbee, A. (2000). In a world without borders: The impact of taxes on Internet commerce. *Quarterly Journal of Economics, 115*(2), 561–577.

Grabher, G. (1993). Rediscovering the social in the economics of interfirm relations. In G. Grabher (Ed.), *The embedded firm: On the socioeconomics of industrial networks* (pp. 1–32). New York: Routledge.

Greguras, F. M. (1980). Tax-exempt bond financing of computer telecommunications and other high technology facilities. *Computer/Law Journal, 2*, 805–828.

Griffiths, J.-M. (1982). The value of information and related systems, products, and services. *Annual Review of Information Science and Technology, 17*, 269–284.

Griliches, Z., & Mairesse J. (Eds.). (1993). *Productivity issues in services at the micro level*. Dordrecht, NL: Kluwer.

Grossman, S. (1977). The existence of futures markets, noisy rational expectations and informational externalities. *Review of Economic Studies, 44*, 431–449.

Grossman, S., & Stiglitz, J. E. (1976). Information and competitive price systems. *American Economic Review, 64*, 246–253.

Grossman, S., & Stiglitz, J. E. (1980). On the impossibility of informationally efficient markets. *American Economic Review, 70*, 393–408.

Grossman, S. J. (1981). The informational role of warranties and private disclosure about product quality. *Journal of Law and Economics, 24*(3), 461–483.

Grunig, J. L. (1971). Communication and the economic decision-making processes of Columbian peasants. *Economic Development & Social Change, 19*, 580–597.

Guerin-Calvert, M. E., & Wildman, S. S. (Eds.). (1993). *Electronic services networks: Functions, structures, and public policy.* New York: Praeger.

Haas, E. B. (1990). *When knowledge is power: Three models of change in international organizations.* Berkeley: University of California Press.

Hall, P., & Preston, P. (1988). *The carrier wave: New information technology and the geography of information 1866–2003.* London: Unwin Hyman.

Hatchuel, A., Le Masson, P., & Weil, B. (2002). From knowledge management to design-oriented organisations. *International Social Science Journal, 54*(171), 25–38.

Hayek, F. A. (1945). The use of knowledge in society. *American Economic Review, 35,* 519–530.

Hayek, F. A. (1973). *Law, legislation, and liberty: A new statement of the liberal principles of justice and political economy.* Chicago: University of Chicago Press.

Hayes, R. M., & Erickson, T. (1982). Added value as a function of purchases of information services. *The Information Society, 1*(4), 307–339.

Hayward, T., & Preston, J. (1999). Chaos theory, economics and information: The implications for strategic decision-making. *Journal of Information Science, 25*(3), 173–182.

Hazlett, T. W. (1990). The rationality of US regulation of the broadcast spectrum. *Journal of Law and Economics, 33*(1), 133–175.

Hechter, M. (1984). When actors comply: Monitoring costs and the production of social order. *Acta Sociologica, 27*(3), 161–183.

Held, D. (1980). *Introduction to critical theory: Horkheimer to Habermas.* Berkeley: University of California Press.

Hepworth, M. (1989). *Geography of the information economy.* New York: Guilford Press.

Hindle, A., & Raper, D. (1976). The economics of information. *Annual Review of Information Science and Technology, 11,* 27–54.

Hirshleifer, J. (1973). Where are we in the theory of information? *American Economic Review, 63*(2), 31–51.

Hirshleifer, J., & Riley, J. G. (1979). The analytics of uncertainty and information: An expository survey. *Journal of Economic Literature, 17,* 1375–1421.

Hopwood, A. G., & Miller, P. (Eds.). (1994). *Accounting as social and institutional practice.* Cambridge, UK: Cambridge University Press.

Hulten, C. R. (1990). *Productivity growth in Japan and the United States.* Chicago: University of Chicago Press.

Hunter, J. A. (1984). What price information? *Information Services & Use, 4*(4), 217–223.

Hyde, A. (2003). *Working in Silicon Valley: Economic and legal analysis of a high-velocity labor market.* Armonk, NY: M. E. Sharpe.

INS v AP, 248 US 215 (1918).

Ito, Y. (1991). *Johoka* as a driving force of change. *KEIO Communication Review, 12,* 33–58.

Iyer, G. R., & Shapiro, J. M. (1999). Ethnic entrepreneurial and marketing systems: Implications for the global economy. *Journal of International Marketing, 7*(4), 83–111.

Jensen, M. C., & Meckling, W. H. (1992). Specific and general knowledge, and organizational structure. In L. Werin & H. Wijkander (Eds.), *Contract economics* (pp. 251–274). Oxford, UK: Blackwell.

Jonscher, C. (1982). The economic causes of information growth. *InterMedia, 10*(6), 34–37.

Jonscher, C. (1986). Information technology and the United States economy: Modeling and measurement. In G. Faulhaber, E. Noam, & R. Tasley (Eds.), *Services in transition: The impact of information technology on the service sector* (pp. 119–131). Cambridge, MA: Ballinger.

Jussawalla, M. (1984). The economic implications of satellite technology and the industrialization of space. *Telecommunications Policy, 8*(3), 237–248.

Jussawalla, M., & Cheah, C. (1983). Towards an information economy: A case study of Singapore. *Information Economics and Policy, 1*(2), 161–176.

Kahin, B. (2004). Codification in context. In S. Braman (Ed.), *The emergent global information policy regime* (pp. 39–61). Houndsmills, UK: Palgrave Macmillan.

Karunaratne, N. D. (1984). Planning for the Australian information economy. *Information Economics and Policy, 1,* 345–367.

Katsenelinboigen, A. (1992). *Indeterministic economics.* New York: Praeger.

Katz, R. L. (1986). Explaining information sector growth in developing countries. *Telecommunications Policy, 10*(3), 209–228.

Katzer, J., & Fletcher, P. T. (1992). The information environment of managers. *Annual Review of Information Science and Technology, 27,* 227–263.

Kaynak, E. (Ed.). (1986). *Service industries in developing countries.* London: Frank Cass.

Kihlstrom, R. E., & Riordan, M. H. (1984). Advertising as a signal. *Journal of Political Economy, 92*(3), 427–450.

King, D. W., Roderer, N. K., & Olsen, H. A. (Eds.). (1983). *Key papers in the economics of information.* White Plains, NY: Knowledge Industry Publications.

King, J. L., & Kraemer, K. L. (1995). *Information infrastructure, national policy, and global competitiveness.* Irvine: University of California-Irvine Center for Research on Information Technology and Organizations Working Paper.

Kissinger, H. A. (1971). Domestic structure and foreign policy. In W. F. Hanrieder (Ed.), *Comparative foreign policy: Theoretical essays* (pp. 22–50). New York: David McKay Company.

Klemperer, P. (1999). Auction theory: A guide to the literature. *Journal of Economic Surveys, 13,* 227–286.

Kling, R., & Allen, J. P. (1993). *How the marriage of management and computing intensifies the struggle for personal privacy.* Unpublished manuscript.

Knight, F. (1921). *Risk, uncertainty, and profit.* Boston, MA: Houghton Mifflin.

Koenig, M. E. D. (1990). Information services and downstream productivity. *Annual Review of Information Science and Technology, 25,* 55–86.

Kogut, B. (Ed.). (1993). *Country competitiveness: Technology and the organizing of work.* New York: Oxford University Press.

Kondratieff, N. D. (1984). *The long wave cycle* (G. Daniels, Trans.). New York: Richardson and Snyder.

Lamberton, D. M. (Ed.). (1971). *Economics of information and knowledge.* Harmondsworth, UK: Penguin Books.

Lamberton, D. M. (1984a). The economics of information and organization. *Annual Review of Information Science and Technology, 19,* 3–30.

Lamberton, D. M. (1984b). The emergence of information economics. In M. Jussawalla & H. Ebenfield (Eds.), *Communication and information economics: New perspectives* (pp. 7–22). New York: North-Holland.

Lamberton, D. M. (1992). Information economics: "Threatened wreckage" or new paradigm? In U. Himmelstrand (Ed.), *Interfaces in Economic and Social Analysis* (113–123). London: Routledge.

Lamberton, D. M. (1994). The information economy revisited. In R. E. Babe (Ed.), *Information and communication economics* (pp. 1–33). Amsterdam: North Holland.

Lamberton, D. M. (1995). Technology, information and institutions. In D. M. Lamberton (Ed.), *Beyond competition: The future of telecommunications* (pp. 3–12). Amsterdam: Elsevier Science.

Lamberton, D. M. (1998a). Information economics research: Points of departure. *Information Economics and Policy, 10*(3), 325–330.

Lamberton, D. M. (1998b). Information economics: Research strategies. *International Journal of Social Economics, 25*(2–4), 338–357.

Landreth, H., & Colander, D. C. (1994). *History of economic thought* (3rd ed.). Boston: Houghton Mifflin.

Lash, S. (1993). Pierre Bourdieu: Cultural economy and social change. In C. Calhoun, E. LiPuma, & M. Postone (Eds.), *Bourdieu: Critical perspectives* (pp. 193–211). Chicago: University of Chicago Press.

Leeson, K. W. (1981). *Trade issues in telecommunications and information.* Washington, DC: National Telecommunications and Information Association.

Lipinski, T. A. (1998). Information ownership and control. *Annual Review of Information Science and Technology, 33*, 3–38.

Locksley, G. (1988). On social software. In G. Muskens & J. Gruppelaar (Eds.), *Global telecommunication networks: Strategic considerations* (pp. 139–158). Dordrecht, NL: Kluwer.

Lucas, R., Jr. (1975). An equilibrium model of the business cycle. *Journal of Political Economy, 83*, 1113–1144.

Lury, C. (1993). *Cultural rights: Technology, legality and personality.* New York: Routledge.

Lytle, R. H. (1986). Information resources management. *Annual Review of Information Science and Technology, 21*, 309–336.

Machlup, F. (1962). *The production and distribution of knowledge in the United States.* Princeton, NJ: Princeton University Press.

Machlup, F. (1979). Uses, value, and benefits of knowledge. *Knowledge: Creation, Diffusion, Utilization, 1*(1), 62–81.

Machlup, F. (1980). *Knowledge: Its creation, distribution, and economic significance. Vol. I: Knowledge and knowledge production.* Princeton, NJ: Princeton University Press.

Machlup, F. (1982). *Knowledge: Its creation, distribution, and economic significance. Vol. II: The branches of learning.* Princeton, NJ: Princeton University Press.

Machlup, F. (1983). The economics of information: A new classification. *InterMedia, 11*(2), 28–37.

Macho-Stadler, I., & Perez-Castrillo, D. (1997). *An introduction to the economics of information: Incentives and contracts* (R. Watt, Trans.). Oxford, UK: Oxford University Press.

Mac Morrow, N. (2001). Knowledge management: An introduction. *Annual Review of Information Science and Technology, 35*, 381–422.

Mansfield, E. (1988). Intellectual property rights, technological change, and economic growth. In C. E. Walker & M. A. Bloomfield (Eds.), *Intellectual property rights and capital formation in the next decade* (pp. 3–26). Boston: University Press of America.

March, J. G., & Simon, H. A. (1963). *Organizations* (2nd ed.). New York: Wiley.

Marcus, G. E., & Fischer, M. M. J. (1986). *Anthropology as cultural critique: An experimental moment in the human sciences.* Chicago: University of Chicago Press.

Markillie, P. (2004, May 15). A perfect market: A survey of e-commerce [Special section]. *The Economist,* 1–20.

Marschak, J. (1954). Towards an economic theory of organization and information. In R. M. Thrall, C. H. Coombs, & R. L. Davis (Eds.), *Decision processes* (pp. 187–220). New York: Wiley.

Marschak, J. (1959). *Remarks on the economics of information: Contributions to scientific research in management.* Los Angeles, CA: University of California-Los Angeles.

Marschak, J. (1968). Economics of enquiring, communicating, deciding. *American Economic Review, 58*(2), 1–18.

Marschak, J. (1971). Economics of information systems. *Journal of the American Statistical Association, 64,* 192–219.

Marshall, A. (1890). *Principles of economics.* London: Macmillan.

Mattelart, A., & Mattelart, M. (1992). On new uses of media in times of crisis. In M. Raboy & B. Dagenais (Eds.), *Media, crisis and democracy: Mass communication and the disruption of social order* (pp. 162–180). London: Sage.

Mattelart, M., & Mattelart, A. (1990). *The carnival of images.* New York: Bergin & Garvey.

McKee, D. L. (1988). *Growth, development, and the service economy in the Third World.* New York: Praeger.

McLaughlin, J. F., & Birinyi, A. E. (1984). Mapping the information business. In B. Compaine (Ed.), *Understanding new media: Trends and issues in electronic distribution of information* (pp. 19–67). Cambridge, MA: Ballinger.

McNamara, J. R. (1991). *The economics of innovation in the telecommunications industry.* Westport, CT: Quorum Books.

Metcalfe, B. (1995, October 2). From the ether: Metcalfe's law: A network becomes more valuable as it reaches more users. *Infoworld,* p. 53.

Meyer, J. R., Wilson, R. W., Baughcum, M. A., Burton, E., & Caouette, L. (1980). *The economics of competition in the telecommunications industry.* Cambridge, MA: Oelgeschlager, Gunn & Hain.

Mick, C. K. (1979). Cost analysis of information systems and services. *Annual Review of Information Science and Technology, 14,* 37–64.

Middleton, K., & Jussawalla, M. (1981). *Economics of communication: An annotated bibliography with abstracts.* New York: Pergamon Press.

Miege, B. (1987). The logics at work in the new cultural industries. *Media, Culture & Society, 9*(3), 273–290.

Milgrom, P., & Roberts, J. (1986). Price and advertising signals of product quality. *Journal of Political Economy, 94*(4), 796–821.

Mirrlees, J. (1971). An exploration in the theory of optimum income taxation. *Review of Economic Studies, 38*(2), 175–208.

Mokyr, J. (1990). *The lever of riches: Technological creativity and economic progress.* New York: Oxford University Press.

Molho, I. (1997). *The economics of information: Lying and cheating in markets and organizations.* Malden, MA: Blackwell.

Monk, P. (1992). *Economic aspects of infrastructural IT systems.* Glasgow, Scotland: University of Strathclyde.

Morgan, E. S., & Morgan, H. M. (1953). *The Stamp Act crisis.* Chapel Hill, NC: University of North Carolina Press.

Morgan, G. (1986). *Images of organizations.* Beverly Hills, CA: Sage Publications.

Mosco, V. (1989). *The pay-per society: Computers and communication in the information age.* Norwood, NJ: Ablex.

Murdock, G., & Golding, P. (1990). Information poverty and political inequality: Citizenship in the age of privatized communications. *Journal of Communication, 39*(3), 180–195.

Muth, J. F. (1961), Rational expectations and the theory of price movements. Econometrica, 29, 315–335.

Nass, C. (1987). Following the money trail: 25 years of measuring the information economy. *Communication Research, 14*(6), 698–708.

Negt, O. (1978). Mass media: Tools of domination or instruments of liberation? Aspects of the Frankfurt School's communications analysis. *New German Critique, 14,* 61–80.

Nelson, P. (1970). Information and consumer behavior. *Journal of Political Economy, 78*(2), 311–329.

Nelson, P. (1974). Advertising as information. *Journal of Political Economy, 82*(4), 729–754.

Nelson, R., & Winter, S. (1982). *An evolutionary theory of economic change.* Cambridge, MA: Belknap Press.

Nissen, H. J., Damerow, P., & Englund, R. K. (1994). *Archaic bookkeeping: Writing and techniques of economic administration in the ancient Near East.* Chicago: University of Chicago Press.

Nitzan, J. (1998). Differential accumulation: Toward a new political economy of capital. *Review of International Political Economy, 5*(2), 169–217.

North, D. C. (1981). *Structure and change in economic history.* New York: Norton.

Overman, E. S. (1989). Continuities in the development of the physical and social sciences: Principles of a new social physics. *Knowledge in Society, 2*(2), 80–93.

Owen, B. M., & Wildman, S. S. (1992). *Video economics.* Cambridge, MA: Harvard University Press.

Palumbo-Liu, D. (1997). Introduction: Unhabituated habituses. In D. Palumbo-Liu & H. U. Gumbrecht (Eds.), *Streams of cultural capital: Transnational cultural studies* (pp. 1–22). Stanford, CA: Stanford University Press.

Parker, I. (1993). Commodities as sign-systems. In R. Babe (Ed.), *Information and communication in economics* (pp. 69–91). Boston/Amsterdam: Kluwer.

Pau, L.-F. (2002). The communications and information economy: Issues, tariffs and economics research areas. *Journal of Economic Dynamics and Control, 26,* 1651–1675.

Pearton, M. (1984). *Diplomacy, war and technology since 1830.* Lawrence: University of Kansas Press.

Pemberton, J. M. (1995). The information economy: A context for records and information management. *Records Management Quarterly, 29*(3), 54–59.

Peters, J. D. (1988). Information: Notes toward a critical history. *Journal of Communication Inquiry, 12*(2), 9–23.

Peters, J. D. (2004). The "marketplace of ideas": History of a concept. In A. Calabrese & C. Sparks (Eds.), *Toward a political economy of culture: Capitalism and communication in the twenty-first century* (pp. 65–82). Boulder, CO: Rowman and Littlefield.

Persico, N. (1996). *Acquiring information: Three essays on the economics of information.* Evanston, IL: Unpublished dissertation, Northwestern University.

Pigou, A. C. (1920). *The economics of welfare.* London: Macmillan.

Polanyi, K. (1957). Marketless trading in Hammurabi's time. In K. Polanyi, C. M. Arensberg, & H. W. Pearson (Eds.), *Trade and markets in early empires: Economies in history and theory* (pp. 12–26). Glencoe, IL: Free Press.

Porat, M. U. (1977). *The information economy: Definition and measurement* [OT Special Publication 77–12(1)]. Washington, DC: Office of Telecommunications, U.S. Department of Commerce.

Porter, M. E., & Millar, V. E. (1985). How information gives you competitive advantage. *Harvard Business Review, 63*(4), 2–13.

Posner, R. A. (1984). An economic theory of privacy. In F. D. Schoeman (Ed.), *Philosophical dimensions of privacy* (pp. 333–345). Cambridge, UK: Cambridge University Press.

Posner, R. A. (1986). *Economic analysis of the law* (4th ed.) (pp. 627–638). Boston: Little, Brown & Co.

Preston, P. (1989). *The information economy and the international standard industrial classification (ISIC): Proposals for updating the ISIC.* Brunel, UK: Economic and Social Research Council Programme on Information and Communication Technology.

Putnam, R. D. (2000). *Bowling alone: The collapse and revival of American community.* New York: Simon & Schuster.

Radner, R. (1989). Competitive equilibrium under uncertainty. In P. A. Diamond & M. Rothschild (Eds.), *Uncertainty in economics: Readings and exercises* (Rev. ed., pp. 177–204). San Diego, CA: Academic Press.

Rayport, J. F., & Sviokla, J. J. (1995). Exploiting the virtual value chain. *Harvard Business Review, 73*(6), 75–85.

Repo, A. J. (1986). *Analysis of the value of information: A study of some approaches taken in the literature of economics, accounting and management science.* Sheffield, UK: The University of Sheffield.

Repo, A. J. (1987). Economics of information. *Annual Review of Information Science and Technology, 22*, 3–35.

Richardson, G. B. (1959). Equilibrium, expectations and information. *The Economic Journal, 69*(274), 223–237.

Riley, J. G. (2001). Silver signals: Twenty-five years of screening and signaling. *Journal of Economic Literature, 39*(2), 432–473.

Rodrik, D. (1998). Promises, promises: Credible policy reform via signaling. *Economic Journal, 99*(3), 756–772.

Roeh, I., & Cohen, A. (1992). One of the bloodiest days: A comparative analysis of open and closed TV news. *Journal of Communication, 42*(2): 42-55.

Rogoff, K. (1990). Equilibrium political budget cycles. *American Economic Review, 80*(1), 21–36.

Rogoff, K., & Siebert, A. (1988). Elections and macroeconomic policy cycles. *Review of Economic Studies, 55*(1), 1–16.

Romer, P. M. (1994). The origins of endogenous growth. *Journal of Economic Perspectives, 8*(1), 3–22.

Rosenau, J. N., & Durfee, M. (1995). *Thinking theory thoroughly.* Boulder, CO: Westview Press.

Rostow, W. (1975). *How it all began: Origins of the modern economy.* New York: McGraw-Hill.

Rothschild, M. (1974). Searching for the lowest price when the distribution of prices is unknown. *Journal of Political Economy, 82*(4), 589–711.

Rothschild, M., & Stiglitz, J. (1976). Equilibrium in competitive insurance markets: An essay on the economics of imperfect information. *Quarterly Journal of Economics, 90,* 629–650.

Rubin, M. R., & Huber, M. T. (1984). *The knowledge industry in the United States, 1960–1980.* Princeton, NJ: Princeton University Press.

Sabel, C. (1991). Moebius-strip organizations and open labor markets: Some consequences of the reintegration of conception and execution in a volatile economy. In P. Bourdieu & J. Coleman (Eds.), *Social theory for a changing society* (pp. 23–54). Boulder, CO: Westview Press.

Salvatore, D. (2003). *Microeconomics: Theory and applications* (4th ed.). New York: Oxford University Press.

Sandler, T. (2001). *Economic concepts for the social sciences.* Cambridge, UK: Cambridge University Press.

Saunders, R. J. (1983). Telecommunications in the developing world: Investment decision and performance monitoring. *Telecommunications Policy, 7*(4), 277–284.

Scazzieri, R. (1993). *A theory of production: Tasks, processes, and technical practices.* Oxford, UK: Clarendon Press.

Schement, J. R. (1990). Porat, Bell, and the information society reconsidered: The growth of information work in the early twentieth century. *Information Processing & Management, 26*(4), 449–465.

Schiller, H. I. (1984). *Information and the crisis economy.* Norwood, NJ: Ablex.

Schmoranz, I. (1981). Macroeconomic sectors of the information economy. In H.-P. Gassmann (Ed.), *Information, computer and communications policies for the 80's* (pp. 75–82). Amsterdam: Organisation for Economic Co-operation and Development.

Schumpeter, J. (1939). *Business cycles.* New York: McGraw Hill.

Sengupta, A. (1987). *Three essays in economic theory.* Unpublished doctoral dissertation, State University of New York at Stony Brook.

Shapiro, C., & Varian, H. R. (1999). *Information rules: A strategic guide to the network economy.* Cambridge, MA: Harvard Business School Press.

Shi, Y. (2001). *The economics of scientific knowledge: A rational choice institutionalist theory of science.* Cheltenham: Edward Elgar.

Simon, H. (1957). *Administrative behavior: A study of decision-making processes in administrative organization.* New York: Macmillan.

Smith, A. (1776/1937). *The wealth of nations.* New York: Collier.

Smythe, D. (1979). Communications: Blindspot of Western Marxism. *Canadian Journal of Political and Social Theory, 1*(3), 1–27.

Snyder, H. W., & Pierce, J. B. (2002). Intellectual capital. *Annual Review of Information Science and Technology, 36*, 467–500.

Solow, R. M. (1957). Technical change and the aggregate production function. *Review of Economics and Statistics, 39*, 312–320.

Spence, A. M. (1973). Job market signaling. *Quarterly Journal of Economics, 87*(3), 355–379.

Spence, A. M. (1974). An economist's view of information. *Annual Review of Information Science and Technology, 9*, 57–78.

Spigai, F. (1991). Information pricing. *Annual Review of Information Science and Technology, 26*, 39–73.

Stein, A. A. (1982). Coordination and collaboration: Regimes in an anarchic world. *International Organization, 36*(2), 299–324.

Steinmueller, W. E. (2002). Knowledge-based economies and information and communication technologies. *International Social Science Journal, 54*(171), 141–154.

Stephen, L. (1907). *English literature and society in the eighteenth century.* New York: Putnam.

Stevenson, N. (1995). *Understanding media cultures: Social theory and mass communication.* Thousand Oaks, CA: Sage.

Stewart, T. A. (1998). *Intellectual capital: The new wealth of organizations.* New York: Doubleday.

Stigler, G. (1961). The economics of information. *Journal of Political Economy, 69*(3), 213–225.

Stigler, G. J. (1962). Information in the labor market. *Journal of Political Economy, 70*(5), 94–105.

Stigler, G. J. (1983). Nobel lecture: The process and progress of economics. *Journal of Political Economy, 91*(4), 529–545.

Stiglitz, J. E. (1982). Information and capital markets. In W. F. Sharpe & C. Cootner (Eds.), *Financial economics* (pp. 118–158). Englewood, NJ: Prentice-Hall.

Stiglitz, J. E. (1985). Information and economic analysis: A perspective. *Economic Journal, 95*, 21–41.

Stiglitz, J. E. (2000). The contributions of the economics of information to twentieth century economics. *The Quarterly Journal of Economics, 115*(4), 1441–1478.

Taylor, R. S. (1982). Value-added processes in the information life cycle. *Journal of the American Society for Information Science, 33*(5), 341–346.

Taylor, R. S. (1986). *Value-added processes in information systems.* Norwood, NJ: Ablex.

Telser, L. G. (1973). Searching for the lowest price. *American Economic Review, 63*(2), 40–49.

Temin, P. (Ed.). (1991). *Inside the business enterprise: Historical perspectives on the use of information.* Chicago: University of Chicago Press.

Tucci, V. K. (1988). Information marketing for libraries. *Annual Review of Information Science and Technology, 23*, 59–82.

U.S. National Telecommunications and Information Administration. (2000). *Falling through the net: Towards digital inclusion.* Washington, DC: U.S. Department of Commerce.

Veblen, T. (1899). *The theory of the leisure class: An economic study of institutions.* London: Allan & Unwin.

Vickrey, W. (1961). Counterspeculation, auctions and competitive sealed tenders. *Journal of Finance, 16*(1), 41–50.

Vishwanath, K., & Finnegan, J. R., Jr. (1996). The knowledge gap hypothesis: Twenty-five years later. *Communication Yearbook, 19*, 187–227.

Voge, J. (1983). The political economics of complexity: From the information economy to the "complexity economy." *Information Economics and Policy, 1*, 97–114.

Wagner, P., Weiss, C. H., Wittrock, B., & Wollman, H. (Eds.). (1991). *Social sciences and modern states: National experiences and theoretical crossroads.* Cambridge, UK: Cambridge University Press.

Webber, S. A. E. (1998). Pricing and marketing online information services. *Annual Review of Information Science and Technology, 33*, 39–83.

Webster, F. (1995). *Theories of the information society.* New York: Routledge.

Webster, F., & Robins, K. (1986). *Information technology: A Luddite analysis.* Norwood, NJ: Ablex.

Weech, T. (1995). The teaching of economics of information in schools of library and information science in the US: A preliminary analysis. *Proceedings of the Annual Meeting of the American Society for Information Science, 70*–75.

Wheelan, C. (2002). *Naked economics: Undressing the dismal science.* New York: W. W. Norton.

Wiley, R. E., & Leibowitz, D. E. (1987). The Electronic Communications Privacy Act of 1986 moves privacy protection towards the 21st century. *Telematics, 4*(2), 1–12.

Williams, S. R., & Radner, R. (1968). *Informational externalities and the scope of efficient strategy mechanisms* (Discussion Paper #761). Chicago: Northwestern University, Center for Mathematical Studies in Economics and Management Science.

Williamson, O. E. (1975). *Markets and hierarchies—Analysis and antitrust implications: A study in the economics of internal organization.* New York: Simon & Schuster.

Winter, H. J. (1988). A view from the US State Department. In C. E. Walker & M. A. Bloomfield (Eds.), *Intellectual property rights and capital formation in the next decade* (pp. 99–106). Boston: University Press of America.

Woodford, M. (2001, December). *Monetary policy in the information economy* (National Bureau of Economic Research Working Paper #W8674). Cambridge, MA: The Bureau.

Wriston, W. B. (1992). *The twilight of sovereignty: How the information revolution is transforming our world.* New York: Scribner.

Yates, J. (1989). *Control through communication: The rise of system in American management.* Baltimore: Johns Hopkins University Press.

Yates, J. (1994). Evolving information use in firms, 1850–1920: Ideology and information techniques and technologies. In L. Bud-Frierman (Ed.), *Information acumen: The understanding and use of knowledge in modern business* (pp. 7–25). New York: Routledge.

Zappia, C. (1999). The economics of information: Market socialism and Hayek's legacy. *History of Economic Ideas, 7*(1–2), 105–138.

CHAPTER 2

The Geographies of the Internet

Matthew Zook
University of Kentucky

Introduction

The word "geography" derives from the Greek term meaning "to write about the Earth," and geography has been a dynamic field of study throughout recorded history. Although many equate geography with simply memorizing names on a map, it is better defined as "the study of the Earth's surface as the space within which the human population lives" with particular attention to the "spatial variation that can occur" (Haggett, 1993, p. 220). The definition of the Earth's surface may seem self-explanatory, in other words, the physical landscape around us, but communication media and technologies have created new nonphysical spaces for human interaction. Although these communication spaces have long existed and shaped the development of civilizations (Innis, 1950, 1951), the rise of electronic information technologies has created a communication space that is categorically richer, more flexible, and more accessible for larger numbers of people. Of particular interest for geographers is the creation—via the widespread diffusion of the Internet—of complex new geographies of interaction and connection between people and places, both near and distant, that blend virtual spaces and physical places.

Although the technical capability of the Internet often promotes an image of uniform and utopian connectivity, these new geographies of the Internet are exceedingly complex. The Internet is far from being a uniform process or system; it is not simply overlaid on existing patterns nor does it entail the end of geography as some have claimed. Rather, it provides new geographies of connection and exclusion (e.g., the difference in access available in physically proximate downtown business districts vs. downtown slums); invites experimentation; and opens the possibility for contests between differing visions of the world. These geographies—ranging from the convergence of hitherto isolated individuals into social and cultural movements to the offshoring of computer programming jobs—are not homogeneous. Instead, the manifestation of Internet geographies turns on issues of cost, usability, reliability, culture, and power to name but a few. Simple expectations of a uniform geography or ubiquitous access are simply unreflective of the reality of the Internet.

This chapter discusses geographies of the Internet rather than the "Geography of the Internet" to emphasize the complexity of the changes in spatial organization engendered by this network. In short, there is no Geography (capital G) of the Internet, but rather a large number of geographies of the Internet created through the interaction of this technology with the places in which and the people by whom it is used. These Internet geographies are both immensely empowering and potentially overpowering depending upon one's location and economic, cultural, and social background. This chapter first provides a brief background to the history of geographers' research in communications, as well as relevant concepts and theories applied to the topic. It then focuses specifically on the ways in which the Internet has been analyzed as an extension of human habitation on the Earth's surface, namely, expanding the core question of geography to electronic spaces. Concentrating primarily on research by geographers (although referencing relevant works outside the discipline), this review explores the geography of Internet technology (where are the technical components of the Internet and how do they vary over space?); the Internet as a recombinant space for political, cultural, and economic interaction (how does the Internet blend with existing human activity embedded in physical places?); and the Internet as a new space to explore and visualize (how do we map the Internet?).

Geographers and Communications Geographies

Communications in general, and more particularly electronic communications, have been the subject of relatively little geographic study and have been characterized by Hillis (1998, p. 559) as the "poor step sister in the 'family' of human geographies."[1] Nevertheless, those interested in space have long explored how changes in communication technologies could bring about "community without propinquity" (Webber, 1965, p. 25) or "time/space convergence" (Janelle, 1973, p. 9) or "time–space compression" (Harvey, 1989, pp. 305–307). Although these formulations differ on a number of theoretical points, they all stress the importance of perceptions (and how these perceptions affect action) of the relative relation between space and time (particularly the reduction of the time required to move through space) as it is embedded in networks of social and economic relations.

This interest in changing time/space relationships is tempered by geographers' deep suspicion toward notions of technological determinism (Graham, 1998; Graham & Marvin, 1996). Even the concept of a "technological impact" per se is problematic for geographers, given the often linear expectations projected onto technology compared with the complex differences between the places that geographers study. Innovations in communication technology (ranging from clay tablets to the Internet) are inevitably shaped as much by the places they connect as they themselves shape these places (Thrift, 1996). Geographers are intensely aware that the use of technologies is part of ongoing social and political struggles embedded in complex power and social networks.

In contrast to this attentiveness to difference, much of the rhetoric surrounding the Internet in the 1990s focused on the transcendence of technology over geography with the widely held assumption (heavily promoted by technology companies) that the Internet meant a world in which everyone and every place was connected (see Cairncross [1997] and Negroponte [1995] for two examples of this). Closely tied to a neoclassical preoccupation with perfect information access and lowering of trade barriers, many argued that the Internet would bring about a perfectly competitive era (or "new economy") in which each individual (through the power of ubiquitous information delivered via the Internet) could achieve his or her potential (see Kelly, 1998).

But this belief in a ubiquitous Internet is belied by a fundamental (if not *the* fundamental) tenet of geography; variation over space. Geographers have historically been concerned with questions of how landscapes, resources, climate, cultures, and economies differ across space, and it is with this sensibility that much of the geographic research on the Internet has been undertaken. As Batty (1993, p. 615) argues in one of the earliest articles by a geographer specifically on the Internet, "this idea of one product or process substituting for another is basic to neoclassical economics, but it increasingly appears to be wrongheaded. It is based on the notion that life is a zero-sum game in which new things must take away the old, in which there are ultimate limits on our capacity to communicate and interact, that one technology replaces another." Without denying the increased reach across physical space that the technologies of the Internet provide, geographers focus on how this technology creates complexity—how it enhances or challenges rather than simply replaces. A technology as fast-moving as the Internet constantly recreates differences—sometimes minor, sometimes significant—between places, peoples, and—in the case of the Internet—within the new virtual spaces that it engenders. Thus a ubiquitous and uniform global digital geography is more rhetoric than reality (Warf, 2001). Even in its densest parts, the Internet is accessed, adapted, and appropriated differently depending on individual and societal imagination, culture, and history.

Space, Place, and the Internet

Central to understanding geographies of the Internet are the concepts of space and place. Although instantly recognizable and already appropriated by William Gibson (1984) in his formulation of cyberspace, these terms have specific meanings and uses for geographers. Space is abstract and geometric—like an impersonal location on a grid—but place is "seen as an extended locale of human activity imbued with the heritage, identity, and commitment of people and institutions ... the meaning of place is subject to transformation through social and technological innovation, and through various levels and means of association and experience" (Janelle & Hodge, 2000, p. 3). Place is a particularly important concept for understanding the geographies of the Internet, because it allows for the construction of places that encompass both the "real" and the "virtual" (Hillis, 1998) and differentiates between spatial scales, such as the local nature of access (or lack thereof) vs. the global nature of Internet interaction.

The Internet (and, more generally, all space-transcending technologies) challenges the historical definition of place, a bounded physical location of human activity, and encourages theories about new hybrid spaces, combining physical and electronic elements, which in turn create new definitions of "near" and "far" (Couclelis, 1996). Distance decay, the fact that nearby things interact more strongly than distant things, is an axiom of geography, but the Internet challenges our very notions of what "distant" and "near" mean. In many ways Euclidean distance (i.e., physical distance) retains its role, but Massey (1993, p. 66) argues that, increasingly, geographers view places as "articulated moments in networks of social relations and understandings" in contrast to historical notions of physical areas surrounded by boundaries.

It is largely through this expanded notion of place that geographers have approached the study of the Internet. As Kitchin (1998a, 1998b) argues, "cyberspaces coexist with geographic spaces, providing a new layer of virtual sites superimposed over geographic spaces" (1998b, p. 403). This layering of the Internet onto places, however, is by no means uniform but filtered through their social, cultural, economic, and political composition. Dodge (2001, p. 1) argues that the Internet "should be treated as an extension of the geographic realm, not as some disembodied, parallel universe. Nevertheless, cyberspace is changing geography; it is warping space, shrinking distance, and modifying our sense of place. Understanding these complex warpings and distortions is very much at the heart of cybergeography." [2]

Space of Flows and Space of Places

Arguably the most appropriate metatheory (a thorny issue for geographers, given their interest in difference over space and their use of multiple theoretical perspectives) for understanding the relationship between space, place, and the Internet was developed by Castells over the course of the 1990s, starting with his *Informational City* (Castells, 1989) and continuing through his *Network Society* trilogy (Castells, 1996, 1997, 1998) to his most recent work, *Internet Galaxy* (Castells, 2001). He theorizes that the interaction between places and new communications technologies is creating a network society in which an electronically formulated "space of flows" connects places to a global network of social and economic interactions (Castells, 1996, p. 423). He further argues that "the space of flows is not placeless, although its structural logic is. It is based on an electronic network, but this network links up specific places, with well-defined social, cultural, physical, and functional characteristics. ... Both nodes and hubs are hierarchically organized according to their relative weight in the network. But such a hierarchy may change depending upon the evolution of activities processed through the network" (Castells, 1996, p. 413). This formulation balances the global and the local and affirms the multiplicity of emerging Internet geographies.

In short, the fact that the Internet allows for the disintermediation of existing (and often locally based) activities into recombined virtually based relations does not mean that this will necessarily occur or that it will occur in the same way in all places. The Web-based interface to the Internet provides the illusion of a realm divorced from socially constructed cultures and markets built over centuries. This perception, however, belies the continual influence of places and their cultural, regulatory, and economic systems on Internet-mediated interactions. Rather, the characteristics of localities, which Castells (1996) refers to as the "space of places," determine the structure of the network and each place's relation (or irrelevance) to it. Not every geographer will formally adopt Castells' theories, but his emphasis on the interaction between the globalizing/totalizing tendencies of information space (i.e., the space of flows) and the differentiating effect of places (i.e., the space of places) is at the heart of most studies of the geographies of the Internet (see Adams & Ghose, 2003; Dodge & Kitchin, 2000b; Zook, 2005).

Analyzing Multiple Internet Geographies

The effort to evaluate the effects of the global and the local is also reflected in Internet geography's antecedents: geographic studies of communications and information use. These range from being as Internet-specific as Kellerman's (1986) examination of BITNET to more general studies of telecommunications and cities (Gillespie, 1992; Langdale, 1989; Moss, 1987) to interactions between information/communications and geography more broadly defined (Brunn & Leinbach, 1991; Castells, 1989; Hepworth, 1990; Kellerman, 1993). For the most part, these studies were conceived more broadly than the Internet per se because, at this early date, few geographers predicted (or perhaps believed) that this network would burgeon into the global and seemingly ubiquitous system that exists today. Batty (1993), Batty and Barr (1994), and Brunn, Jones, and Purcell (1994) offer some of the earliest work specifically on Internet geographies, although others outside the discipline, most notably Quarterman (1990), had begun to pursue the subject.[3]

This early interest in the geographies of the Internet by those outside geography mirrors the continued and regular interaction by Internet geographers with researchers beyond their immediate discipline. For example, geographers focused on visualizing and representing informational space or virtual places share much with computer scientists. In contrast, those interested in cultural or economic issues interact with architects, anthropologists, urban planners, sociologists, or economists. This necessarily makes the scope of Internet geography wide, but, at the core, all work shares an interest in the spatial component of human interaction and how the fortunes of specific places have co-evolved alongside the use of this new technology.

As the field of Internet geography (or cybergeography) developed, a distinction emerged between research on the *technology* and that on the *use* of the Internet, mirroring a partition within the larger discipline. Although geographers emphasize the fundamental interconnection between nature and society, a delineation is often made (both externally and internally) between physical geography (Earth processes such as weather and geomorphology) and human geography (the spatial component of human culture, society, and economy). Likewise, Internet geography developed lines of research focused on the *technology and infrastructure* of the Internet (bandwidth and fiber networks, hosts, etc.) and how *use* of this technology has blended with existing cultural, political, and economic structures manifest in physical places (virtual communities and movements, e-commerce, etc.). Although there is a fundamental and reflexive bond between Internet technology and use (just as there is between nature and society), this review employs the distinction as a means to understand the multiplicity of ways in which the spatial component of human activity has been extended to the Internet.

Building largely upon work by geographers (although including a range of influential research outside the discipline), this chapter first reviews the geography of Internet technology (e.g., where are the technical components of the Internet and how does this vary over space?) before turning to the geography of the use of the Internet (e.g., how does the Internet blend with existing human activity embedded in physical places?). This second topic (the human geography of the Internet) is further divided into topical concentrations of cultural and political geographies and economic geographies of the Internet, reflecting a common distinction within the discipline of geography. A third and final topic—visualizations of the Internet—employs a basic tool of geography (maps) to address the question, "How do we map the Internet?". Technical and human Internet geographies are the source of many useful representations of the Internet; this final topic expands into issues of representation of data from electronic space that may (or may not) have a spatial association in the physical world. This survey of these three main points demonstrates both the breadth of geographic research and the complexity of spatial transformation resulting in multiple geographies of the Internet.

Technical Geographies of the Internet

At first blush, the technical geographies of the Internet (locating the infrastructure, nodes, and data flows of the network) seem the most straightforward both to understand and research. It quickly becomes apparent, however, that this initial impression is overly simplistic. First, unlike most other infrastructure such as roads or sewers, the physical components of the Internet are largely invisible to the casual observer. Graham and Marvin (1996) argue that this basic fact is in large part responsible for the relative dearth (until recently) of studies in communications geography (see also Hillis, 1998). Nevertheless, Internet infrastructure exists and can be mapped and analyzed as any other piece of infrastructure. But as Hayes (1997) demonstrates and any number of researchers discuss (Moss & Townsend, 1997; Zook, 2000), obtaining accurate and comprehensive data on this infrastructure is difficult, costly, time-consuming, or a combination of all three.[4] Data with geographic relevance are increasingly available, but much of the data used in this work around the end of the 20th century was laboriously collected by the researchers themselves.[5]

Despite these difficulties (or perhaps inspired by them), several researchers and institutions have both collected and analyzed geographically marked Internet infrastructure data.[6] Commonly used indicators are amounts of bandwidth going to a city or country (Townsend, 2001; Wheeler & O'Kelley, 1999), Internet fiber backbone (Gorman & Malecki, 2002; Malecki, 2002; O'Kelley & Grubesic, 2002), points of presence (POPs) (Grubesic & O'Kelley, 2002), or broadband deployment (Grubesic, 2002; Grubesic & Murray, 2004). In general, these studies document the role of geography in producing a significantly uneven rollout of Internet infrastructure (even when controlling for size) with key metropolitan areas (both within the U.S. and abroad) retaining a disproportional share of infrastructure. At the same time, however, this mapping of infrastructure has also highlighted some emerging metropolitan nodes. As Malecki (2002, p. 419) argues, "to a large degree, the evolving infrastructure of the Internet is reinforcing old patterns of agglomeration: the world cities are alive and well. At the same time, new technologies cause new 'disturbances' that can result in the emergence of new clusters. … the prominence of Amsterdam and Stockholm in Europe and Salt Lake City and Atlanta in the United States suggests that new clusters can emerge."

This differentiated implementation of infrastructure highlights the importance of geography in understanding the infrastructure of the Internet and how data flow across it. Even physical distance between an Internet user and the computer upon which a Web page is hosted (theoretically a moot point because electrons move across computer networks at speeds imperceptible to humans) is relevant. Murnion and Healey (1998) used latency values (the time required for packets to travel across the Internet) as a measure of Internet distance and found that Web interaction (i.e., the number of visitors to a Web page) declines with distance (as measured by latency). Thelwall (2002) found a similar trend when he analyzed the hyperlink connections between academic Web sites compared to the physical distance between them. Although it is arguably an artifact of social networks rather than infrastructure (despite the variable of distance), this finding nevertheless shows that, rather than simply removing the "tyranny of distance," the Internet is dynamically shaped by distance and geography.

Often because of the types of indicators used, efforts to map the technical geographies of the Internet spill over into questions dealing more broadly with diffusion, for example, distribution of users or domain names (Kellerman, 2000; Moss & Townsend, 1997; Zook, 2001). When these markers are treated as simple indicators (as opposed to an in-depth analysis of how people are using the Internet), they help address questions about where the Internet is located. This is an area of considerable interest to geographers (Batty & Barr, 1994; Warf, 1995; Zook, 2001); it has also been a topic on which non-geographers have produced a number of important papers, particularly early on, describing the levels of use and diffusion of the Internet globally (Coffman & Odlyzko, 1998; Goodman, Press, Ruth, & Rutkowski, 1994; Petrazzini & Kibati, 1999; Press, 1997, 2000). Other researchers have focused on the distribution of Internet indicators within countries including Norway (Steineke, 2000); South Korea (Huh & Kim, 2003); Israel, Palestine, and Jordan (Ein-Dor, Goodman, & Wolcott, 2000); China (Loo, 2003a, 2003b); Japan (Rimmer & Morris-Suzuki, 1999); and the U.S. (Moss & Townsend, 1997, 2000; U.S. Department of Commerce, National Telecommunications and Information Administration, 1995, 1998, 1999; Zook, 2000).

An equally intriguing but significantly more difficult topic is that of Internet connections between places. Using a combination of hyperlinks on specific Web pages and search engine results, Brunn and Dodge's (2001) worldwide matrix of hyperlinks shows a distinct pattern of interconnections related to income. The analyses by Thelwall, Tang, and Price (2003) and Thelwall and Smith (2002) of interlinkages of academic Web sites in Europe and Asia reveal a similar pattern.

Although important in understanding the distribution of Internet use and interconnection, descriptive geographies lack the ability to identify the factors driving Internet adoption. As a result, several researchers have examined this causal question across a wide range of countries. Zook (2000), using the distribution of .com domain names within the U.S., argued for a causal connection between information-intensive industries and Internet location. Sternberg and Krymalowski (2002), looking at the distribution of .de domains within Germany, found that economic factors, including external economies, knowledge generation, and skilled labor, were the best predictors of domain distribution. Barnett, Chon, and Rosen (2001) studied Organisation for Economic Co-operation and Development (OECD) data on Internet hypertext links between various top level domains (TLDs). They found that the structure of the Internet (circa 1998) was significantly related to existing world-wide networks including trade, telecommunications, and also the co-authorship patterns among scientific and technical research. Hargittai (1999) examined the distribution of Internet hosts in the European Union and showed that economic factors such as per capita income and telecommunication pricing were key in understanding the diffusion of Internet use. Using a combination of Internet indicators, Loo and Wong (2002) found that per capita income, Internet pricing policies, and level of foreign direct investment were important explanatory variables for Internet usage in Asian and Pacific Rim countries.

To date, research has focused primarily on static measures of the Internet, such as number of hosts, amount of bandwidth connectivity, number of users, or domains. Currently lacking (and constituting an important direction for future research) are studies of more dynamic and detailed indicators of flows between places that move beyond simple counts of data packet traffic or hyperlinks. A downloaded flash program takes up significantly more space than a text-based e-mail, but the former could represent an annoying advertisement and the latter could be a job offer or love letter. Thus, it is clear that the nature of the connection between places, as opposed to a simple measure of the amount of connection, is key to improving our understanding of the technical geographies of the Internet.

Human Geographies of the Internet

The human dimension of Internet geography focuses upon how this new technology interacts with existing polities, cultures, economies, and individuals. Space is a fundamental facet of human experience (our bodies live in so-called meat-space), but the electronic spaces of the Internet have expanded the realm of human interaction. Although nonphysical space such as telephone-space, media-space, or video game-space existed previously, the immediacy, bandwidth, and scope of the Internet vastly expand our control and ability to participate virtually. As Mitchell (1995, p. 167) argues in his treatise on digital networks and cities, "we are entering an era of electronically extended bodies living at the intersection points of the physical and virtual worlds, of occupation and interaction through telepresence as well as through physical presence." In other words, the Internet is contributing to the ongoing process (begun with earlier communications technology) of extending human interaction beyond physical co-presence.

One obvious manifestation of the declining relative importance of physical co-presence (particularly from the perspective of the average Internet user) is the emergence of "virtual community," a notion first popularized by Rheingold (1993). Although presaged by Webber (1963), the concept is based on the idea of communities and groups that coalesce and communicate not through physical proximity but via a variety of information technologies. Wellman (1999, 2001) notes that this practice has served to strengthen personal networking around shared interests (at the expense of door-to-door and place-to-place communities) and to disperse interaction spatially. However, he is quick to argue against a simple and false dichotomy of cyber- and physical-space communities. Instead, he emphasizes the fluidity of communication choices and how individuals' abilities to use technology shape the scope and scale of their interaction.[7] Geographers have contributed and extended this argument to larger changes in existing political, cultural, and economic geographies.

Political and Cultural Geographies of the Internet

Although the 1990s saw much punditry on how the Internet would necessarily lead to a "borderless world," geographers have noted that the implementation of some of its most basic building blocks has reproduced notions of sovereign and territoriality. Wilson (2001) notes the continued relevance of geography and sovereignty in his analysis of the development of laws governing domain ownership and disputes. Steinberg and McDowell (2003), however, demonstrate that the affirmation of sovereignty has in turn been undermined by attempts to associated new meanings and identities with top-level domains such as marketing Niue (.nu) as alternatives to .com domains.

The edited volume of Crang, Crang, and May (1999) examines the changing nature of our relations to one another, our cultures, and space, exploring ideas about how the Internet interacts with the boundary between the real and the virtual. Central to many of the chapters is the "socially constructed" nature of technology (and the Internet) and how the interaction of online and offline users constructs places as diverse as an Internet café or a virtual world. Geographers have been intensely interested in how this new electronic space is used and how the Internet interacts with existing places, cultures, and people (Haythornthwaite & Hagar, 2005). Many other disciplines, such as psychology (e.g., Turkle, 1995), have made key contributions in this topic. I focus primarily on geographic research as is appropriate for this chapter.

A key approach in the study of these Internet geographies is the analysis of power relations and contestation. Although recognizing the ability for marginalized groups to contest and present competing discourse, Warf (2001) argues that the uneven distribution of the Internet has increased the advantages of an elite (see also Adams & Warf, 1997; Crampton, 2003). This strand of research focuses upon inherent power differences within society and the ability of the Internet to reinforce existing inequalities or, more optimistically, provide a means for contesting existing power relations (see Froehling's [1997] study of the Zapatista rebels' use of the Internet). Other examples of geo-cultural contestation on the Internet include conflicting efforts to represent the former republics of Yugoslavia (Jackson & Purcell, 1997), cyberactivism by socially marginalized groups (Warf & Grimes, 1997), and the redefinition of community membership among widely scattered actors (Brunn, Jones, & Purcell, 1994). Adams and Ghose (2003, p. 433) provide a particularly good example in their study of the use of the Internet by temporary and permanent immigrants to the U.S. from India. Using the concept of ethnicity, they (2003, p. 433) examine the creation of "bridgespace," or a life built upon the Internet's facilitation of creating roots in multiple places, thus generating "different types of space in technologies."

Of course, the ability to contest existing power structures also makes the Internet extremely useful for criminals (Barrett, 1998) and terrorists (Wade, 2003). Moreover, it facilitates surveillance and censorship of online activities (measures increasingly prevalent in a post-9/11 world). Privacy International (2003) notes that censoring of the Internet is increasingly commonplace worldwide (China and Burma are arguably the most extreme examples) and governments are simultaneously more protective of information that was previously widely available (Blumenfeld, 2003) and more emboldened to expand their monitoring of online activity. This increased scrutiny (for commercial as well as governmental purposes) is troubling from a privacy standpoint; Lyon (2001) alerts us to the tendency of this trend to reinforce already existing divisions within society, while reminding us that the amount of data passing through any surveillance system challenges a system's ability to process it adequately.

Kwan (2001, 2002) changes the scale of the analysis to individual experiences in cyberspace. He employs a cognitive/behavioral approach to understand how people construct and use electronic environments and how this, in turn, affects their ability to access and interact with physical places. This echoes Mitchell's (1995) and Couclelis's (1996) argument that we increasingly live in hybrid space at the intersection of the virtual and the physical. Valentine and Holloway (2002) are likewise interested in this intersection and examine how the "real" and "virtual" worlds of children are mutually constituted and argue that one world (be it virtual or physical) cannot be understood without the other. They again emphasize that there are no inherent properties associated with an electronic world and that technology is differentially used by different communities of practice.

Geographers could benefit, however, from other disciplines that have developed techniques for visualizing and highlighting these interactions, such as those offered by Donath (2002) and Smith (2002). Conversely, cartographic techniques taken from geography could inform the ways in which these visualizations are created. Moreover, efforts to expand social navigation (i.e., markers and tools resulting from interactions with other people), such as that offered by Munro, Höök, and Benyon (1999), could greatly assist efforts by geographers to analyze the social and relational contours of Internet interaction. Just as we may be tempted to pick up a well-thumbed book or listen to a street musician who has attracted a crowd, elements of social navigation actualized online can create a sense of place and agglomeration in electronic networks and are therefore intriguing topics for geographic research.

Economic Geographies of the Internet

Economic geographers were among the first to study the broad topic of telecommunications and geography. The rise of multinational corporations and changes in the global economy demonstrated the role of information technology in the location of economic activity and did much to spur this interest. As Castells (1996) theorizes, the emerging network society, defined as an interdependent space of flows interacting through digital exchange, is increasingly relevant to the organization (and control) of the economies of places. Thus, the Internet and economic places recursively and interactively exert pressure on one another to bring about new economic geographies.

Urban Economic Geographies

A particularly strong strand in this research is the effect of the Internet on urban geography.[8] Some of the earliest work within communications geography dealt with the role of information and urban growth (Abler, 1977; Pred, 1973). This is due in part to the important role that telecommunication has played in the development of cities, particularly so-called "global cities" (Graham & Marvin, 1996; Hall, 1998; Moss, 1987), and in part to the urban bias that has been a hallmark of Internet diffusion and penetration (Drennan, 2002; Kolko, 2000).

Although exceptions to the pattern exist, the Internet has followed the standard diffusion model of a new technology. Places specialized in a particular activity that can best take advantage of the innovation will be the first to take it on. In the case of the Internet, intense users of information in the economy were among the earliest adopters and their location in certain cities (Sassen, 1991) contributed to this clustering. A number of indicators of Internet activity such as bandwidth (Malecki, 2002; Townsend, 2001) or domain names (Kolko, 2000; Zook, 2000) were disproportionately located in urban centers. This geography was marked by significant regional variation and reflected the history and fortunes of specific regions, conforming in some ways to existing hierarchies, for example, New York, while suggesting new hubs in others, for example, Austin, Texas (Zook, 2005). Moreover, arguments advanced by Leamer and Storper (2001) and Zook (2002, 2005) point to the role of knowledge and co-presence in economic (particularly innovative) activity. Although the Internet does make information more readily available, physical co-presence—particularly for activities that require trust and/or involve complex and changing concepts—vastly out-performs any electronic means of communication so far created. Thus, Leamer and Storper (2002) argue that the Internet will tend to produce economic geographies with an increased number of "conversations" (via e-mail and other electronic media) between distant locations but will still require localized clusters where face-to-face interaction for "handshake"-level activities can take place.

These empirical and theoretical findings argue strongly against notions of uniform deployment of the Internet and instead point to variable outcomes in different cities and different parts of the same city. Graham and Marvin (2001) characterize this uneven distribution of electronic technologies in space (within and between cities) as "splintered urbanism," extending Mitchell's (1995) vision of living at the intersection of the physical and the virtual to its logical extreme. As Graham (1999, p. 948) argued earlier,

> the geostrategic "roll out" of planetary optic-fibre grids ... seem[s] likely to generate very high degrees of uneven development internally and ever-more problematic relations with their [cities'] traditional hinterlands and peripheries. Once relatively cohesive, homogeneous and equalizing infrastructure grids at the national level are thus clearly "splintering" into tailored, customised and global-local grids, designed to meet the needs of hegemonic economic and social actors.

This focus on cities has its counterpart in research on the possible opportunities and pitfalls involving the use of the Internet in more rural locations. Beyond the basic question of accessibility—a real and ongoing issue for most low-density locations (Grubesic, 2003; Grubesic & Murray, 2004)—are questions about whether Internet access will automatically bring development (Malecki, 2003). Others, including Grimes (2003) and Gillespie, Richardson, and Cornford (2001), echo this concern, and Richardson and Gillespie (2000, p. 201) remind us that the introduction of any technology (in this case, the Internet) "does not automatically result in the decentralization of activities." There are some indications that, at least on the issue of accessibility, wireless technology may be closing the gap between urban and less urban locations (Gorman & McIntee, 2003); larger questions remain as to whether this will spur economic development.

E-Commerce Geographies

Although the urban (and rural) dimension is a crucial variable in understanding the economic geographies of the Internet, a potentially more significant issue is the creation of complex linkages (value chains) between and within firms facilitated by the Internet. In addition to broadening the scale and scope of the marketplace, cyberspace is transforming the spatial organization of the market in ways that were previously not possible. The Internet, in the form of electronic commerce (e-commerce), offers the means to recast the geographies of production and consumption of firms and regions, forming new e-commerce geographies. Currah (2002, p. 1434), following the lead of Graham (1998), argues for a "co-evolutionary perspective" on the process of e-commerce development in which "technological systems and material economic-geographical landscapes are recursively produced together."

The manner in which these changes are implemented varies at the firm level, but the more important ones have specific sectoral and geographical characteristics (see Currah [2003a], Leyshon [2001], and Murphy [2003] for analyses of the geographic implications for the film, music, and online grocery industries, respectively).[9] This latter category—variation in the nature and form of e-commerce implementation between places—is understandably of particular interest to geographers (Aoyama, 2003; Gibbs, Kraemer, & Dedrick, 2002; Leinbach & Brunn, 2001; Organisation for Economic Co-operation and Development, 2001). As Kling, Kraemer, and Dedrick (2002, p. 3) argue, "the diffusion of e-commerce is varied and complex across countries. It is significantly influenced by features of each country's national environment and national policy...as with earlier ICTs, e-commerce is occurring in a more evolutionary than revolutionary manner, as it is closely intertwined with broader business, socioeconomic and political change."

Economic behavior is governed by convenience, familiarity, and social habits, which in turn are shaped by the historical evolution of the economy. Thus, countries with comparable levels of wealth and education can follow very different technology adoption trajectories. The most telling example of these differences, particularly when compared to U.S. experiences, is documented by Aoyama (2001, 2003) in her study of e-commerce in Japan. Due to a lack of history of nonstore retailing (i.e., mail order), high population densities, the prevalence of local convenience stores, and a preference for mobile Web devices (tied to the Japanese-language character set), e-commerce in Japan was built around convenience stores. Rather than placing an order on a personal computer and having an item delivered to one's home, the Japanese utilized stores as the point of purchase and delivery in marked contrast to U.S. practice.

Although rarely noted, e-commerce has also extended beyond traditional venues to provide so called "gray" and/or illegal economies with the means to reorganize their geographies of production and consumption, particularly in the provision of completely digital products such as pornography, copyrighted materials (software, movies, and music), and gambling. Capitalizing on an ability to conduct geographic arbitrage, in other words, locating in hospitable or indifferent regulatory environments, can be an attractive proposition. Wilson (2003) and Zook (2003) have identified the Caribbean and Eastern Europe as important nodes in the economic geography of Internet gambling and pornography, respectively. Thus, the Internet has both increased the ability for isolated businesses or individuals to access (and be accessed by) the rest of the world and also strengthened the ability of organizations (although not always those in the mainstream) to extend the scope and reach of their markets and consumption.

Visualized Geographies of the Internet

Common to the Internet geographies discussed so far is the effort to map the phenomena under analysis. Ranging from the reference maps of Internet bandwidth presented by Townsend (2001) to graphical representations of the topography of the Gnutella protocol used by Gorman and Malecki (2002) through the graphical depiction of the Web sites used by immigrant Indians in the U.S. by Adams and Ghose (2003), to the location maps of urban Japan provided by Aoyama (2001), these maps all provide insight into a single aspect of Internet geography. Additional efforts to map the technical, cultural, political, and economic spheres of the Internet are presented and critiqued by Dodge and Kitchin (2000a, 2000b, 2001).[10] Long a tool of geography, maps can be characterized as efforts to represent complex locations, situations, and phenomena in a condensed and simplified way. One must recognize the fundamental cartographic fact that, in map making (as with any representation), one subjectively selects and excludes, and in so doing alters one's (and one's readers') perspective on the mapped phenomenon. Graphical representations of the Internet (or any information space) are no different. It is ironic that, even as the expansion of the Internet led people to declare the "end of geography," the Internet continued to be understood largely through metaphors of geographic place, for example, superhighways, teleports, server farms, home pages (Graham, 1998). Even as these geographic metaphors were being popularized, geographers were grappling with the extent to which these cartographic tools, created largely to represent Euclidean geography, were relevant to our understanding of Internet geographies (Dodge & Kitchin, 2000b).

The topic of visualized geographies is larger than the Internet and extends easily into the question of representing data that may (or may not) have any spatial association in the physical world. This is referred to as "spatialization"; Fabrikant (2000, p. 67) defines it as visualizations that "rely on the use of spatial metaphors to represent data that are not necessarily spatial."[11] Thus, spatialization constitutes a potential point of convergence for geography and information science, although information scientists have only sporadically used the tools and techniques of geographers (Skupin & Fabrikant, 2003).[12] The efforts of Skupin (2000), Skupin and Fabrikant (2003), and Fabrikant and Buttenfield (2001) are all examples of using cartographic concepts to visualize diverse types of information. The representation of knowledge domains and their interconnections have been described by Börner, Chen, and Boyack (2003). Conversely, a lesson that geographers can take from this collaboration is the use of maps as dynamic tools for understanding complex datasets rather than as static final products, which was the proclivity of predigital cartography. A particularly intriguing possibility is the ability to create real-time visualizations of social geographies. As Zook, Dodge, Aoyama, and Townsend (2004, p. 174) argue, "the fusion of fine-scale individual activities patterns that are automatically logged and novel forms of geo-visualization could give rise to fully dynamic time-space diagrams."

A final element of visualized geographies of the Internet is the representation of virtual spaces themselves, ranging from text-based online communities (Donath, 2002; Smith, 2002) to graphically depicted virtual spaces. The latter are particularly intriguing for geographers because, as Taylor (1997, p. 189) notes, "to be within a virtual world is to have an intrinsically geographical experience, as virtual worlds are experienced fundamentally as places." In their edited volume, Fisher and Unwin (2002) explore how geography in general and, more specifically, cartography can inform our understanding of virtual reality environments. Particularly interesting is the section discussing "other worlds" and explorations in using geography to interact with, understand, and manipulate a wide range of data (Fisher & Unwin, 2002, pp. 293–400). In this vein, Dodge (2002) analyzes the geography of one such virtual space (Alphaworld), including observations on avatars, travel time versus teleportation, the establishment of community norms, and the creation of private or public spaces in these virtual settings. Of equal interest to geographers, because of the effort to replicate key aspects of physical places, are inhabited information spaces (IIS), which combine virtual reality and information visualization to support collaboration (Snowdon, Churchill, & Frécon, 2004). Although much information can be transmitted electronically, there has hitherto been a premium placed on physical co-presence (particularly in economic geography work). Simply put, the level of communication afforded by co-presence (body language, dress, etc.) is orders of magnitude greater than any yet offered by electronic means.[13]

An intriguing (and potentially troubling) aspect of these geographies of virtual places is, as Taylor (1997, p. 189) observes, the desire "to occupy, to produce, and to utilize new spaces. ... I view it as a component of a much larger and older vision or narrative of territorial expansion." Although this concern may at first glance appear overstated, socially constructed virtual places are ever subject to exploitation, e.g., the infamous posting of law service advertisements to Usenet groups in 1994 or the current wave of spam e-mail. These actions are generally perpetrated by individuals, but some governments are actively engaged in controlling and censoring their citizens' online activities (Privacy International, 2003). Attempts to exploit or censor virtual places, however, are often met with a vigorous defense by their "netizens," reflecting the inherently contested nature of the Internet.

Conclusions

This review of the geographies of the Internet shows the multiple ways in which physical and electronic space overlap and the range of outcomes that result from their interaction. We may live in a global village, but even the smallest community (virtual or not) has unmistakable spatial demarcations that are paralleled (although not precisely) in the technical, human, and visualized geographies of the Internet. As Goodchild (2004, p. 168) argues, "distance is not disappearing as a basis for human organization; but the parameters that define its importance are changing, and new scales of organization are emerging as a consequence." The Internet geographies discussed in this chapter focus on uncovering and understanding the parameters upon which distance matters and the relevant scale of analysis in understanding the organization of human activity concurrently in physical and electronic space.

Common to all Internet geographies is the centrality of place; in Castells' formulation, the space of flows cannot be understood without reference to the space of places. The widespread use of the Internet has altered the geographic organization of our economy, sense of identity, and how we visualize the places in which we live and work; it has also served to reconnect us to place. Consider the example of a wireless Internet-enabled telephone equipped with a global positioning device (GPS) and a camera (a work of science fiction 20 years ago, but increasingly available and utilized today) and how it both grounds and networks. It locates a person in a physical place via the x,y coordinates of location (the very epitome of Euclidean geography) while at the same time allowing real-time or asynchronous interaction with other people and with virtual places. One can be simultaneously located in a physical place (particularly important if you are trying to get your body from one location to another during rush hour) while connected to social networks and the global web of information. Thus, just as geographers view the recursive link between nature and society as the source of the variation of human experience over the Earth's surface, Internet geographers look to the complexity of the interaction between electronic technology and human use as the origin of the multiplicity of Internet geographies.

Acknowledgments

The author would like to thank Martin Dodge and three anonymous reviewers for their suggestions and critique of this chapter.

Endnotes

1. As this chapter documents, there has been a dramatic increase in attention to communications geography, but the tendency to discount communication—or electronic—space remains. For example, a Communication Geography specialty group of the Association of American Geographers (AAG) was not formed until 2003.

2. Although often espousing the ability of the Internet to transcend geography during the dot-com boom of the 1990s, the weekly magazine, *The Economist*, has begun to acknowledge the relevance of geography to the Internet, addressing the issues of accessibility, speed, and regulation (Geography and the Net, 2001; The Internet's new borders, 2001; The revenge of geography, 2003).

3. As an outcome of the Internet's origins and relatively more recent study by geographers, most if not all historical maps of the Internet were designed by computer scientists. See Dodge and Kitchin (2001) for an excellent collection of these and other maps of the Internet and Abbate (1999) for an excellent history of the Internet, which, with careful reading, implicitly provides a geohistorical overview of the development of the Internet.

4. The scarcity and value of geographic data on the Internet's infrastructure were made abundantly clear when a geography dissertation by Sean Gorman based on a comprehensive mapping of the fiber-optic network in the U.S. was considered a possible security risk by the U.S. government (Blumenfeld, 2003).

5. The ability of researchers to use the Internet to measure itself contrasts sharply with the difficulty in gathering geographic measures of broadcast television and other proprietary networks. It has, however, become increasingly difficult to collect Internet infrastructure data as the initially relaxed and open systems of the Internet have been hardened.

6. Of special note is the survey of Internet hosts begun by Network Wizards (http://www.nw.com) and continued by the Internet Software Consortium (http://www.isc.org), which provides the longest running freely available metric of Internet growth with some geographic reference. Also worthy of note are country-level data available from the International Telecommunication Union (ITU) and United Nations Development Programme (UNDP). Other examples of Internet data used by Internet geography researchers can be seen in *Hubs and spokes: A telegeography Internet reader* (2000), Paltridge (1998), and Claffy (2001).

7. Of course, not all are as optimistic about the benefits of this type of community (see Robins, 1999).

8. The inclusion of urban geography as a subset of economic geography reflects the author's background as an economic geographer rather than a consensus of geographers. The spatial economy is a major focus within urban geography (particularly as it relates to the Internet), which includes many other noneconomic topics such as social inequality and sociospatial polarization.

9. Even the virtual space represented by a Web site can be an important site for analysis (Currah, 2003b).

10. Of special note is the map created by Cheswick and Burch (see Branigan, Burch, Cheswick, & Wojcik, 2001) which represents the topography of the Internet and has been referred to as a "smashed peacock." This map has circulated widely, appearing both as a wall poster and on the cover of Castells's (2001) book.

11. Information scientists have long been interested in the related question of mapping intellectual interconnections using citations as one means to do this (Cronin, 1984; Cronin & Shaw, 2001).

12. This overlap between data visualization and Geography has implications beyond academic realms as the discussion by *The Economist* (Grokking the infoviz, 2003b) on the growing importance of information visualization demonstrates.

13. Despite a general skepticism on the part of geographers regarding the extent to which technological presence can substitute for physical presence, the concept remains intriguing. The International Society for Presence Research (ISPR) is a good example of research efforts in this direction (see http://www.ispr.info for more information).

References

Abbate, J. (1999). *Inventing the Internet*. Cambridge, MA: MIT Press.

Abler, R. (1975). Effects of space-adjusting technologies on the human geography of the future. In R. Abler, D. Janelle, A. Philbrick, & J. Sommer (Eds.), *Human geography in a shrinking world* (pp. 36–56). North Scituate, MA: Duxbury Press.

Abler, R. (1977). The telephone and the evolution of the American Metropolitan system. In I. d. S. Pool (Ed.), *The social impact of the telephone* (pp. 318–341). Cambridge, MA: MIT Press.

Adams, P., & Ghose, R. (2003). India.com: The construction of a space between. *Progress in Human Geography, 27*(4), 414–437.

Adams, P., & Warf, B. (1997). Cyberspace and geographical space. *Geographical Review, 87*(2), 139–145.

Aoyama, Y. (2001). The information society, Japanese style: Corner stores as hubs for e-commerce access. In T. Leinbach & S. Brunn (Eds.), *Worlds of electronic commerce* (pp. 109–128). New York: Wiley.

Aoyama, Y. (2003). Sociospatial dimensions of technology adoption: Recent m-commerce and e-commerce developments. *Environment and Planning A, 35*(7), 1201–1221.

Barnett, G. A., Chon, B. S., & Rosen, D. (2001). The structure of international Internet flows in cyberspace. *NETCOM (Network and Communication Studies), 15*(1–2), 61–80.

Barrett, N. (1998). *Digital crime: Policing the cybernation*. Dover, NH: Kogan Page.

Batty, M. (1993). The geography of cyberspace. *Environment and Planning B, 20*, 615–616.

Batty, M., & Barr, B. (1994). The electronic frontier: Exploring and mapping cyberspace. *Futures, 26*(7), 699–712.

Blumenfeld, L. (2003, July 8). Dissertation could be security threat: Student's maps illustrate concerns about public information. *Washington Post*, A01.

Branigan, S., Burch, H., Cheswick, B., & Wojcik, F., (2001). What can you do with Traceroute? *Internet Computing, 5*(5), 96.

Börner, K., Chen, C., & Boyack, K. (2003). Visualizing knowledge domains. *Annual Review of Information Science and Technology, 37*, 179–255.

Brunn, S., & Dodge, M. (2001). Mapping the 'worlds' of the World-Wide Web: (Re)structuring global commerce through hyperlinks. *American Behavioral Scientist, 44*(10), 1717–1739.

Brunn, S., Jones, J., & Purcell, D. (1994). Ethnic communities in the evolving 'electronic' state: Cyberplaces in cyberspace. In W. Galluser, M. Brügin, & W. Leimgruber (Eds.), *Political boundaries and co-existence* (pp. 415–424). New York: Peter Lang.

Brunn S., & Leinbach, T. (Eds.). (1991). *Collapsing space and time: Geographic aspects of communications and information*. London: HarperCollins Academic.

Cairncross, F. (1997). *The death of distance: How the communications revolution will change our lives*. Cambridge, MA: Harvard Business School Press.

Castells, M. (1989). *The informational city*. Cambridge, MA: Blackwell.

Castells, M. (1996). *The rise of the network society*. Cambridge, MA: Blackwell.

Castells, M. (1997). *The power of identity*. Cambridge, MA: Blackwell.

Castells, M. (1998). *The end of millennium*. Cambridge, MA: Blackwell.

Castells, M. (2001). *Internet galaxy: Reflections on the Internet, business, and society.* Oxford, UK: Oxford University Press.

Claffy, K. C. (2001). CAIDA: Visualizing the Internet. *Internet Computing Online, 1.* Retrieved January 9, 2005, from http://www.computer.org/internet/v5n1/caida.htm

Coffman, K. G., & Odlyzko, A. (1998). The size and growth rate of the Internet. *First Monday, 3*(10). Retrieved January 9, 2005, from http://www.firstmonday.org/issues/issue3_10/coffman

Couclelis, H. (1996). The death of distance. *Environment and Planning B, 23,* 387–389.

Crampton, J. W. (2003). *The political mapping of cyberspace.* Chicago: University of Chicago Press.

Crang, M., Crang, P., & May, J. (1999). *Virtual geographies: Bodies, space and relations.* London: Routledge.

Cronin, B. (1984). *The citation process: The role and significance of citations in scientific communication.* London: Taylor Graham.

Cronin, B., & Shaw, D. (2001). Identity-creators and image-makers: Using citation analysis and thick description to put authors in their place. *Scientometrics, 54*(1), 31–49.

Currah, A. (2002). Behind the Web store: The organizational and spatial evolution of multi-channel retailing in Toronto. *Environment and Planning A, 34*(8), 1411–1441.

Currah, A. (2003a). Digital effects in the spatial economy of film: Towards a research agenda. *Area, 35*(1), 64–73.

Currah, A. (2003b). The virtual geographies of retail display. *Journal of Consumer Culture, 3*(1), 5–37.

Dodge, M. (2001). Cybergeography. *Environment and Planning B, 28,* 1–2.

Dodge, M. (2002). Explorations in AlphaWorld: The geography of 3D virtual worlds on the Internet. In P. Fisher & D. Unwin (Eds.), *Virtual reality in geography* (pp. 305–331). London: Taylor and Francis.

Dodge, M., & Kitchin, R. (2000a). Exposing the 'second text' of maps of the Net. *Journal of Computer-Mediated Communication, 5*(4). Retrieved January 9, 2005, from http://www.ascusc.org/jcmc/vol5/issue4/dodge_kitchin.htm

Dodge, M., & Kitchin, R. (2000b). *Mapping cyberspace.* London: Routledge.

Dodge, M., & Kitchin, R. (2001). *Atlas of cyberspace.* London: Addison-Wesley.

Donath J. (2002). A semantic approach to visualizing online conversations. *Communications of the ACM, 45*(4), 45–49.

Drennan, M. (2002). *The information economy and American cities.* Baltimore: Johns Hopkins University Press.

Ein-Dor, P., Goodman, S. E., & Wolcott, P. (2000). From Via Maris to electronic highway: The Internet in Canaan. *Communications of the ACM, 43*(7), 19–23.

Fabrikant, S. I. (2000). Spatialized browsing in large data archives. *Transactions in GIS, 4*(1), 65–78.

Fabrikant, S. I., & Buttenfield, B. P. (2001). Formalizing semantic spaces for information access. *Annals of the Association of American Geographers, 91*(2), 263–280.

Fisher, P., & Unwin, D. (Eds.). (2002). *Virtual reality in geography.* London: Taylor and Francis.

Froehling, O. (1997). The cyberspace war of ink and Internet in Chiapas, Mexico. *The Geographical Review, 87,* 291–307.

Geography and the Net: Putting it in its place. (2001, August 9). *Economist*, 18–20.

Gibbs, J., Kraemer, K. L., & Dedrick, J. (2002). *Environment and policy factors shaping e-commerce diffusion: A cross-country comparison* (Global IT paper 311). Irvine, CA: Center for Research on Information Technology and Organizations. Retrieved May 1, 2005, from http://repositories.cdlib.org/crito/globalization/311

Gibson, W. (1984). *Neuromancer*. New York: Berkley Publishing.

Gillespie, A. (1992). Communications technologies and the future of the city. In M. J. Breheny (Ed.), *Sustainable development and urban form* (pp. 67–77). London: Pion.

Gillespie, A., Richardson, R., & Cornford, J. (2001). Regional development and the new economy. *European Investment Bank Papers, 6*(1), 109–131.

Goodchild, M. F. (2004). Scales of cybergeography. In E. Sheppard & R. B. McMaster (Eds.), *Scale and geographic enquiry: Nature, society and method* (pp. 154–169). Cambridge, MA: Blackwell.

Goodman, S. E., Press, L. I., Ruth, S. R., & Rutkowski, A. M. (1994). The global diffusion of the Internet: Patterns and problems. *Communications of the ACM, 37*(8), 27–31.

Gorman, S. P., & Malecki, E. J. (2002). Fixed and fluid: Stability and change in the geography of the Internet. *Telecommunications Policy, 26*(7–8), 389–413.

Gorman, S. P., & McIntee, A. (2003). Tethered connectivity? The spatial distribution of wireless infrastructure. *Environment and Planning A, 35*(7), 1157–1171.

Graham, S. (1998). The end of geography or the explosion of place? Conceptualizing space, place and information technology. *Progress in Human Geography, 22*(2), 165–185.

Graham, S. (1999). Global grids of glass: On global cities, telecommunications, and planetary urban networks. *Urban Studies, 36*(5), 929–949.

Graham, S., & Marvin, S. (1996). *Telecommunications and city: Electronic spaces, urban places*. London: Routledge.

Graham, S., & Marvin, S. (2001). *Splintering urbanism: Networked infrastructures, technological mobilities, and the urban condition*. London: Routledge.

Grimes, S. (2003). The digital economy challenge facing peripheral rural areas. *Progress in Human Geography, 27*(2), 174–193.

Grokking the infoviz. (2003, June 19). *Economist*, 28–30.

Grubesic, T. H. (2002). Spatial dimensions of Internet activity. *Telecommunications Policy, 26*(7–8), 363–387.

Grubesic, T. H. (2003). Inequities in the broadband revolution. *Annals of Regional Science, 37*, 263–289.

Grubesic, T. H., & Murray, A. (2004). Waiting for broadband: Local competition and the spatial distribution of advanced telecommunication services in the United States. *Growth and Change, 35*(2), 139–165.

Grubesic, T. H., & O'Kelly, M. E. (2002). Using points of presence to measure city accessibility to the commercial Internet. *The Professional Geographer, 54*(2), 259–278.

Haggett, P. (1993). Geography. In R. J. Johnson, D. Gregory, & D. Smith (Eds.), *The dictionary of human geography* (pp. 220–223). Oxford, UK: Blackwell Reference.

Hall, P. (1998). *Cities in civilization*. New York: Pantheon Books.

Hargittai, E. (1999). Weaving the western Web: Explaining differences in Internet connectivity among OECD countries. *Telecommunications Policy, 23*(10/11), 701–718.

Harvey, D. (1989). *The condition of post-modernity*. Cambridge, MA: Cambridge Press.

Hayes, B. (1997). The infrastructure of the information infrastructure. *American Scientist, 85*(3), 214–218.

Haythornthwaite, C., & Hagar, C. (2005). The social worlds of the Web. *Annual Review of Information Science and Technology,* 39, 311–346.

Hepworth, M. (1990). *Geography of the information economy.* New York: Guilford Press.

Hillis, K. (1998). On the margins: The invisibility of communications in geography. *Progress in Human Geography, 22*(4), 543–566.

Hubs and spokes: A telegeography Internet reader. (2000). Washington, DC: Telegeography Inc.

Huh, W. K., & Kim, H. (2003). Information flows on the Internet of Korea. *Journal of Urban Technology, 10*(1), 61–87.

Innis, H. (1950). *Empire and communications.* Oxford, UK: Clarendon Press.

Innis, H. (1951). *The bias of communication.* Toronto: University of Toronto Press.

The Internet's new borders. (2001, August 9). *Economist,* 9–10.

Jackson, M., & Purcell, D. (1997). Politics and media richness in World Wide Web representations of the former Yugoslavia. *Geographical Review, 87*(2), 219–239.

Janelle, D. (1973). Measuring human extensibility in a shrinking world. *Journal of Geography, 72*(5), 8–15.

Janelle, D., & Hodge, D. (2000). Information, place, cyberspace and accessibility. In D. Janelle & D. Hodge (Eds.), *Information, place, and cyberspace* (pp. 3–11). Berlin: Springer-Verlag.

Kellerman, A. (1986). The diffusion of BITNET: A communications system for universities. *Telecommunications Policy, 10,* 88–92.

Kellerman, A. (1993). *Telecommunications and geography.* London: Belhaven Press.

Kellerman, A. (2000). Where does it happen? The location of production, consumption and contents of Web information. *Journal of Urban Technology, 7*(1), 45–61.

Kelly, K. (1998). *New rules for the new economy.* New York: Penguin Books.

Kitchin, R. (1998a). *Cyberspace: The world in wires.* Chichester, MA: Wiley.

Kitchin, R. (1998b). Towards geographies of cyberspace. *Progress in Human Geography, 22*(3), 385–406.

Kling, R., Kraemer, K., & Dedrick, J. (2003). Introduction. *The Information Society, 19*(1), 1–3.

Kolko, J. (2000). The death of cities? The death of distance? Evidence from the geography of commercial Internet usage. In I. Vogelsang & B. Compaine (Eds.), *The Internet upheaval* (pp. 73–98). Cambridge, MA: MIT Press.

Kwan, M.-P. (2001). Cyberspatial cognition and individual access to information: The behavioral foundation of cybergeography. *Environment and Planning B, 28,* 21–38.

Kwan, M.-P. (2002). Time, information technologies and the geographies of everyday life. *Urban Geography, 23*(5), 471–482.

Langdale, J. V. (1989). The geography of international business telecommunications: The role of leased networks. *Annals of the Association of American Geographers, 79,* 501–522.

Leamer, E. E., & Storper, M. (2001). *The economic geography of the Internet age* (NBER Working Paper No. W8450). Cambridge, MA: National Bureau of Economic Research.

Leinbach, T. R., & Brunn, S. (Eds.). (2001). *Worlds of e-commerce: Economic, geographical and social dimensions*. Chichester, MA: Wiley.

Leyshon, A. (2001). Time–space (and digital) compression: Software formats, musical networks, and the reorganisation of the music industry. *Environment and Planning A, 33*, 49–77.

Loo, B. (2003a). Internet development in China: Geographic distribution of Internet production and consumption. *GIM International, 17*(12), 68–71.

Loo, B. (2003b). The rise of a [sic] digital communities in the People's Republic of China. *Journal of Urban Technology, 10*(1), 1–21.

Loo, B., & Wong, A. (2002). Internet development in Asia: Spatial patterns and underlying locational factors. *Network and Communications, 16*(3–4), 113–134.

Lyon, D. (2001). *Surveillance society*. London: Open University Press.

Malecki, E. J. (2002). The economic geography of the Internet's infrastructure. *Economic Geography, 78*(4), 399–424.

Malecki, E. J. (2003). Digital development in rural areas: Potentials and pitfalls. *Journal of Rural Studies, 19*(2), 201–214.

Massey, D. (1993). Power-geometry and a progressive sense of place. In J. Bird, B. Curtis, T. Putman, G. Robertson, & L. Tickner (Eds.), *Mapping the futures: Local cultures, global change* (pp. 59–69). London: Routledge.

Mitchell, W. (1995). *City of bits*. Cambridge, MA: MIT Press.

Moss, M. (1987). Telecommunications, world cities and urban policy. *Urban Studies, 24*, 534–546.

Moss, M., & Townsend, A. (1997). Tracking the Net: Using domain names to measure the growth of the Internet in U.S. cities. *Journal of Urban Technology, 4*(3), 47–60.

Moss, M., & Townsend, A. (2000). The Internet backbone and the American metropolis. *The Information Society, 16*, 35–47.

Munro, A. J., Höök, K., & Benyon, D. (Eds.). (1999). *Social navigation of information space*. Berlin: Springer.

Murnion, S., & Healey, R. G. (1998). Modeling distance decay effects in Web server information flows. *Geographical Analysis, 30*(4), 285–303.

Murphy, A. J. (2003). (Re)solving space and time: Fulfilment issues in online grocery retailing. *Environment and Planning A, 35*(7), 1173–1200.

Negroponte, N. (1995). *Being digital*. New York: Knopf.

O'Kelly, M. E., & Grubesic, T. (2002). Backbone topology, access, and the commercial Internet. *Environment and Planning B, 29*(4), 533–552.

Organisation for Economic Co-operation and Development. (2001). *The Internet and business performance* (Business and Industry Policy Forum Series). Paris: Organisation for Economic Co-operation and Development.

Paltridge, S. (1998). *Internet infrastructure indicators: Report by the Directorate for Science, Technology and Industry*. Paris: Organisation for Economic Co-operation and Development.

Petrazzini, B., & Kibati, M. (1999). The Internet in developing countries. *Communications of the ACM, 42*(6), 31–36.

Pred, A. (1973). *Urban growth and the circulation of information*. Cambridge, MA: Harvard University Press.

Press, L. (1997). Tracking the global diffusion of the Internet. *Communications of the ACM, 40*(11), 11–17.

Press, L. (2000, July). *The state of the Internet: Growth and gaps.* Paper presented at the INET 2000 Conference, Yokohama, Japan.

Privacy International. (2003). *Silenced: An international report on censorship and control of the Internet.* London: Privacy International. Retrieved January 4, 2004, from http://www.privacyinternational.org/survey/censorship/silenced.pdf

Quarterman, J. S. (1990). *The matrix: Computer networks and conferencing systems worldwide.* Bedford, MA: Digital Press.

The revenge of geography. (2003, March 13). *Economist,* 22–27.

Rheingold, H. (1993). *The virtual community.* Cambridge, MA: MIT Press.

Richardson, R., & Gillespie, A. (2000). The economic development of peripheral rural areas in the information age. In M. Wilson & K. Corey (Eds.), *Information tectonics* (pp. 199–218). New York: Wiley.

Rimmer, P. J., & Morris-Suzuki, T. (1999). The Japanese Internet: Visionaries and virtual democracy. *Environment and Planning A, 31*(7), 1189–1206.

Robins, K. (1999). Against virtual community: For a politics of distance. *Angelaki, 4*(2), 163–170.

Sassen, S. (1991). *The global city: New York, London, Tokyo.* Princeton, NJ: Princeton University Press.

Skupin, A. (2000). From metaphor to method: Cartographic perspectives on information visualization. *Proceedings IEEE Symposium on Information Visualization (InfoVis 2000),* 91–97.

Skupin, A., & Fabrikant, S. (2003). Spatialization methods: A cartographic research agenda for non-geographic information visualization. *Cartography and Geographic Information Science, 30*(2), 95–115.

Smith, M. (2002). Tools for navigating large social cyberspace. *Communications of the ACM, 45*(4), 51–55.

Snowdon, D. N., Churchill, E. F., & Frécon, E. (2004). *Inhabited information spaces: Living with your data.* Berlin: Springer.

Steinberg, P., & McDowell, S. (2003). Mutiny on the bandwidth: The semiotics of statehood in the Internet domain name registries of Pitcairn Island and Niue. *New Media & Society, 5*(1), 47–67.

Steineke, J. M. (2000). *The Web and the cities: Explaining spatial patterns of Internet accessibility and use in Norway* (Working Paper RF-2000/116). Stavanger, Norway: Rogaland Research. Retrieved January 9, 2005, from http://www.rf.no/internet/student.nsf/199f312efd2a0cacc125680e00635b85/8f0160b7cd298035c12569360057ee9d/$FILE/rf-2000-116.pdf

Sternberg, R., & Krymalowski, M. (2002). Internet domains and the innovativeness of cities/regions: Evidence from Germany and Munich. *European Planning Studies, 10*(2), 251–274.

Taylor, J. (1997). The emerging geographies of virtual worlds. *Geographical Review, 87*(2), 172–192.

Thelwall, M. (2002). Evidence for the existence of geographic trends in university Web site interlinking. *Journal of Documentation, 58*(2), 563–574.

Thelwall, M., Tang, R., & Price, L. (2003). Linguistic patterns of academic Web use in Western Europe. *Scientometrics, 56*(3), 417–432.

Thelwall, M., & Smith, A. (2002). Interlinking between Asia-Pacific university Web sites. *Scientometrics, 56*(3), 335–348.

Thrift, N. (1996). New urban eras and old technological fears: Reconfiguring the goodwill of electronic things. *Urban Studies, 33*(8), 1463–1493.

Townsend, A. (2001). The Internet and the rise of the new network cities, 1969–1999. *Environment and Planning B, 28*(1), 39–58.

Turkle, S. (1995). *Life on the screen.* Cambridge, MA: MIT Press.

U.S. Department of Commerce. National Telecommunications and Information Administration (NTIA). (1995). *Falling through the Net: A survey of the "have nots" in rural and urban America.* Washington, DC: The Department.

U.S. Department of Commerce. National Telecommunications and Information Administration (NTIA). (1998). *Falling through the Net II: New data on the digital divide.* Washington, DC: The Department.

U.S. Department of Commerce. National Telecommunications and Information Administration (NTIA). (1999). *Falling through the Net: Defining the digital divide.* Washington, DC: The Department.

Valentine, G., & Holloway, S. L. (2002). Cyberkids? Exploring children's identities and social networks in on-line and off-line worlds. *Annals of the Association of American Geographers, 92*, 296–315.

Wade, L. (2003). Terrorism and the Internet: Resistance in the information age. *Knowledge, Technology & Policy, 16*(1), 104–127.

Warf, B. (1995). Telecommunications and the changing geographies of knowledge transmission in the late 20th century. *Urban Studies, 32*, 361–378.

Warf, B. (2001). Segueways into cyberspace: Multiple geographies of the digital divide. *Environment and Planning B, 28*(1), 3–19.

Warf, B., & Grimes, J. (1997). Counterhegemonic discourses and the Internet. *Geographical Review, 87*(2), 259–274.

Webber, M. (1963). Order in diversity: Community without propinquity. In L. Wingo (Ed.), *Cities and space: The future use of urban land* (pp. 23–54). Baltimore: Johns Hopkins University Press.

Wellman, B. (1999). The network community. In B. Wellman (Ed.), *Networks in the global village* (pp. 1–48). Boulder, CO: Westview.

Wellman, B. (2001). Physical place and cyberplace: The rise of personalized networking. *International Journal of Urban and Regional Research, 25*(2), 227–252.

Wheeler, D. C., & O'Kelly, M. E. (1999). Network topology and city accessibility of the commercial Internet. *The Professional Geographer, 51*(3), 327–339.

Wilson, M. I. (2001). Location, location, location: The geography of the dot com problem. *Environment and Planning B, 28*(1), 59–71.

Wilson, M. (2003). Chips, bits, and the law: An economic geography of Internet gambling. *Environment and Planning A, 35*(7), 1245–1260.

Zook, M. A. (2000). The Web of production: The economic geography of commercial Internet content production in the United States. *Environment and Planning A, 32*, 411–426.

Zook, M. A. (2001). Old hierarchies or new networks of centrality? The global geography of the Internet content market. *American Behavioral Scientist, 44*(10), 1679–1696.

Zook, M. A. (2002). Grounded capital: Venture financing and the geography of the Internet industry, 1994–2000. *Journal of Economic Geography, 2*(2), 151–177.

Zook, M. A. (2003). Underground globalization: Mapping the space of flows of the Internet adult industry. *Environment and Planning A, 35*(7), 1261–1286.

Zook, M. A. (2005). *The Geography of the Internet industry: Venture capital, dot-coms and local knowledge.* Cambridge, MA: Blackwell.

Zook, M. A., Dodge, M., Aoyama, Y., & Townsend, A. (2004). New digital geographies: Information, communication, and place. In S. Brunn, S. Cutter, & J. W. Harrington (Eds.), *TechnoEarth: Geography and technology* (pp. 155–176). Boston: Kluwer Academic.

CHAPTER 3

Open Access

M. Carl Drott
Drexel University

Introduction

An important role of library and information science has been the identification of better ways to understand and support the "effectual transmission of information" (Weiner, 1948, p. 156) in an ever-changing intellectual and technological context. One of the important vehicles of scholarly information for the past 300-plus years has been the journal. The fundamental nature of this long-established organ may seem destined for reshaping in the light of the recent so-called open access (OA) movement. At the heart of this phenomenon in publishing and distributing information to the community of researchers is the principle that all those who want to read articles published in scholarly periodicals should be able to do so at no cost. In our current model, access costs are borne by libraries and other publicly and privately supported institutions, whereas open access implies broader access without institutional or technical constraints.

One of the widely influential definitions of open access comes from the Budapest Open Access Initiative (BOAI) convened in Budapest by the Open Society Institute (OSI, 2004):

> By "open access" to this literature, we mean its free availability on the public internet, permitting any users to read, download, copy, distribute, print, search, or link to the full texts of these articles, crawl them for indexing, pass them as data to software, or use them for any other lawful purpose without financial, legal, or technical barriers other than those inseparable from gaining access to the internet itself. The only constraint on reproduction and distribution, and the only role for copyright in this domain, should be to give authors control over the integrity of their work and the right to be properly acknowledged and cited.

The Public Library of Science (http://www.plos.org/faq.html) adds:

> all material is also deposited in an archival public reposi-
> tory [...], which enhances the utility of all deposited papers by
> allowing sophisticated searching, manipulation, and mining
> of the literature, using new and emerging tools.

Almost all definitions agree that open access should be viewed as a property of individual articles. Thus, by extension, an open access journal is one in which all content is open access.

Some definitions expand the materials to be made available through open access to include preprints, revisions or commentaries, raw or partially structured data, and, indeed, anything that an author wishes to make freely available. For the purposes of this chapter I will focus mainly on the issue of open access to scholarly journal articles.

These definitions are very liberal in providing researchers and scholars the broadest possible access to materials they might find useful. Others favor much more modest rights, while nonetheless liberalizing present standards of access. These groups and individuals raise serious questions about both the protection of ownership and the cost of supporting the publication of even a Web-based journal. The following is a list of limitations that have been proposed—sometimes in various combinations.

- Limit the permitted use to read and print for personal use only

- Impose a delay between publication for paid subscribers and free access

- Institute system of "micro-payments," a small charge for each access

- Make only selected articles open access

- Make only research articles open access

- Allow authors to make individual articles open access through the payment of a fee

- Provide free or low-cost access only to scholars in poor nations

There is a wide range of variation in the economic models developed to undergird the complex of rights usually posited in these varied open-access models. The most popular model is that of "author pays." In this model, publication and distribution costs are covered by fees assessed to the author, although the assumption is that the fees may actually be paid by research funders or the author's employer.

Some have argued that open access can be achieved without change to the present subscription-supported journal model by the creation of repositories by individual authors or their institutions. Under this model, authors publish their articles through traditional journals but also make a copy available on a Web server. Supporters of this model point to the long acceptance of preprint servers in some areas of physics. Skeptics note that author posting requires either journal policy changes or negotiation of rights by individual authors, because there are so many variations in the transfer of copyright required by most journals as a condition of publication.

In evaluating the various models for open access, it is important to recognize that the issue of open access is as much about social-political issues as it is about scholarly communication, technology, or economics. Discussion thus moves into an arena in which the rhetoric may be more focused on proselytizing than on careful articulation of scholarly argument. In such an atmosphere, even authors who show impeccable semantic hygiene in their own research work are given to expressing themselves in a more populist style. Thus, there can be no completely satisfactory definition of "open access." Representatives of different points of view seek to characterize the term in ways most agreeable to their own conclusions. Similarly, there can be no *neutral* view of the situation. In this respect, I have attempted to provide a *balanced* view. As part of this effort, I have used the first person voice as a reminder that any position is subjective in the heated conflict that surrounds the topic of open access. Because the debate about open access is not strictly a matter of scholarship, much that is written appears in online discussion groups and individual Web sites. Such sources may not have the complete scholarly apparatus that allows recognition of the originator of a particular view. Rather than risk misattribution, I have tended not to identify these sources. I apologize to those whom I have inadvertently denied credit for their ideas.

Finally, I want to note that issues involving open access include many topics that have been the subject of chapters in past volumes of *ARIST*. Readers who wish to explore any of these in greater depth will find a wealth of supporting references. I would especially note: "The Internet and Unrefereed Scholarly Publishing" (Kling, 2004), "Preservation of Digital Objects" (Galloway, 2004), "Digital Libraries" (Fox & Urs, 2002), and "Legal Aspects of the Web" (Borrull & Oppenheim, 2004).

Overview

The emergence of the discussion of open access as a viable alternative to traditional publishing rests on developments in three main areas: economics, technology, and social justice. The issues in these areas are complex and intertwined, but the foregoing division will serve to set out the main themes to be discussed.

For many, especially those in the library community, the continuing and substantial rise in journal subscription costs over several decades has been a great burden. In addition, the costs of binding and storing back issues of journals have continued to climb. Moreover, the cost of computing has dropped dramatically and the growth of extremely cheap electronic networking has been remarkable. At the same time, advances in technology have meant that almost all authors produce their original manuscripts in digital form. Word processing and page layout programs have simplified editing and proofreading and have made it much easier to prepare manuscripts for publication. Web servers and browsers along with network technology make the dissemination and retrieval of electronic documents a largely transparent task. This means that most manuscripts are easily available via electronic access systems, whether on the Web or as part of more limited services.

In addition, the press for open access satisfies a growing awareness of issues that are best labeled matters of ethics or even social justice. Many advocates would explain the philosophical ramifications of the current system as follows: The public, through government funding of both research and universities, pays a significant part of the cost of producing the research that underlies scholarly articles. The people are then charged additional amounts, through support of libraries, for access to the published product that they have already subsidized. An additional unfairness is seen to lie in the growth of information-dependent activities in the developing world. This is evident both in the emergence of strong indigenous universities and educational institutions and in the creation of new industries that are highly information-dependent. In these countries, the vast majority of research is published in journals that are priced according to Western economic models, making them almost unaffordable in local economic terms.

Financial Pressures on Libraries

One of the driving forces of the open access movement has been the increase in subscription costs of scholarly journals and the resultant pressure on academic library budgets. Regardless of how the data is analyzed, the average costs of subscriptions to scholarly journals have risen faster than the average inflation rate. This is true even when one takes into account the increase in journal size measured by the number of pages published annually. One estimate places the overall annual rate of increase for journals at 12 percent, only half of which can be attributed to average inflation and increased size. The increases have been particularly troublesome in science, technology, and medicine (usually abbreviated as STM), both because the rates of increase have been higher and because journals in those areas account for a large fraction of the serials budgets of many major libraries. Buckholtz (2001) summarizes the situation by noting that in the 15 years prior to 2001, serial unit cost rose by 207 percent, general inflation was 52 percent, faculty salaries rose 68 percent, and healthcare costs increased by 107 percent.

The growth in journal subscription cost has varied with the nature of the journal publisher. Journals published by commercial for-profit publishers have had steeper average increases than have those produced by professional societies or nonprofit publishers such as university presses. To some of the advocates of open access journals, this has been seen as evidence that commercial publishers are making excessive price demands on the essentially captive customer base of academic and research libraries. However, averages are misleading, and there are examples of steep price increases for every type of publisher.

Publishers have argued that they have made large financial investments in providing electronic accessibility and that the extra income allows them to take risks in supporting unprofitable journals. In particular, nonprofit publishers, such as learned societies and academic presses, note that surpluses from journal sales are used to support projects that advance both scholarship and the benefits to the public that derive therefrom. The *Washington DC Principles for Free Access to Science* (http://dcprinciples.org) was written by a coalition of nonprofit publishers who see their mission to "enhance the independence, rigor, trust, and visibility" of scholarly journals. They argue that their revenues support "scholarships, scientific meetings, grants, educational outreach, [and] advocacy for research funding." Another report on the benefits of nonprofit publishing has been prepared by the Association of Learned and Professional Society Publishers (2004).

A complicating factor in assessing the rise in journal costs has been the growth of journals that are available online. Such journals may be distributed in both print and online formats, or increasingly, in online only versions. Online availability has changed both the nature of the financial transactions between libraries and publishers and the services that the library can provide to users. Instead of purchasing the issues of a journal produced in a single year, libraries contract for one year's access to all available current and back issues of one or more journal titles. In some cases, for example *Lancet*, this amounts to a run of over 150 years. Further, this access is available to every member of the community served by the library around the clock and, in many cases, anyplace in the world from which they have Internet access. This is clearly an expansion of library services that, without electronic journals, would be far beyond the budget dreams of librarians. Thus, one reasonable position is that, with electronic access, increased subscription costs have purchased vastly increased user service.

A complicating factor in acquiring electronic journals is the fondness of some publishers for offering multiyear packages consisting of a large number of titles for a fixed price. The process of negotiating contracts for electronic access has become a significant burden for some librarians. This so-called "big deal" often combines journals that the library wants to acquire along with many that would be considered marginal. Frazier

(2001) warns against such deals on the grounds that they add to costs and tie up a disproportionate share of library funds with a small number of publishers. In 2003, the Massachusetts Institute of Technology Libraries (2003) explained that they were refusing three-year packages from Elsevier and Wiley because such a commitment would inappropriately limit changes in title holdings. They have since been joined by other libraries.

In a sense, it does not matter what the objective truth of journal price increases is. It suffices to note that rising journal prices and relatively static budgets have galvanized librarians, individually, in collective groups, and through their professional organizations, to assume a significant leadership role in the open access movement. It is clear that in any discussion of journal cost, benefits must also be weighed. As noted earlier, electronic journals offer significant additional benefits to the user. When subscription prices are viewed in terms of cost-per-access, there may be dramatic changes in the perception of which journals are truly "expensive." For example, Morgan Stanley Research (Morgan Stanley & Co., 2002) used 1999 data from the University of Wisconsin to calculate cost per use (University of Wisconsin-Madison Libraries, n.d.). *Brain Research*, which at that time had a subscription price of $14,669, cost $8.25 per article use. On the other hand, *Hospital Medicine*, with a subscription price of $398, had a per use cost of $66.33. However, such thinking may disadvantage researchers in fields with a small readership that are nonetheless vital to the synergy of scholarly work in larger fields.

One cannot leave a discussion of journal costs without looking at the polemics of the issue. The *San Francisco Chronicle* (2004), in a comparison more sensational than illuminating, equated the cost of a year's subscription to *Nuclear Physics A & B* to the price of a new Toyota Camry. Similarly, Elsevier, the for-profit publisher of both titles and some 1,700 other scholarly journals, is frequently excoriated because their margin of return (before taxes) on STM journal publishing approaches 38 percent. In response, Crispin Davis of Elsevier (U.K. House of Commons. Science and Technology Committee, 2004c) has explained that nearly half of this amount is invested in research and development. Plutchak (2004, online), discussing the difficulty of assessing "fairness" in journal pricing, observes that a "fair price" for a journal may simply be a substitute for "a price that is so low that I am happy to pay it." Public Knowledge is an organization devoted to the political aspects of open access and many other intellectual property issues. Its mission is to serve as an "advocacy group working to defend your rights in the emerging digital culture" (Public Knowledge, 2004).

It is beyond the scope of this chapter to review the literature on serials costs, but it is worthwhile to identify some important sources. The 2000 monograph by Tenopir and King presents a comprehensive study of scholarly journals and their users and, although their focus is on print publishing, they also include some information on electronic publication.

King and Montgomery (2002) and Montgomery and King (2002) examine cost per use and the economics of a shift to an electronic environment. Both Holmström (2004) and Cox (2003) add information on the cost per article-reading. Cox notes that the cost per use of the electronic-only Emerald Fulltext is lower than that for the nonprofit Institute of Physics Publishing subscription, which offers a combination of print and online journals.

Journal Quality and Reputation

The enterprise of academic journal publishing has been characterized as a "gift economy." Authors write articles and submit them for publication with no expectation of direct payment. The peculiarity of scholarly authorship is that, unlike other forms of property, almost none of the benefits to the author derive from actual ownership of the work. The benefits derive from acknowledgment of the work by others. In its most abstract form, authors achieve the intangible benefit of the recognition and respect of their colleagues. But the very real reward structure of tenure, promotion, salary raises, and better employment and research opportunities is strongly dependent on the researcher's publication record. The ways in which those who evaluate authors recognize the "quality" of research publications is a complex and largely unstudied one. Evaluation ranges from knowledgeable senior researchers who carefully read and judge a work, to administrators who simply compare the journals in which the author has published against some list of "good" titles. In the U.K., the evaluation of research publications is a part of the Research Assessment Exercise that has a direct bearing on government funding for individual universities (U.K. House of Commons. Science and Technology Committee, 2004b).

With respect to what constitutes a "good title," the impact factor developed by Garfield and published by ISI may well be the single most widely accepted measure of journal quality. Simply put, impact factor is the number of citations to articles published two and three years ago in a particular journal divided by the number of articles published by that journal. Impact factor has been subject to a great deal of both praise and criticism, the discussion of which is beyond the scope of this chapter. However, two observations are particularly germane: The impact factor measure favors established journals and works against new ones; and, for good or ill, impact factor is soundly entrenched in the evaluation strategies of a great many administrative bodies. A hope for the future expressed by Harnad (2001) is that the creation of open archives will increase both the availability and the impact of the papers included. Web-based citation services such as ResearchIndex (also called CiteSeer) will offer broader citation counts than the ISI database and thus may serve to validate open access materials.

A recent study by McVeigh (2004) using ISI's *Journal Citation Reports* compared the impact factors of open access journals with traditional journals in the same subject fields. There are many factors to consider in evaluating this analysis, but, in general, OA journals tended to have somewhat lower impact factors. On the immediacy index, which counts citations to articles in the calendar year of their publication, OA journals performed better, but were still slightly below traditional journals. These data are sufficiently close and subject to interpretation that either side could claim their support. ISI attempts to select the most important journals for coverage so the increasing number of OA journals (now 239 titles out of about 9,000 covered) is a sign of the growth of this model. It is also instructive to note that the fastest growth of OA journals has been in South America and the Asian Pacific regions.

The increasing "citedness" of open access journals is an important indicator of their quality. Even more intriguing is the possibility that the ready availability of open access articles may actually increase the number of citations that they receive. A study of the relation between online availability and citedness by Lawrence (2001b) has provided significant support for open access. From a sample of nearly 120,000 conference articles in computer science whose publication spanned the decade of the 1990s, he found that articles available online averaged 7.03 citations each, whereas those not so available averaged only 2.74. This is an increase in citedness of 157 percent. This difference held even when online and offline articles from the same conference in the same year were matched. Lawrence (2001b) reasons that articles from the same conference are likely to be similar in quality and finds that in this comparison the advantage to articles available online is even greater (an increase of 336 percent) than when articles are compared across conferences. Unfortunately, some who cite Lawrence commit the fallacy of *post hoc propter hoc*. They assume that online availability is the cause of the increase in citations. This ignores possibilities such as biases in the citation data that he derived from ResearchIndex or that "better" authors have greater access to online storage for their papers or greater incentives to provide such availability. Oddly, this research illustrates a problem with self-archiving of preprints. Lawrence's self-archived version (2001a) is titled "Online or invisible?" The published version (2001b) changes the title to "Free online availability substantially increases a paper's impact." This change seems to lead the reader to only one of a number of possible conclusions to be derived from the study.

Journal quality and reputation are important issues for both authors and evaluators. It should be no surprise that subscription-based journals argue that their long history of developing and preserving journal quality should not be taken lightly. Open access advocates reply that there is no reason that their journals cannot achieve or maintain high levels of quality.

Peer Review—Quality Control

The process of peer review, or refereeing, is generally considered the gold standard of quality control for academic publishing. In its simplest form, peer review is the judgment of a submitted article by experienced researchers (generally two or three reviewers per article) concerning whether the work should be published. In explaining the importance of rigorous peer review in a high-quality journal O'Nions said,

> the importance of a piece of work sometimes takes many, many years to establish, but people can feel assured that the piece of work will have been rigorously peer reviewed, and therefore has a high chance of being free of error, and will be timely and will not be just a repetitive piece of research. It is that sort of assurance that people feel. What its real impact is on science may be minimal or may take another 30 years to discover. (U.K. House of Commons. Science and Technology Committee, 2004c, p.18)

As O'Nions notes, the quality of peer review in a particular journal must be judged over many articles and a long period of time. This presents a serious problem for any new journal. Because new open access journals have sought to establish credible claims to quality by emulating the practices of established journals, in assessing these claims, it is worthwhile to consider the process of manuscript acceptance and some of the recommendations for establishing open access journals of high quality.

The quality control of a journal begins with an editorial board. The board is ideally composed of senior scholars with well-established reputations within the field. Such a board is especially important for a newly created journal, since board members are expected to submit some of their own papers to the journal and to encourage prestigious colleagues to do the same. PubMed Central (http://www.pubmedcentral.gov), described later, will accept articles from new journals only if the board members are funded researchers. The new open access journal *PLoS Biology* published by the Public Library of Science has gained in prestige from having Nobel laureate Harold Varmus and many other scientists of international repute as active members of its board.

An editor, again a prestigious scholar is best, is appointed to handle the regular running of the enterprise. One of the primary responsibilities of the editor (in the case of a popular journal there may also be several associate editors) is to receive article submissions, reject those that are clearly unacceptable, and to select appropriate referees to judge the remainder. The referee may judge the importance of the topic to the journal and to the field, the conformance of the methodology to the field's research paradigms, the clarity of the presentation, and even the way the submission links to the previous research in the field through citation.

Finally, it is the job of the editor to combine the judgments of the reviewers and to decide for publication, against publication, or to request that the author revise the submission subject either to re-refereeing or to the editor's satisfaction that the reviewers' comments have been addressed. These functions are considered important to maintaining the quality of a journal, and a journal of any publishing model must have the resources to carry them out.

Up to this point, most of the effort has been either uncompensated, for most referees and many editors, or under-compensated for some editors. There is, however, clerical effort that involves tracking manuscripts and correspondence with authors and referees. A number of free or low-cost software packages are available to support editors in this task and thus reduce both the cost and effort of these quality control steps. For example eFirst XML, eprints, Electronic Submission and Peer Review (ESPERE), and myICAAP. The Open Journal Systems software from the Public Knowledge Project (n.d.) supports both the refereeing and the editing phases of the process. DSpace software supports digital archives (http://www.dspace.org).

The clerical effort of handling manuscripts is proportional to the number of manuscripts submitted, so that highly selective journals, those with high rejection rates, incur substantially greater costs per article actually published. Journals of high reputation report rejection rates of over 90 percent, so that each article published represents the accumulated costs of 10 articles refereed. Any business plan for a journal, whether open access or not, must have a way of covering such expenses.

One of the areas of uncertainty in the production of open access journals is the issue of how much the review process is likely to cost. This point is important because OA advocates tend to present low numbers whereas traditional journals present higher ones. Data on the compensation of editors is rare and scattered. It is probably the case that most editors receive little or nothing in direct compensation and that payments for clerical support, if any, are likely to be small. On the other hand, it is not uncommon for universities to partially support editors with clerical time, computers, office space, and time off from other duties. In terms of compensation from the publisher, Birman (2000) reports on four journal editors who received $6,000, $12,000, $14,000, and $22,500 per year. In another case, she reports that an editor for the journal of a professional society received $12,000 for clerical support. Knuth (2003) reports receiving $1,000 per year as editor of *The Journal of Algorithms* published by Elsevier. In addition, he received $1,667 for clerical support and two complimentary copies. Guédon reports (private communication) that one editor of a commercially published journal received $50 for each manuscript sent out for review. It is not clear if this was compensation for editorial effort or if it was intended to support clerical costs. Editors and referees may also receive other perks such as support for attendance at conferences.

Even when a journal is owned by a commercial publisher, the compensation of editors and referees is largely the intangible benefit of being seen as a gatekeeper protecting the quality of research in a discipline. The prestige of serving on the board of a respected journal may give individuals leverage in making decisions about the future of open access. Knuth and the editorial board of the commercial *The Journal of Algorithms* (Elsevier) resigned en masse to found the open access *Transactions on Algorithms* (ACM). Similarly, the board of *The Journal of Logic Programming* (Elsevier) resigned to found *Theory & Practice of Logic Programming* (Cambridge University Press). Individuals, too, have taken stands against excessive journal subscription rates. Economist Ted Bergstrom (2002), in a widely circulated letter, announced that he would refuse to review articles for journals whose subscription charges exceeded certain limits. On the other hand, the political nature of the debate over open access may lead to divisive characterizations of those who remain on the boards of subscription journals. Guédon (2001, online) chastises professors who have agreed to become editors of commercial journals; "but I must admit, alas, scientists often place the enhancement of their personal career ahead of the collective good."

As I mentioned earlier, there are no clear cost figures for this quality assessment stage of the publication process. The Wellcome Trust (2004) suggests that a conservative (that is to say, high) estimate for refereeing costs for a journal with an acceptance rate of 12 percent, would be $600 per article published, and for a journal with a 50 percent acceptance rate, the estimated refereeing cost would be $300 per article accepted. These numbers seem high compared to Knuth's clerical remuneration of under $2,000 a year reported earlier. A professional society confirmed informally that a figure of $1,000 to $2,000 per year would be reasonable clerical support for an editor. Rowland (2002) put the cost of refereeing at $200 per paper published, and Tenopir and King (2000) give a figure of $20 per page reviewed. If we assume that the prestige of office allows editors to obtain a reasonable fraction of the refereeing cost from their employers, then, not counting the labor of the referees or the editor, refereeing may have a cost to the journal of below $50 per article published.

Editorial Processing

After an article is accepted for publication, it must go through a process of editorial correction and formatting. In traditional journal publishing, this is the point at which the editor sends the accepted manuscript to the publisher. The cost of the effort required to edit and format an article are referred to as "first copy costs." Values for these costs reported in the literature vary greatly. In part, the effort is field-dependent—articles that are largely text are easier to prepare than those with equations, figures, and pictures. One issue with formatting is the extent to which a journal wishes to impose a consistent "look and feel" on all articles, but it is generally assumed that readers expect that any reasonably good journal will have well edited copy that is clearly formatted. Regardless of the standards of the journal, there is nearly universal agreement among editors that authors are largely incapable of preparing acceptable publication-ready copy. Boyce and Dalterio (1996) discuss the difficulty of turning manuscripts into journal copy. They point out that future searchability depends on strict adherence to formatting standards. They suggest that as long as publishers prepare both electronic and paper copies, improvements in technology are not likely to save more than 25 percent.

It is first-copy costs that make up the most significant portion of the cost of producing any journal. Data from the Wellcome Trust (2004) suggest that first-copy costs are about 33 percent of the total cost for a print journal. This is also the kind of effort that is most likely to require hiring skilled staff—both editors and proofreaders as well as those who do page layout. The Public Knowledge Project (n.d.) estimates the cost of freelance editing to be $20,000 for 1,000 pages. This puts the cost of editing a 10-page article at $200. The Entomological Society of America (2004, online) imposes an "editorial review charge" of $48 per page for members and $75 per page for nonmembers, but it is unclear to what extent either of these figures reflects actual costs.

One way of trying to estimate first-copy costs would be to work backward from total income per article. According to the Wellcome Trust (2004), Blackwell Publishing, with over 600 journals in a wide variety of scholarly fields, reported generating an average revenue *from libraries* of $1,425 per published article in 2003. On the other hand, Odlyzko (1999) says, without support, that in general each article brings the publisher $4,000. Applying the 33 percent factor yields first-copy costs of $470 to $1,320. These costs presumably still include profit, so let us give a range of $350 to $1,100. If you add in something like $100 for refereeing, the result is very close to the going rate for most open access journals. Along the same line, Harnad (1997, online) quotes "brave souls who have launched electronic-only journals" as reporting that their costs are about 25 percent of the cost of a print journal. His estimate results in somewhat lower numbers—but scholarship is diverse and it would be fairly conservative to assume that the costs for different journals vary by a factor of two or three.

The discussion of the actual costs involved in producing a scholarly journal has suffered both from lack of data and from lack of agreement on how to make data from various sources comparable. The development of a formalized economic model of the scientific publishing process by Björk and Hedlund (2004) offers hope.

I should perhaps conclude this discussion by noting some of the costs that are not incurred by open access journals. The most obvious are printing and distribution. There are additional costs of maintaining subscription lists, sending bills, and handling payments. These all apply to subscription journals whether they are online or print.

Subscription journals have sales costs. The demand for the top journals in a field is relatively inelastic, but lower-tier journals are in competition for the remaining subscription funds. Further, as the complexity of the "big deal" packages marketed to libraries has increased, so contract negotiations have become more time consuming and, thus, expensive. Online journals of whatever publishing model incur costs for servers, computers, storage, and Internet access. Subscription journals that are available online must maintain security arrangements to limit access. On the other hand, open access journals must track and collect author payments.

Archiving

Given the growth of electronic publishing, archiving is an important factor in the context of open access. As more publications shift to electronic form, the archives themselves can become vehicles for materials to be more widely available. No journal publication scheme, open access or traditional, can meet the needs of scholars unless it is supported by an archiving function. In the case of print journals, archiving has been largely the province of libraries. Libraries bind and store back issues of journals—for less-used materials, they may coordinate their efforts with other libraries. They also provide access, both on site and through interlibrary loan. In the U.S., special provisions have been written into the copyright law to facilitate both preservation and lending. In the U.K., government-established Depository Libraries receive, store, and lend materials that are published in print form. The development of electronic publishing has presented a challenge both to national libraries and to national laws requiring the deposit of published materials.

The archiving of electronic publications presents special problems. In the first place, rapid changes in computer technology make it very likely that any current archives will have to be converted to match new standards. This is less of a problem for text, but may be significant as electronic-only journals add nontext features such as multimedia, video clips, and three-dimensional renderings. The all-electronic *New Journal of Physics* is one that incorporates such materials. Another problem for many academic libraries is not only the potential cost of archiving electronic materials, but the issue of ownership. In any scheme in which libraries are not owners, but licensees, governed by contractual terms, the options for them to continue to perform the function of archiving may be limited or nonexistent.

To some extent, the move to archiving of electronic journals may simply be a matter of the time it takes to shift long-standing library practice. Both the British Library and several European National Libraries have agreed to accept electronic archiving of the *New Journal of Physics*. Elsevier has entered into agreements with several national depositories to provide copies of all of its electronic journal files. The agreement is that, if the material were no longer available in any other way, the libraries would have the right to distribute it.

The positive side of electronic archiving is represented by Ginsparg's (1996) pioneering work in developing arXiv.org as a preprint server in high-energy physics. Since 1991 this project has grown to include other physics specialties and other fields. Because developments in high-energy physics move faster than the journal-refereeing and publication process, the research community in that field has long used preprint distribution. The server is heavily used both by authors who deposit their preprints and by users who generate over 100,000 connections per day to the archive. This electronic archive has proved to be a sustainable enterprise even though it depends on contributed effort and resources. Preprints are accepted by the scholarly community even though they have not yet been refereed—indeed, arXiv.org data show that preprint access continues even after the refereed versions are published. In encouraging much wider preprint archiving, Ginsparg notes that the academic leaders in physics have become comfortable with evaluating candidates using evidence based in part on preprint publication. He believes that absolute consistency of format among papers is not necessary, and, in a memorable phrase, characterizes print publication as the "chemicals adsorbed onto sliced processed dead trees" format (Ginsparg, 1996, online).

Although arXiv.org is an example of one of a number of subject or *discipline-based* archives, a growing number of universities including Cornell, Harvard, MIT, Stanford, and others have begun their own *institutional* archives. To support other universities considering the creation of electronic archives, the Digital Library Federation (2003) has collected case studies and supporting materials.

Archiving as an Alternative to Open Access Journals

The intriguing argument that open access can be achieved through author archiving rather than the form of publication has come from Harnad (1999). He has been a strong leader in promoting open access and especially in encouraging authors to electronically archive their manuscripts (preprints) when they are submitted for publication. He then proposes that, after the manuscript is accepted, the author either append a list of the changes that were made, or substitute a revised version of the paper. He argues that the language in many copyright assignment forms allows authors to create collections of their own works without violating the terms of the agreement and thus the archiving of the finished article is legal. Where the author does not have such rights, Harnad argues that posting the preprint and list of corrections would avoid copyright issues. Others have noted that, no matter what the legal situation, young scholars especially could place their careers in jeopardy by behaving in ways that angered editorial boards or publishers.

I should note that the open access community has come to use the word "archive" as a synonym for "store." This reasonably annoys archivists who would prefer to think of their duties as a far step beyond storage, including preserving and providing access for an indefinitely long future. With apologies, I use the terminology as it appears in the open access literature.

To simplify information exchange on this topic, Harnad, Brody, Vallieres, Carr, Hitchcock, Gingras, et al. (2004) have proposed a classification system for journals according to their policies on making free copies of articles available. According this scheme, *gold* journals provide open access as a matter of policy. *Green* journals explicitly permit authors to self-archive materials. Harnad et al. (2004) estimate that 5 percent of the world's journals are gold and 90 percent are green. This seems high when compared to estimates of others who typically estimate that less than 1 percent of scholarly journals are open access. But it points up the difficulty stated at the outset with respect to lack of agreement on what constitutes open access.

The advantage of a policy of self-archiving with the permission of the journal (green) is that it does not require that a journal make the admittedly risky shift to full open access. More than 10 years ago, Harnad (1995) labeled his proposal for self-archiving "subversive." Unfortunately, open access advocates can no longer claim such a radical title. In June 2004 Elsevier, the world's largest commercial publisher of STM journals, turned "green" by announcing that that it would explicitly permit authors to post final versions of their articles on personal or institutional Web sites.

One of the concerns raised by self-archiving, or any archiving not under the control of the original publisher, is identity, or whether the archived document is the same as the published version. Such a concern has apparently not affected the use of preprint collections even after the material has been published. Now that Elsevier (Elsevier Ltd., 2004, online) has agreed to permit self-archiving of final article versions by authors, rehearsing its former arguments against the practice would be churlish. However, because the arguments continue to be used by others, they remain relevant to the issue of open access being considered here. This is Elsevier's former argument:

> The scientific communication process revolves around the peer review process and the question of what the scientific record is. Researchers need to know when they obtain an Elsevier journal article that it is the article as published, that is, as having been edited and peer reviewed in conformity with the quality which the researcher associates with that particular journal. Having the article on an Elsevier server provides the integrity seal of approval for researchers. Permitting the same article to be published elsewhere on public servers, with researchers unsure about which version was actually peer reviewed, is confusing and potentially harmful to science.

It might be tempting to categorize such an argument as crass commercial justification rather than deeply held commitment to the integrity of science. On the other hand, concerns about the veracity and reliability of documents found on the Web are widespread. I would argue that any author who considers it necessary to include a date of access in a citation to a Web document, especially to an item that is an archived version of an article, is expressing exactly such concerns about integrity. Unfortunately, electronic documents are easily mutable and no widely accepted standard for version control has yet emerged.

It remains to be seen whether creating author or institutional archives can ever capture more than a scattered fraction of the published literature. Advocates believe that as universities see the publicity benefits of institutional archives, and as deposit in archives becomes accepted by trend-setting faculty, the movement will grow. Cynics view the acquiescence of commercial publishers to self-archiving as an indication that it represents no threat to their existing business models.

Author Fees for Open Access

The predominant model for the support of open access publication is one of "author pays." That is, for each accepted article the author is charged a fee of from $300 to over $3,000 depending on the journal. These numbers seem comparable to the cost estimates for a journal of average selectivity of $470 to $1,320, which were presented earlier. The large spread of these numbers is in part due to differences in what services the fee must cover—support for the refereeing phase, the extent of editorial processing, and the number of formats supported, sophistication of online access including special search features, and long-term archiving. For journals with high rejection rates, an alternate author-pays model is one in which each submission is subject to a fee regardless of the referees' decision. So far, this has been a much less popular option. The two major sources of debate over the author-pays model are: "Who really pays?" and "Are costs fairly borne in proportion to benefits?"

Proponents of "author pays" argue that, in the case of funded research, publication should reasonably be covered by the research grant. In support of this mode of funding, the Howard Hughes Institute, an important supporter of medical research in the U.S., has announced a policy that it will support up to $3,000 per year in open access fees, beyond the grant amount, for any researcher it supports. In the U.K., the Wellcome Trust, a major research funder, has taken a number of initiatives to support open access fees. Other funders, both government and private, have indicated a willingness to allow the inclusion of publication fees in research grants.

Another source of author fees can be the author's institution—especially if it is a university or research organization. One line of reasoning is that open access journals significantly cut journal subscription costs and hence author fees represent a transfer of funds rather than an additional cost. Plutchak (2004) observes that a shift to open access could result in a shift in university budgets—moving money from libraries to support membership fees for OA publishing groups. It is not surprising that librarians, who consider their present budgets "under siege," are not enthusiastic about this argument. Quint (2004) argues that in the short term, libraries must form purchasing consortia to combat the power of publishers, but that as open access grows, libraries that wish to keep their budgets will have to emphasize their service focus to their institutions.

Some open access publishers offer institutions "memberships" so that, in return for an annual fee, all authors who belong to that institution may publish without additional fee. In Europe, both the U.K. and Finland have announced membership participations that are national in scope. Memberships have proven to be an attractive option because the costs are known in advance and can be included in an annual budget. The problem that faces open access publishers is an institutional preference for a small number of agreements that cover a large number of journals. There is thus a clear advantage for publishers to form consortia—at least for marketing purposes.

Proponents of author charges for open access frequently argue that the major benefit of the publication of an article is increased prestige for authors, their institutions, and their funding agencies. They also note that, in some fields, authors who publish in subscription journals already bear the costs of page charges or added charges for color in pictures or diagrams. Thus, some authors already pay publication fees and seem to find them acceptable.

Author payments clearly impose a special burden on authors who do not have institutional support, and especially on authors from developing countries. Almost every open access journal has a policy of permitting authors to request that fees be waived. Some of these journals are careful to keep the author's payment status unknown to referees to avoid introducing any bias. Clearly, fees must be set so that paying authors bear sufficient charges to cover those for whom the fee is waived.

The author-pays model of journal publishing raises the specter of the vanity press in which willingness to pay is the only criterion for acceptance. Journals that adopt an author-pays model tend to claim scrupulous attention to the refereeing and editorial control process. Critics warn that the temptation to increase income by accepting marginal papers presents a danger to scholarly integrity. OA supporters point out that subscription journals are susceptible to similar temptations to maintain their page counts or to use growth in the number of articles published to justify price increases.

If fees shift from information consumers to information producers, the economic burden of publication may become more concentrated in a small number of universities and research centers. Plutchak (2004) provides a hypothetical example suggesting that a shift to an author-pays system could concentrate the costs currently spread over 1,000 institutions plus 1,000 individual subscribers to as few as 200 research institutions. Arguments about such extra charges have even been raised on a national level. Policy makers in the U.K. have noted that the country is a net exporter of scientific research. The implication is that many of those "exports" go to well-off countries that would disproportionately benefit from author/sponsor-funded publication (U.K. House of Commons. Committee on Science and Technology, 2004b). A similar fairness issue involves companies, especially those in pharmaceuticals and chemicals, who are heavy users of the journal literature. These organizations are well able to pay subscription fees, but under open access would be relieved of such expenses.

The literature on open access suggests additional sources of income besides author fees (see, for example, the publications of the Open Society Institute), some of which have already been mentioned. Even though these sources may be viable for some journals, there are very few examples among existing open access journals of anything except the fee-for-publication model and its cousin, the flat rate membership fee for all institutional authors. A variation on the author fee is to charge for each submission rather than each publication. Although this would have the effect of reducing fees for the most selective journals, it could be discriminatory against younger researchers who might have higher rejection rates.

Additional possibilities for income include advertisements or corporate sponsorships. Income could be generated from the sale of offprints or CD versions of the journal. But all of these possibilities represent special situations for particular journals in particular fields. It may also be that print subscriptions can coexist with online open access. As early as 1996, the Entomological Society of America began experimenting with electronic access to articles. Authors publishing in any of its four journals could voluntarily purchase immediate free Web access to other articles. The intriguing aspect of this proposal was that the price for immediate free access was set at 75 percent of the cost of 100 reprints— about $90. The idea was that if the number of subscriptions fell, the fee would be raised. Thomas Walker (2001), a strong advocate for open access, announced the policy in *Nature* in 2001. Walker's (2003) Web site presents data showing that between January 2000 and January 2004, the fraction of authors taking advantage of this opportunity rose from 13 percent to 66 percent.

The collection of micropayments, a small fee for each reading, from readers is contrary to most of the open access definitions that have been published. Thus, it is little discussed in the open access literature. Graczynski and Moses (2004) argue that author costs are too high and propose the use of such payments from readers to reduce them. Odlyzko (2003) summarizes a number of reasons why it is unreasonable to expect micropayments to come into wide use. Odlyzko's arguments about the high costs of making multiple small payments may apply to employers who support author payments.

Key Perspectives Ltd. (2004) reported a survey of 311 authors split almost evenly between those who had published in open access journals and those who had published in conventional journals. The study presents a good history of open access and a very complete analysis of the survey responses. Commentators on both sides of the OA debate have already found support in this study. I would note the relatively high error in so small a sample (±4.6 percent). The two author groups differ mainly in terms of: (1) their knowledge of OA publishing, and (2) their comfort with OA journal quality. An interesting finding was that only 4 percent of the OA authors paid their own fees. For the rest, fees were either waived, paid out of grants, paid by their institutions, or covered under membership agreements.

Concerns About Open Access

No one involved in the open access movement denies that, if carried to all scholarly journals, it would produce great economic, technical, and social shifts. The discussion between those in favor of open access and those concerned about the effects of changing the status quo has grown so heated at times that Kaiser (2003, p. 16) compares it to "estranged lovers fighting over child custody." The following authors whom I present here are not alone in their concerns, but reflect some of the major issues that have been raised. Considerable debate has also arisen about whether open access would actually improve matters for developing countries. Some scientists have argued that they would feel stigmatized in asking to have author fees waived. Others have noted that, in many regions of the world, Internet access is expensive, undependable, or even nonexistent—so open access would represent no access in these regions.

In an editorial in the *Journal of the American Medical Association (JAMA)*, DeAngelis and Musacchio (2004) argue against the author-pay system, noting that *JAMA* accepts only 8 percent of submissions, thus placing the cost per article published well over the frequently quoted estimate of $1,500 per article. They express concern that such a system favors authors with the means to pay and might tempt journals to accept more papers (thus lowering quality) in order to increase income. They also set out *JAMA*'s plan to increase access under the current subscription model by (1) providing immediate free access to one major article per issue; (2) providing free access to all major articles and editorials that are between six months and five years old; (3) unlocking online article PDF files so that readers can highlight and annotate the copies that they read; (4) providing 25 free accesses (a kind of online offprint) for authors who reside outside of the United States; and (5) Participating in HINARI, the Health InterNetwork, which provides free and low cost access in the poorest countries (http://www.healthinternetwork.org/src/eligibility.php).

Held (2003), in an editorial in the *Journal of Cell Biology*, warns that attempts to pressure all journals into the open access format are premature. In particular, he is responding to the "Public Access to Science Act" (H.R. 2613) introduced into the U.S. House of Representatives by Martin O. Sabo (and since withdrawn). Held considers the open access model economically untested and perhaps unsustainable. He notes that many journals, in addition to scholarly reports, also publish news and commentary, which are valued by the scholarly community. The authors of these portions are paid for their work, but most proposals for open access ignore this cost component. He notes that the Rockefeller University Press, of which he is executive director, works with other medium-sized publishers to make content available electronically through HighWire Press. Finally, he warns that allowing authors to retain copyright would complicate obtaining rights for further dissemination and might lead to "misuse of the materials by third parties or commercial organizations."

Babbitt, in 1997, predicted the demise of open access journals, noting that many of the current open access journals are supported by subsidies either from foundations or from universities. He also expressed concern that the enthusiasm that leads creators of open access journals to contribute great amounts of time and effort will not be sustained as the founders need to be replaced. Manuscript preparation by authors is likely to be erratic, and this will either require significant editing and formatting costs or compromise the appearance of the journals.

Although generally in favor of open access, Björk (2004) concludes that progress toward open access has not been as fast as many had earlier believed. In particular, he identifies the most difficult problems facing OA journals as the academic reward system, business models, and marketing and critical mass. The latter two can be seen as related. He notes that many business plans depend on volunteer labor and informal employer support. Both of these will be stressed if a journal grows, thus inhibiting strong marketing and economies of scale. He finds that there have been fewer barriers to the growth of repositories, although he notes that individual archiving is still the most common form of open access.

Ewing (2002) argues that present publishers are too well entrenched and that they will be able to make adjustments to their pricing and access policies that will allow them to retain their dominant role. One of the difficulties with any balanced presentation of open access is that many of those raising objections to OA are directly involved in the publication of subscription model journals. If we wish to believe that all such commentators are self-serving and seek only to preserve their lucrative positions in the status quo, they can be easily dismissed. On the other hand, it would seem unwise to so glibly dismiss the views of individuals with so much experience in the journal publication process.

Finding Archived Material

Fundamental to the success of individual, organizational, or topic-focused archives is the ability of potential readers to find the materials they want. In this respect, the most commonly used resource, the Web search engine, is not adequate. In the first place, there may be a considerable delay before the site is revisited by the search service's robot. In the second, the vast size of the Web means that unless the readers have at least an author and title, they will be unable to find the item. Even worse, a subject search, rather than one for a specific item, will be overwhelmed by pages that are not scholarly publications. Google's (2004) introduction of the Google Scholar search service, which is aimed at locating scholarly materials on the Web, is a strong positive achievement with respect to this problem.

The Open Archives Initiative (OAI) has been a leader in setting standards to facilitate indexing and retrieval of materials in scholarly archives. Its approach has been twofold: First, it has worked cooperatively with other organizations to develop standards for including descriptive metadata (think of indexing information: author, title, etc.) as a part of each deposited document. Second, it has developed standards (protocols) describing how a Web server that hosts archival papers can interact with a program that collects metadata to use for creating indexes. Such a collection program is called a "harvester." An important aspect of the OAI approach is that archive operators may make metadata available (the term used is "expose") without the requirement of providing free access to the actual document. In the words of the Open Archives Initiative, "open is not the same as free."

If search engines for archived materials are to work effectively, the metadata for the individual articles must be consistent and of high quality. The metadata standards are based on a limited subset of the Dublin Core. These are standards familiar to catalogers and so, in many universities, the library participates in preparing materials for the institutional online archive.

The OAI standards have made possible a number of indexing projects to locate scholarly material on the Web. The University of Michigan Library (2004) runs the OAIster Project, which shares its indexing with Yahoo!. The most comprehensive source for locating open access journals is the Directory of Open Access Journals run by Lund University Libraries (2004). J-STAGE (http://www.jstage.jst.go.jp/browse) serves as both a search service and an electronic publisher for materials from Japan. Similarly, SciELO (The Scientific Electronic Library Online; http://www.scielo.br) provides indexing and access for scholarly work from Brazil as well as other South American and Caribbean countries.

Copyright and Ownership

The principal question involving copyright for both open access publishers and archives designed to support open access is, how can the legitimate rights of the author be protected while advancing the intention of open access to provide the maximum possible utility to the scholarly community? The obvious answer, and probably the one most frequently used, is to allow the author to retain the copyright and to give the journal or archive a nonexclusive perpetual right of distribution.

The problem of leaving copyright in the hands of individual authors is that it may limit future use of scholarly works because it may be hard to locate living authors and even harder to locate heirs to obtain permission for further use. Thus, if one accepts the broadest definition of open access, a document that is under copyright is not a completely open access document.

One of the organizations that has been addressing this problem is the Creative Commons. This organization has used the Free Software Foundation's GNU General Public License as an inspiration. The Creative Commons' slogan "Some Rights Reserved" is a focus of their efforts to provide more user-friendly forms of authorial protection than copyright law. By developing standard language and clear, nonlegalese explanations, the Creative Commons makes it easy for authors to automatically "give away" rights such as nonprofit copying, while still retaining some copyright control. Authors can link to the appropriate Creative Commons' Web page, or provide the link as metadata, to give readers a standard statement of which rights are automatically licensed and which are reserved. This makes it much easier for metadata harvesters to track the rights information for each document. More recently, Creative Commons has begun to explore the possibility of a Science Commons (http://science.creativecommons.org), which would encourage sharing not only of published articles, but a great deal of other information produced through scholarly research.

A developing issue in copyright is the ownership of databases. Legislation has been proposed in the U.S., but is further developed in the European Union, to protect database content. This has caused concern, such as that articulated by Elliott (1997), that legislation that is too broad may prevent the open sharing of metadata from scholarly archives. The RoMEO project (Gadd, Oppenheim, & Probets, 2003) addresses both the issue of protecting metadata and the articles themselves.

Organizations Supporting Open Access

Many organizations have lent their support to open access and the movement continues to grow. Any attempt to list even the major supporters is certain to have omissions. I apologize for these in advance.

The *Open Society Institute*, a philanthropy of George Soros, has committed to a multimillion-dollar investment in the promotion of open access journals. The institute organized the conference that produced the Budapest Open Access Initiative (BOAI). The BOAI model (whose definition of open access is quoted here earlier) views open access journals as completely free to readers and thus subsidized by a combination of author charges, advertising sales, sponsorships, and other support.

One of the significant contributions of the Open Society Institute (2003a, 2003b, 2004) is the set of three documents that are guides to creating business plans for open access journals. These documents, prepared by Raym Crow and Howard Goldstein of the SPARC (Scholarly Publishing and Academic Resources Coalition) Consulting Group, are an extremely comprehensive discussion of a wide range of alternatives and recommendations for those creating and managing open access journals. The strong point of these reports is their attempt to be comprehensive in laying out funding alternatives and discussing the realities of planning and operating a successful business venture. On the other hand, recognizing the vast differences among scholarly fields and the many national economic situations, the guides have little in the way of actual figures for either costs or income. Although the three documents total over 160 pages, there is considerable redundancy within them, and a reader can skip considerable portions. The guides provide extensive links to other available resources.

The Wellcome Trust (2003) has produced a thorough discussion of the current state of scientific journal publishing. This report covers not only economic issues, but the roles of scholarly journals, guidance for those who wish to develop advocacy organizations for open access, and summaries of the varying views of the many OA participants. One interesting point made in this report is the description of scientific research as a "public good." A public good, as defined in this document, is something that is of value to the public but whose value is difficult either to assess or to allocate to individual people. For example, medical research benefits a great number of people in largely unpredictable ways. Further, the benefit that a person receives from medical research is not traceable to how much they contribute to it. In such a situation, individuals may benefit as "free-riders" by not providing any support. The report notes that in situations of public good, the costs are frequently distributed widely over the population through taxation.

The *Public Library of Science* (PLoS) approaches the idea of open access by trying to create new journals of the highest possible caliber in biology and medicine. Under the chairmanship of Nobel laureate Harold Varmus, this organization has launched *PLoS Biology* and *PLoS Medicine*. These journals have editorial boards of high international standing as well as the ability to attract papers from some of the best researchers. The principal objective is to help to dispel any doubts that scientists may have about the inherent quality of journals published under open access. In addition, PLoS hopes to lead by example, producing financially viable journals based on an author-pays model. Currently the fee is set at $1,500 per accepted article. Although it explicitly denies any desire to compete with journals published by professional societies, PLoS is clearly sending a message to commercial publishers that there may be an effective competitor to journals with high subscription prices.

SPARC (http://www.arl.org/sparc) (Case, 2002) is not an open access organization. Rather, it is an umbrella group for publishers—especially those wishing to start new journals. Its aim is to assist with pooling resources and experience in order to keep journal costs as low as possible. Johnson (2002, p. 648) characterized the formation of SPARC as "built as a response to market dysfunctions in the scholarly communication system, which have reduced dissemination of scholarship and crippled libraries."

Within this organization some members do offer open access journals, whereas others use the traditional subscription model. SPARC encourages the creation of both electronic journals and repositories. It has produced a detailed manual to aid in planning such ventures (Scholarly Publishing and Academic Resources Coalition, 2002). SPARC has also supported and encouraged the formation of electronic depositories for use by journals that are mainly distributed in print. BioOne aggregates articles from dozens of journals in the biological, environmental, and ecological sciences. The strong ties between SPARC and the Association of Research Libraries builds confidence in associated fledgling publishers that their new journals will be noticed by libraries and considered for acquisition in a timely way.

PubMed Central is an electronic archive of journal literature in the life sciences. It is operated by a division of the U.S. National Library of Medicine (NLM). The service is offered to journals, not yet to individual authors, and a journal must include at a minimum all of its primary research content. Journals may delay public release of their material for a year or more in order to preserve value to their subscriber base. To qualify, an established life sciences journal must be currently indexed by a major abstracting and indexing service. New journals may qualify if at least three editorial board members are funded as principal investigators on research grants from a major funding agency. Only English language materials are accepted at present.

The principal value to participating journals is that PubMed Central assumes responsibility for the long-term preservation and accessibility of articles. A second significant advantage is that PubMed Central journals are indexed in the NLM's popular PubMed search service and in the other search services that make up Entrez. PubMed Central's use of standardized formats makes it easy for metadata harvesters to collect indexing information for display on other search services. Clearly, both functions would be of significant value to an open access journal. The major cost of participation is the development of the ability to submit text in an acceptable Extensible Markup Language (XML) of Standard

Generalized Markup Language (SGML) format and to transmit images as either Tagged Image File Format (TIFF) or Encapsulated PostScript.

Developing Trends

Any author trying to judge the important events in the history of a movement while that history is still unfolding is likely to provide ample space for his own future chagrin. Yet, it seems to me that two events in 2004 have lent considerable momentum to the open access movement. These are: the evidentiary hearings and subsequent report on open access of the U.K. House of Commons Science and Technology Committee (2004a, 2004b, 2004c, 2004d, 2004e, 2004f), and the proposal by the U.S. National Institutes of Health (NIH) (2004) that all research publications funded by it should become open access within six months.

The extensive hearings and final report of the House of Commons Science and Technology Committee gathered information on many aspects of the OA debate. Some of the evidence submitted to the committee has been mentioned earlier. BioMed Central (2004) responded to some of the testimony against open access, characterizing it as "myth." The report of the House of Commons Committee agrees that the current subscription model restricts access to research and recommends that the government create a network of freely accessible repositories and require that all publicly funded researchers deposit copies of their articles. The report, however, is not an unqualified endorsement of open access journals, noting that the change to open access would have uncertain and perhaps negative consequences, perhaps weakening learned societies and decreasing the number of high quality scientific publications. The response of the U.K. government could be generally characterized as negative. In most cases the government does not see either a serious problem with the present publication system or the need for increased government support of open access or archiving (U.K. House of Commons. Science and Technology Committee, 2004b).

The debate on open access became considerably more intense when the U.S. National Institutes of Health proposed that electronic versions of all publications based on NIH funded research should be submitted to NIH and then be made freely available through PubMed Central not more than six months after publication. The NIH notice goes on to suggest that those seeking new or extended grants provide links to their work as archived in PubMed. One paragraph also warns investigators not to incur unreasonable or disproportionate charges from publishers. I take this also to be a warning to subscription-based publishers not to institute high-price open access options. Language mandating this policy was written into the appropriations bill for 2004 and passed by the House of Representatives, but the Senate did not concur. It remains to be seen whether the traditional subscription publishers have sufficient political capital to turn back this movement. So far, the main function of both of these developments has been to greatly increase the visibility of the open access movement and to provide proposals that can be a concrete focus for both sides of the debate.

Concluding Remarks

It is not possible to conclude this chapter by synthesizing both the promises and problems of open access and coming to a conclusion about its future. That future is being created even as this is written, and the final outcome remains uncertain. It is clear that the necessary conditions for a publishing revolution are in place. There are organized battalions of angry librarians. Editors, board members, and referees have thrown their shoes into the machinery of for-profit publishing. Subversive authors are archiving their articles. Organizations in many countries provide advice and economic support. The first free enclaves of open access publishing have been established and have raised their banners as rallying points.

Against this, we must admit that, in its present form, the journal has served the cause of scholarly communication long and well, and continues to do so today. Publishers, both for-profit and nonprofit, not only have strong vested interests in the status quo—they have the economic resources to defend those interests. Editors and editorial boards see the value of their own contributions and find that these outweigh somewhat inflated journal prices. Successful scholars have come to depend on the present system for recognition and new scholars hope to do the same. University administrators and research funders understand present financial allocations and are not anxious to venture into the confusion of reallocation.

There is an old environmentalist slogan, "Think globally, act locally." The success or failure of the open access movement will depend on the local actions of individual researchers. Scholars will vote with their paper submissions, with their archives, and with their participation in the publication process.

References

Association of Learned and Professional Society Publishers. (2004). *What do societies do with their publishing surpluses?* Retrieved December 10, 2004, from http://www.alpsp.org/news/NFPsurvey-summaryofresults.pdf

Babbitt, D. (1997). Mathematical journals: Past present and future: A personal view. *Notices of the American Mathematical Society, 44*(1), 29–32.

Bergstrom, T. (2002). *A Lysistratian scheme.* Retrieved May 5, 2005, from http://www.econ.ucsb.edu/%7Etedb/Journals/lysistrata.html

Bethesda Statement on Open Access Publishing (2003, April 11). Retrieved December 8, 2004, from http://www.earlham.edu/~peters/fos/bethesda.htm

BioMed Central. (2004). *(Mis)leading open access myths.* Retrieved December 8, 2004, from http://www.biomedcentral.com/openaccess/inquiry/myths

Birman, J. (2000). Scientific publishing: A mathematician's viewpoint. *Notices of the American Mathematical Society, 47*(7), 770–774.

Björk, B.-C. (2004). Open access to scientific publications: An analysis of the barriers to change? *Information Research, 9*(2). Retrieved December 8, 2004, from http://InformationR.net/ir/9-2/paper170.html

Björk, B.-C., & Hedlund, T. (2004). A formalised model of the scientific publication process. *Online Information Review, 28*(1), 8–21. Retrieved December 8, 2004, from http://oacs.shh.fi/publications/Bjork&HedlundOIR2004.pdf

Borrull, A. L., & Oppenheim, C. (2004). Legal aspects of the Web. *Annual Review of Information Science and Technology, 38*, 483–548.

Boyce, P. B., & Dalterio, H. (1996). Electronic publishing of scientific journals. *Physics Today, 49*(42). Retrieved December 8, 2004, from http://www.aas.org/~pboyce/epubs/pt-art.htm

Buckholtz, A. (2001). Returning scientific publishing to scientists. *Journal of Electronic Publishing, 7*(1). Retrieved December 8, 2004, from http://www.press.umich.edu/jep/07-01/buckholtz.html

Case, M. M. (2002). *Capitalizing on competition: The economic underpinnings of SPARC.* Retrieved December 8, 2004, from http://www.arl.org/sparc/core/index.asp?page=f41

Cox, J. (2003). Value for money in electronic journals: A survey of the early evidence and some preliminary conclusions. *Serials Review, 29*(2), 83–88.

Crow, R. (2002, August 27). *The case for institutional repositories: A SPARC position paper.* Washington, DC: The Scholarly Publishing and Academic Resources Coalition. Retrieved December 8, 2004, from http://www.arl.org/sparc/IR/ir.html

DeAngelis, C. D., & Musacchio, R. A. (2004). Access to *JAMA. Journal of the American Medical Association, 291*(3), 370–371.

Digital Library Federation. (2003). *Archiving electronic journals.* Retrieved December 8, 2004, from http://www.diglib.org/preserve/ejp.htm

Elliott, R. (1997, September). *The impact of electronic publishing in the scientific information chain.* Oxford, UK: International Network for the Availability of Scientific Publications. Retrieved December 8, 2004, from http://www.inasp.info/psi/ejp/elliot.html

Elsevier, Ltd. (2004). *Author gateway for Elsevier journals.* Retrieved December 8, 2004, from http://authors.elsevier.com

Entomological Society of America. (2004). *Publish with ESA.* Retrieved December 8, 2004, from http://www.entsoc.org/pubs/PUBLISH/index.html

Ewing, J. (2002). Predicting the future of scholarly publishing. *Conference on Electronic Information and Communication*, Tsinghau University, China, August 29–31, 2002. Retrieved December 8, 2004, from http://www.ams.org/ewing/Documents/Predicting25.pdf

Fox, E. A., & Urs, S. R. (2002). Digital libraries. *Annual Review of Information Science and Technology, 36*, 503–589.

Frazier, K. (2001). The librarian's dilemma: Contemplating the costs of the "Big Deal." *D-Lib Magazine, 7*(3). Retrieved December 8, 2004, from http://www.dlib.org/dlib/march01/frazier/03frazier.html

Gadd, E., Oppenheim, C., & Probets, S. (2003, July). The RoMEO Project: Protecting metadata in an open access environment. *Ariadne, 36.* Retrieved December 8, 2004, from http://www.ariadne.ac.uk/issue36/romeo

Galloway, P. (2004). Preservation of digital objects. *Annual Review of Information Science and Technology, 38*, 549–590.

Ginsparg, P. (1996, February). Winners and losers in the global research village. Paper presented at a conference held at UNESCO Headquarters. Retrieved December 8, 2004, from http://arXiv.org/blurb/pg96unesco.html

Google. (2004). *About Google Scholar.* Retrieved December 15, 2004, from http://scholar.google.com/scholar/about.html#about

Graczynski, M. R., & Moses, L. (2004). Open access publishing: Panacea or Trojan horse? *Medical Science Monitor, 10*(1), ED1–ED3.

Guédon, J.-C. (2001). In Oldenburg's long shadow: Librarians, research scientists, publishers, and the control of scientific publishing. *Proceedings of the 138th Annual Meeting of the Association of Research Libraries.* Retrieved December 8, 2004, from http://www.arl.org/arl/proceedings/138/guedon.html

Guédon, J.-C. (2003). Open access archives: From scientific plutocracy to the republic of science. *IFLA Journal, 29*(2), 129–140.

Harnad, S. (1995). Universal FTP archives for esoteric science and scholarship: A subversive proposal. In: A. Okerson & J. O'Donnell (Eds.), Scholarly journals at the crossroads: A subversive proposal for electronic publishing. Washington, DC: Association of Research Libraries. Retrieved May 10, 2005, from http://www.arl.org/scomm/subversive/toc.html

Harnad, S. (1997, December). Learned inquiry and the Net: The role of peer review, peer commentary and copyright. *Antiquity, 71*(274), 1042–1048. Retrieved May 5, 2005, from http://www.ecs.soton.ac.uk/~harnad/Papers/Harnad/harnad97.antiquity.html

Harnad, S. (1999). Free at last: The future of peer-reviewed journals. *D-Lib Magazine, 5*(12). Retrieved December 8, 2004, from http://www.dlib.org/dlib/december99/12harnad.html

Harnad, S. (2000). E-knowledge: Freeing the refereed journal corpus online. *Computer Law & Security Report, 16*(2), 78–87. Retrieved December 15, 2004, from http://cogprints.org/1701/00/harnad00.scinejm.htm

Harnad, S. (2001, May 18). Why I think research access, impact and assessment are linked. *Times Higher Education Supplement*, 16.

Harnad, S., Brody, T., Vallieres, F., Carr, L., Hitchcock, S., Gingras, et al. (2004). The access/impact problem and the green and gold roads to open access. *Serials Review, 30*(4), 310–314.

Held, M. J. (2003). Proposed legislation supports an untested publishing model. *Journal of Cell Biology, 162*(2), 171–172. Retrieved December 15, 2004, from http://www.jcb.org/cgi/content/full/162/2/171

Holmström, J. (2004). The cost per article reading of open access articles. *D-Lib Magazine, 10*(1). Retrieved December 8, 2004, from http://www.dlib.org/dlib/january04/holmstrom/01holmstrom.html

Johnson, R. K. (2002). SPARC and ACRL working together to reform scholarly communication. *College & Research Libraries News, 63*(9), 648–651.

Kaiser, D. (2003). The politics of open access. *Information Today, 20*(11), 16.

Key Perspectives Ltd. (2004). *JISC / OSI journal authors survey report.* Retrieved December 8, 2004, from http://www.jisc.ac.uk/uploaded_documents/JISCOAreport1.pdf

King, D. W., & Montgomery, C. H. (2002). After migration to an electronic journal collection: Impact on faculty and doctoral students. *D-Lib Magazine, 8*(12). Retrieved December 8, 2004, from http://www.dlib.org/dlib/december02/king/12king.html

Kling, R. (2004). The Internet and unrefereed scholarly publishing. *Annual Review of Information Science and Technology, 38*, 591–631.

Knuth, D. E. (2003, October 27). Letter to the editorial board. *Journal of Algorithms.* Retrieved December 8, 2004, from http://www-ca-faculty.stanford.edu/~Knuth/joalet.pdf

Lawrence, S. (2001a). Online or invisible? [preprint]. Retrieved December 8, 2004, from http://www.neci.nec.com/~lawrence/papers/online-nature01

Lawrence, S. (2001b, May 31). Free online availability substantially increases a paper's impact. *Nature, 411*(6837), 521–522

Lund University Libraries. (2004). *Directory of open access journals.* Retrieved December 8, 2004, from http://www.doaj.org

McVeigh, M. E. (2004). *Open access journals in the ISI citation databases: Analysis of impact factors and citation patterns.* Stamford, CT: Thomson Scientific. Retrieved December 8, 2004, from http://www.isinet.com/media/presentrep/essayspdf/openaccesscitations2.pdf

Massachusetts Institute of Technology Libraries. (2003). *Wiley and Elsevier subscription packages.* Retrieved December 8, 2004, from http://libraries.mit.edu/about/journals/packages.html

Montgomery, C. H., & King, D. W. (2002). Comparing library and user related costs of print and electronic journal collections. *D-Lib Magazine, 8*(10). Retrieved December 8, 2004, from http://www.dlib.org/dlib/october02/montgomery/10montgomery.html

Morgan Stanley & Co. (2002). *Scientific publishing: Knowledge is power.* Retrieved December 8, 2004, from http://www.alpsp.org/MorgStan300902.pdf

Odlyzko, A. (1999). Competition and cooperation: Libraries and publishers in the transition to electronic scholarly journals. *Journal of Electronic Publishing, 4*(4). Retrieved December 8, 2004, from http://www.press.umich.edu/jep

Odlyzko, A. (2003). The case against micropayments. In R. N. Wright (Ed.), *Proceedings of the 7th International Conference on Financial Cryptography* (Lecture Notes in Computer Science, No.2742, pp. 77–83). Berlin: Springer. Retrieved December 8, 2004, from http://www.dtc.umn.edu/~odlyzko/doc/complete.html

Open Society Institute. (2002, February 14). *Budapest Open Access Initiative.* Retrieved December 8, 2004, from http://www.soros.org/openaccess/read.shtml

Open Society Institute. (2003a). *Guide to business planning for launching a new open access journal* (2nd ed.). Retrieved December 15, 2004, from http://www.soros.org/openaccess/oajguides/html/business_planning.htm

Open Society Institute. (2003b). *Model business plan: A supplemental guide for open access journal developers & publishers.* Retrieved December 15, 2004, from http://www.soros.org/openaccess/oajguides/html/OAJGuideBPSuppl_Ed.1.htm

Open Society Institute. (2004). *Guide to business planning for converting a subscription-based journal to open access* (3rd ed.). Retrieved December 15, 2004, from http://www.soros.org/openaccess/oajguides/html/business_converting.htm

Plutchak, T. S. (2004). Embracing open access. *Journal of the Medical Library Association, 92*(1), 1–3. Retrieved December 8, 2004, from http://www.pubmedcentral.gov

Public Knowledge. (2004). *Public Knowledge home.* Retrieved December 8, 2004, from http://www.publicknowledge.org

Public Knowledge Project. (n.d.). *Open journal systems*. Retrieved December 8, 2004, from http://pkp.ubc.ca/ojs

Quint, B. (2004, February). The great divide. *Searcher Magazine, 12*(2). Retrieved December 8, 2004, from http://www.infotoday.com/searcher/feb04/voice.shtm

Rowland, F. (1997). Print journals: Fit for the future? *Ariadne, 7.* Retrieved December 8, 2004, from http://www.ariadne.ac.uk/issue7/fytton

Rowland, F. (2002). The peer review process. *Learned Publishing, 15*(4), 247–258.

San Francisco Chronicle. (2004, March 28). The staggering price of the world's best research, B1.

Scholarly Publishing and Academic Resources Coalition. (2002). *Gaining independence: A manual for planning the launch of a nonprofit electronic publishing venture.* Washington, DC: Scholarly Publishing and Academic Resources Coalition. Retrieved December 8, 2004, from http://www.arl.org/sparc/GI

Tenopir, C., & King, D. W. (2000). *Towards electronic journals.* Washington, DC: Special Libraries Association.

U.K. House of Commons. Science and Technology Committee. (2004a, November 1). *Responses to the Committee's Tenth Report, Session 2003–04, Scientific publications: Free for all? Fourteenth Report of Session 2003–04.* Retrieved December 8, 2004, from http://www. publications.parliament.uk/pa/cm200304/cmselect/cmsctech/1200/120002.htm

U.K. House of Commons. Science and Technology Committee. (2004b, July 7). *Scientific publications: Free for all?* Tenth Report of Session 2003–04. Retrieved December 8, 2004, from   http://www.publications.parliament.uk/pa/cm200304/cmselect/cmsctech/399/ 39902.htm

U.K. House of Commons. Science and Technology Committee. (2004c, March 1). *Uncorrected transcript of oral evidence to be published as HC 399-iv.* Retrieved December 8, 2004, from   http://www.publications.parliament.uk/pa/cm200304/cmselect/cmsctech/uc399-i/ uc39902.htm

U.K. House of Commons. Science and Technology Committee. (2004d, April 21). *Uncorrected transcript of oral evidence to be published as HC 399-iv.* Retrieved December 8, 2004, from   http://www.publications.parliament.uk/pa/cm200304/cmselect/cmsctech/uc399-iii/ uc39902.htm

U.K. House of Commons. Science and Technology Committee. (2004e, May 5). *Uncorrected transcript of oral evidence to be published as HC 399-iv.* Retrieved December 8, 2004, http://www.publications.parliament.uk/pa/cm200304/cmselect/cmsctech/uc399-iv/ uc39902.htm

U.K. House of Commons. Science and Technology Committee. (2004f, July 7). *Written evidence.* Retrieved December 8, 2004, from http://www.publications.parliament.uk/pa/ cm200304/cmselect/cmsctech/399/399we01.htm

University of Michigan Library. (2004, March 10). *U-M expands access to hidden electronic resources with OAIster.* Retrieved December 8, 2004, from http://www.umich.edu/news/? Releases/2004/Mar04/r031004

University of Wisconsin-Madison Libraries. (n.d.). *Journal cost per use statistics University of Wisconsin-Madison Libraries, 1998–1999.* Retrieved December 8, 2004, from http:// www.wisc.edu/wendt/journals/costben.html

University of Wisconsin-Madison Libraries. (n.d.). *More information about journal cost per use statistics University of Wisconsin-Madison Libraries.* Retrieved December 8, 2004, from http://www.wisc.edu/wendt/journals/costben/mcostben.html

U.S. National Institutes of Health. (2004, September 3). *Notice: Enhanced public access to NIH research information* (NOT-OD-04-064). Retrieved December 8, 2004, from http://grants.nih.gov/grants/guide/notice-files/NOT-OD-04-064.html

Walker, T. J. (2001, May 31). Authors willing to pay for instant Web access. *Nature, 411*(6837), 521–522.

Walker, T. J. (2003). *Electronic publication of journals by the Entomological Society of America*. Retrieved December 8, 2004, from http://csssrvr.entnem.ufl.edu/~walker/epub/esaepub.htm

Wiener, N. (1948). Cybernetics; or, Control and communication in the animal and the machine. Cambridge, MA: Technology Press.

Wellcome Trust. (2003). *Economic analysis of scientific research publishing*. Retrieved December 8, 2004, from http://www.wellcome.ac.uk/en/1/awtpubrepeas.html

Wellcome Trust. (2004). *Costs and business models in scientific and research publishing*. Retrieved December 8, 2004, from http://www.wellcome.ac.uk/en/1/awtpubrepcos.html

Technologies and Systems

CHAPTER 4

TREC: An Overview

Donna K. Harman and Ellen M. Voorhees
National Institute of Standards and Technology

Introduction

The pervasiveness of digital information and the ease with which it can be found via sophisticated search engines make it hard to believe that this was not always the case. But back in 1992, when the Text REtrieval Conference (TREC) began, little digital information was publicly available, and it could only be found by knowing its exact location and using ftp to download the information. Commercial services such as Chemical Abstracts, BRS, and Dialog were available, but these operated on large-scale, custom-built databases of information culled from scientific journals and other such sources. These services were not available to the general public, except via subscribing libraries; furthermore, they could only be accessed via complex (Boolean) search mechanisms requiring a skilled user. Today's simple searching techniques, which return a ranked list of results, were developed in information retrieval research laboratories many years ago, but there was no way to test these on large-scale data; the only data available were collections of abstracts and/or titles on the order of two megabytes of text.

In 1990 the National Institute of Standards and Technology (NIST) was asked to build a new, very large test collection for use in evaluating the text retrieval technology being developed as part of the U.S. Department of Defense, Advanced Research Projects Agency (DARPA) TIPSTER project (for more on the TIPSTER project, see Merchant, 1994). This collection was to be on the order of one million full-text documents, about 100 times larger than existing non-proprietary test collections. The following year, NIST proposed that this large test collection be made available to the full information retrieval community by the formation of TREC.

So, in early 1992, the 25 research groups in TREC-1 undertook to scale their prototype retrieval systems from searching two megabytes of text to searching two gigabytes of text. Large disk drives were expensive in 1992 (over $5,000 for two gigabytes), typical research computers were SPARC 2s with as little as 16 MB of RAM and speeds of 16 MIPS, and most groups made Herculean efforts to finish the task. Yet, despite these technological constraints, a truly significant event had occurred: It had been shown that the statistical methods used by the participating research groups were capable of handling operational amounts of text, and that research on these large test collections could lead to new insights in text retrieval.

By the end of 2004, 13 more TREC conferences had been held, co-sponsored by the NIST, DARPA, and the U.S. intelligence community's Advanced Research and Development Activity (ARDA). The work being evaluated in TREC has expanded far beyond the initial tasks in TREC-1, including retrieval from Web data of a terabyte. The number of participating systems has grown from 25 in TREC-1 to over 100 in TREC 2004, including participants from 22 different countries. The TREC effort represents literally thousands of experiments and many person-hours.

This chapter is an overview of the first 12 years of TREC. Each TREC has published a set of proceedings (Harman, 1993b, 1994b, 1995b, 1996b; Voorhees & Buckland, 2003, 2004; Voorhees & Harman, 1997b, 1998b, 1999b, 2000b, 2001, 2002), both in paper and online at http://trec.nist.gov. Additionally, Voorhees and Harman's (in press) book contains detailed chapters on most of the tasks (tracks) included in TREC and on the experiences of several research groups during their years in the conference. This overview is intended to provide a sense of what has been accomplished in TREC, including some of the major issues in evaluation, and to point interested readers both to further information and to the test material that is available for use by the general research community.

This chapter starts with some background on TREC, including the evaluation model on which TREC is based. Subsequent sections deal with a description of the test collections and overviews of what has been accomplished in the various evaluation tasks that have been investigated in the 12 years of TREC's existence. It should be noted that the numbering scheme for the TRECs was changed in 2001. Until then, the conferences were numbered consecutively (i.e., TREC-1, TREC-2, etc.), but from 2001 on, the TRECs have been labeled by the year of occurrence.

The Basic TREC Paradigm

The Cranfield tests (Cleverdon, Mills, & Keen, 1966) were an investigation of appropriate indexing languages for information retrieval. The conclusion from the tests was that using the words of the texts themselves was very effective—a conclusion that was highly controversial for suggesting that full-text automatic indexing (as opposed to laborious manual indexing) was a viable option for retrieval systems.

Today, full-text indexing is an accepted practice and the tests are better known for the experimental methodology they introduced. Cleverdon created a test collection—a set of 1,400 abstracts and 225 information requests in the field of aerodynamics, with a list of which abstracts should be retrieved for each request—which he used to compare the effectiveness of the different indexing languages. The Cranfield test collection was subsequently used by other groups to compare other retrieval devices. Although this testing paradigm received much criticism (Harter & Hert, 1997), mainly because of its emphasis on batch mode evaluation as opposed to evaluation involving users, it was widely accepted by the information retrieval community for use in batch mode testing and led to important insights in the field.

The Cranfield testing paradigm was the obvious choice to guide TREC, beginning with the creation of the new test collection, because of its widespread acceptance within the information retrieval community. This testing paradigm has received many modifications over the TREC years in order to handle new evaluation situations (discussed throughout this chapter); nevertheless, the basic Cranfield paradigm has proven to be the correct choice.

TREC was planned from the beginning to be more than just a mechanism for making a new test collection available to the community. It included clear specifications of the tasks to be performed, uniform scoring procedures, and the requirement that all groups attending the yearly meetings undergo prior evaluation in one or more of these tasks. By providing the test collection, scoring procedures, and a forum for organizations interested in comparing results, it was hoped that comparisons would be possible across a much wider variety of techniques than any one research group could tackle alone.

From the beginning, TRECs have had four main goals:

1. To encourage research in text retrieval based on large (reusable) test collections

2. To increase communication among industry, academia, and government by developing common tasks that allow system comparisons and by creating an open forum for the discussion of the results and the research that lay behind those results

3. To speed the transfer of technology from research labs into commercial products by demonstrating substantial improvements in retrieval performance on real-world problems

4. To increase the availability of appropriate evaluation techniques for use by industry and academia, including development of new evaluation techniques more applicable to current systems

The Test Collections

The creation of a set of large, unbiased test collections has been critical to the success of TREC. As with the Cranfield collection, these collections have three distinct parts: the documents, the information requests, and the relevance judgments or "right answers." It was important to match all three components (documents, requests, and relevance judgments) of the collection to the TIPSTER application, so that the collection matched at least one real application of text retrieval.

Central in the early TRECs was the ad hoc task. This task tests the ability of systems to retrieve accurate and complete ranked lists of documents in response to 50 information need statements called "topics." The task is similar to how a researcher might use a library—the collection is known but the questions likely to be asked are not known—or to doing a Web search for information. NIST provided the participants with approximately two gigabytes' worth of documents and a set of 50 natural language topic statements. Using either automatic or manual query construction, the participants produced a set of queries from the topic statements and ran those queries against the documents. The output from this run, consisting of the top 1,000 documents retrieved for each topic, was the official test result returned to NIST for the ad hoc task.

The initial collections used for the ad hoc task are described here in order to illustrate the process; many more details can be found in Harman (in press b). Summaries of collections for other tasks are presented in the following section.

The Ad Hoc Collections

The document collection needed to reflect the types of corpora used by information analysts, not only because this was a requirement of the funding, but also because it was desirable to build a collection that would model real information needs. This meant that a very large collection was needed to test the ability of the algorithms to handle huge numbers of full-text documents. The documents were to cover many different subject areas in order to test the domain independence of the algorithms. Additionally, the documents needed to mirror the different types of documents used in the TIPSTER application; specifically, they had to be of varied length, writing style, level of editing, and vocabulary. As a final requirement, the documents had to cover information from different years to show the effects of document date.

There are currently five CD-ROMs of documents in the ad hoc collection, with approximately one gigabyte of text per disk. These documents consist primarily of news articles (taken from, *inter alia*, the *Wall Street Journal*, the Associated Press newswire, the *Financial Times*, the *San Jose Mercury News*, and the *Los Angeles Times*) and government documents (the *Federal Register*, the *Congressional Record*, patent applications, and abstracts from the U.S. Department of Energy publications). Document selection was based on availability and also on congruence with the TIPSTER criteria. Only two disks (two gigabytes of data) were used for each TREC ad hoc test, along with 50 new topics each year.

The topics (requests) for the new test collection were also designed to model some of the analysts' needs. It was assumed that the typical user of these retrieval systems was a dedicated (not a novice) searcher. It was also assumed that the users needed the ability to do both high precision and high recall searches, and were willing to look at many documents and to modify queries repeatedly in order to achieve high recall. Therefore, the topics were created to be very specific, but required both broad and narrow searching capacity.

The topics used in TREC have consistently been the most difficult part of the test collection to control. The designers of the TREC task have made a conscious decision to provide "user need" statements rather than the more traditional queries. Starting in TREC-3, different lengths (and component parts) of topics have been used in each TREC to explore the effects of topic length, such as the use of short titles versus sentence length descriptions versus full user narratives (which include all parts of the topic).

Also starting in TREC-3, topics have generally been created by the same person (or assessor) who performed the relevance assessments for that topic. Each assessor comes to NIST with ideas for topics based on his or her own interests, and searches the document collection (looking at approximately 100 documents per topic) to estimate the likely number of relevant documents per candidate topic. NIST personnel select the final 50 topics from among the candidates, attempting to provide a range of estimated numbers of relevant documents, while balancing the load across assessors. A sample TREC-8 topic is shown in Figure 4.1.

For each topic it is necessary to compile a list of relevant documents, ideally as comprehensive a list as possible. With well over one million documents as possible targets, relevance assessment for all documents for each topic would have been impossible. TREC uses a sampling

---

<num> Number: 409
<title> legal, Pan Am, 103

<desc> Description:
What legal actions have resulted from the destruction of Pan Am Flight 103 over Lockerbie, Scotland, on December 21, 1998?

<narr> Narrative:
Documents describing any charges, claims, or fines presented to or imposed by any court or tribunal are relevant, but documents that discuss charges made in diplomatic jousting are not relevant.

Figure 4.1   Sample TREC-8 topic.

method known as pooling (Sparck Jones & van Rijsbergen, 1975) that takes the top 100 retrieved by each system for a given topic and merges them into a pool for relevance assessment. This is a valid sampling method because all the systems use ranked retrieval methods, placing first those documents most likely to be relevant. The merged list of results is then shown to the human assessors, with each topic judged by a single assessor to ensure consistency of judgment. The assessors are instructed to view the task as if they were addressing a real information need and to mark a document as relevant if they would have used the document in writing a report on this need. For TREC-8 there was an average of 1,736 documents judged per topic, with about 5 percent, or 94, of the documents found to be relevant.

Tests for completeness of the relevance judgments were made in TRECs 2 and 3. For TREC-3 testing, where all runs (a run is the output of a system for a given set of parameters) had a second set of 100 documents judged (thus yielding a total of 200 top documents), a median of 30 new relevant documents was found per topic across all runs. This averages to less than one new relevant document per run. As there were initially 196 relevant documents (on average) for a topic, the addition of one more relevant document was not significant. (For more details on the specific tests, see Harman, 1995a.) These levels of completeness are quite acceptable. Furthermore, it was demonstrated that there was no relationship between the number of new relevant documents found and the number of documents judged within the top 200, implying that additional judgments were not necessary. Instead, the number of new relevant documents found was shown to be correlated with the original number of relevant documents—in other words, topics with many relevant documents were more likely to have additional ones that have not been found. These findings were independently verified by Zobel (1998) at the Royal Melbourne Institute of Technology (RMIT). He also found that any lack of completeness did not bias the results of particular systems, and that systems that did not contribute documents to the pool could still be evaluated fairly using the pooled judgments.

A second issue important to any set of relevance judgments is consistency—specifically, how stable are the judgments and how does their stability or lack thereof affect comparison of the performance of systems using that test collection? In TREC-4, all the ad hoc topics had samples re-judged by two additional assessors, with 72 percent agreement among all three judges and 88 percent agreement between the initial judge and one of the two additional judges. This remarkably high level of agreement in relevance assessment probably is due to the similar backgrounds and training of the judges and a general lack of ambiguity in the topics as represented by the narrative section.

When one looks more deeply into the inconsistencies, however, it becomes clear that most of the agreement had to do with the large numbers of documents that are clearly not relevant. On average, 30 percent of the documents judged relevant by the initial judge were marked as nonrelevant by both additional judges, whereas less than 3 percent of the initial nonrelevant documents were marked as relevant by the secondary judges. This average masks high variability across topics; for 12 of the 50 topics, the disagreement on documents initially marked relevant is higher than 50 percent.

A critical question addresses how all this variation affects system comparisons. Voorhees (2000b) investigated this question by using different subsets of the relevance judgments from TREC-4. She found that although the absolute scores of a given set of system results did change, the changes were highly correlated across systems and the relative ranking of different system runs did not change significantly. Even when the two runs were from the same organization (and therefore more likely to be similar), the two systems were ranked in the same order by all subsets of relevance judgments. This clearly demonstrates the stability of the TREC relevance judgments in the sense that groups can test two different algorithms and be reasonably assured that results reflect a true difference between those algorithms. These results were independently verified as a result of the University of Waterloo's work in TREC-6 (Cormack, Clarke, Palmer, & To, 1998). Although even less agreement was found between the NIST assessors and the Waterloo assessors because of their very different backgrounds and training, the changes in system rankings were still not significant.

As a final comment on TREC relevance assessments, the relationship of the TREC definition of relevance to the wider area of relevance judgments needs to be addressed. The TREC relevance judgments for the ad hoc task should be viewed as the broadest type of judgments—the fact that a document contained *any* information about a topic/question was enough to make it relevant. This was important because of the perceived definition of the TREC task as aiming for high recall. But it also was important in terms of creating the most complete set of relevance judgments possible. It is hoped that others will take these judgments as the starting point for other types of relevance judgments, such as removal of "duplicate" documents, the use of graded relevance judgments (Järvelin & Kekäläinen, 2000; Sormunen, 2002), or even the measurement of some type of learning effect.

Other Collections and Data Availability

The expansion of TREC into multiple tasks (tracks) in TREC-4 led to the design and building of many specialized test collections. None of these is as large or as heavily used as the ad hoc collections described earlier, but they demonstrate the wide range of test collections built for the TREC evaluations. The test collections for the tracks all bear some resemblance to the ad hoc collections, ranging from slight modifications to major evolutions for tracks in the more recent TRECs. Table 4.1 gives some idea of the range of these collections, but later sections describing the various tracks should be read for more information. Details of how

Table 4.1 TREC test collections

TREC	Task	Collection type	Size of collection	Topics
1-8	Ad hoc	English newspapers/wires government documents	5 1-GB CDs	English, 10+ sets of 50
5	Ad hoc	OCR of 1994 *Federal Register*	395 MB	50 known-item searches
6-9	Ad hoc	English broadcast news with truth data and ASR	50, 100, 550 hours	50 known-item searches 23, 50, and 50 ad hoc style
2001, 2002	Ad hoc	Video clips	11, 73 hours	74 mixed known item and ad hoc style, 25 ad hoc
3-6	Ad hoc	Spanish *El Norte* and *Agence France Presse*	200 MB 308 MB	Spanish, 3 sets of 25
5-6	Ad hoc	Chinese *Peoples Daily*, *Xinhua*	178 MB	Chinese, sets of 28 and 26
6-8	Ad hoc CLIR	English, French, German, Italian newswires	750 MB, 250 MB 330 MB, 90 MB	2 sets of 25, 1 of 28 in all four languages
9	Ad hoc CLIR	Chinese Hong Kong newspapers	188 MB	25 in English and Chinese
2001, 2002	Ad hoc CLIR	Arabic *Agence France Presse*	896 MB	2 sets of 25 and 50 in English and Arabic 25 in first set also in French
8	Ad hoc	English Web	2 GB	50 TREC-8 ad hoc
9, 2001	Ad hoc	English Web	10 GB	2 sets of 50 built for Web
1-8	Routing/filtering	English newspapers/ newswires government documents	5-1 GB CDs	8 sets of 50
9	Routing/filtering	OSHUMED (medical)	350,000 abstracts	101 topics
9	Categorization	OSHUMED (medical)	350,000 abstracts	About 5,000 MeSH headings
2001	Categorization	English Reuters newswire	2.5 GB	84 categories
2002	Routing/filtering	English Reuters newswire	2.5 GB	50 topics
2001-2003	Home/name page finding	English Web	10, 18, 18 GB	145, 150, 150 known item
2002-2003	Topic distillation	English Web	18 GB	50, 50 topics

to obtain all the data are on the NIST Web site, http://trec.nist.gov, under the data section.

Evaluation

Retrieval runs on a test collection can be evaluated in a number of ways. In TREC, all ad hoc tasks are evaluated using the trec_eval package (Buckley, 2004). This package reports some 85 different numbers for a run, including recall and precision at various cut-off levels plus single-valued summary measures that are derived from recall and precision. Precision is the proportion of retrieved documents that are relevant, whereas recall is the proportion of relevant documents that are retrieved. A cut-off level is a rank that defines the retrieved set; for example, a cut-off level of 10 defines the retrieved set as the top 10 documents in the ranked list. The trec_eval program reports the scores as averages over the set of topics, with each topic being equally weighted.

Precision reaches its maximal value of 1.0 when only relevant documents are retrieved, and recall reaches its maximal value (also 1.0) when all the relevant documents are retrieved. Note, however, that these theoretical maximum values are not obtainable as an average over a set of topics at a single cut-off level because different topics have different numbers of relevant documents. For example, a topic that has fewer than 10 relevant documents will have a precision score less than

1.0 after 10 documents are retrieved, regardless of how the documents are ranked. Similarly, a topic with more than 10 relevant documents must have a recall score less than 1.0 after 10 documents are retrieved. At a single cut-off level, recall and precision reflect the same information, namely the number of relevant documents retrieved. At varying cut-off levels, recall and precision tend to be inversely related because retrieving more documents usually increases recall while degrading precision and vice versa.

Of all the numbers reported by trec_eval, the recall–precision curve and mean (non-interpolated) average precision are the most commonly used measures to describe TREC retrieval results. A recall–precision curve plots precision as a function of recall. The actual recall values obtained for a topic depend on the number of relevant documents, so that the average recall–precision curve for a set of topics must be interpolated to a set of standard recall values. Recall–precision graphs show the behavior of a retrieval run over the entire recall spectrum.

Mean average precision is the single-valued summary measure used when an entire graph is too cumbersome. The average precision for a single topic is the mean of the precision obtained after each relevant document is retrieved (using zero as the precision for relevant documents that are not retrieved). The mean average precision for a run consisting of multiple topics is the mean of the average precision scores of each of the individual topics in the run. The average precision measure has a recall component in that it reflects the performance of a retrieval run across all relevant documents, and a precision component in that it weights documents retrieved earlier more heavily than documents retrieved later. Geometrically, mean average precision is the area underneath a non-interpolated recall–precision curve.

For more details on the evaluation measures used in TREC, plus discussion of many of the issues surrounding these metrics, see Buckley and Voorhees (in press).

The Ad Hoc Task

From the beginning, the basic TREC ad hoc paradigm presented three major challenges to search engine technology. The first is the vast scale-up in terms of the number of documents to be searched, from several megabytes of documents to two gigabytes of documents. This system-engineering problem occupied most systems in TREC-1, and has continued to be the initial work for most new groups entering TREC. The second challenge is that the documents used are predominantly full-text and, therefore, much longer than most algorithms in TREC-1 were designed to handle. The document length issue resulted in major changes to the basic term weighting algorithms starting in TREC-2. The third challenge has been the idea that a test question or topic contains multiple fields, each representing either facets of a user's question or ways the user might express the question using different amounts of text. The particular fields and their respective lengths have changed across the various TRECs, resulting in different research issues as the basic environment has changed.

Table 4.2 summarizes the research trends in the ad hoc task from TREC-2 to TREC-7. It illustrates some of the common issues that have affected all groups and also shows the initial use and subsequent spread of some of the now-standard techniques that have emerged from TREC. Further details for each of the eight ad hoc task years can be found in the TREC overviews (Harman, 1993a, 1994a, 1995a, 1996a; Voorhees & Harman, 1997a, 1998a, 1999a, 2000a).

Five different research areas are shown in the table, with research in many of these triggered by changes in the TREC evaluation environment. For example, the use of subdocuments, or passages, was caused by the initial difficulties in handling full-text documents—in particular, excessively long ones. Better term weighting, including correct length normalization procedures, reduced the use of this technique in TRECs 4 and 5, but it resurfaced in TREC-6 to facilitate better input to relevance feedback.

Term weighting is the first research area shown in the table. Most initial participants in TREC used term weighting that had been developed and tested on very small test collections with short documents (abstracts). Many of these algorithms were modified to handle longer documents in simple ways; however, some algorithms were not amenable to this approach, resulting in new fundamental research. The group from the Okapi system, City University, London (Robertson &

Table 4.2 Use of new techniques in the ad hoc task

	TREC-2	TREC-3	TREC-4	TREC-5	TREC-6	TREC-7
Term weighting	Baseline for most systems. Beginning of Okapi weighting experiments	Okapi perfects BM25 algorithm	New weighting algorithms in SMART, INQUERY, and PIRCS systems	Use of Okapi/ SMART weighting algorithms by other groups	Adaptations of Okapi/ SMART weighting algorithms in most systems	New retrieval models by TNO and BBN
Passages	Use of sub-documents by PIRCS system	Heavy use of passages/ subdocuments	Decline in use of passages	Decline in use of passages	Use of passages in relevance feedback	Multiple use of passages
Automatic query expansion		Beginning of expansion using top X documents	Heavy use of expansion using top X documents	Beginning of more complex expansion schemes	More sophisticated expansion experiments by many groups	More sophisticated expansion experiments by many groups
Manual query modification		Beginning of manual expansion using other sources	Major experiments in manual editing, user-in-the-loop	Extensive user-in-the-loop experiments	Simpler user-specific strategies tested	Simpler user-specific strategies tested
Other new areas		Initial use of "data fusion"		Start of more concentration on initial topic	More complex use of data fusion. Continued focus on initial topic, especially the title	More complex use of data fusion. Continued focus on initial topic, especially the title

Walker, 1994; Robertson, Walker, Hancock-Beaulieu, & Gatford, 1994) decided to experiment with a completely new term weighting algorithm that was both theoretically and practically based on term distribution within longer documents. By TREC-3, this algorithm had been "perfected" into the BM25 algorithm. Continuing along this same row in Table 4.2, three other systems, the SMART system from Cornell (Singhal, Buckley, & Mitra, 1996), the PIRCS system from the City University of New York (CUNY) (Kwok, 1996), and the INQUERY system from the University of Massachusetts (Allan, Ballesteros, Callan, Croft, & Lu, 1996) changed their weighting algorithms in TREC-4 on the basis of analysis comparing their old algorithms to the new BM25 algorithm. By TREC-5, many of the groups had adopted these new weighting algorithms, with the early adopters being those systems with similar structural models. The common thread in all these algorithms was a complex approach to dealing with document length, either by normalizing with respect to the average document lengths in a given document collection or by working with more subtle methods of measuring term probability.

TREC-6 saw even further expansion of the use of these new weighting algorithms. In particular, many groups adapted these algorithms to new models, often involving considerable experimentation to find the correct fit. For example, IRIT (Boughanem & Soulé-Dupuy, 1998) modified the Okapi algorithm to fit a spreading activation model, IBM[1] (Brown & Chong, 1998) modified it to deal with unigrams and trigrams, and the Australian National University (Hawking, Thistlewaite, & Craswell, 1998) and the University of Waterloo (Cormack, et al., 1998) used it in conjunction with various types of proximity measures. City University also ran major experiments (Walker, Robertson, Boughanem, Jones, & Sparck Jones, 1998) with the BM25 weighting algorithm in TREC-6, including extensive exploration of the existing parameters and the addition of some new ones involving the use of non-relevant documents.

It might be expected that six years of term weighting experiments would lead to a convergence of the algorithms. However, a snapshot of the top eight systems in TREC-7 shows that these systems were derived from many retrieval models and used differing term weighting algorithms and similarity measures. Especially to be noted here is that new models and term weighting algorithms were still being developed (Hiemstra & Kraaij, 1999; Miller, Leek, & Schwartz, 1999) and that these were competitive with the more established methods.

The second new technique, initiated in TREC-2 (the second line of Table 4.2), was the use of smaller sections of documents, called subdocuments, by the PIRCS system at Queens College, City University of New York (Kwok & Grunfeld, 1994). This issue was forced by the difficulty of using the PIRCS spreading activation model for documents having a wide variety of lengths. By TREC-3, many other groups were also using subdocuments, or passages, to help with retrieval. But, as has already been mentioned, TRECs 4 and 5 saw far less use of this technique, as many groups dropped the use of passages due to better term weighting algorithms. In TREC-6, there was a revival of the use of passages, but generally only for specific uses, such as topic expansion. This diverse use of passages has continued in TREC, clearly serving as one of the standard tools for experimentation.

The query expansion/modification techniques shown in the third and fourth lines of Table 4.2 were started when the topics were substantially shortened in TREC-3. In the search for a technique that would automatically expand the topic, several groups revived the old approach of assuming that the top retrieved documents are relevant and then using them in relevance feedback. This technique, which had not worked on smaller collections, turned out to work very well in the TREC environment.

By TREC-6, almost all groups were using variations on expanding queries using information from the top retrieved documents (often called pseudo-relevance feedback or blind feedback). Here, success requires many parameters, such as how many top documents to use for mining terms, how many terms to select, how to rank the terms for selection, and how to weight those terms. There has been general convergence on some of these parameters, with four basic models being used in TREC-7:

1. Rocchio (Buckley, Mitra, Walz, & Cardie, 1999; Kwok & Grunfeld, 1994)

2. Probabilistic (Sparck Jones, Walker, & Robertson, 2000)

3. Local Context Analysis (LCA) (Xu & Croft, 1996)

4. Hidden Markov Model (HMM) (Miller et al., 1999)

TRECs 5, 6, and 7 also saw many varied and successful experiments in the query expansion area. The Open Text Corporation (Fitzpatrick & Dent, 1997) gathered terms for expansion by looking at relevant documents from past topics that were loosely similar to the TREC-5 topics. Several groups, such as LexisNexis (Lu, Ayoub, & Dong, 1997) and Fujitsi Laboratories (Namba, Igata, Horai, Nitta, & Matsui, 1999), tried clustering the top retrieved documents in order to select expansion terms more accurately; and, in TREC-6, three groups (City University, AT&T, and IRIT) successfully generated information from negative feedback, using nonrelevant documents to modify the expansion process. NEC (Mandala, Tokunaga, Tanaka, Okumura, & Satoh, 1999) compared the use of three different thesauri for expansion (Word Net, a simple co-occurrence thesaurus, and an automatically built thesaurus using predicate–argument structures) and AT&T (Singhal, Choi, Hindle, Lewis, & Pereira, 1999) investigated "conservative enrichment" to avoid the additional noise caused by using larger corpora (all five disks) for query expansion.

From TREC-3 on, groups that built their queries manually also looked into better query expansion techniques (see fourth line of Table 4.2). At first, these expansions involved manual editing or using other sources to expand the initial query manually. However, the rules governing manual query building changed in TREC-5 to allow unrestricted interactions with the systems. This change caused a major evolution in the manual query expansion, with most participating groups not only manually expanding the initial queries, but then looking at retrieved documents in order to further expand the queries, much in the manner of human users of potential systems. Both CLARITECH (Milic-Frayling, Evans, Tong, & Zhai, 1997) and General Electric Research (Strzalkowski, Lin, Wang, Guthrie, Leistensnider, Wilding, et al., 1997) ran experiments with complex user interaction scenarios. But by TREC-6, the manual experiments moved back to the simpler practice of having users edit the automatically generated query or having users select documents to be used in automatic relevance feedback.

The final line in Table 4.2 shows some of the other areas that have seen concentrated research in the ad hoc task. Data fusion has been used in TREC by many groups in various ways and has increased in complexity over the years. For example, a project involving four teams led by Strzalkowski continued the investigation of merging results from multiple streams of input using different indexing methods (Strzalkowski, Lin, & Perez-Carballo, 1998; Strzalkowski, Stein, Wise, Perez-Carballo, Tapananinen, Järvinen, et al., 1999). In TREC-6, several groups, such as LexisNexis (Lu, Meier, Rao, Miller, & Pliske, 1998) and MDS (Fuller, Kaszkiel, Ng, Vines, Wilkinson, & Zobel, 1998), used multiple stages of data fusion, including merging results from different term weighting schemes, various mixtures of documents and passages, and different query expansion schemes.

The INQUERY system from the University of Massachusetts worked in all TRECs to build more structure into queries automatically on the basis of information "mined" from the topics (Brown, 1995). From TREC-5 on, there were experiments by other groups to use more information from the initial topic. LexisNexis (Lu, Ayoub, & Dong, 1997) used the interterm distance between nouns in the topic. Several other groups, such as Australian National University (Hawking, Thistlewaite, & Bailey, 1997), University of Waterloo (Clarke & Cormack, 1997), and IBM (Chan, Garcia, & Roukos, 1997) have made use of term proximity features to improve retrieval scores, whereas others, such as CUNY (Kwok & Grunfeld, 1997), AT&T (Singhal,1998), and INQUERY (Allan, Callan, Sanderson, Xu, & Wegmann, 1999) have used the initial topic to look for clues that would suggest a need for more emphasis on certain topic terms. TREC-7 had two additional groups, ETH (Braschler, Wechsler, Mateev, Mittendorf, & Schäuble, 1999) and NTT (Nakajima, Takaki, Hirao, & Kitauchi, 1999) working with the use of term co-occurrence and proximity as alternative methods for ranking.

The creation of two formal topic lengths in TREC-5 (a short version using the title and description sections only and the full topic as given) inspired many experiments comparing results using those different topic lengths; the addition of a formal "title" in TREC-6 increased these investigations. It should be noted that although most of the best runs used the full topic, there was now a smaller performance difference between runs that used the full topic and those that used only the title and description sections than was seen in earlier TRECs. The improvement attributable to the full topic was only 1 percent in TREC-7 for several groups. This is most likely due to improved query expansion methods, but variations across topic sets could also account for the change.

The graph in Figure 4.2 shows that retrieval effectiveness approximately doubled in the first seven years of TREC. The figure plots retrieval effectiveness for one well-known retrieval engine, the SMART system of Cornell University. The SMART system has consistently been one of the more effective systems in the ad hoc task in TREC, but other systems perform comparably; thus, the graph is representative of the increase in effectiveness for the field as a whole. Researchers at Cornell ran the version of SMART used in each of the first seven TREC conferences against each of the first seven ad hoc test sets (Buckley, et al., 1999). Each line in the graph connects the mean average precision scores produced by each version of the system for a single test. For each test, the TREC-7 system has a markedly higher mean average precision than the TREC-1 system.

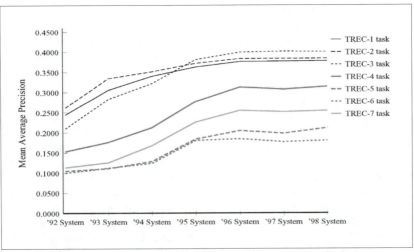

Figure 4.2 Retrieval effectiveness improvement for Cornell's SMART system, TREC-1 through TREC-7.

Figure 4.2 also shows a flattening of the improvements by TREC-7. Possible explanations for this include a dilution of effort from the ad hoc task to all the new (and more exciting) tracks. However, it may be that there are limits as to how far current technology can be pushed in the difficult test environment that systems face at TREC.

The first limitation concerns realistic performance expectations for the systems, given the known conflicts in relevance judgments. Voorhees (2000b) claimed a practical upper bound on retrieval system performance of 65 percent precision at 65 percent recall on average, with a wide variation across topics, on the basis of her studies of agreement between relevance assessors. Note that this limitation on performance should not be considered as an evaluation problem in TREC. The ability to operate within such a noisy environment is critical to operational systems and, therefore, an important part of the TREC environment.

The second limitation is the obvious lack of user interaction. TREC was originally conceived as a test-collection-based evaluation in the Cranfield tradition. What is basically measured are the initial results a user would see after entering a query, but before any interaction. Although this point of measurement is definitely important, and many users will be satisfied with these initial results, the average precision measure shown in Figure 4.2 has a strong recall component. The recall performance can be further improved only by user interaction and the use of appropriate new tools. These tools are generally not being tested in TREC, except by the groups operating in a manual mode or participating in the interactive track. Note that these types of tools are the specific focus of the TREC 2003 HARD track (Allan, 2004).

Comparing the results from manual and automatic systems can provide a sense of how much further improvement might be obtained from user interaction. CLARITECH Corporation has consistently done extensive user-in-the-loop experiments since TREC-5, using both complex and simple interaction models. Table 4.3 shows the average precision for both the Cornell system shown in Figure 4.2 and the best CLARITECH manual run. Note that the average precision for the manual results is only about 10 percentage points above that for the automatic systems, and clearly well below the ceiling suggested by Voorhees.

One major problem is that the use of average performance across the 50 topics hides a wide variation in how the systems perform on the different topics. For any system, there is a wide variation across topics in the effectiveness of particular devices such as relevance feedback. When one compares systems, there is a wide variation in which system, or even which run within a given system, does best on a given topic.

More work needs to be done on customizing methods for each topic; this was taken up in the first running of the robust track in TREC 2003 (Voorhees, 2004b). Although this track used the same task as the ad hoc task, the metrics used emphasized good performance for every topic rather than just good performance on average.

Table 4.3 Comparison of performance using precision at 30 and average precision for automatic and manual systems

	TREC-5	TREC-6	TREC-7
Cornell (automatic)	0.29 / 0.21	0.32 / 0.21	0.39 / 0.27
CLARITECH (manual)	0.36 / 0.25	0.46 / 0.37	0.57 / 0.37

The Tracks

One of TREC's goals has been to provide a common task evaluation that allows cross-system comparisons; this has proven to be a key strength in TREC. A second major strength is the loose definition of the tasks, which allows a wide range of experiments. The addition of secondary tasks (called tracks) in TREC-4 combined these strengths by creating a common evaluation for retrieval sub-problems. The tracks invigorate TREC by focusing research on new areas or particular aspects of text retrieval. To the extent that the same retrieval techniques are used for different tasks, the tracks also validate the findings of the ad hoc task. Each track has a set of guidelines, developed under the direction of the track coordinator, and participants are free to choose which of the tracks they will join.

Figure 4.3 shows various tracks that have been run in TREC over the 12-year period. The left-hand column lists the domain or area of the particular information retrieval aspect being tested, the middle columns give the TREC years in which this aspect was tested, and the right-hand column gives the track names for those tests. Each of the left-hand areas is reviewed in the sections that follow.

Streamed Text (the Routing/Filtering Track)

The routing or filtering problem can be viewed as the inverse of the ad hoc retrieval task in that the question is assumed to be known but the document set changes. These searches are similar to those required

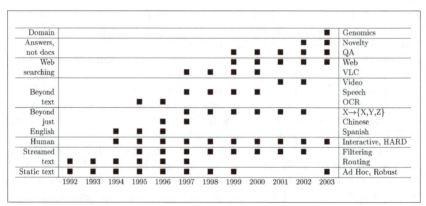

Figure 4.3 Tasks performed in TREC.

by news clipping services and library profiling systems. This section contains information about the tasks that were run and about some of the issues—for a more extensive overview of this track, see Robertson and Callan (in press).

As the routing task was initially defined in TREC, the goal was to use training data to build an "ideal" query that would produce extremely good ranked lists of documents from new data (streaming data). Participants used old ad hoc topics, along with their relevance judgments, to form routing queries. These queries were then run against a previously unseen document collection to produce a ranked document list. This is how the routing task was run in TRECs 1–3 (Harman, 1993a, 1994a, 1995a). The majority of the participants built the routing queries by using some method of relevance feedback; for examples see Robertson, Walker, Hancock-Beaulieu, and Gatford (1994), Broglio, Callan, Croft, and Nachbar (1995), and Buckley, Salton, Allan, and Singhal (1995). However, real routing applications generally require a system to make a binary decision whether to retrieve the current document or not, not to form the ranking of a document set. The filtering track was started in TREC-4 to address this more difficult version of the task (Lewis, 1996).

A slow evolution in the filtering tasks has occurred, with TREC-7 (Hull, 1999) and TREC-8 (Hull & Robertson, 2000) containing three tasks of increasing difficulty (and realism). The first task was the original routing task. The second was a batch filtering task in which systems were given topics and relevance judgments as in the routing task, but then had to decide whether to retrieve each document in the test portion of the collection (in time-received order, not in ranked order). The third was an adaptive filtering task. In this task, a filtering system started with only the query derived from the topic statement (in TREC-8 a few training documents were also used) and processed documents, one at a time, in date order. If the system decided to retrieve a document, it obtained the relevance judgment for it and could modify its query as desired. Most groups continued to use variations on relevance feedback for the adaptive filtering task; for examples, see Evans, Huettner, Tong, Jansen, and Bennett (1999), Hiemstra and Kraaij (1999), and Robertson, Walker, and Beaulieu (1999). Some minor improvements in performance occurred in the second year of the adaptive task, notably from the University of Iowa (Eichmann & Srinivasan, 2000) and the CLARIT group (Zhai, Jansen, Roma, Stoica, & Evans, 2000), but, in general, it was decided that the adaptive task was too hard as posed and, in particular, that new data were needed (the TREC ad hoc data had been used for eight years).

For TREC-9 (Robertson & Hull, 2001), the tasks stayed the same, but the data and the use thereof changed. Although work on getting new data from Reuters had started, it was not completed in time for TREC-9. Instead, the OSHUMED collection (Hersh, 1994) was used in two different tasks. The first used 63 of the 101 topics as filtering topics, with training material consisting of relevant documents from the first year of the collection. This was similar to the TREC-8 work. The second, however, used the MeSH subject headings as the topics, with the documents having those headings being considered the relevant set. Note that this task was very close to a categorization task—specifically, given a set of documents and a set of training material, it sought to categorize those documents with respect to the subject headings. Because this was an adaptive filtering task, however, the time element was important. Therefore, the categorization was assumed to occur one document at a time, as the documents arrived, with only a few early documents used as training. Although this new definition of the task encouraged machine learning approaches (Ault & Yang, 2001), most of the systems still used variations on relevance feedback, but with careful attention to thresholding (for examples, see Arampatzis, Beney, Koster, & van der Weide, 2001; Robertson & Walker, 2001; Zhang & Callan, 2001).

The new Reuters collection arrived in time for TREC 2001 and was used for two TRECs. This collection has about 800,000 Reuters news articles from 1996 and 1997, along with hand-classified categories for those documents. In TREC 2001 (Robertson & Soboroff, 2002), the filtering task was done in a similar manner to the MeSH heading task, with 84 of the Reuters categories used as filtering targets. A wide variety of approaches was used, including a comparison of machine learning with Rocchio feedback (Ault & Yang, 2002), the use of a thesaurus to assign concepts that were "mapped" to the categories (Alpha, Dixon, Liao, & Yang, 2002) and use of both positive and negative feedback (Wu, Huang, Niu, Guo, Xia, & Feng, 2002).

The Reuters collection was again used in TREC 2002 (Soboroff & Robertson, 2003), this time as part of a new test collection expressly designed for tests of filtering. This collection consisted of 50 topics created in a manner similar to the ad hoc task, and 50 topics automatically created using intersections of the Reuters categories. For the manually created topics, extensive relevance assessments were made using successive rounds of user relevance feedback. In general, the automatically created intersection topics did not work well, but the manual collection was quite useful. Many more of the systems used some type of machine learning approach (see Cancedda, Goutte, Renders, Ces-Bianchi, Conconi, Li, et al., 2003; McNamee, Piatko, & Mayfield, 2003; Wu, Huang, Niu, Xia, Feng, & Zhou, 2003), although some groups continued to use Rocchio-type feedback methods (Collin-Thompson, Ogilvie, Zhang, & Callan, 2003; Robertson, Walker, Zaragoza, & Herbrich, 2003).

The question of how to evaluate filtering runs has been a major focus of the filtering track since its inception. Because filtering results are an unordered set of documents, the rank-based measures used in the ad hoc and routing tasks are not appropriate. The main approach has been to try utility functions as measures of the quality of the retrieved set—the quality is computed as a function of the benefit of retrieving a relevant document and the cost of retrieving an irrelevant document. Each TREC since TREC-4 has tried different implementations of utility functions in search of the elusive "ideal" measure.

Robertson and Callan (in press) provide an excellent discussion of the evaluation problem and sum up 11 years of the routing/filtering task by noting that the research community has moved from mostly theoretical work to modeling real filtering situations and creating systems that adapt rapidly to changing conditions. The filtering tasks, and especially the adaptive filtering task, have pushed solid research in thresholding and brought together the traditional information retrieval and machine learning communities. Although there was no filtering track in TREC 2003, it is expected that some new filtering challenge will be proposed for future TRECs.

Human-in-the-Loop (the Interactive Track)

The interactive track, one of the first tracks to be started in TREC, studied text retrieval systems in interaction with users and has been interested in the process as well as the results. Dumais and Belkin (in press) provide both an extensive summary of the work done in the track and discussion of many of the issues raised in interactive testing in general. A special issue of *Information Processing & Management* (Hersh & Over, 2001a) contains a collection of papers based on the TREC interactive track.

TRECs 3 and 4 attempted to mimic the ad hoc task, but in an interactive mode. In TREC-3 (Harman, 1995a), it was decided to use the TREC-3 routing topics because these would be available long before the TREC-3 ad hoc ones, and this earlier availability would allow time for experimentation. Four groups participated in TREC-3, all using the opportunity to gain insights into tools needed by their systems or how online searchers handled the new techniques. Verity (Tong, 1995) found a 24 percent improvement in results obtained by humans using the training material after their initial query. But the major surprise from the track was that the interactive results were notably worse than those for automatic routing. This was the first confirmation of the power exhibited by automatic systems that had been given reasonable amounts of training data.

Eleven groups took part in the interactive track in TREC-4 (Harman, 1996a), using a subset of the ad hoc topics. Many different experiments were run, but the common thread was that all groups used the same topics, performed the same task(s), and recorded the same information about how the searches were done. Many of the systems tested new interfaces, such as Georgia Tech (Veerasamy, 1996), City University, London (Robertson, Walker, Beaulieu, Gatford, & Payne, 1996), and the University of Toronto (Charoenkitkarn, Chignell, & Golovchinsky, 1996), whereas others performed specific experiments, such as the LexisNexis group (Lu, Holt, & Miller, 1996), who used expert Boolean searchers to compare retrieval performance between Boolean and non-Boolean systems, and Xerox (Hearst, Pederson, Pirolli, Schutze, Grefenstette, & Hull, 1996), who tested three modes of search interfaces. But a major track result for TREC-4 was that there was no way to compare across sites in a manner similar to other TREC tracks because the differences between the users and the user groups were much larger than the differences between the systems, a common problem in interactive evaluations.

TREC-5 (Over, 1997) and TREC-6 (Over, 1998) concentrated on a new experimental design comparing the particular retrieval system used at a site (an experimental system) to a common control system that was also run at each site. The direct comparison between the experimental and control systems was used to derive a measure of how much better the experimental system was than the control, independent of topic, searcher, and any other site-specific effects. Different experimental systems could then be indirectly compared across sites relative to the common control. Two groups piloted the new method in TREC-5, and 12 systems took part in TREC-6. Additional experiments before and after TREC-6 addressed the effectiveness of the control (i.e., the equivalence of the direct and indirect comparison of systems) but neither confirmed nor refuted its effectiveness (Lagergren & Over, 1998; Swan & Allan, 1998). As a practical matter, it is difficult to justify the cost of adding a control system to an experimental design in the absence of clear positive evidence for its effectiveness. The decision was made to run future interactive tasks using a similar experimental framework, but without the requirement to use the single control system. The framework both defined a common task for participants to perform and prescribed an experimental matrix for running experiments with a minimum of eight searchers.

TREC-7 (Over, 1999) and TREC-8 (Hersh & Over, 2000) used a small subset of the ad hoc topics for doing "instances" searching. The searchers' job was to save documents covering as many distinct answers to the topic as possible in a short (15–20 minute) timeframe. Eight groups participated in TREC-7, and seven in TREC-8. Some of these groups used TREC to test new capabilities in their systems (Larson, 2000); others continued specific experimental interests, such as relevance feedback variations (Belkin, Head, Jeng, Kelly, Lin, Lobash, et al., 2000). Participants were also required to collect demographic and psychometric data from the searchers and to report extensive data on each searcher's interactions with the search systems.

TREC-9 (Hersh & Over, 2001b) had six participants who tried a fact-finding task across multiple documents. Several of the groups experimented with different document presentation interfaces, such as using the titles and passages (Belkin, Keller, Kelly, Perez-Carballo, Sikora, & Sun, 2001) or using query-based summaries (Alexander, Brown, Jose, Ruthven, & Tombros, 2001). TREC 2001 (Hersh & Over, 2002) and TREC 2002 (Hersh, 2003) worked with Web searching. Again, each of the groups tried experiments of their own, such as different result presentation interfaces (Craswell, Hawking, Wilkinson, & Wu, 2002), using query vs. browsing mode (Toms, Barlett, Freund, & Kopak, 2002), and looking at differences in search expertise and domain expertise (Bhavnani, 2002). In an attempt at more comparability of results, TREC 2002 used a common collection (the TREC Web-track collection) and a common index using the Panoptic search engine (Craswell, Hawking, Wilkinson, Wu, Thom, & Upstill, 2003). Two of the six groups used the Panoptic-ranked document interface as a baseline to test against an interface using categories based on domain names (Craswell, Hawking, Wilkinson, Wu, Thom, & Upstill, 2003) or categories coming from a dynamic clustering algorithm (Osdin, Ounis, & White, 2003).

Although the track abandoned efforts at cross-site comparisons, all groups ran individual experiments using the same task, and this provided a common focus for experimentation. The results of the interactive track need to be understood in the context of the particular research goals of the individual research groups. In general, however, many of the results have been inconclusive, illustrating the extreme difficulty with user testing of systems (as opposed to usability testing). Part of the problem here has been finding a suitable task or scenario for evaluation. For example, if users spend a majority of their limited time reading documents and not using a browsing system, it is hard to evaluate the differences between two browsing systems.

In TREC 2003, the interactive track became the interactive part of the Web track, and some groups participated in a new track on user interaction, the HARD track (Allan, 2004).

Beyond English (the Spanish, Chinese, and Cross-Language Tracks)

The first non-English track was started in TREC-3 (Harman, 1995a). Four groups worked with 25 topics in Spanish, using a document collection from a Monterey, California, Mexican newspaper (*El Norte*) consisting of about 200 megabytes (58,000 documents). Because no training data was available for testing (similar to the startup problems for TREC-1), the groups used simple techniques. The major result from this very preliminary experiment in a second language was the ease of porting the retrieval techniques across languages. Cornell (Buckley, et al., 1995) reported that only five to six hours of system changes were necessary (beyond creation of any stemmers or stop word lists).

Ten groups took part in TREC-4 (Harman, 1996a), using the same document collection and 25 new topics. The final round of Spanish retrieval took place in TREC-5 (Smeaton & Wilkinson, 1997), again with 25 new topics and also with additional text (1994 newswire from Agence France Presse, including 308 megabytes or 173,950 documents). Seven groups took part, with several of them building more elaborate procedures for testing, such as Spanish part-of-speech taggers (Hull, Grefenstette, Schulze, Gaussier, & Schütze, 1997). But in the main, these did not improve performance and the major outcome of the Spanish track was that most of the techniques used in English retrieval, including the advanced ones used in the ad hoc task, could be applied successfully to Spanish (for examples, see Allan, Callan, Croft, Ballesteros, Broglio, Xu, et al., 1997, and Gey, Chen, He, Xu, & Meggs, 1997).

TREC then switched to Chinese in order to investigate retrieval performance for a language with an orthography that is not word-oriented. The document set was a collection of articles selected from the *People's Daily* newspaper and the Xinhua newswire, a total of 168,811 documents in 170 megabytes. Twenty-eight topics were created in Chinese for the track in TREC-5, with an additional 26 topics for TREC-6.

Nine groups submitted Chinese runs in TREC-5 (Smeaton & Wilkinson, 1997), and since it was the first year for Chinese in TREC, most groups concentrated on segmentation issues (Allan, et al., 1997; Kwok & Grunfeld, 1997). TREC-6 (Wilkinson, 1998) had 12 participating groups, which found that approaches using single characters or bigrams as features were competitive with word-based approaches and had the advantage of not requiring complicated segmentation schemes. Several groups, including Fuller et al. (1998) and Kwok, Grunfeld, and Xu (1998), tried combining word and character approaches.

TREC-6 also saw the inclusion of a new task, that of cross-language retrieval of documents (CLIR) (Schäuble & Sheridan, 1998). This task focused on retrieving documents that were written in different languages (English, French, and German) using topics that are in only one language. NIST piloted the task the first year with 25 topics created by trilingual assessors, but this was not satisfactory for many reasons (Harman, Braschler, Hess, Kluck, Peters, Schäuble, et al., 2001). The decision was made to run the track in cooperation with European institutions working in their native languages: the University of Zurich, Switzerland (working on the French portion); Informationszentrum Sozialwissenschaften, Bonn and the University of Koblenz (working on the German portion); and CNR, Pisa, Italy. Twenty-five new topics were created for TREC-7, with an additional 28 topics for TREC-8. Note that for TRECs 7 and 8, these topics were created by each of the cooperating institutions in their native languages and relevance judgments for these topics were made separately for each language. The topics were translated (carefully) into all languages and groups could choose the topic language in which to work.

The track used a document set composed of 250 MB of French documents from the Swiss news agency Schweizerische Depeschen Agentur (SDA); 330 MB of German documents from SDA plus 200 MB from the newspaper *Neue Zürcher Zeitung* (NZZ); 90 MB of Italian documents from SDA; and 750 MB of English documents from the Associated Press newswire. All of the document sets contained news stories from approximately the same time period, but were not aligned or specially coordinated with one another.

The task in TREC-6 was to retrieve both in a monolingual and a cross-lingual manner. For later TRECs, the additional task of merging documents from multiple languages was added—for example, one might start with a topic in English and seek to retrieve a ranked list of documents in English, French, German, and Italian. During the three years that TREC did European CLIR, four major approaches were used: machine translation, where either the topics or the documents were translated into the target language (Gey, Jiang, Chen, & Larson, 1999), the use of machine-readable bilingual dictionaries or other existing linguistic resources (Hiemstra & Kraaij, 1999), the use of n-gram techniques (Mayfield, McNamee, & Piatko, 2000), and the use of corpus resources to train or otherwise enable the cross-language retrieval mechanism (Braschler & Schäuble, 1998). The approaches all behaved similarly in that some groups obtained good cross-language performance for each method. In general, the best cross-language performance was between 50 percent and 75 percent as effective as a high-quality monolingual run in TREC-6. The task of merging results across different languages turned out to be a particularly difficult research issue (Franz, McCarley, & Roukos, 1999; Hiemstra & Kraaij, 1999).

In 2000, the European CLIR task moved to the new Cross Language Evaluation Forum (CLEF) (http://www.iei.pi.cnr.it/DELOS/CLEF), where it continues to run (and grow). NIST moved on to CLIR for English to Chinese in TREC-9 (Gey & Chen, 2001), and then to two years of English to Arabic in TREC 2001 (Gey & Oard, 2002) and TREC 2002 (Oard & Gey, 2003).

Chinese CLIR was run only one year, using data from Hong Kong and 25 topics. The research was more sophisticated than the earlier Chinese track, attracting a group from Microsoft Asia (Gao, Xun, Zhou, Nie, Zhang, & Su, 2001) who performed many experiments using both words and characters, and BBN (Xu & Weischedel, 2001), who modeled the word translation probabilities with a hidden Markov method and used both lexicons and parallel corpora for the translations. The Chinese CLIR then moved to a new Asian CLIR workshop (http://research. nii.ac.jp/ntcir/workshop), where it expanded to include Japanese, Chinese, Korean, and English, plus other tasks in later years.

The TREC CLIR Arabic track used data from the Agence France Press Arabic Newswire (1994–2000), with over 800,000 stories. There were 25 English topics built by native Arabic speakers at the Linguistic Data Consortium in 2002 and another 50 for 2003. Nine groups took part in both years, in particular concentrating on dealing with the rich morphology in Arabic. IIT (Chowdhury, Aljlayl, Jensen, Beitzel, Grossman, & Frieder, 2003) worked with several different stemming approaches, BBN (Fraser, Xu, & Weischedel, 2003) continued their work with probabilistic translation, and APL/JHU (McNamee et al., 2003) used n-grams. Most groups used machine translation of the topics to Arabic and then worked on Arabic retrieval. It should be noted that by the end of both the Chinese and the Arabic CLIR tasks, several systems had cross-lingual performances that equaled (and sometimes surpassed) their monolingual performance, showing the improvements in cross-language retrieval techniques since 1997 (and probably improvements in a type of query expansion resulting from the translation of the query [or document] during the CLIR process).

The TREC tasks in non-English languages produced an extensive set of test collections, all available to the public. A new distributed method of building test collections for multiple languages was created and is still being used in the European and Asian CLIR workshops. Equally important, the nine years of TREC non-English tasks formed a community of researchers interested in working with other languages. For more discussion of the non-English TREC research, including evaluation issues, see Harman (in press a).

Beyond Text (the OCR, Speech, and Video Tracks)

A confusion (or data corruption) track was run in TREC-4 and TREC-5 (Kantor & Voorhees, 2000) to investigate the problems with using "corrupted" data such as would come from optical character recognition (OCR) or speech input. The TREC-4 track did the ad hoc task, using data randomly corrupted at NIST through character deletions, substitutions, and additions to create data with 10 percent and 20 percent error rates (i.e., 10 percent or 20 percent of the characters were affected). Note that this process is neutral in that it does not model OCR or speech input. Four groups used the baseline and 10-percent corruption level; only two groups tried the 20-percent level. As some had expected, the 10-percent error rate did not hurt performance in general and the track results were considered inconclusive.

In TREC-5, the test data were actual OCR output of scanned images of the 1994 *Federal Register*. This time a new task was tried: known-item searching, where the goal was to retrieve a single specific document, rather than a set of relevant documents. Three versions of the documents were used, including the original documents, the documents that resulted after the originals were subjected to an OCR process with a character error rate of approximately 5 percent, and the documents produced through OCR with a 20-percent error rate (caused by downsampling the image before doing the OCR). Five groups tried very different methods, with the group from the Swiss Federal Institute of Technology (ETH) (Ballerini, Büchel, Domenig, Knaus, Mateev, Mittendorf, et al., 1997) performing the best by expanding possible candidate words to improve the best match score.

It was decided to migrate the confusion track to the speech area in TREC-6, where it was called the Spoken Document Retrieval (SDR) track. This track fostered research on retrieval methods for spoken documents (i.e., recordings of speech). It was run in TRECs 6, 7, 8, and 9 using different document sets and different tasks. The TREC-6 (Garofolo, Voorhees, Stanford, & Sparck Jones, 1998) document set was a set of transcripts from 50 hours of broadcast news originally collected by the Linguistic Data Consortium for DARPA Hub-4 speech recognition evaluations (Garofolo, Fiscus, & Fisher, 1997). Three versions of the transcripts were used: a "truth" transcript that was hand-produced, a transcript produced by an IBM baseline speech recognition system, and the transcript produced by each participant's own speech recognition system. Although 50 hours of news presented a serious challenge to the speech systems, the resulting document set was small by retrieval standards, consisting of only 1,451 stories.

As with the earlier confusion track, the task in the TREC-6 SDR track was a known-item search. Thirteen groups submitted SDR track runs and the results suggested that speech recognition and information retrieval technologies were sufficiently advanced to do a credible job of retrieving specific documents. The better systems were able to retrieve the target document at rank 1.0 over 70 percent of the time using their own recognizer transcripts, compared to the best performance on the truth transcripts of 78.7 percent. Search performance was a bigger factor than recognition accuracy in the overall results, with the best results obtained by groups that included both speech and information retrieval experts (Allan, Callan, Croft, Ballesteros, Byrd, Swan, et al., 1998; Milic-Frayling, Zhai, Tong, Jansen, & Evans, 1998).

The TREC-7 track (Garofolo, Voorhees, Auzanne, Stanford, & Lund, 1999) implemented a full-ranked retrieval task, with 23 ad hoc style topics. The document collection was doubled to approximately 100 hours, representing about 3,000 news stories. Different versions of the transcripts (similar to TREC-6, but with two baselines having 35-percent and 50-percent error rates) allowed participants to observe the effect of recognizer errors on their retrieval strategy. Eleven groups participated; the results of the track displayed a linear correlation between the error rates of the recognition and a decrease in retrieval effectiveness. Not surprisingly, the correlation was stronger when recognizer error rate was computed over content-based words (e.g., named entities) rather than all words. AT&T (Singhal, et al., 1999) used contemporaneous newswire for query expansion, allowing them to avoid problems with error-rich speech expansion.

The TREC-8 track (Garofolo, Auzanne, & Voorhees, 2000) made a major jump in collection size, with more than 550 hours of news broadcasts (21,500 stories) and 50 topics. The 10 participating groups found that the larger collection did not affect results and that spoken document retrieval was effective even in very "large" speech corpora. The best speech recognition (and retrieval) was from Cambridge University (Johnson, Woodland, Jourlin, & Sparck Jones, 2000), with a 20.5-percent word error rate. For a summary of both the OCR and the speech tracks, including discussion of the issues, see Voorhees and Garofolo (in press).

TREC 2001 saw the introduction of a new track in video retrieval (Smeaton & Over, 2002). Some 12 groups worked on retrieval of video clips from 5.4 hours of donated video (2.96 gigabytes) that included old government videos, some NIST public video, and 27 minutes of donated BBC stock videos. Seventy-four search topics were built by the participating groups; each topic contained not only text but possibly video, audio, or images of what was needed. Additionally, there was a shot boundary task. The major outcomes of this new track were the creation of a community interested in video retrieval and the beginning of a video test collection.

The video retrieval track continued in TREC 2002 (Smeaton & Over, 2003), this time using 73.3 hours of digital video from the Internet Archive (http://www.archive.org) and the Open Video Project (Marchionini, 2002). The group grew to 17 teams who worked on the shot boundary task, 25 new search topics, and a semantic feature extraction task. For some examples of the type of research involved, see the Informedia project group (Hauptmann, Yan, Qi, Jin, Christel, Derthick, et al., 2003), the CLIPS-IMAG group (Quenot, Moran, Besacier, & Mulhern, 2003) from France, and the Dublin City University group (Browne, Czirjek, Gurrin, Jarina, Lee, Marlow, et al., 2003). This track split off from TREC in 2003, forming a new evaluation project called TRECVID (http://www-nlpir.nist.gov/projects/trecvid).

Web Searching (Very Large Corpus and Web Tracks)

The Very Large Corpus (VLC) track in TREC-6 (Hawking & Thistlewaite, 1998) and TREC-7 (Hawking, Craswell, & Thistlewaite, 1999b) explored how well retrieval algorithms scale to larger document collections. In contrast to the ad hoc task that used a two-gigabyte document collection, the first running of the VLC track in TREC-6 tried a 20-gigabyte collection and the TREC-7 track scaled up to a 100-gigabyte document collection. For both years, the TREC ad hoc topics (for that year) were used, with a set of relevance judgments produced by assessors at the Australian National University (ANU) for the top 20 documents retrieved (precision at 20 was the major effectiveness measures for the task). Also reported were query response time, data structure (e.g., inverted index) building time, and a cost measure of the number of queries processed per minute per hardware dollar. To measure the effect that size had on the retrieval systems used by the participants, the track provided sub-samples of the full data as baselines: one of 10 percent for TREC-6 and two of 1 percent and 10 percent for TREC-7.

The TREC-7 collection consisted of a 100-gigabyte sample of World Wide Web data that was collected by the Internet Archive. Six groups were able to process the entire corpus, with three groups able to process queries at a rate of two seconds or less and to index in 10 hours or less (Cormack, Palmer, Biesbrouck, & Clarke, 1999; Hawking, Craswell & Thistlewaite, 1999a; Singhal, et al., 1999), thus showing that processing a 100-gigabyte corpus was well within the capabilities of the TREC-7 retrieval systems. Note that these systems are not operational systems, they are tuned for flexibility rather than speed, and they are running on suboptimal equipment.

TREC-8 (Hawking, Voorhees, Craswell, & Bailey, 2000) used the 100-gigabyte Web corpus in a new Web track—that is, the use of the links was included. The purpose of the track was to provide the infrastructure required to evaluate new search techniques reliably and to perform repeatable experiments in the context of the World Wide Web. The track defined two subtasks, the small Web and the large Web tasks: the small Web task used a two-gigabyte, 250,000-document subset of the VLC2 collection; the large Web task used the entire collection.

The focus of the small Web task was on answering two questions: do the best methods in the TREC ad hoc task also work best on Web data, and can link information in Web data be utilized to obtain more effective search rankings than can be obtained from page content alone? The TREC-8 ad hoc topics were also the Web track topics and the NIST relevance assessors who judged the ad hoc pools also judged the corresponding Web pools. Results from the 17 participants showed little difference between performance of special Web searching techniques and "normal" ad hoc techniques. Questions about the effects of the size and structure of the two-gigabyte Web data led to a much larger (10-gigabyte) Web track in TREC-9 featuring a more controlled structure.

The large Web task was also a traditional ad hoc retrieval task. In this case, however, the full VLC2 collection of documents was searched using 10,000 queries extracted from logs provided by the AltaVista and Electric Monk search engines. Eight participants submitted the top 20 documents for all 10,000 queries to the Cooperative Research Centre for Advanced Computational Systems (ACSys). ACSys selected 50 of the 10,000 queries to judge and evaluated all 20 documents for each run for those 50 queries. Results again verified the ability of these systems to handle the large amount of Web data.

A special-purpose Web collection (Bailey, Craswell, & Hawking, 2003) was built for TREC-9—the WT10g corpus (a 10-gigabyte sub-sample of the 100-gigabyte corpus) was carefully constructed to increase the density of interserver links. Fifty topics were built for the Web track at NIST by reverse engineering from queries selected from real Web engine logs, and graded relevance judgments (highly relevant, relevant, and nonrelevant) were used (Voorhees, 2001a). Despite all this effort, no conclusive benefit could be found for using link information in TREC ad hoc style topics. For some examples of the techniques tried, see Fujita (2001), Tomlinson and Blackwell (2001), and Singhal and Kaszkiel (2001).

The WT10g corpus was again used in TREC 2001 (Hawking & Craswell, 2002), with 50 more ad hoc style topics (this time the title held the actual Web log query, with misspellings corrected). Once again, no conclusive benefit could be found for utilizing link information. However, in a second task, that of finding 145 home pages, the groups working with the page content only achieved half of the performance of the best run, which incorporated both links and URL structure (Kraaij, Westerveld, & Hiemstra, 2002; Westerveld, Hiemstra, & Kraaij, 2002).

For TREC 2002 (Craswell & Hawking, 2003) and TREC 2003 (Craswell & Hawking, 2004), another new corpus was built and new tasks were designed to model Web searching tasks. The new corpus was a crawl of Web sites in the ".gov" domain from early 2002, capturing 18 gigabytes, or over 1.2 million Web pages. The two new tasks in TREC 2002 were a "named page" task (similar to known-item searches), where the goal was to find a particular page given the name of a specific item, and a topic distillation task, where the goal was to find the best set of pages about a specific broad area (similar to creating bookmarks). TREC 2003 used a revised version of the topic distillation task and a combined named page/home page finding task (in order to ascertain if different techniques were needed for these two tasks). Those completing the topic distillation task generally found that traditional content-based methods were helped by use of stemming and the anchor text, but not helped by use of expansion (Craswell, Hawking, Upstill, McLean, Wilkinson, & Wu, 2004; Tomlinson, 2004). For the named page/home page finding task, groups formed document surrogates based on titles and anchor text, text from other referring documents, and other noncontent items, and then used these surrogates in language modeling modes (Ogilvie & Callan, 2004) or in other combinations (Savoy, Rasolofo, & Perret, 2004). Although the Web track started mainly as an efficiency evaluation effort, the move to a Web-searching task produced a series of important results, as well as providing the infrastructure for testing in the Web environment. The initial Web tasks used ad hoc topics and found no improvement from Web-centric techniques. But when the tests moved into tasks more commonly performed on the Web, techniques such as the use of anchor text, link measures, and site structure became critical to good performance. In an extensive overview of this track, Hawking and Craswell (in press) conclude that the track illustrated the critical relationship between the nature of the task and the specific pieces of Web evidence that should be used to accomplish that task well.

Answers, Not Documents (Question Answering)

TREC-8 (Voorhees, 2000a) saw the first running of the Question Answering track. The goal was to encourage research into systems that return actual answers, as opposed to ranked lists of documents, in response to a question. The track used the TREC-8 ad hoc document collection (two gigabytes of text) and 198 fact-based, short-answer questions such as "How many calories are there in a Big Mac?" Often these questions had been back-formulated from the answers and, therefore, contained similar vocabulary. Each question was guaranteed to have at least one document in the collection that provided the answer. Participants were to return a ranked list of five strings per question on the supposition that each string contained an answer to the question.

Depending on the run type, answer strings were limited to either 50 or 250 bytes. Human assessors read each string and made a binary decision as to whether the string actually did contain an answer to the question. Individual questions received a score equal to the reciprocal of the rank at which the first correct response was returned (or zero if none of the five responses contained a correct answer). The score for a run was the mean of the individual questions' reciprocal ranks.

For TREC-9 (Voorhees, 2001b), the task was kept the same, but the document set was expanded to include all the newspaper and newswire articles from the first five TREC disks, amounting to approximately 979,000 documents. More importantly, the question set consisted of 693 questions: 500 questions drawn from log files (Encarta and Excite), and 193 linguistic variants of those questions. Using naturally occurring questions made the task much more difficult for the systems—but more realistic. There were 28 participants in TREC-9, with the best system able to answer 66 percent of the questions correctly (Harabagiu, Moldovan, Pasca, Mihalcea, Surdeanu, Bunescu, et al., 2001). Most systems first attempted to classify a question according to the type of its answer suggested by the question word—for example, "who" expects a person; "when" expects a day, time, or date. Then the systems used standard information retrieval techniques to select a good target set of documents, and finally performed a shallow parse looking for entities of the same type as the expected answer. Additionally, some groups used sophisticated pattern matching techniques (Hovy, Gerber, Hermjakob, Junk, & Lin, 2001) whereas others worked with standard techniques derived from information retrieval (Clarke, Cormack, Kisman, & Lynam, 2001).

This testing paradigm was continued for TREC 2001 (Voorhees, 2002) with 500 new questions coming from the logs of MSNSearch and AskJeeves. These raw logs were filtered (at Microsoft and AskJeeves) to select queries that contained a question word (e.g., what, when, where, which) anywhere in the query, began with modals or the verb "to be" (e.g., are, can, could, define, describe, does, do), or ended with a question mark. Thirty-six participants worked on the task, which was also modified to allow only a 50-byte response string and removed the guarantee of an answer (groups returned "NIL" as the correct response if there were no answers found). Some of the participating groups continued to use even more sophisticated language processing (Harabagiu, Moldovan, Pasca, Surdeanu, Mihalcea, Girju, et al., 2002) and others built (by hand) very complex patterns (Soubboutin, 2002). Of particular interest was the group from Microsoft (Brill, Lin, Banko, Dumais, & Ng, 2002), who exploited the redundancy in extremely large corpora.

NIST made no attempt to control the relative number of different types of questions; instead, the distribution of question types in the final set reflected the distribution in the source of questions. Definition questions ("Who is ...?" and "What is...?") appear very frequently in query logs and approximately one-quarter of the TREC 2001 test set consisted of definition questions. This resulted in a very challenging test set: it was difficult to return short answers to definition questions because the system had no context in which to frame a suitable response.

For TREC 2002 (Voorhees, 2003), the definition questions were dropped, but the task was made harder by requiring that exact answers be returned, not just the 50-byte string. Accepting only exact answers forced the systems to demonstrate that they knew precisely where the answer lay, even if any reasonable interface would provide a user with some context. Groups were allowed only one answer, not five, and new documents were used. Questions were again drawn from MSNSearch and AskJeeves logs. TRECs 2001 and 2002 also contained a list task in which questions asked for lists of items, such as "Name four countries that can produce synthetic diamonds." There was not much participation in these tasks, so in TREC 2003 (Voorhees, 2004a) the list task, the short answer task, and a reformulated definition task were all combined into a single task with a single score (weighted by each component task).

The reformulation of the definition task forced a major change in the evaluation paradigm because judgment of answers as simply right or wrong does not work when the questions are more complex and require longer answers. However, it was clear that this was needed if TREC were ever to tackle more complex questions. The evaluation of the TREC 2003 definition questions, therefore, was seen as a pilot test of how to evaluate longer, more complex answers.

Systems were asked to return an unordered set of answer strings as the response to a definition question, with each string presumed to be a facet of that definition. The quality of the system response was judged in two steps. First, all answer strings from all responses were presented to the assessor in a single (long) list. Using these responses and his or her own research done during question development, the assessor created a list of "information nuggets" about the target of the question. An information nugget was defined as a fact for which the assessor could make a binary decision as to whether a response contained the nugget. Second, the assessor went through each of the system responses in turn and marked where each nugget appeared in the response. This allowed recall to be calculated based on the number of nuggets found; precision was based on the length of the answer strings. This pilot test worked well; at least it was the start of a new path for evaluation that will be continued for the track. For some examples of how systems tackled the reformulated definition task, see Harabagiu, Moldovan, Clark, Bowden, Williams, and Bensley (2004) and Xu, Licuanan, and Weischedel (2004). For more discussion of this issue and for more detail on the track, see Voorhees (in press).

Conclusions

It is difficult to summarize all results from 12 years of TREC work, comprising thousands of major experiments conducted by all the participating groups. This chapter serves only as an introduction to TREC. Each of the conferences has produced proceedings containing papers from all the participating groups with the details of these experiments and highlights of what was accomplished.

It is equally hard to encapsulate the impact of these 12 years of research on the information retrieval field. When TREC began, there was real doubt as to whether the statistical systems that had been developed in the research labs could effectively retrieve documents from the "large" collections. Clearly, they can. The effectiveness has been demonstrated not only in the laboratory on test collections, but also by today's operational systems that incorporate these techniques. Further, the techniques are routinely used on collections far larger than what was considered large in 1992. Web search engines are a prime example of the power of the statistical techniques; the ability of search engines to point users to the information they seek has been fundamental to the success of the Web.

The various test collections built through the TREC tasks are used by many more groups than actually participate in the conference. This is because TREC has provided the test collections of choice for the community. Additionally, the existence of these collections and the open participation policy have created a major expansion in groups working in the information retrieval area (as evidenced by the growth in TREC). The range of problems that are considered information retrieval problems has been extended by joining with other communities in new tasks, such as the question-answering task, and by focusing on new, highly specialized areas, such as cross-language information retrieval.

TREC has also provided a forum where a large community of research groups, each of which has attempted the same task with the same data, openly discusses their retrieval techniques. This was one of the original goals of TREC and it has proven to be extremely beneficial. By defining a common set of tasks, TREC focuses retrieval research on problems that have a significant impact throughout the community. The meeting itself provides a forum in which researchers can learn efficiently from one another and thus facilitates technology transfer.

Finally, TREC has both validated and standardized evaluation methods, including the importance of building and using test collections. The large number of retrieval results accumulated over the years has provided the means for characterizing the stability of the different evaluation measures and the difficulty of the particular tests. The use of pooling that was first operationalized in TREC has been the key for working with these larger collections and has become one of the standard test collection methods. The TREC model has been followed in the formation of other long-term research evaluation programs, such as CLEF, the Japanese National Center for Science Information Systems (NACSIS), the Test Collection for evaluation of Information Retrieval Systems (NTCIR), and the Initiative for the Evaluation of XML Retrieval (INEX). There are many new areas for exploration, including retrieval of domain-specific information (a new genomics track [Hersh, 2004] started in 2003), retrieval of documents that contain significant amounts of usable metadata, and retrieval of more heterogeneous documents, such as the mix in enterprise searching (an enterprise track is scheduled for TREC 2005). Innumerable challenges also have presented themselves to the research community in terms of filling in the gaps where more analysis is needed. Sparck Jones (1995, 2000, in press) has raised many interesting questions in her reflections on TREC. The results from TREC experiments are publicly available and could serve as a rich resource for analysis on topic variability and other issues. The test collections are available not only for further system experiments but for useful "add-ons," such as user experiments with the core set of relevance judgments to examine learning effects.

TREC continues to be successful in advancing the state of the art in text retrieval, providing a forum for cross-system evaluation that uses common data and evaluation methods and acting as a focal point for discussion of methodological questions on how retrieval research evaluation should be conducted.

Acknowledgments

The authors gratefully acknowledge the continued support of the TREC conferences by ARDA. Thanks also go to the TREC program committee and the staff at NIST. The TREC tracks could not happen without the efforts of the track coordinators; a special thanks to them.

Endnote

1. The inclusion or omission of a particular company or product implies neither endorsement nor criticism by NIST.

References

Alexander, N., Brown, C., Jose, J., Ruthven, I., & Tombros, A. (2001). Question-answering, relevance feedback and summarisation: TREC-9 Interactive Track Report. *Proceedings of the Ninth Text REtrieval Conference (TREC-9)*, 523–532.

Allan, J. (2004). Hard track overview in the TREC 2003 high accuracy retrieval from documents. *Proceedings of the Twelfth Text REtrieval Conference (TREC 2003)*, 24–37.

Allan, J., Ballesteros, L., Callan, J., Croft, B., & Lu, Z. (1996). Recent experiments with INQUERY. *Proceedings of the Fourth Text REtrieval Conference (TREC-4)*, 49–63.

Allan, J., Callan, J., Croft, B., Ballesteros, L., Broglio, J., Xu, J., et al. (1997). INQUERY at TREC-5. *Proceedings of the Fifth Text REtrieval Conference (TREC-5)*, 119–133.

Allan, J., Callan, J., Croft, W., Ballesteros, L., Byrd, D., Swan, R., et al. (1998). INQUERY does battle with TREC-6. *Proceedings of the Sixth Text REtrieval Conference (TREC-6)*, 169–206.

Allan, J., Callan, J., Sanderson, M., Xu, J., & Wegmann, S. (1999). INQUERY and TREC-7. *Proceedings of the Seventh Text REtrieval Conference (TREC-7)*, 201–216.

Alpha, S., Dixon, P., Liao, C., & Yang, C. (2002). Oracle at TREC-10: Filtering and question answering. *Proceedings of the Tenth Text REtrieval Conference (TREC 2001)*, 423–433.

Arampatzis, A., Beney, J., Koster, C., & van der Weide, T. (2001). Incrementality, half-life, and threshold optimization for adaptive document filtering. *Proceedings of the Ninth Text REtrieval Conference (TREC-9)*, 589–600.

Ault, T., & Yang, Y. (2001). kNN at TREC-9. *Proceedings of the Ninth Text Retrieval Conference (TREC-9)*, 127–134.

Ault, T., & Yang, Y. (2002). kNN, Rocchio and metrics for information filtering at TREC-10. *Proceedings of the Tenth Text REtrieval Conference (TREC 2001)*, 84–93.

Bailey, P., Craswell, N., & Hawking, D. (2003). Engineering a multi-purpose test collection for Web retrieval experiments. *Information Processing & Management, 39*, 853–872.

Ballerini, J.-P., Büchel, M., Domenig, R., Knaus, D., Mateev, B., Mittendorf, E., et al. (1997). SPIDER retrieval system at TREC-5. *Proceedings of the Fifth Text REtrieval Conference (TREC-5)*, 217–228.

Belkin, N., Head, J., Jeng, J., Kelly, D., Lin, S., Lobash, L., et al. (2000). Relevance feedback versus local context analysis as term suggestion devices: Rutgers TREC-8 interactive track experience. *Proceedings of the Eighth Text REtrieval Conference (TREC-8)*, 565–574.

Belkin, N., Keller, A., Kelly, D., Perez-Carballo, J., Sikora, C., & Sun, Y. (2001). Support for question-answering in interactive information retrieval: Rutgers TREC-9 interactive track experience. *Proceedings of the Ninth Text REtrieval Conference (TREC-9)*, 463–474.

Bhavnani, S. (2002). Important cognitive components of domain-specific search knowledge. *Proceedings of the Tenth Text REtrieval Conference (TREC 2001)*, 571–578.

Boughanem, M., & Soulé-Dupuy, C. (1998). Mercure at TREC-6. *Proceedings of the Sixth Text REtrieval Conference (TREC-6)*, 321–328.

Braschler, M., & Schäuble, P. (1998). Multilingual information retrieval based on document alignment techniques. In P. Constantopoulos & I. T. Sölvberg (Eds.), *Research and Advanced Technology for Digital Libraries, Second European Conference, ECDL '98*, (Lecture Notes in Computer Science No. 1513, pp.183–197). Berlin: Springer Verlag.

Braschler, M., Wechsler, M., Mateev, B., Mittendorf, E., & Schäuble, P. (1999). SPIDER retrieval system at TREC7. *Proceedings of the Seventh Text REtrieval Conference (TREC-7)*, 509–518.

Brill, E., Lin, J., Banko, M., Dumais, S., & Ng, A. (2002). Data intensive question answering. *Proceedings of the Tenth Text REtrieval Conference (TREC 2001)*, 393–400.

Broglio, J., Callan, J. P., Croft, W. B., & Nachbar, D.(1995). Document retrieval and routing using the INQUERY system. *Proceedings of the Third Text REtrieval Conference (TREC-3)*, 29–38.

Brown, E. (1995). Fast evaluation of structured queries for information retrieval. *Proceedings of the Eighteenth Annual International ACM SIGIR Conference on Research and Development in Information Retrieval*, 30–38.

Brown, E., & Chong, H. (1998). The GURU System in TREC-6. *Proceedings of the Sixth Text REtrieval Conference (TREC-6)*, 535–540.

Browne, P., Czirjek, C., Gurrin, C., Jarina, R., Lee, H., Marlow, S., et al. (2003). Dublin City University video track experiments for TREC 2002. *Proceedings of the Eleventh Text REtrieval Conference (TREC 2002)*, 217–226.

Buckley, C. (2004). trec_eval IR evaluation package. Retrieved January 1, 2005, from ftp://ftp.cs.cornell.edu/pub/smart

Buckley, C., Mitra, M., Walz, J., & Cardie, C. (1999). SMART high precision: TREC 7. *Proceedings of the Seventh Text REtrieval Conference (TREC-7)*, 285–298.

Buckley, C., Salton, G., Allan, J., & Singhal, A. (1995). Automatic query expansion using SMART. *Proceedings of the Third Text REtrieval Conference (TREC-3)*, 69–80.

Buckley, C., & Voorhees, E. (in press). Retrieval system evaluation. In E. M. Voorhees & D. K. Harman (Eds.), *TREC: Experiment and Evaluation in Information Retrieval.* Boston: MIT Press.

Cancedda, N., Goutte, C., Renders, J.-M., Ces-Bianchi, N., Conconi, A., Li, Y., et al. (2003). Kernel methods for document filtering. *Proceedings of the Eleventh Text REtrieval Conference (TREC 2002)*, 373–382.

Chan, E., Garcia, S., & Roukos, S. (1997). TREC-5 ad-hoc retrieval using K nearest-neighbors re-scoring. *Proceedings of the Fifth Text REtrieval Conference (TREC-5)*, 415–426.

Charoenkitkarn, N., Chignell, M., & Golovchinsky, G. (1996). Is recall relevant? An analysis of how user interface conditions affect strategies and performance in large scale text retrieval. *Proceedings of the Fourth Text REtrieval Conference (TREC-4)*, 211–232.

Chowdhury, A., Aljlayl, M., Jensen, E., Beitzel, S., Grossman, D., & Frieder, O. (2003). IIT at TREC 2002: Linear combinations based on document structure and varied stemming for Arabic retrieval. *Proceedings of the Eleventh Text REtrieval Conference (TREC 2002)*, 299–310.

Clarke, C., & Cormack, G. (1997). Interactive substring retrieval (multitext experiments for TREC-5). *Proceedings of the Fifth Text REtrieval Conference (TREC-5)*, 267–278.

Clarke, C., Cormack, G., Kisman, D., & Lynam, T. (2001). Question-answering by passage selection (multitext experiments for TREC-9). *Proceedings of the Ninth Text REtrieval Conference (TREC-9)*, 673–682.

Cleverdon, C., Mills, J., & Keen, E. (1966). *Factors determining the performance of indexing systems.* Cranfield, UK: Aslib Cranfield Research Project.

Collin-Thompson, K., Ogilvie, P., Zhang, Y., & Callan, J. (2003). Information filtering, novelty detection and named-page finding. *Proceedings of the Eleventh Text REtrieval Conference (TREC 2002)*, 107–118.

Cormack, G., Clarke, C., Palmer, C., & To, S. (1998). Passage-based refinement (multitext) experiments for TREC-6. *Proceedings of the Sixth Text REtrieval Conference (TREC-6)*, 303–319.

Cormack, G., Palmer, C., Biesbrouck, M. V., & Clarke, C. (1999). Deriving very short queries for high precision and recall (multitext experiments for TREC-7). *Proceedings of the Seventh Text REtrieval Conference (TREC-7)*, 121–132.

Craswell, N., & Hawking, D. (2003). Overview of TREC 2002 Web track. *Proceedings of the Eleventh Text REtrieval Conference (TREC 2002)*, 86–95.

Craswell, N., & Hawking, D. (2004). Overview of TREC 2003 Web track. *Proceedings of the Twelfth Text REtrieval Conference (TREC 2003)*, 78–92.

Craswell, N., Hawking, D., Upstill, T., McLean, A., Wilkinson, R., & Wu, M. (2004). Trec 12 Web and interactive tracks at CSIRO. *Proceedings of the Twelfth Text REtrieval Conference (TREC 2003)*, 193–203.

Craswell, N., Hawking, D., Wilkinson, R., & Wu, M. (2002). Trec 10 Web and interactive tracks at CSIRO. *Proceedings of the Tenth Text REtrieval Conference (TREC 2001)*, 151–158.

Craswell, N., Hawking, D., Wilkinson, R., Wu, M., Thom, J., & Upstill, T. (2003). TREC 11 Web and interactive tracks at CSIRO. *Proceedings of the Eleventh Text REtrieval Conference (TREC 2002)*, 197–206.

Dumais, S., & Belkin, N. (in press). The TREC interactive tracks: Putting the user into search. In E. M. Voorhees & D. K. Harman (Eds.), *TREC: Experiment and evaluation in information retrieval*. Boston: MIT Press.

Eichmann, D., & Srinivasan, P. (2000). Filters, webs and answers: The University of Iowa TREC-8 results. *Proceedings of the Eighth Text REtrieval Conference (TREC-8)*, 259–266.

Evans, D., Huettner, A., Tong, X., Jansen, P., & Bennett, J. (1999). Effectiveness of clustering in ad hoc retrieval. *Proceedings of the Seventh Text REtrieval Conference (TREC-7)*, 143–148.

Fitzpatrick, L., & Dent, M. (1997). Automatic feedback using past queries: Social searching? *Proceedings of the Twentieth Annual International ACM SIGIR Conference on Research and Development in Information Retrieval*, 306–313.

Franz, M., McCarley, J., & Roukos, S. (1999). Ad hoc and multilingual information retrieval at IBM. *Proceedings of the Seventh Text REtrieval Conference (TREC-7)*, 157–168.

Fraser, A., Xu, J., & Weischedel, R. (2003). TREC 2002 cross-lingual retrieval at BBN. *Proceedings of the Eleventh Text REtrieval Conference (TREC 2002)*, 102–106.

Fujita, S. (2001). Reflections on "aboutness": TREC-9 evaluation experiments. *Proceedings of the Ninth Text REtrieval Conference (TREC-9)*, 281–288.

Fuller, M., Kaszkiel, M., Ng, C., Vines, P., Wilkinson, R., & Zobel, J. (1998). MDS TREC-6 report. *Proceedings of the Sixth Text REtrieval Conference (TREC-6)*, 241–257.

Gao, J., Xun, E., Zhou, M., Nie, J.-Y., Zhang, J., & Su, Y. (2001). TREC-9 CLIR experiments at MSRCN. *Proceedings of the Ninth Text REtrieval Conference (TREC-9)*, 343–354.

Garofolo, J., Auzanne, C., & Voorhees, E. (2000). The TREC spoken document retrieval track: A success story. *Proceedings of the Eighth Text REtrieval Conference (TREC-8)*, 107–130.

Garofolo, J., Fiscus, J., & Fisher, W. (1997). Design and preparation of the 1996 hub-4 broadcast news benchmark test corpora. *Proceedings of the DARPA Speech Recognition Workshop*, 15–21.

Garofolo, J., Voorhees, E., Auzanne, C., Stanford, V., & Lund, B. (1999). 1998 TREC-7 spoken document retrieval track overview and results. *Proceedings of the Seventh Text REtrieval Conference (TREC-7)*, 79–90.

Garofolo, J., Voorhees, E., Stanford, V., & Sparck Jones, K. (1998). 1997 TREC-6 spoken document retrieval track overview and results. *Proceedings of the Sixth Text REtrieval Conference (TREC-6)*, 83–92.

Gey, F., & Chen, A. (2001). The TREC-9 cross-language information retrieval (English-Chinese) overview. *Proceedings of the Ninth Text REtrieval Conference (TREC-9)*, 15–24.

Gey, F., Chen, A., He, J., Xu, L., & Meggs, J. (1997). Term importance, Boolean conjunct training, negative terms, and foreign language retrieval: Probabilistic algorithms at TREC-5. *Proceedings of the Fifth Text REtrieval Conference (TREC-5)*, 187–191.

Gey, F., Jiang, H., Chen, A., & Larson, R. (1999). Manual queries and machine translation in cross-language retrieval and interactive retrieval with Cheshire II at TREC-7. *Proceedings of the Seventh Text REtrieval Conference (TREC-7)*, 527–540.

Gey, F., & Oard, D. (2002). The TREC 2001 Cross-language information retrieval track: Searching Arabic using English, French or Arabic queries. *Proceedings of the Tenth Text REtrieval Conference (TREC 2001)*, 16–25.

Harabagiu, S., Moldovan, D., Clark, C., Bowden, M., Williams, J., & Bensley, J. (2004). Answer mining by combining extraction techniques with abductive reasoning. *Proceedings of the Twelfth Text REtrieval Conference (TREC 2003)*, 375–382.

Harabagiu, S., Moldovan, D., Pasca, M., Mihalcea, R., Surdeanu, M., Bunescu, R., et al. (2001). Falcon: Boosting knowledge for answer engines. *Proceedings of the Ninth Text REtrieval Conference (TREC-9)*, 479–488.

Harabagiu, S., Moldovan, D., Pasca, M., Surdeanu, M., Mihalcea, R., Girju, R., et al. (2002). Answering complex, list and context questions with LCC's question-answering server. *Proceedings of the Tenth Text REtrieval Conference (TREC 2001)*, 355–361.

Harman, D. (1993a). Overview of the First Text REtrieval Conference (TREC-1). *Proceedings of the First Text REtrieval Conference (TREC-1)*, 1–20.

Harman, D. (1993b*). Proceedings of the First Text REtrieval Conference (TREC-1)*. National Institute of Standards and Technology Special Publication 500-207.

Harman, D. (1994a). Overview of the Second Text REtrieval Conference (TREC-2). *Proceedings of the Second Text REtrieval Conference (TREC-2)*, 1–20.

Harman, D. (1994b). *Proceedings of the Second Text REtrieval Conference (TREC-2)*. National Institute of Standards and Technology Special Publication 500–215.

Harman, D. (1995a). Overview of the Third Text REtrieval Conference. *Proceedings of the Third Text REtrieval Conference (TREC-3)*, 1–20.

Harman, D. (1995b). *Proceedings of the Third Text REtrieval Conference (TREC-3)*. National Institute of Standards and Technology Special Publication 500–225.

Harman, D. (1996a). Overview of the Fourth Text REtrieval Conference (TREC-4). *Proceedings of the Fourth Text REtrieval Conference (TREC-4)*, 1–23.

Harman, D. (1996b). *Proceedings of the Fourth Text REtrieval Conference (TREC-4)*. National Institute of Standards and Technology Special Publication 500–236.

Harman, D. (in press a). Beyond English. In E. M. Voorhees & D. K. Harman (Eds.), *TREC: Experiment and Evaluation in Information Retrieval*. Boston: MIT Press.

Harman, D. (in press b). The TREC test collections. In E. M. Voorhees & D. K. Harman (Eds.), *TREC: Experiment and Evaluation in Information Retrieval*. Boston: MIT Press.

Harman, D., Braschler, M., Hess, M., Kluck, M., Peters, C., Schäuble, P., et al. (2001). CLIR evaluation at TREC. In C. Peters, M. Braschler, J. Gonzalo, & M. Kluck (Eds.), *Cross-language information retrieval and evaluation* (Lecture Notes in Computer Science No. 2069, pp. 7–23). Berlin: Springer Verlag.

Harter, S. P., & Hert, C. A. (1997). Evaluation of information retrieval systems: Approaches, issues, and methods. *Annual Review of Information Science and Technology, 32*, 3–94.

Hauptmann, A., Yan, R., Qi, Y., Jin, R., Christel, M., Derthick, M., et al. (2003). Video classification and retrieval with the Informedia Digital Library System. *Proceedings of the Eleventh Text REtrieval Conference (TREC 2002)*, 119–127.

Hawking, D., Bailey, P., & Craswell, N. (2000). *Efficient and flexible search using text and metadata* (Technical Report TR2000-8. Canberra, Australia: CSIRO Mathematical and Information Sciences. Panoptic system retrieved December 30, 2004, from http://www.panopticsearch.com

Hawking, D., & Craswell, N. (2002). Overview of TREC 2001 Web track. *Proceedings of the Tenth Text REtrieval Conference (TREC 2001)*, 61–67.

Hawking, D., & Craswell, N. (in press). The very large collection and Web tracks. In E. M. Voorhees & D. K. Harman (Eds.), *TREC: Experiment and evaluation in information retrieval*. Boston: MIT Press.

Hawking, D., Craswell, N., & Thistlewaite, P. (1999a). ACSys TREC-7 experiments. *Proceedings of the Seventh Text REtrieval Conference (TREC-7)*, 299–312.

Hawking, D., Craswell, N., & Thistlewaite, P. (1999b). Overview of TREC-7 very large collection track. *Proceedings of the Seventh Text REtrieval Conference (TREC-7)*, 91–104.

Hawking, D., & Thistlewaite, P. (1998). Overview of TREC-6 very large collection track. *Proceedings of the Sixth Text REtrieval Conference (TREC-6)*, 93–106.

Hawking, D., Thistlewaite, P., & Bailey, P. (1997). ANU/ACSys TREC-5 experiments. *Proceedings of the Fifth Text REtrieval Conference (TREC-5)*, 359–376.

Hawking, D., Thistlewaite, P., & Craswell, N. (1998). ANU/ACSys TREC-6 experiments. *Proceedings of the Sixth Text REtrieval Conference (TREC-6)*, 275–290.

Hawking, D., Voorhees, E., Craswell, N., & Bailey, P. (2000). Overview of TREC-8 Web track. *Proceedings of the Eighth Text REtrieval Conference (TREC-8)*, 131–150.

Hearst, M., Pederson, J., Pirolli, P., Schutze, H., Grefenstette, G., & Hull, D. (1996). Xerox Site TREC-4 Report: Four TREC-4 tracks. *Proceedings of the Fourth Text REtrieval Conference (TREC-4)*, 97–120.

Hersh, W. (1994). Ohsumed: An interactive retrieval evaluation and new large test collection. In W. B. Croft & C. J. van Rijsbergen (Eds.), *Proceedings of the Seventeenth SIGIR Conference on Research and Development in Information Retrieval*, 192–201.

Hersh, W. (2003). TREC-2002 interactive track report. *Proceedings of the Eleventh Text REtrieval Conference (TREC 2002)*, 40–45.

Hersh, W. (2004). TREC GENOMICS track overview. *Proceedings of the Twelfth Text REtrieval Conference (TREC 2003)*, 14–23.

Hersh, W., & Over, P. (2000). TREC-8 interactive track report. *Proceedings of the Eighth Text REtrieval Conference (TREC-8)*, 57–64.

Hersh, W., & Over, P. (2001a). Interactivity at the Text Retrieval Conference (TREC) [Special Issue]. *Information Processing & Management, 37*(3).

Hersh, W., & Over, P. (2001b). TREC-9 interactive track report. *Proceedings of the Ninth Text REtrieval Conference (TREC-9)*, 41–50.

Hersh, W., & Over, P. (2002). TREC-2001 interactive track report. *Proceedings of the Tenth Text REtrieval Conference (TREC 2001)*, 38–41.

Hiemstra, D., & Kraaij, W. (1999). Twenty-one at TREC-7: Ad-hoc and cross-language track. *Proceedings of the Seventh Text REtrieval Conference (TREC-7)*, 227–238.

Hovy, E., Gerber, L., Hermjakob, U., Junk, M., & Lin, C. Y. (2001). Question answering in Webclopedia. *Proceedings of the Ninth Text REtrieval Conference (TREC-9)*, 655–664.

Hull, D. (1999). The TREC-7 filtering track: Description and analysis. *Proceedings of the Seventh Text REtrieval Conference (TREC-7)*, 33–56.

Hull, D., Grefenstette, G., Schulze, B., Gaussier, E., & Schütze, H. (1997). Xerox Site TREC-5 report: Routing, filtering, NLP and Spanish tracks. *Proceedings of the Fifth Text REtrieval Conference (TREC-5)*, 167–180.

Hull, D., & Robertson, S. (2000). The TREC-8 filtering track final report. *Proceedings of the Eighth Text REtrieval Conference (TREC-8)*, 35–56.

Järvelin, K., & Kekäläinen, J. (2000). IR evaluation methods for retrieving highly relevant documents. *Proceedings of the Twenty-Third Annual International ACM SIGIR Conference on Research and Development in Information Retrieval*, 41–48.

Johnson, S., Woodland, P., Jourlin, P., & Sparck Jones, K. (2000). Spoken document retrieval for TREC-8 at Cambridge University. *Proceedings of the Eighth Text REtrieval Conference (TREC-8)*, 197–208.

Kantor, P., & Voorhees, E. M. (2000). The TREC-5 confusion track: Comparing retrieval methods for scanned text. *Information Retrieval, 2*, 165–176.

Kraaij, W., Westerveld, T., & Hiemstra, D. (2002). The importance of prior probabilities for entry page search. *Proceedings of the Twenty-Fifth Annual International ACM SIGIR Conference on Research and Development in Information Retrieval*, 27–34.

Kwok, K. (1996). A new method of weighting query terms for ad-hoc retrieval. *Proceedings of the Nineteenth Annual International ACM SIGIR Conference on Research and Development in Information Retrieval*, 187–196.

Kwok, K., & Grunfeld, L. (1994). TREC-2 document retrieval experiments using PIRCS. *Proceedings of the Second Text REtrieval Conference (TREC-2)*, 233–242.

Kwok, K., & Grunfeld, L. (1997). TREC-5 English and Chinese retrieval experiments using PIRCS. *Proceedings of the Fifth Text REtrieval Conference (TREC-5)*, 133–142.

Kwok, K., Grunfeld, L., & Xu, J. (1998). TREC-6 English and Chinese retrieval experiments using PIRCS. *Proceedings of the Sixth Text REtrieval Conference (TREC-6)*, 207–215.

Lagergren, E., & Over, P. (1998). Comparing interactive information retrieval systems across sites: The TREC-6 interactive track matrix experiment. *Proceedings of the Twenty-First Annual International ACM SIGIR Conference on Research and Development in Information Retrieval*, 164–172.

Larson, R. (2000). Berkeley's TREC-8 interactive track entry: Cheshire II and Zprise. *Proceedings of the Eighth Text REtrieval Conference (TREC-8)*, 613–622.

Lewis, D. (1996). The TREC-4 filtering track. *Proceedings of the Fourth Text REtrieval Conference (TREC-4)*, 165–180.

Lu, A., Ayoub, M., & Dong, J. (1997). Ad hoc experiments using EUREKA. *Proceedings of the Fifth Text REtrieval Conference (TREC-5)*, 229–240.

Lu, A., Holt, J., & Miller, D. (1996). Boolean system revisited: Its performance and its behavior. *Proceedings of the Fourth Text REtrieval Conference (TREC-4)*, 459–475.

Lu, A., Meier, E., Rao, A., Miller, D., & Pliske, D. (1998). Query processing in TREC-6. *Proceedings of the Sixth Text REtrieval Conference (TREC-6)*, 567–576.

Mandala, R., Tokunaga, T., Tanaka, H., Okumura, A., & Satoh, K. (1999). Ad hoc retrieval experiments using WordNet and automatically constructed thesauri. *Proceedings of the Seventh Text REtrieval Conference (TREC-7)*, 475–480.

Marchionini, G. (2002). *The open video project*. Retrieved December 30, 2004, from http://www.open-video.org

Mayfield, J., McNamee, P., & Piatko, C. (2000). JHU/APL HAIRCUT System at TREC-8. *Proceedings of the Eighth Text REtrieval Conference (TREC-8)*, 445–452.

McNamee, P., Piatko, C., & Mayfield, J. (2003). JHU/APL at TREC 2002. *Proceedings of the Eleventh Text REtrieval Conference (TREC 2002)*, 358–363.

Merchant, R. (Ed.) (1994). *The proceedings of the TIPSTER text program: Phase I*. San Mateo, CA: Morgan Kaufmann.

Milic-Frayling, N., Evans, D., Tong, X., & Zhai, C. (1997). CLARIT compound queries and constraint-controlled feedback in TREC-5. *Proceedings of the Fifth Text REtrieval Conference (TREC-5)*, 315–334.

Milic-Frayling, N., Zhai, C., Tong, X., Jansen, P., & Evans, D. (1998). Experiments in query optimization. *Proceedings of the Sixth Text REtrieval Conference (TREC-6)*, 415–454.

Miller, D., Leek, T., & Schwartz, R. (1999). A hidden Markov model information retrieval system. *Proceedings of the Twenty-Second Annual International ACM SIGIR Conference on Research and Development in Information Retrieval*, 214–221.

Nakajima, H., Takaki, T., Hirao, T., & Kitauchi, A. (1999). NTT DATA at TREC-7: System approach for ad hoc and filtering. *Proceedings of the Seventh Text REtrieval Conference (TREC-7)*, 481–490.

Namba, I., Igata, N., Horai, H., Nitta, K., & Matsui, K. (1999). Fujitsu Laboratories TREC7 report. *Proceedings of the Seventh Text REtrieval Conference (TREC-7)*, 383–392.

Oard, D., & Gey, F. (2003). The TREC 2002 Arabic/English CLIR track. *Proceedings of the Eleventh Text REtrieval Conference (TREC 2002)*, 17–26.

Ogilvie, P., & Callan, J. (2004). Combining structural information and the use of priors in mixed named-page and homepage finding. *Proceedings of the Twelfth Text REtrieval Conference (TREC 2003)*, 177–184.

Osdin, R., Ounis, I., & White, R. (2003). Using hierarchical clustering and summarisation approaches for Web retrieval: Glasgow at the TREC 2002 interactive track. *Proceedings of the Eleventh Text REtrieval Conference (TREC 2002)*, 640–644.

Over, P. (1997). TREC-5 interactive track report. *Proceedings of the Fifth Text REtrieval Conference (TREC-5)*, 29–56.

Over, P. (1998). TREC-6 interactive track report. *Proceedings of the Sixth Text REtrieval Conference (TREC-6)*, 73–81.

Over, P. (1999). TREC-7 interactive track report. *Proceedings of the Seventh Text REtrieval Conference (TREC-7)*, 65–72.

Quenot, G., Moran, D., Besacier, L., & Mulhern, P. (2003). CLIPS at TREC 11: Experiments in video retrieval. *Proceedings of the Eleventh Text REtrieval Conference (TREC 2002)*, 181–187.

Robertson, S., & Callan, J. (in press). Routing and filtering. In E. M. Voorhees & D. K. Harman (Eds.), *TREC: Experiment and evaluation in information retrieval*. Boston: MIT Press.

Robertson, S., & Hull, D. (2001). The TREC-9 filtering track final report. *Proceedings of the Ninth Text REtrieval Conference (TREC-9)*, 25–40.

Robertson, S., & Soboroff, I. (2002). The TREC 2001 filtering track report. *Proceedings of the Tenth Text REtrieval Conference (TREC 2001)*, 26–37.

Robertson, S., & Walker, S. (1994). Some simple effective approximations to the 2-Poisson model for probabilistic weighted retrieval. *Proceedings of the Seventeenth Annual International ACM SIGIR Conference on Research and Development in Information Retrieval*, 232–241.

Robertson, S., & Walker, S. (2001). Microsoft Cambridge at TREC-9: Filtering track. *Proceedings of the Ninth Text REtrieval Conference (TREC-9)*, 361–368.

Robertson, S., Walker, S., & Beaulieu, M. (1999). Okapi at TREC-7: Automatic ad-hoc, filtering, VLC, and interactive track. *Proceedings of the Seventh Text REtrieval Conference (TREC-7)*, 253–264.

Robertson, S., Walker, S., Beaulieu, M., Gatford, M., & Payne, A. (1996). Okapi at TREC-4. *Proceedings of the Fourth Text REtrieval Conference (TREC-4)*, 73–96.

Robertson, S., Walker, S., Hancock-Beaulieu, M., & Gatford, M. (1994). Okapi and TREC-2. *Proceedings of the Second Text REtrieval Conference (TREC-2)*, 21–34.

Robertson, S., Walker, S., Jones, S., Hancock-Beaulieu, M., & Gatford, M. (1995). OKAPI at TREC-3. *Proceedings of the Third Text REtrieval Conference (TREC-3)*, 109–126.

Robertson, S., Walker, S., Zaragoza, H., & Herbrich, R. (2003). Microsoft Cambridge at TREC 2002: Filtering track. *Proceedings of the Eleventh Text REtrieval Conference (TREC 2002)*, 439–446.

Savoy, J., Rasolofo, Y., & Perret, L. (2004). Report on the TREC 2003 experiment: Genomic and Web searches. *Proceedings of the Twelfth Text REtrieval Conference (TREC 2003)*, 739–750.

Schäuble, P., & Sheridan, P. (1998). Cross-language information retrieval (CLIR) track overview. *Proceedings of the Sixth Text REtrieval Conference (TREC-6)*, 31–43.

Singhal, A. (1998). AT&T at TREC-6. *Proceedings of the Sixth Text REtrieval Conference (TREC-6)*, 215–225.

Singhal, A., Buckley, C., & Mitra, M. (1996). Pivoted document length normalization. *Proceedings of the Nineteenth Annual International ACM SIGIR Conference on Research and Development in Information Retrieval*, 21–29.

Singhal, A., Choi, J., Hindle, D., Lewis, D., & Pereira, F. (1999). AT&T at TREC-7. *Proceedings of the Seventh Text REtrieval Conference (TREC-7)*, 239–252.

Singhal, A., & Kaszkiel, M. (2001). AT&T at TREC-9. *Proceedings of the Ninth Text REtrieval Conference (TREC-9)*, 103–105.

Smeaton, A., & Over, P. (2002). TREC-2001 video track report. *Proceedings of the Tenth Text REtrieval Conference (TREC 2001)*, 52–60.

Smeaton, A., & Over, P. (2003). TREC-2002 video track report. *Proceedings of the Eleventh Text REtrieval Conference (TREC 2002)*, 69–85.

Smeaton, A., & Wilkinson, R. (1997). Spanish and Chinese document retrieval in TREC-5. *Proceedings of the Fifth Text REtrieval Conference (TREC-5)*, 57–64.

Soboroff, I., & Robertson, S. (2003). Building a filtering test collection for TREC 2002. *Proceedings of the Twenty-Sixth Annual International ACM SIGIR Conference on Research and Development in Information Retrieval*, 243–250.

Sormunen, E. (2002). Liberal relevance criteria of TREC: Counting on negligible documents?. *Proceedings of the Twenty-Fifth Annual International ACM SIGIR Conference on Research and Development in Information Retrieval*, 324–330.

Soubboutin, M. (2002). Patterns of potential answer expressions as clues to the right answers. *Proceedings of the Tenth Text REtrieval Conference (TREC 2001)*, 293–302.

Sparck Jones, K. (1995). Reflections on TREC. *Information Processing & Management, 31*, 291–314.

Sparck Jones, K. (2000). Further reflections on TREC. *Information Processing & Management, 36*, 37–86.

Sparck Jones, K. (in press). Meta-reflections on TREC. In E. M. Voorhees & D. K. Harman (Eds.), *TREC: Experiment and evaluation in information retrieval*. Boston: MIT Press.

Sparck Jones, K., & van Rijsbergen, C. (1975). *Report on the need for and provision of an "ideal" information retrieval test collection* (British Library Research and Development Report 5266). Cambridge, UK: University of Cambridge Computer Laboratory.

Sparck Jones, K., Walker, S., & Robertson, S. (2000). A probabilistic model of information retrieval: Development and comparative experiments part I and part II. *Information Processing & Management, 36,* 779–840.

Strzalkowski, T., Lin, F., & Perez-Carballo, J. (1998). Natural language information retrieval: TREC-6 report. *Proceedings of the Sixth Text REtrieval Conference (TREC-6),* 347–366.

Strzalkowski, T., Lin, F., Wang, J., Guthrie, L., Leistensnider, J., Wilding, J., et al. (1997). Natural language information retrieval: TREC-5 report. *Proceedings of the Fifth Text REtrieval Conference (TREC-5),* 291–314.

Strzalkowski, T., Stein, G., Wise, G. B., Perez-Carballo, J., Tapananinen, P., Jarvinen, T., et al. (1999). Natural language information retrieval: TREC-7 report. *Proceedings of the Seventh Text REtrieval Conference (TREC-7),* 217–226.

Swan, R., & Allan, J. (1998). Aspect windows, 3-D visualizations, and indirect comparisons of information retrieval systems. *Proceedings of the Twenty-First Annual International ACM SIGIR Conference on Research and Development in Information Retrieval,* 173–181.

Tomlinson, S. (2004). Robust, Web and genomic retrieval with Hummingbird SearchServer at TREC 2003. *Proceedings of the Twelfth Text REtrieval Conference (TREC 2003),* 254–267.

Tomlinson, S., & Blackwell, T. (2001). Hummingbird's fulcrum SearchServer at TREC-9. *Proceedings of the Ninth Text REtrieval Conference (TREC-9),* 209–222.

Toms, E., Barlett, J., Freund, L., & Kopak, R. (2002). Selecting versus describing: A preliminary analysis of the efficacy of categories in exploring the Web. *Proceedings of the Tenth Text REtrieval Conference (TREC 2001),* 653–662.

Tong, R. (1995). Interactive document retrieval using TOPIC. *Proceedings of the Third Text REtrieval Conference (TREC-3),* 201–210.

Veerasamy, A. (1996). Interactive TREC-4 at Georgia Tech. *Proceedings of the Fourth Text REtrieval Conference (TREC-4),* 421–432.

Voorhees, E. (2000a). The TREC-8 question answering track report. *Proceedings of the Eighth Text REtrieval Conference (TREC-8),* 77–82.

Voorhees, E. (2000b). Variations in relevance judgments and the measurement of retrieval effectiveness. *Information Processing & Management, 36,* 697–716.

Voorhees, E. (2001a). Evaluation by highly relevant documents. *Proceedings of the Twenty-Fourth Annual International ACM SIGIR Conference on Research and Development in Information Retrieval,* 74–82.

Voorhees, E. (2001b). Overview of the TREC-9 question answering track. *Proceedings of the Ninth Text REtrieval Conference (TREC-9),* 71–80.

Voorhees, E. (2002). Overview of the TREC-2001 question answering track. *Proceedings of the Tenth Text REtrieval Conference (TREC 2001),* 42–51.

Voorhees, E. (2003). Overview of the TREC-2002 question answering track. *Proceedings of the Eleventh Text REtrieval Conference (TREC 2002),* 57–68.

Voorhees, E. (2004a). Overview of the TREC-2003 question answering track. *Proceedings of the Twelfth Text REtrieval Conference (TREC 2003),* 54–68.

Voorhees, E. (2004b). Overview of the TREC 2003 robust retrieval track. *Proceedings of the Twelfth Text REtrieval Conference (TREC 2003),* 69–77.

Voorhees, E. M. (in press). Question answering in TREC. In E. M. Voorhees & D. K. Harman (Eds.), *TREC: Experiment and evaluation in information retrieval*. Boston: MIT Press.

Voorhees, E., & Buckland, L. (Eds.). (2003). *Proceedings of the Eleventh Text REtrieval Conference (TREC 2002)* (NIST Special Publication 500-251).

Voorhees, E., & Buckland, L. (Eds.). (2004). *Proceedings of the Twelfth Text REtrieval Conference (TREC 2003)* (NIST Special Publication 500-255).

Voorhees, E. M., & Buckley, C. (in press). Retrieval system evaluation. In E. M. Voorhees & D. K. Harman (Eds.), *TREC: Experiment and evaluation in information retrieval*. Boston: MIT Press.

Voorhees, E. M., & Garofolo, J. (in press). Retrieving noisy text. In E. M. Voorhees & D. K. Harman (Eds.), *TREC: Experiment and evaluation in information retrieval*. Boston: MIT Press.

Voorhees, E., & Harman, D. (1997a). Overview of the Fifth Text REtrieval Conference (TREC-5). *Proceedings of the Fifth Text REtrieval Conference (TREC-5)*, 1–28.

Voorhees, E. & Harman, D. (Eds.). (1997b). *Proceedings of the Fifth Text REtrieval Conference (TREC-5)* (NIST Special Publication 500-238).

Voorhees, E., & Harman, D. (1998a). Overview of the Sixth Text REtrieval Conference (TREC-6). *Proceedings of the Sixth Text REtrieval Conference (TREC-6)*, 1–24.

Voorhees, E., & Harman, D. (Eds.). (1998b). *Proceedings of the Sixth Text REtrieval Conference (TREC-6)* (NIST Special Publication 500–240).

Voorhees, E., & Harman, D. (1999a). Overview of the Seventh Text REtrieval Conference (TREC-7). *Proceedings of the Seventh Text REtrieval Conference (TREC-7)*, 1–24.

Voorhees, E., & Harman, D. (Eds.). (1999b). *Proceedings of the Seventh Text REtrieval Conference (TREC-7)* (NIST Special Publication 500–242).

Voorhees, E., & Harman, D. (2000a). Overview of the Eighth Text REtrieval Conference (TREC-8). *Proceedings of the Eighth Text REtrieval Conference (TREC-8)*, 1–24.

Voorhees, E., & Harman, D. (Eds.). (2000b). *Proceedings of the Eighth Text REtrieval Conference (TREC-8)* (NIST Special Publication 500–246).

Voorhees, E., & Harman, D. (Eds.). (2001). *Proceedings of the Ninth Text REtrieval Conference (TREC-9)* (NIST Special Publication 500–249).

Voorhees, E., & Harman, D. (Eds.). (2002). *Proceedings of the Tenth Text REtrieval Conference(TREC 2001)* (NIST Special Publication 500–250).

Voorhees, E. M., & Harman, D. (Eds.). (in press). *TREC: Experiment and evaluation in information retrieval*. Boston: MIT Press.

Walker, S., Robertson, S., Boughanem, M., Jones, G., Sparck Jones, K. (1998). Okapi at TREC-6: Automatic ad hoc, VLC, routing, filtering and QSDR. *Proceedings of the Sixth Text REtrieval Conference (TREC-6)*, 125–136.

Westerveld, T., Hiemstra, D., & Kraaij, W. (2002). Retrieving Web pages using content, links, URLs, and anchors. *Proceedings of the Tenth Text REtrieval Conference (TREC 2001)*, 663–672.

Wilkinson, R. (1998). Chinese document retrieval at TREC-6. *Proceedings of the Sixth Text REtrieval Conference (TREC-6)*, 25–30.

Wu, L., Huang, X., Niu, J., Guo, Y., Xia, Y., & Feng, Z. (2002). FDU at TREC-10: Filtering, QA, and video tasks. *Proceedings of the Tenth Text REtrieval Conference (TREC 2001)*, 192–207.

Wu, L., Huang, X., Niu, J., Xia, Y., Feng, Z., & Zhou, Y. (2003). FDU at TREC 2002: Filtering, Q&A, Web and video tasks. *Proceedings of the Eleventh Text REtrieval Conference (TREC 2002)*, 232–247.

Xu, J., & Croft, W. (1996). Query expansion using local and global document analysis. *Proceedings of the Nineteenth Annual International ACM SIGIR Conference on Research and Development in Information Retrieval*, 4–11.

Xu, J., Licuanan, A., & Weischedel, R. (2004). TREC 2003 QA at BBN: Answering definitional questions. *Proceedings of the Twelfth Text REtrieval Conference (TREC 2003)*, 98–106.

Xu, J., & Weischedel, R. (2001). TREC-9 cross-lingual retrieval at BBN. *Proceedings of the Ninth Text REtrieval Conference (TREC-9)*, 106–116.

Zhai, C., Jansen, P., Roma, N., Stoica, E., & Evans, D. (2000). Optimization in CLARIT TREC-8 filtering. *Proceedings of the Eighth Text REtrieval Conference (TREC-8)*, 253–258.

Zhang, Y., & Callan, J. (2001). YFilter at TREC-9. *Proceedings of the Ninth Text REtrieval Conference (TREC-9)*, 135–140.

Zobel, J. (1998). How reliable are the results of large-scale information retrieval experiments. *Proceedings of the Twenty-First Annual International ACM SIGIR Conference on Research and Development in Information Retrieval*, 307–314.

CHAPTER 5

Semantic Relations in Information Science

Christopher S. G. Khoo and Jin-Cheon Na
Nanyang Technological University

Introduction

This chapter examines the nature of semantic relations and their main applications in information science. The nature and types of semantic relations are discussed from the perspectives of linguistics and psychology. An overview of the semantic relations used in knowledge structures such as thesauri and ontologies is provided, as well as the main techniques used in the automatic extraction of semantic relations from text. The chapter then reviews the use of semantic relations in information extraction, information retrieval, question-answering, and automatic text summarization applications.

Concepts and relations are the foundation of knowledge and thought. When we look at the world, we perceive not a mass of colors but objects to which we automatically assign category labels. Our perceptual system automatically segments the world into concepts and categories.[1] Concepts are the building blocks of knowledge; relations act as the cement that links concepts into knowledge structures. We spend much of our lives identifying regular associations and relations between objects, events, and processes so that the world has an understandable structure and predictability. Our lives and work depend on the accuracy and richness of this knowledge structure and its web of relations. Relations are needed for reasoning and inferencing.

Chaffin and Herrmann (1988b, p. 290) noted that "relations between ideas have long been viewed as basic to thought, language, comprehension, and memory." Aristotle's *Metaphysics* (Aristotle, 1961; McKeon, 1941/2001) expounded on several types of relations. The majority of the 30 entries in a section of the *Metaphysics* known today as the *Philosophical Lexicon* referred to relations and attributes, including cause, part-whole, same and opposite, quality (i.e., attribute) and kind-of, and defined different types of each relation. Hume (1955) pointed out that there is a connection between successive ideas in our minds, even in our dreams, and that the introduction of an idea in our mind automatically recalls an associated idea. He argued that all the objects of human reasoning are divided into relations of ideas and matters of fact and that factual reasoning is founded on the cause-effect relation. His *Treatise of Human Nature* identified seven kinds of relations: resemblance, identity, relations of time and place, proportion in quantity or number, degrees in quality, contrariety, and causation. Mill (1974, pp. 989–1004) discoursed on several types of relations, claiming that all things are either feelings, substances, or attributes, and that attributes can be a quality (which belongs to one object) or a relation to other objects.

Linguists in the structuralist tradition (e.g., Lyons, 1977; Saussure, 1959) have asserted that concepts cannot be defined on their own but only in relation to other concepts. Semantic relations appear to reflect a logical structure in the fundamental nature of thought (Caplan & Herrmann, 1993). Green, Bean, and Myaeng (2002) noted that semantic relations play a critical role in how we represent knowledge psychologically, linguistically, and computationally, and that many systems of knowledge representation start with a basic distinction between entities and relations. Green (2001, p. 3) said that "relationships are involved as we combine simple entities to form more complex entities, as we compare entities, as we group entities, as one entity performs a process on another entity, and so forth. Indeed, many things that we might initially regard as basic and elemental are revealed upon further examination to involve internal structure, or in other words, internal relationships."

Concepts and relations are often expressed in language and text. Language is used not just for communicating concepts and relations, but also for representing, storing, and reasoning with concepts and relations. We shall examine the nature of semantic relations from a linguistic and psychological perspective, with an emphasis on relations expressed in text. The usefulness of semantic relations in information science, especially in ontology construction, information extraction, information retrieval, question-answering, and text summarization is discussed.

Research and development in information science have focused on concepts and terms, but the focus will increasingly shift to the identification, processing, and management of relations to achieve greater effectiveness and refinement in information science techniques. Previous chapters in *ARIST* on natural language processing (Chowdhury, 2003), text mining (Trybula, 1999), information retrieval and the philosophy of language (Blair, 2003), and query expansion (Efthimiadis, 1996) provide a background for this discussion, as semantic relations are an important part of these applications.

What Are Semantic Relations?

Semantic Relations in Language and Logic

Semantic relations are meaningful associations between two or more concepts, entities, or sets of entities. They can be viewed as directional links between the concepts/entities that participate in the relation. The concepts/entities are an integral part of the relation as a relation cannot exist by itself. Associations between concepts/entities can be categorized into different types, abstracted, conceptualized and distinguished from other associations, and can thus be assigned meaning. The meaning or type of an association can sometimes but not always be derived from the meanings of the concepts involved. Psychologists and philosophers have attempted to identify the main types of relations and their features.

Two concepts connected by a relation are often represented as a concept-relation-concept triplet: [concept1] → (relation) → [concept2].[2] The link is labeled to indicate the type or meaning of the relation. A relation can thus be viewed as containing two places or slots that need to be filled. A relation exerts selectional restrictions on the slots that constrain the kind of concepts or entities that can occupy them. A valid participant of a relation may need to have certain semantic features or belong to a semantic category. For example, in the relation [John] → (is-father-of) → [Mary], the entity represented by "John" has to belong to the category of human beings and have the gender feature *male*. A relation can also constrain the slot filler to a concept, an entity (i.e., instance of a concept), a set of entities, or a mass concept (denoting a set of entities).

Although most relations are binary relations having two slots, a relation may have three or more slots. For example, the *buy* relation may relate four participants: the buyer, the seller, the thing that is bought, and the price. The number of slots of a relation is called its arity or valence. *Buy* is a 4-ary relation, and the four participants in the relation are assigned the roles *agent* (buyer), *source* (seller), *patient* (thing bought), and *price* to distinguish between them. It is, however, well known that relations with arity higher than two can be decomposed into a set of more primitive binary relations. For example, the *buy* relation can be converted to a *buy* concept, which can be linked to the four participants with the binary relations agent, source, patient, and price. Sowa (1984) proposed the generic *link* relation as the most primitive relation. All other relations can be defined in terms of concepts combined with the *link* relation. For example, the *eat* relation in [John] → (eat) → [apple], can be decomposed into the concept *eat* and the case relations *agent* and *patient*: [John] → (agent) → [eat] → (patient) → [apple]. The *agent* relation can be further reduced to the concept *agent* and the *link* relation: [John] → (link) → [agent] → (link) → [eat].

Sowa (1984) further suggested that tenses and modalities, such as *possibility*, *necessity*, *permission*, and *negation*, be treated as 1-ary or "monadic" relations. For example, the PAST relation can indicate that a proposition was true in the past: (PAST) → [PROPOSITION].

Semantic relations can refer to relations *between concepts in the mind* (called conceptual relations), or relations *between words* (lexical relations) or text segments. However, concepts and relations are inextricably bound with language and text and it is difficult to analyze the meaning of concepts and relations apart from the language that expresses them. Wittgenstein (1953, p. 107) said, "When I think in language, there aren't 'meanings' going through my mind in addition to the verbal expressions: the language is itself the vehicle of thought." Often the distinction between conceptual relations and lexical relations are unimportant and authors use the term *lexical-semantic relations* (Evens, 1988, p. 2) to refer to relations between lexical concepts—concepts denoted by words. They are also sometimes called *sense relations*, as some linguists maintain that they relate particular senses of words (Lyons, 1977).

In addition to words, semantic relations can occur at higher levels of text—between phrases, clauses, sentences, and larger text segments, as well as between documents and sets of documents. The analysis of semantic relations can be carried out at the textual level, close to the words that express the meaning, or at a logical level, focusing on the meaning expressed by the text or concepts in the mind.

Let us now consider some properties of relations. Murphy (2003) listed the following general properties of lexical-semantic relations that have been identified by linguists:

1. Productivity—new relations can be created easily.

2. Binarity—some relations, for example *antonymy*, are binary in the sense that a word can have only one true antonym, whereas other relations, for example *synonymy*, can relate a set of words (i.e., a word can have many synonyms).

3. Variability—relations between words vary with the sense of the word used and the context.

4. Prototypicality and canonicity—some word pairs are better exemplars of a relation than others, and some word pairs have special status as canonical examples of a relation (particularly for *antonyms*).

5. Semisemanticity—nonsemantic properties, such as grammatical category, co-occurrence in text, and similarity in morphological form, can affect whether a particular relation is considered to hold between two words.

6. Uncountability—semantic relations are an open class and they cannot all be listed or counted.

7. Predictability—semantic relations follow certain general patterns and rules.

8. Universality—the same types of semantic relations are used in any language and the same concepts are related by the same semantic relations in different languages.

A semantic relation can have one or more of the following logical properties (Cruse, 2004; Sowa, 1984, p. 381):

- Reflexivity: A relation R is reflexive if it can relate an entity to itself; $[x] \rightarrow (R) \rightarrow [x]$ is true for every x (e.g., the part-whole relation).

- Symmetry: A relation R is symmetric if the two participants of the relation can occupy either slot; $[x] \rightarrow (R) \rightarrow [y]$ implies $[y] \rightarrow (R) \rightarrow [x]$ (e.g., synonymy).

- Transitivity: A relation R is transitive if $[x] \rightarrow (R) \rightarrow [y]$ and $[y] \rightarrow (R) \rightarrow [z]$ implies $[x] \rightarrow (R) \rightarrow [z]$ (e.g., IS-A relation, and ancestor-descendent relation).

- One-to-one relation: A relation R is one-to-one if, when one participant of the relation is known, the other participant is fixed; $[x] \rightarrow (R) \rightarrow [y]$ and $[z] \rightarrow (R) \rightarrow [y]$ implies $x = z$.

A relation can be related to another relation by *similarity* (i.e., the two relations are the same) or by an *inverse* relation. A relation R is the inverse of a relation S if both can accept the same pair of participants or slot fillers but the direction of the two relations is reversed; $[x] \rightarrow (R) \rightarrow [y]$ implies $[y] \rightarrow (S) \rightarrow [x]$ (e.g., *broader* versus *narrower* relation, *parent* versus *child* relation). One relation can be a *subrelation*, or more specific type of relation, of another, and relations can thus be organized into a relation hierarchy.

The variety of semantic relations and their properties plays an important role in human comprehension and reasoning. Spellman, Holyoak, and Morrison (2001) said that conceptual relations and the role bindings that they impose on the participant objects are central to such cognitive tasks as discourse comprehension, inference, problem solving, and analogical reasoning. Chaffin and Herrmann (1984) noted that the variety of relations is important both to general models of comprehension and to semantic models. For general models of comprehension, the relations differ in their logical properties and thus permit different kinds of inferences. The different relations also call into play different sets of decision criteria in decision making (Herrmann, Chaffin, Conti, Peters, & Robbins, 1979). Relations have also been found to be important in analogical reasoning and in the use of metaphors, which involve cross-domain mapping within a conceptual system (Lakoff, 1993, p. 203). In analogical reasoning, people map connected systems of relations, in particular cause-effect relations, rather than individual features (Gentner, 1983, 1989; Holyoak & Thagard, 1995; Lakoff, 1993; Turner, 1993).

Comprehensive treatments of semantic relations in language and text can be found in Cruse (1986, 2004), Lyons (1977, 1995), and Murphy (2003).

The Psychological Reality of Semantic Relations

Are semantic relations real, or are they just an abstract theoretical construct of linguists and psychologists? Do people really perceive, recognize, and process semantic relations? There is substantial evidence from experimental psychology that, to human beings, semantic relations are endowed with psychological reality.

Chaffin and Herrmann (1984, 1987, 1988b) and Glass, Holyoak, and Kiger (1979) carried out a series of studies to demonstrate that people can distinguish between different types of relations, identify instances of similar relations, express relations in words, recognize instances of relation ambiguity, and create new relations. The evidence comes from sorting experiments in which subjects were asked to sort relations (represented by pairs of terms) into groups of similar relations, analogy tests in which subjects were asked to assess the similarity of pairs of terms representing different relations, and tasks of relating term pairs to relation names indicating the type of relation exemplified by each term pair.

Psychologists have determined that some types of semantic relations, for example antonymy, are easier for adults and children to comprehend and process than others (Chaffin & Herrmann, 1987; Herrmann & Chaffin, 1986). Landis, Herrmann, and Chaffin (1987) studied children's developmental rates in understanding five types of semantic relations (antonymy, class inclusion, part-whole, syntactic relations, and synonymy), reaching the conclusion that the ability to match relations developed faster for antonymy and part-whole relations than for others and that comprehension of class inclusion developed least rapidly.

Researchers in anthropology and psychology have also found substantial cross-cultural agreement on the meanings, and in the use, of semantic relations (Chaffin & Herrmann, 1984; Hermann & Raybeck, 1981; Hudon, 2001; Romney, Moore, & Rusch, 1997). Raybeck and Herrmann (1990) found that some types of relations (particularly antonymy, part-whole, and cause-effect relations) are recognized equally easily and used with equal frequency and accuracy by diverse groups of people from different cultural backgrounds.

Psychologists consider semantic relations to be important in explaining the coherence and structure of concepts and categories. A *category* is not just a random set of entities—the entities in a category must belong together in some way. A category or concept is *coherent*—it must make meaningful sense. Psychologists have investigated several theoretical models for explaining conceptual coherence and structure. Initial studies focused on similarity of features, but this was found to be inadequate in explaining why certain features are more important than others in determining category membership. Researchers now believe that relations between the features of the category members, the functions of the features, and the configuration of features are important. For example, Markowitz (1988) has suggested that the *modification, part-whole, function, agent,* and *object* relations are important in determining category membership ranking. *Modification,* particularly *size,* is used in the definitions of most categories and many categories have a specific range of acceptable sizes. The *part-whole* relation is important in natural categories, whereas *function* is important in those pertaining to manufactured objects.

Some psychologists have espoused an explanation-based or theory-based model of categorization that accounts for conceptual coherence in terms of theories people have about the relations between attributes in a concept and about the relations between concepts (Ahn & Kim, 2001; Keil, 1989, 2003; Murphy & Medin, 1985). Wattenmaker, Nakamura, and Medin (1988) argued that categories derive their coherence not from overlapping attributes but from the complex web of causal and theoretical relationships in which these attributes participate. Ahn (1999) and Rehder (2003) found that causal relations appear to determine the importance of specific attributes in human evaluation of category membership. Rehder and Hastie (2001) showed that attributes occupying a central position in a network of causal relationships (either as a common cause or a common effect) dominate category membership judgment. Ahn and Kim (2001) found that the deeper an attribute is in a causal chain, the more dominant it is in category membership judgments.

Are semantic relations *concepts*? Chaffin and Herrmann (1988b) and Chaffin (1992) found that relations have the main characteristics of concepts and concluded that they are abstract concepts. They identified four characteristics that relational concepts share with concrete concepts: (a) relations can be analyzed into more basic elements or features; (b) a new relation may be an elaboration or combination of other relations; (c) relations have graded structure (i.e., some instances of relations, represented by word pairs, are more typical of a particular relation than others); and (d) relations vary in the ease with which they can be expressed.

Linguists and psychologists have shown that the antonym, synonym, IS-A, part-whole, and case relations, often taken as primitive relations, can be decomposed into simpler relational elements (Chaffin, 1992; Chaffin & Herrmann, 1987, 1988a, 1988b; Cruse, 1986; Klix, 1986; Lyons, 1977). Murphy (2003) stated that most lexical-semantic relations have some kind of similarity and contrast element. For example, synonyms are similar in meaning but different in lexical form, and antonyms have contrasting positions on the same dimension. Chaffin and Herrmann (1984) found that subjects distinguished relations in terms of three features: contrasting/noncontrasting, logical/pragmatic, and inclusion/noninclusion. Shared features can also account for perceptions of similarity between relations (Caplan & Herrmann, 1993; Chaffin, 1992; Chaffin & Herrmann, 1984, 1987, 1988a, 1988b).

Categories of relation instances (expressed as word pairs) also differ in the extent to which their memberships are graded (Caplan & Barr, 1991). Some relations can be defined "classically" in terms of necessary and sufficient features, whereas others have "fuzzy" boundaries with many partial members. Semantic relations, like concepts, can be organized into taxonomies with broader and narrower relations (Chaffin & Herrmann, 1987; Green, 2002; Stasio, Herrman, & Chaffin, 1985).

Semantic Relations in Semantic Memory

In addition to semantic relations expressed in text, semantic relations are also encoded in knowledge structures in our brains. Psychologists working in the area of *semantic memory* have attempted to characterize the nature of these knowledge structures and the semantic relations that support them. *Semantic memory* has been characterized as our mental storehouse of knowledge about language as well as general knowledge about the world (McNamara & Holbrook, 2003; Smith, 1978).

Semantic memory is usually modeled as a network, with nodes representing concepts and labeled directional links representing relations. This semantic network model was first proposed by Quillian (1967) and Collins and Quillian (1969). In Quillian's theory (1967, 1968), words are stored in memory as configurations of pointers to other words and each configuration of pointers represents the meaning of a word. The use of semantic memory for memory recall and comprehension is modeled as spreading activation—activation that spreads from one node to neighboring nodes along the links (Collins & Loftus, 1975). A major debate in semantic memory research is the structure vs. process question: Are semantic relations prestored in semantic memory or computed dynamically from the representation of concepts (Kounios & Holcomb, 1992)? Experimental evidence suggests that at least some relations, for example the *ownership* relation, are computed as needed (Kounios, Montgomery, & Smith, 1994).

Klix (1980, 1986) distinguished between *intraconcept relations* and *interconcept relations*. Interconcept relations, also called *event relations*, are based on associations between words, concepts, and events that have been observed and experienced (e.g., *knife* is for *cutting*), and are hypothesized as being stored directly in memory. Intraconcept relations, or feature-based relations between concepts, are based on common features or feature relationships within the concepts. These relations are not stored explicitly in memory but are hypothesized to be computed from concept features using cognitive procedures stored in the brain (Kukla, 1980).

These two types of relations have been found to have different effects on memory recall and analogy recognition (Hoffmann & Trettin, 1980). Murphy (2003) argued that paradigmatic relations (discussed later), which are mainly feature-based relations, are generated using cognitive rules because new instances of the relations can be easily produced at any time. She hypothesized that paradigmatic relations are represented as "metalinguistic knowledge" about words rather than hard-coded in the lexicon and that this explains why semantic relations are determined partly by context.

Herrmann (1987) suggested another possibility: A relation between two words may be represented in semantic memory as simpler relations or relation elements between aspects of the meanings of the two words. He further proposed an *alternative-form model* of relation comprehension in which different ways of representing relations in semantic memory are employed in relation comprehension, each form providing an alternative way of processing relations under different conditions.

General knowledge in human memory has also been modeled as being organized into structures of relations called a *schema* (Alba & Hasher, 1983). One implementation of the schema introduced by Minsky (1975) is a *frame*—basically a set of labeled slots, each indicating the role of a participant in the frame. Frames with a temporal element indicating a sequence of subevents in an event type are called *scripts* (Rumelhart & Ortony, 1977; Schank, 1982; Schank & Abelson, 1977). Frames, scripts, and story schemas play a major role in models of human comprehension (Brewer & Nakamura, 1984; Butcher & Kintsch, 2003; Whitney, Budd, Bramucci, & Crane, 1995).

Types of Semantic Relations

Overview

This section surveys the types of semantic relations that have been identified by researchers: lexical-semantic relations, case relations, and relations at a higher level of text.

Can a comprehensive list of semantic relations be constructed? What are the main types of relations? There are two broad approaches to constructing a list of semantic relations: the minimalist approach and the elaborate approach. Evens (1988) referred to the two groups of researchers as "lumpers" and "splitters." The lumpers, or minimalists, define a small number of general relations based on philosophical or logical principles (e.g., Sowa, 1984, 2000; Werner, 1988). Werner (1988) used only three relations: *modification*, *taxonomy*, and *queuing*. Other researchers have a much more elaborate list of specific relations, often based on lexical-semantic relations and words found in a text (e.g., Calzolari, 1988). Lexical-oriented models often group relations into families of relations with the same core meaning or function.

Most researchers recognize two broad categories of relations: paradigmatic and syntagmatic relations. This distinction can be traced to Ferdinand de Saussure (1959). [3] Paradigmatic relations are relations between pairs of words or phrases that can occur in the same position in the same sentence (Asher, 1994, v.10, p. 5153). The words often are instances of the same part of speech, belong to the same semantic class, and are to some extent grammatically substitutable. Examples include IS-A (broader-narrower), part-whole, and synonym relations. These relations tend to be part of our semantic memory, and are typically used in thesauri. Lancaster (1986) characterized paradigmatic relations as *a priori* or permanent relations.

Syntagmatic relations refer to relations between words that co-occur (often in close syntactic positions) in the same sentence or text (Asher, 1994, v. 10, p. 5178). It is a linear or sequence relation that is synthesized and expressed between two words or phrases when we construct a sentence. The relations are governed partly by the syntactic and grammatical rules of a language. Lancaster (1986) characterized syntagmatic relations as *a posteriori* or transient relations. Green (2001) suggested that paradigmatic relations are a closed, enumerable class of relations, whereas syntagmatic relations are an open class that cannot be fully enumerated, as a new relation is invented whenever a new verb is coined.

The distinction between paradigmatic and syntagmatic relations is fuzzy. Evens, Litowitz, Markowitz, Smith, and Werner (1980) pointed out that paradigmatic relations can be expressed syntagmatically. However, they also noted that

> we seem to receive paradigmatic information typically in generic (always true) sentences, while syntagmatic relationships come to us in occasional sentences. A generic or standing sentence contains a piece of permanent information about the world, such as 'Food is edible.' An occasional sentence contains information about a particular context. (Evens et al., 1980, pp. 10–11)

Syntagmatic relations between two words can become part of our semantic memory if the words co-occur frequently enough in text or discourse to be associated (Harris, 1987). Gardin (1965) argued that paradigmatic data should be derived from accumulated syntagmatic data. As we shall see later, researchers performing corpus-based linguistic analysis have found that paradigmatically related words, especially antonyms, do, indeed, often co-occur in text.

Many authors have attempted to enumerate semantic relations—generally, either of a particular type or for a particular purpose. Warren (1921) identified 13 other classification systems proposed before 1911. Evens et al. (1980) surveyed the sets of lexical-semantic relations that had been studied by researchers in anthropology, linguistics, psychology, and computer science before 1980. Lists of semantic relations can be found in Chaffin and Herrmann (1987, 1988b), Myaeng and McHale (1992), Neelameghan (1998, 2001), Neelameghan and Maitra (1978), Smith (1981), and Sowa (1984, 2000). Vickery (1996) has provided a summary history of associative relationships in information retrieval over the past few decades.

Lexical-Semantic Relations

Lexical-semantic relations are an important group of relations because they provide structure to lexicons, thesauri, taxonomies, and ontologies. The *structure* vs. *process* debate in semantic memory research is also present in lexical semantics. Are semantic relations stored in semantic memory as part of the meaning of a word, or are words defined in terms of their features and relations between words inferred dynamically from word meanings?

Lyons (1995) and other structural linguists hold that words cannot be defined independently of other words. A word's relationship with other words is a part of its meaning. The vocabulary of a language is thus viewed as a web of nodes, each representing a sense of a word and labeled links representing relations between the word senses. As Lyons (1977, pp. 231–232) put it,

> We cannot first identify the units [i.e., words] and then, at a subsequent stage of the analysis, enquire what combinatorial or other relations hold between them: we simultaneously identify both the units and their interrelations. Linguistic units are but points in a system, or network, of relations; they are the terminals of these relations, and they have no prior and independent existence.

Ferdinand de Saussure, generally regarded as the founder of modern structural linguistics, argued that "language is a system of interdependent terms in which the value of each term results solely from the simultaneous presence of the others" (Saussure, 1959, pp. 114–116). Other linguists maintain that the lexical representation of a word is mainly a set of semantic features based on semantic primitives and that semantic relations are derivable from the semantic features of the words through the use of some basic relational rules (Clark, 1973; Katz, 1972; Murphy, 2003).

The main lexical-semantic relations are the paradigmatic relations of hyponymy (IS-A or broader-narrower term), part-whole relation, synonymy, and antonymy, which are discussed later. However, frequently occurring syntagmatic relations between a pair of words can be part of our linguistic knowledge and considered lexical-semantic relations. As Firth (1957, p. 195; 1968, p. 179) put it, "you shall know a word by the company it keeps." Pairs of words that co-occur in a sentence more often than pure chance would allow are referred to, broadly, as *collocations* (Smadja, 1993), although some writers prefer narrower definitions of the term.

There are different degrees of syntagmatic word association. At the extreme are idioms (e.g., "kick the bucket") whose meanings cannot be derived from the meanings of the component words. Other word sequences are less strongly associated—their meaning is related to the meaning of the component words but not completely derivable from them. Hausmann (1985) divided word associations into fixed (i.e., idiom) and nonfixed combinations, subdividing the latter into counter-affine, affine, and free combinations.

Some word pairs are so strongly associated that the presence of one word almost determines that the other word will also appear in a particular context. Mel'cuk (1988) introduced the idea of *lexical functions* (LFs) in the framework of his Meaning-Text Theory. Wanner (1996) has referred to lexical functions as "institutionalized" lexical relations. A lexical function is a mapping or relation between two terms—term1 and term2—denoted "LF(term1) = term2" for a particular meaning context. So, if a term, *term2*, is to be selected to express a particular meaning or relation, the choice of term2 is predetermined if term1 is given. An example is LF("aircraft") = "crew." The value of a lexical function can also be a set of words, e.g., LF("flock") = {"birds", "sheep"}. Institutionalized lexical relations are directed and asymmetrical, as well as language-specific. For example, LF("aircraft") = "crew" does not imply LF("crew") = "aircraft."

There are many LF relations. Mel'cuk (1996) listed 27 paradigmatic and 37 syntagmatic lexical functions. Examples of paradigmatic lexical functions are: *Syn* (synonym), *Anti* (antonym), *Conv* (converse), *Contr* (contrastive), and *Gener* (genus). Syntagmatic lexical functions include:

- Center/culmination: Centr("crisis") = "the peak" [of the crisis]

- Very/intensely: Magn("naked") = "stark"

- More: Plus("prices") = {"soar", "skyrocket"}

- Less: Minus("pressure") = "decreases"

A good introduction to lexical functions and Meaning-Text Theory is given by Wanner (1996).

The most extensive lexical-semantic network that has been constructed for the English language is WordNet (http://www.cogsci.princeton.edu/~wn) (Fellbaum, 1998; Miller, 1995; Miller & Fellbaum, 1991). WordNet is a lexical database comprising about 150,000 English nouns, verbs, adjectives, and adverbs, organized into sets of synonymous words called *synsets*, each of which represents a lexical concept. Its design is based on psycholinguistic theories of human lexical memory. Its construction has given insight into how the lexicon is structured by lexical-semantic relations. For example, nouns are structured mainly by IS-A and part-whole relations; nouns are linked to adjectives with the *attribute* link and to verbs with the *function* link; adjectives are linked primarily by *antonymy*; the most frequent relation among verbs is *troponymy*, which expresses a *manner* elaboration. Other relations among verbs encoded in WordNet are lexical entailment (e.g., *snoring* entails *sleeping*), causal relation (e.g., show/see, feed/eat, have/own), and antonymy.

Following the success of WordNet, EuroWordNet (http://www.illc.uva.nl/EuroWordNet)—a multilingual lexical database covering several European languages—was constructed (Alonge, Calzolari, Vossen, Bloksma, Castellon, Marti, et al., 1998; Vossen, 1998). EuroWordNet is patterned after WordNet but uses a richer set of lexical-semantic relations. For example, causal relations are divided into *non-factive causal relations* (i.e., one event is likely to cause another event but not necessarily so, e.g., *search* → *find*) and *factive causal relations* (the causal relation necessarily holds, e.g., *kill* → *die*). A causal relation can also be labeled with the property of *intention* to cause the result (e.g., *search* → *find*) in order to distinguish it from inadvertent causal relations. Near-synonymy, near-antonymy, and five types of part-whole relations are also used in EuroWordNet. Furthermore, sets of relations can be labeled with the properties of *conjunction* and *disjunction* to indicate relationships among sets of concepts: for example, an airplane is typically composed of several parts—wings, nose, tail AND door (conjunction), but may have only one of several possible means of propulsive power—propellers OR jet engines (disjunction). WordNets for other languages are being constructed, and these projects are listed on the Global WordNet Association Web site (http://www.globalwordnet.org).

Case Relations

Case relations, also called case roles, thematic relations, and theta roles, are the primary syntagmatic relations between the main verb in a clause and the other syntactic constituents of the clause (Fillmore, 1968; Somers, 1987). According to case grammar theory, verbs assign semantic roles to the various clause constituents—subject, direct object, indirect object, prepositional phrase, and so forth—which are sometimes termed the arguments of verbs. For example, in the sentence "Mary bought a watch for John," the case relations between the verb *buy* and the other clause constituents are:

buy → (agent) → [Mary]

→ (patient) → [watch]

→ (recipient) → [John]

Each verb sense is associated with a case frame with slots, each slot having a case role. A case frame specifies the number of entities the verb expects in the clause, the case roles assigned to these entities, whether each role is obligatory (i.e., must be filled) or optional, selectional restrictions specifying the semantic category of the entity filling a role, and the syntactic realization of each role in the clause (whether expressed as subject, direct object, etc.).

Somers (1987, p. 111) said that "a recurring problem for Case grammarians has always been the definition of a 'comfortable' set of cases." Rosner and Somers (1980) stressed that a case system should be tailored to the particular application. The rationale for using case roles is to classify and generalize the semantic roles between a verb and its arguments, and so the set of case roles should be at a level of abstraction that is appropriate for the application.

Fillmore (1971) produced a "case hierarchy" with eight roles: *agent, experiencer, instrument, object, source, goal, location,* and *time.* Cook's (1989) case frame matrix had five case roles: *agent, experiencer, benefactive, object,* and *locative.* He also listed additional "modal cases": *time, manner, instrument, cause, result,* and *purpose.* Somers's (1987) case grid defined 24 case roles using a combination of two dimensions: a spatial/temporal-orientation dimension comprising the values *source, path, goal,* and *neutral* and a second, mostly verb-type dimension with the values *active, objective, dative psychological/possessive, locative, temporal,* and *ambient.* A case role is thus considered a bundle of more primitive features. The *experiencer* role, for example, is represented as a combination of *dative psychological + goal* features. Sets of case roles have been constructed by many authors. Longacre (1996) presented 10 case roles. Myaeng, Khoo, and Li (1994) identified 46 case roles in the process of constructing case frames for all the verb senses in the *Longman Dictionary of Contemporary English* (1987). Various case grammar systems have been reviewed by Cook (1989) and Somers (1987).

Dowty (1991), however, has argued that case roles are not discrete roles but cluster concepts that have fuzzy boundaries. An individual verb-specific semantic role can belong to a case role to a greater or lesser extent. A case role is thus seen as a category or type of semantic role, which includes a cluster of more specific roles with overlapping sets of features. Each semantic role can be decomposed into features that Dowty called verbal entailments. He proposed two large clusters of case roles called *proto-agent* and *proto-patient* roles. Examples of entailments for the proto-agent role include *volitional involvement, perception, causing an event or change of state,* and *movement relative to the position of another participant.*

Case grammar theory can be extended to other parts of speech, such as nouns and adjectives. Verb case frames are applicable to nominalized verbs and gerunds, formed by adding one of several possible suffixes, such as -ing and -ion, to verbs. Case frames for these nouns can be derived from the case frames of their associated verbs, although the process is not straightforward. It has been suggested that some adjectives and nouns also have valency, inasmuch as they expect certain prepositional phrases and certain kinds of complements (e.g., Somers, 1987).

Constructing case frames for a comprehensive set of verbs is a difficult task. Automatic construction methods using text mining and corpus statistics are described later in the chapter. A major manual effort to construct a comprehensive set of case frames for English verb senses as well as predicative nouns and adjectives is being undertaken in the Berkeley FrameNet project (http://www.icsi.berkeley.edu/~framenet) (Baker, Fillmore, & Cronin, 2003; Baker, Fillmore, & Lowe, 1998). The project does not define a small number of case roles to use in all the case frames. Instead, a set of case roles called "frame elements" is defined for each "frame." A frame in the FrameNet project is a schematic representation of a particular type of situation involving various participants. Example frames are *action*, *awareness,* and *transaction* frames. To construct case frames for individual word senses, the words are clustered into groups corresponding to situations or frames and the case roles for each word sense are selected from the frame elements defined for the situation.

Many natural language processing applications make use of case frames because they correspond quite closely to the surface structure of clauses and it is thus relatively easy to label clause constituents with case roles by means of a computer program. This serves as a useful intermediate processing step when converting the text to a semantic representation. Indeed, instantiated case frames with slots filled by terms/concepts extracted from the text are often used as the intermediate representation or interlingua in natural language understanding systems (e.g., Chan & Franklin, 2003; Minker, Bennacef, & Gauvain, 1996), question answering and dialogue systems (e.g., Takemura & Ashida, 2002; Xu, Araki, & Niimi, 2003), and machine translation systems (e.g., Dorr, Levow, & Lin, 2002).

Relations Between Larger Text Segments

We turn now to semantic relations between larger units of text. Relations between sentences can be analyzed from a logical or textual perspective. Logical relations between sentences are dealt with in the fields of formal semantics (e.g., Cann, 1993), logic and philosophy (e.g., Quine, 1982), and knowledge representation (e.g., Ringland & Duce, 1988; Sowa, 1984, 2000). Often, sentences and clauses are represented as propositions or predicates, and inferencing is performed using propositional, predicate, and other kinds of logics. The main semantic relations used are entailment (or implication or consequence), presupposition, equivalence, and contradiction (Cann, 1993; Lyons, 1995; Van Dijk, 1972). The most important relation is entailment. When we say that *a sentence* S *entails a sentence* S', we mean that if S is true then S' is true. Van Dijk (1972) presented other semantic relations: time, place, cause, purpose, result, condition, concession, and topic (theme)-comment (rheme). Crombie's (1985) semantic relations between propositions were grouped under the headings *temporal, matching, cause-effect, truth and validity, alternation, bonding, paraphrase, amplification,* and *setting/conduct.* Other lists of propositional relations can be found in Beekman, Callow, and Kopesec (1981), Hobbs (1985), and Longacre (1996).

At the textual level, sentences and clauses are linked by relations of cohesion and coherence. Halliday and Hasan (1976) analyzed relations between adjacent sentences and clauses, which they termed *cohesive relations.* They emphasized that cohesion is a semantic relation and that "cohesion occurs where the interpretation of some element in the discourse is dependent on that of another" (Halliday & Hassan, 1976, p. 4). Their work focused on the linguistic devices that writers use to effect "cohesive ties" between two proximate items, usually words and phrases, in the text. They divided cohesive devices into grammatical devices (anaphoric reference, substitution, ellipsis, and conjunction) and lexical devices (use of vocabulary and repetition of words).

Cohesion is often contrasted with coherence relations. Dooley and Levinsohn (2001, p. 27) characterized text coherence as "in essence, a question of whether the hearer can make it 'hang together' conceptually, that is, interpret it within a single mental representation." Eggins (1994, p. 87) said that coherence refers to the way a group of clauses or sentences relates to the context. Cohesion emphasizes local relations between two adjacent text units, whereas coherence focuses on networks of related units and larger structures as well as on the argumentative and pragmatic purposes of the text unit.

At an even higher level of text are discourse relations and macrostructure. Van Dijk (1988) argued that syntax and semantics can be applied to sequences of clauses, sentences, or whole texts. An influential discourse structure model in information science derives from the *Rhetorical Structure Theory* of Mann and Thompson (1988, 1989; see also Mann, Matthiessen, & Thompson, 1992), which uses a set of rhetorical relations to model text structure. Rhetorical relations include *evidence, elaboration, motivation, volitional cause, evaluation,* and *background.* Each relation links two text segments, one of which is considered the nuclear, or more central, segment and the other, the satellite or peripheral segment. A few relations, for example *sequence* and *contrast,* are "multinuclear" in that the linked text segments are both considered nuclear. The rhetorical structure is recursive—a text is decomposed into a sequence of segments linked by rhetorical relations and each segment can be further decomposed into smaller segments linked by the same or other rhetorical relations.

Van Dijk (1980) maintained that a text has an overall macrolevel syntactic structure, called a *superstructure,* that is governed by a rule-based schema. He suggested the following hierarchical schema for news articles:

Situation

Episode (subdivided into Main Events and Consequences)

Background

Context (circumstances, previous events)

History

Comments

Verbal reactions

Conclusions (subdivided into Expectations and Evaluations)

Though Van Dijk regarded these as syntactic units, the unit labels suggest semantic roles. The segments can perhaps be considered to have a semantic relation to the overall content of the text. In fact, Van Dijk (1988) postulated the existence of summarizing macrorules, which relate lower-level propositions to higher-level macropropositions—topics or themes derived from the meanings of a text. A more recent discussion of the discourse structure of news articles can be found in Bell (1998).

The macrolevel structures of stories, called *story schemas* and *story grammars*, have been studied by several authors (e.g., Mandler, 1987; Mandler & Johnson, 1977; Rumelhart, 1975; Schneider & Winship, 2002), and are used in the teaching of reading comprehension, literary analysis, and story-writing in schools (see Dimino, Taylor, & Gersten, 1995; Olson & Gee, 1988). A recent review of the theory can be found in Lang (2003).

At the document level, relations between documents may be structural (e.g., an article in a journal, a chapter in a book) or associative (e.g., articles by the same author, cited articles, hyperlinked Web pages). The documents can be linked by various kinds of semantic relations—two articles may be on the same topic, one article may be a condensed version of another, an article could report a follow-up study or refute the results of another study, and so forth. Topical semantic relations can be indicated using controlled subject terms taken from a thesaurus or subject headings list, or class numbers taken from a classification scheme. Another type of document-level semantic relation can be derived from an author's citation of other works and his or her reasons for citing them. Liu (1993) reviewed many citation studies and compiled a list of possible reasons for citing another work. Green (2001) noted that little is known about the range of semantic relations between citing and cited documents. Relatively little work has been done on identifying semantic relations between documents. The main semantic relations at the document level appear to be those provided by thesauri and classification schemes.

Finally, an important type of semantic relation in information science is the relevance relation—the relevance of a document to a query or to the information need of a user. Researchers have identified many factors, in addition to topical relevance, that affect a user's judgment of the relevance of a document (Barry, 1994; Park, 1997; Schamber, 1991, 1994; Tang & Solomon, 2001). Green (2001) suggested that there may be several types of semantic relations underlying these factors, which have not been studied in depth. Green and Bean (1995) and Bean and Green (2001) have explored some of the relations underlying topical relevance.

Selected Semantic Relations

This section takes a close look at five well-known paradigmatic relations used in thesauri and ontologies, as well as the cause-effect relation, which is an important syntagmatic relation in human knowledge structures. These relations are often treated as unitary primitive relations. We show that they are complex relations that can be divided into subtypes with different properties.

Hyponym-Hyperonym Relation

The hyponymy relation has been referred to in the literature under various names, including IS-A (is-a), a-kind-of, taxonymic, superordinate-subordinate, genus-species, and class-subclass relations. *Hyponym* refers to the narrower term/concept (e.g., Alsatian) and *hyperonym* to the broader term/concept (dog). The relation implies class inclusion, e.g., all instances of Alsatians are dogs, the set of Alsatian instances is a subset of dogs, and the meaning of "Alsatian" is included in the meaning of "dog" (Cruse, 2000). Different logical definitions of the hyponymy relation are presented and discussed by Cruse (2000). Related to hyponym is the *co-hyponym* or *coordinate*—another hyponym of the same hyperonym, such as siblings with the same parent.

Lyons (1968, p. 453) called the hyponymy relation the most fundamental paradigmatic relation of sense in terms of which the vocabulary is structured. Together with the *part-whole* relation, it is a hierarchical relation often found in thesauri, taxonomies, and ontologies. Cruse (2002) asserted that of all the sense relations, it occurs across the widest range of grammatical categories and domains.

There is some question whether the hyponymy relation relates word senses, lexemes (root words), or concepts. Most linguists take the hyponymy relation to relate word senses. Cruse (2002) has argued that, in some cases, even senses can be subdivided into "facets" (e.g., the physical *book* versus the abstract text of a *book*), and that sense relations relate facets. However, the form of a word has been found to affect human judgment of relations. For example, Cruse found that people considered *cat* to be a better hyponym of *animal* than *pussy*, suggesting that people are influenced by word forms.

The hyponymy relation exhibits different linguistic behavior when expressed by means of different terms. Cruse (2002) pointed out that the expression "An X is *a kind / type of* Y" is more discriminating than "an X *is a* Y." Cruse (1986) called the first relation *taxonymy* and the second relation *simple hyponymy*. He claimed that taxonymy is not just a logical class inclusion relation—the terms used to represent the classes are important. He gave the following examples of logical hyponymy relations that do not sound correct when expressed as "a kind of":[4]

> ?A stallion/mare/foal is a kind/type of horse.
>
> A stallion is a horse.
>
> ?A blonde/queen/actress is a kind of woman.
>
> An actress is a woman.

The expression "a kind/type of" exerts selectional restrictions on the pair of terms. Cruse has suggested the existence of a "principle of taxonomic subdivision" that selects only good categories that are internally cohesive, externally distinctive, and maximally informative. Good taxonyms tend to be natural kinds that cannot be defined in terms of a few necessary and sufficient features. Cruse (1986) suggested that single-feature category division may be the reason that *stallion*, *kitten*, and *blonde* are not satisfactory taxonyms of *horse*, *cat*, and *woman*. Another possible reason is that a term may "highlight" a particular semantic feature. The word *prostitute* highlights the *sexual activity* feature so that "A prostitute is a kind of sex-worker" is better than "A prostitute is a kind of woman." The hyponymy relation is generally taken to be a transitive relation. However, Cruse (2004, p. 152) cited the following example where transitivity breaks down:

> A car seat is a type of seat.
>
> A seat is a type of furniture.
>
> * A car seat is a type of furniture.

Fellbaum (2002) suggested that the hyponymy relation works best between closely related terms, and less well between terms far apart in the hierarchy.

Troponymy Relation

Troponymy refers to broader-narrower relations between verbs. Fellbaum (2002) pointed out that the expressions "a kind of" and "IS-A" sound odd when applied to verbs, for example "(To) yodel is a kind of (to) sing" and "To murmur is to talk." She said that the main relation between verb senses is the *manner* relation, which Fellbaum and Miller

(1991) termed "troponymy." For example, the *Longman Dictionary of Contemporary English* (1995) defines *run* and *fly* as to move in some manner (*to move quickly on foot* in the case of *run*, and *to move through the air* for *fly*). The manner relation involves several dimensions. Motion verbs differ along the dimension of speed (e.g., *walk* vs. *run*) or the means of transportation. Verbs of impact (e.g., *hit*) vary along the dimension of degree of force (e.g., *chop* and *slam*). In addition to the manner relation, troponyms include the *function* and *result* relations.

Fellbaum and Chaffin (1990) determined in a psychological study that people were able to recognize and process troponymy relations: Subjects had no trouble labeling verb pairs with the type of troponymy relation, sorting verbs into related pairs, responding with related verbs in an association task, and accomplishing an analogy task. Finally, Fellbaum (2002) has observed that verb hierarchies are flatter and more "bushy" than noun hierarchies, since they generally do not exceed three or four levels.

Meronym-Holonym Relation

The meronymy relation, also known as the part-whole relation and partonymy, refers to the relation between a concept/entity and its constituent parts. The distinction between meronymy and hyponymy relations is clear for concrete concepts but fuzzy for abstract concepts. Hyponymy relations can be said to exist within concepts, but meronymy relations are between concepts. Pribbenow (2002) has pointed out that both are logically asymmetric and transitive relations. Hyponyms inherit features from hyperonyms, but parts do not inherit features from wholes, although there is upward inheritance for some attributes, such as color, material, and function (Tversky, 1990).

Lyons (1977, v. 1, p. 313) demonstrated that the part-whole relation is intransitive at the level of linguistic expression:

> The door has a handle.
>
> The house has a door.
>
> ? The house has a handle.

Cruse (1979) attempted to resolve the problem by characterizing the functional context of the relation. He claimed that when we say X is a (functional) component of Y, we usually mean that X is a major component of Y.

Iris, Litowitz, and Evens (1988) found that the part-whole relation is really a family of relations, divided into four main types:

- Functional component of a whole (e.g., wheel of a bicycle)

- The segmented whole (the whole divided into pieces like a pie)

- Members of a collection of elements

- Subsets of sets (set inclusion, e.g., fruits and apples)

Winston, Chaffin, and Herrmann (1987) identified six types of part-whole relations, including the following three additional types: stuff-object (steel-car), feature-activity (paying-shopping), and place-area.

Gerstl and Pribbenow (1995) divided part-whole relations broadly into those relating to the natural structure of the whole (e.g., functional components of an object) and partitions of the whole by construction (i.e., artificial partitions based on attributes, e.g., dividing objects by color). These were further divided into subtypes.

Within the Meaning-Text Theory, Wanner (1996) listed the following meronymic relations:

- LF Mult (member-collection), e.g., Mult("dog")="pack"

- LF Equip (social whole-staff), e.g., Equip("aircraft")="crew"

- LF Cap (organization and its head), e.g., Cap("ship")="captain"

- LF Sing (a whole and its uniform unit), e.g., Sing("sand")="grain"

- LF Centr (a whole and its center or culmination), e.g., Centr("mountain")="peak" [of the mountain].

Other classifications of the part-whole relation have been developed by Barriere (1997, 2002); Markowitz, Nutter, and Evens (1992); and Sattler (1995) specifically for an engineering application; Uschold (1996) for ecological information systems; and Bernauer (1996) for the medical domain.

Synonymy

Lyons (1995) has noted that absolute synonymy is very rare. Two expressions are absolutely synonymous if all their meanings are identical in all linguistic contexts. Synonymy can be analyzed from a logical point of view or from the linguistic expression point of view. Logically synonymous terms have been called *logical synonyms* (Murphy, 2003) and *propositional synonyms* (Cruse, 2004).

Common types of synonyms are *sense-synonyms* (which share one or more senses), *near-synonyms* (which have no identical senses but are close in meaning), and *partial synonyms* (which share some senses but differ in some aspect, e.g., in the way they are used or in some dimension of meaning) (Cruse, 1986; Lyons, 1995). Sense-synonyms that share at least one sense and match in every other property for that sense are *complete synonyms* (Lyons, 1981). Church, Gale, Hanks, Hindle, and Moon (1994) discussed *gradient synonyms*—sets of synonyms in which one core term is considered prototypical and the other synonyms differ from the prototype in various ways, often giving additional information.

Synonyms are usually treated as reflexive, symmetrical, and transitive, although Murphy (2003) has argued that they are not always so.

Antonymy

Antonymy, or opposition, is one of the best-studied relations and the one that people find easiest to learn and process (Jones, 2002). Cruse (1986, p. 197) called it the most readily apprehended of sense relations, hedged with magical properties in the eyes of many people:

> Indeed, there is a widespread idea that the power of uniting or reconciling opposites is a magical one, an attribute of the Deity, or a property of states of mind brought about by profound meditation, and so on. … Philosophers and others from Heraclitus to Jung have noted the tendency of things to slip into their opposite states; and many have remarked on the thin dividing line between love and hate, genius and madness, etc.

Evens et al. (1980) observed that antonymy is irreflexive, symmetric, and intransitive. Of the many different types of antonymy, *canonical antonymy* is the best studied. Canonical antonyms constitute a special class of opposites that are stable and enjoy wide cultural currency. For example, *hot/cold* is a better example of antonymy than *steamy/frigid*, even though both pairs indicate opposite ends of the temperature scale (Murphy, 2003). Such antonym pairs, for example *big/small, good/bad, good/evil*, are automatically recalled by subjects in free word association tasks and are taught to children (Murphy, 2003, p. 10).

Justeson and Katz (1991, 1992) and Jones (2002) found that antonymous adjectives tend to co-occur within the same sentences in texts, often linked by the conjunctions *and* and *or*, as in, for example, the phrases "rich or poor" and "large and small." They also often substitute for each other in parallel, essentially identical, phrases such as "am I right or am I wrong?" and "new lamps for old ones." Justeson and Katz (1992, p. 181) concluded that "the patterns [of phrasal substitution] are so pervasive, that there is simply no chance for a genuine antonym pair to fail to show up in them, at a reasonable rate. So those that do not, cannot be antonymic." They suggested that the frequent co-occurrence of antonyms in text and discourse reinforces people's knowledge of antonymous pairs, which partly explains how antonymous pairs are learned and why antonym relations are graded.

Frequently co-occurring antonymous words are more likely to be judged as good antonyms than less frequently co-occurring antonyms.

Many types of antonymy have been identified (Cruse, 1986; Lehrer & Lehrer, 1982; Lyons, 1977; Murphy, 2003; Ogden, 1967). Jones (2002) examined how antonyms are used in a newspaper corpus and identified several antonym classes based on their linguistic behavior.

Cause-Effect Relation

The concept of causation is complex and surprisingly difficult to define. Since Aristotle, philosophers have grappled with the concept (Ehring, 1997; Mellor, 1995; Owens, 1992; Sosa & Tooley, 1993). Reviews of the concept from a philosophical and psychological perspective can be found in Khoo, Chan, and Niu (2002) and Khoo (1995).

One can distinguish between *necessary* and *sufficient* causes. An event A is a *sufficient*, although not a *necessary*, condition for event B if, when A occurs, B always follows, but when A does not occur, B sometimes occurs and sometimes not. A is a *necessary*, although not a *sufficient*, condition for B if, when A does not occur, B never occurs, but when A occurs, B sometimes occurs and sometimes not. An often cited definition of causation is Mackie's (1980) INUS condition, which defined a cause as an *Insufficient* but *Necessary* part of an *Unnecessary* but *Sufficient* condition for an event. Psychologists Jaspars, Hewstone, and Fincham (1983) and Jaspars (1983) found evidence that whether a cause is a necessary and/or sufficient condition varies with the type of entity being considered for causal status. Cause is likely to be attributed to a person if the person is a sufficient condition, whereas cause is likely to be attributed to the circumstances or situation if the situation is a necessary condition. Cause is ascribed to a stimulus when it is both a necessary and a sufficient condition. So, "a personal cause is seen more as a sufficient condition, whereas situational causes are conceived primarily as necessary conditions" (Jaspars et al., 1983, pp. 16–17).

However, Mackie (1980) pointed out that our concept of causation also includes some presumption of continuity from the cause to the effect, a causal mechanism by which the cause generates the effect. The concept of probabilistic causation has also gained popularity (Eells, 1991; Salmon, 1984). This view recognizes the possibility of indeterministic causation—instances where the causal mechanism is inherently probabilistic, as in the field of quantum mechanics.

Aristotle (1996) identified four kinds of cause: material cause (i.e., the material of an object causes its existence), formal cause (i.e., the form or structure of an object causes its existence), efficient cause (i.e., an entity acting mechanically upon an object causes it to change, move, or come to rest), and final or teleological cause (the intended future effect is the ultimate cause of a present action undertaken to bring about that future event).

Recently, Barriere (1997, 2002) has presented a classification of general cause-effect relations:

Existence dependency

Creation

Prevention

Destruction

Maintenance

Influence dependency

Preservation

Modification

Increase

Decrease

Taking a somewhat different approach, Terenziani and Torasso (1995) have categorized cause and effect on the basis of temporal considerations. Their analysis yielded five categories:

- *One-shot causation*: The presence of the cause is required only momentarily to allow the action to begin.

- *Continuous causation*: The continued presence of the cause is required to sustain the effect.

- *Mutually sustaining causation*: Each bit of cause causes a slightly later bit of the effect.

- *Culminated event causation*: The effect comes about only by achieving the culmination of the causal event (e.g., "run a mile in less than 4 minutes" causes "receive a prize").

- *Causal connection with a threshold*: There is a delay between the beginning of the cause and the beginning of the effect, and the effect is triggered only when some kind of threshold is reached.

Warren, Nicholas, and Trabasso (1979) identified four types of cause-effect relations in narrative texts: motivation, psychological causation, physical causation, and enablement. Dick (1997), in attempting to model the causal situation in a legal case, distinguished between the following types of cause and effect: distant vs. direct cause, animate vs. inanimate agent, animate agent vs. instrument, volitive vs. nonvolitive cause, active vs. passive cause, central vs. peripheral (or abstract) cause, explicit vs. implicit cause, and aims vs. actual effect.

Khoo (1995) analyzed the verb entries in the *Longman Dictionary of Contemporary English* (1987) and came up with a total of 2,082 causative verbs (verbs with a causal component in their meaning), which he grouped into 47 types of effects. Levin (1993) provided a systematic and extensive classification of verbs based on their syntactic behavior. Many of the verb classes were found to have a causal component in their meaning.

Semantic Relations in Knowledge Structures

Semantic Relations in Thesauri

A thesaurus is a set of terms structured using a small set of semantic relations to indicate the controlled (or preferred) term for each concept and relationships between the terms/concepts. It is designed to support consistent subject indexing of documents and effective information retrieval. The relations between terms help both indexers and searchers to navigate thesauri in order to identify various kinds of related terms.

The American National Standards Institute/National Information Standards Organization (ANSI/NISO) Z39.15-1993 standard "Guidelines for the Construction, Format, and Management of Monolingual Thesauri" (National Information Standards Organization, 1994) and the International Organization for Standardization (ISO) 2788 standard "Guidelines for the Establishment and Development of Monolingual Thesauri" (International Organization for Standardization, 1986) recognize three types of semantic relations: equivalence (*use* and *use for*), hierarchical (*broader term* and *narrower term*), and associative (*related term*).

The ANSI/NISO standard lists seven types of synonym relations: terms of different linguistic origins, popular term-scientific name, generic noun-trade name, variant names, current-outdated term, common nouns-slang/jargon, and dialectical variants. It also describes other kinds of equivalence relation: lexical variants and quasi-synonyms. Hierarchical relations include generic (IS-A), part-whole, and instantiation relations. Part-whole relations include organs of the body; geographic locations; subject disciplines; and hierarchical organizational, corporate, social, or political structures. Nine types of associative relations are also identified.

Associative relations in thesauri have been analyzed by several authors. Aitchison, Gilchrist, and Bawden (1997) listed 14 categories. Lancaster (1986) and Raitt (1980) each listed 10 categories. In an analysis of hierarchical relations in Medical Subject Headings (MeSH), Bean (1998) identified 67 types of relations other than the generic and instantiation relations. Most of the relations could be considered associative. Aitchison et al. (1997) cited a 1965 study by Perreault, who found 120 types of relations. Discussion of thesaural relations in general can be found in Aitchison et al. (1997), Clarke (2001), and Milstead (2001).

Semantic Relations in Indexing Languages

In the subject indexing of a document, controlled terms from a thesaurus, subject headings list, or classification scheme are assigned to the document to reflect the main subjects and concepts in the content of the document. The index terms are generally not precoordinated, in other words, the index terms are assigned as separate terms and there is no indication whether two or more concepts are related in a particular way in the document. For example, if a document is assigned the terms *information retrieval, user interface,* and *evaluation*, there is no indication whether *evaluation* is related to *information retrieval* or to *user interface*. During retrieval, the user may specify the Boolean query "information retrieval AND evaluation," which requires the system to search for the two terms separately and to combine the two sets of documents retrieved to identify documents containing both terms. There is no assurance that the documents retrieved will discuss *evaluation of information retrieval*—only that these two concepts occur in the same document. Such an indexing approach is called *postcoordinate* indexing.

In some indexing languages, the index terms are *precoordinated*, in other words, the human indexer indicates an association between two or more concepts in the document by using the syntax of the language and placing the terms in a particular order. However, the type of association is not specified explicitly but is implied by the context. Such is the case with the Library of Congress Subject Headings (LCSH) and faceted classification schemes like Ranganathan's Colon Classification (Kishore, 1986; Ranganathan, 1965). Precoordinate indexing allows the user to search for some kind of association between two or more index terms.

Farradane (1967) advocated the use of explicitly specified relations in indexing systems. He pointed out that implied relations in precoordinate indexing are unambiguous only in a narrow domain. More recently, Green (1995a, 1995b) has called for the inclusion of syntagmatic relations in indexing languages, examined the issues involved, and suggested a frame-based representation of syntagmatic relations.

Two indexing systems that make explicit use of relations are Farradane's (1950, 1952, 1967) relational classification system and the SYNTOL model (Gardin, 1965; Levy, 1967). Farradane's system had nine types of relations: concurrence, self-activity, association, equivalence, dimensional (time, space, state), appurtenance, distinctness, reaction, and functional dependence (causation). The SYNTOL project used four main types of relations: coordinative, consecutive, predicative, and associative. The associative relation was subdivided into finer relations. There is no experimental evidence yet that the use of explicitly specified relations in indexing yields better retrieval results compared to postco-ordinate indexing or precoordinate indexing with implied relations.

Semantic Relations in Ontologies

A thesaurus lists the main concepts/terms in a particular domain and specifies relations between the concepts/terms using only a small number of relation types. This small set of relations may be adequate for information retrieval applications because the focus of a thesaurus is on indexing and searching, but it is not sufficient for more complex or intelligent applications that require knowledge-based inferencing and a detailed representation of domain knowledge.

A more detailed representation of domain knowledge is called an ontology. Many definitions of ontology from different perspectives have been put forward. The following definition by Berners-Lee, Hendler, and Lassila (2001, p. 40) alludes to some of the different aspects of ontology:

> In philosophy, an ontology is a theory about the nature of existence, of what types of things exist; ontology as a discipline studies such theories. Artificial-intelligence and Web researchers have co-opted the term for their own jargon, and for them an ontology is a document or file that formally defines the relations among terms. The most typical kind of ontology for the Web has a taxonomy and a set of inference rules.

Ontology, with an uppercase "O," refers to a branch of philosophy dealing with the nature of being or existence—what categories of things exist and what their features are (Guarino & Giaretta, 1995; Sowa, 2000). This is often contrasted with *Epistemology*, which deals with the nature and sources of knowledge. Ontology with a lowercase "o" can refer to the conceptual framework or knowledge of a particular domain shared by a group of people—for instance, something that exists in people's minds. Or it can refer to the symbolic representation of this conceptual frame, perhaps in the form of a "logical theory" that can be used by a computer program (Guarino & Giaretta, 1995).

An often quoted definition is that of Gruber (1993, p. 199): "An ontology is an explicit specification of a conceptualization." In practice, an ontology is expressed as a taxonomy of concepts linked by IS-A, part-whole, and attribute-value relations, sometimes enriched by other kinds of relations as well as additional rules or constraints called *axioms*. One major difference between an ontology and a thesaurus is the richer set of relations used in an ontology. Guarino and Giaretta (1995), Guarino (1997), and Gómez-Pérez, Fernandez-Lopez, and Corcho (2004) analyzed various definitions of ontology. A collection of definitions can be found at http://www.aaai.org/AITopics/html/ontol.html.

Ontologies come in many types and flavors, depending on the domain, application, representation scheme used, philosophical principles adopted by the authors, and the construction method and tools used. Those functioning as online search aids are more lexically oriented and may not contain nontaxonomic relations or axioms, whereas others supporting inferencing may be formally represented in a logic representation and have many axioms. Gómez-Pérez et al. (2004) outlined the different typologies of ontologies that have been proposed and suggested that even thesauri can be considered light-weight ontologies.

There is growing interest in ontologies because of their potential for encoding knowledge in a way that allows computer programs and agent software to perform intelligent tasks on the Web.

> Ontologies provide support in integrating heterogeneous and distributed information sources. This gives them an important role in areas such as knowledge management and electronic commerce. ... Ontologies enable machine-understandable semantics of data, and building this data infrastructure will enable completely new kinds of automated services. (Fensel, 2001, p. 8)

The OWL Web Ontology Language Use Cases and Requirements (World Wide Web Consortium, 2004a) lists the following areas where ontologies are expected to be useful: Web portals, multimedia collections, corporate Web site management, design documentation, agents and services, and ubiquitous computing.

Ontologies are seen as the backbone of the Semantic Web. The Semantic Web was characterized by Berners-Lee et al. (2001, p. 37) as "an extension of the current web in which information is given well-defined meaning, better enabling computers and people to work in cooperation." The World Wide Web Consortium (2004c) views the Semantic Web as providing "a common framework that allows data to be shared and reused across application, enterprise, and community boundaries." A fundamental technology for realizing the Semantic Web is the Web service. Web services are self-contained computer programs that can be accessed on the Internet by other computer programs through public interfaces and bindings that are defined by means of Extensible Markup Language (XML) (World Wide Web Consortium, 2004d). Because the interface definition of a Web service can be discovered by other computer programs, this allows computer programs to dynamically locate and interact with one another in an automated and unattended way. To help people and software agents locate appropriate information, objects, and Web services on the Internet, ontologies are needed.

The World Wide Web Consortium (2004b) has developed the Resource Description Framework (RDF) and the Web Ontology Language (OWL) for encoding ontologies with XML. OWL can be used to specify types of relations between concept instances, called *properties*. The following relations between user-defined *properties* can be specified: *equivalentProperty* (synonymous property), *inverseOf* (e.g., "hasChild" is the inverse of "hasParent"), and *subPropertyOf* (a kind of hyponymy relation). The user-defined *properties* can also be labeled with the following attributes: *TransitiveProperty, SymmetricProperty, Functional Property* (i.e., each instance has no more than one value for this property), and *InverseFunctionalProperty*.

Well-known ontologies include:

- CYC (http://www.cyc.com), which contains about 40,000 concepts and 300,000 axioms (inter-concept relations and constraints) and is built for commonsense reasoning.

- Suggested Upper Merged Ontology (SUMO) (http://suo. ieee.org, http://ontology.teknowledge.com, http://www.ontol ogyportal.org), a standard upper ontology developed by the Institute of Electrical and Electronic Engineers (IEEE) Standard Upper Ontology Working Group. SUMO and its several domain ontologies contain altogether about 20,000 terms and 60,000 axioms.

- Unified Medical Language System (UMLS) (http://www. nlm.nih.gov/research/umls), which contains 135 semantic types, 54 semantic relations, and about 250,000 concepts.

- MIKROKOSMOS (http://crl.nmsu.edu/users/sb/papers/ thesis/node26.html), which contains about 4,800 concepts and is built to support machine translation.

- Descriptive Ontology for Linguistic and Cognitive Engineering (DOLCE) (http://www.loa-cnr.it/DOLCE.html), which aims at capturing ontological categories underlying natural language and human common sense and has been developed to serve as a starting point for comparing and analyzing other ontologies.

- WordNet (http://www.cogsci.princeton.edu/~wn), a lexical database, often considered a lexical or terminological ontology, containing approximately 150,000 English nouns, verbs, adjectives, and adverbs that are grouped into 115,000 synonym sets (synsets), each of which represents an underlying lexical concept.

- The Enterprise Ontology (http://www.aiai.ed.ac.uk/ project/enterprise), a collection of terms and definitions relevant to business enterprises to assist in the acquisition, representation, and manipulation of enterprise knowledge.

- Toronto Virtual Enterprise (TOVE) (http://www.eil. utoronto.ca/enterprise-modelling/tove/index.html), used to model the structure, activities, processes, information, resources, people, behavior, goals, and constraints of an enterprise.

Ontologies vary widely in the number and types of relations used. They can range from a simple taxonomy structured by IS-A and part-whole relations, to a small number of relations as in WordNet, to thousands of relations in CYC (Lenat, Miller, & Yokoi, 1995). Relations in ontologies are often structured in a relation hierarchy or grouped into major categories. Many relation hierarchies have been proposed in the literature. For example, Sowa (2000, 2001) divided his role concepts into two groups: roles pertaining to the *PrehendingEntity* (the subject of the relation, e.g., *whole*) and the *PrehendedEntity* (the object of the relation, e.g., *part*). *PrehendedEntity* is subdivided into *Correlative* and *Component*, the latter further subdivided as follows:

Component
    Part
        Piece
        Participant
        Stage
    Property
        Attribute
        Manner

A role concept is converted to a relation by combining the concept with the *has* relation. For example, the *part* role concept can be converted to the *has-part* relation.

A more extensive relation hierarchy is presented in the Generalized Upper Model (GUM) ontology (Bateman, Fabris, & Magnini, 1995). At the top level, relations are arranged into four categories: *participant, circumstance, process,* and *logical-relation*. The CGKAT system (Martin, 1995, 1996) has a default hierarchy of about 200 relations. Relations are organized into nine classes at the top-level: *attributive_relation, component _relation, constraint_or_measure_relation, relation_from_a_situation,*

relation_to_a_situation, relation_from_a_proposition, relation_refer-
ring_to_a_process, relation_with_a_special_property, and *relation_*
used_by_an_agent. The Unified Medical Language System (UMLS) relation hierarchy (U.S. National Library of Medicine, 2004) contains 54
relations grouped broadly into *IS-A* and *associated_with* relation types,
the latter being subdivided into *physically_related_to, spatially_related_*
to, functionally_related_to, temporally_related_to, and *conceptually_*
related_to. Markowitz et al. (1992) presented a hierarchy of lexical relations containing nearly 100 relations as leaf nodes.

Some researchers have developed methods to formally represent relations in a knowledge representation scheme for use in data modeling
and knowledge-based inferencing. This usually involves explicitly representing the attributes of semantic relations, modeling the hierarchical
relationships between semantic relations, and defining axioms or rules
for reasoning with the relations. Priss (1999) developed a mathematical
formalism for representing a network of semantic relations in a lattice
structure by analyzing the relations using *formal concept analysis*
(Ganter & Wille, 1997) and identifying relational components (see the
chapter by Priss in the present volume). Wille (2003) also showed how
commonsense logical relations between concepts can be represented
using a concept lattice. Methods for representing and reasoning with
semantic relations have been developed in the *conceptual graph* formalism (Sowa, 1984, 2000) as well as in description logics (Baader,
Calvanese, McGuinness, Nardi, & Patel-Schneider, 2003)—a family of
knowledge representation languages that focuses on expressing knowledge about concepts, concept hierarchies, roles and instances, and reasoning about them. A collection of papers describing various formalisms
for modeling concepts and relations can be found in Lehmann (1992).
Several authors have examined the issues involved in formalizing the
part-whole relation in data modeling and inferencing systems (e.g.,
Artale, Franconi, Guarino, & Pazzi, 1996; Lambrix, 2000; Lee, Chan, &
Yeung, 1995). Some issues involved in organizing semantic relations in
a knowledge base were examined by Stephens and Chen (1996).

To our knowledge, no systematic analysis of the types of semantic
relations used in ontologies has been reported in the literature. Such an
analysis should be carried out in the context of the domain and application for which the ontology was constructed. Little is known about what
constitutes an appropriate set of semantic relations for a domain or
application, or the most effective way to structure the relations into a
relation hierarchy. Although much has been written about the potential
uses of ontologies and methods for their construction, and although
small case studies of applications have been reported, there has not been
any systematic evaluation of the effectiveness of ontologies or the various types of semantic relations occurring in real applications.

One possible exception is the part-whole relation. Researchers have
analyzed the different types of part-whole relations used to model data
and objects for various purposes (e.g., Artale et al., 1996; Gerstl &

Pribbenow, 1995). Many ontologies specify a few types of part-whole relations. Nevertheless, in a review of 10 well-known ontologies, Noy and Hafner (1997) found that different ontologies represented the part-whole relation in dissimilar ways and that, in many cases, the ontologies did not deal adequately with the distinctions between different types of part-whole relations.

Automatic Identification of Semantic Relations

Overview

Automatic identification and extraction of semantic relations in text is a difficult task. The accuracy rate varies widely and depends on many factors: the type of semantic relation to be identified, the domain or subject area, the type of text/documents being processed, the amount of training text available, whether knowledge-based inferencing is used, and the accuracy of the syntactic preprocessing of the text. Furthermore, because there are different types of semantic relations at different text levels, no system can identify semantic relations accurately at all levels. This is a major barrier to widespread use of semantic relations in information science.

In this section, we consider the automatic identification and extraction of semantic relations between words/phrases and the concepts that they represent. Identification of higher-level relations such as cohesion relations (including anaphor and co-reference resolution), rhetorical relations, and text macrostructure is important, but the literature is too broad to cover in this survey. A general introduction to information extraction technology is given by Appelt and Israel (1999). The three major applications of automatic identification of relations in text are in information extraction, ontology construction/knowledge acquisition, and information retrieval. This section examines the main techniques used to extract relations in information extraction and ontology construction. Information retrieval applications are discussed later in the chapter.

In information extraction applications, concepts and relations are extracted from text to fill predefined templates representing various kinds of information about an event (e.g., terrorist attack or corporate merger), entity (e.g., company), or process. The slots in a template are labeled and can be considered roles related to the event/entity/process. In the 1980s and early 1990s, artificial intelligence researchers used sophisticated natural language processing and knowledge-based inferencing to extract concepts and relations from text and represent them in a semantic representation or knowledge representation scheme (e.g., Berrut, 1990; Mauldin, 1991; Rau, 1987; Rau, Jacobs, & Zernik, 1989). Unfortunately, such complex systems could be built only for narrow domains. In the 1990s, it was found that simple methods of relation extraction using shallow text processing and pattern matching utilizing many simple patterns were equally effective. However, constructing a good set of extraction patterns for an application still involves considerable manual effort. Current research is focused on automatic pattern construction, which requires a large training set of documents and manually filled templates representing the associated answer key. For information extraction technology to become widely used, automatic pattern construction techniques that are effective with small training sets need to be developed, together with good interfaces that help the end-user to construct the training examples and to guide the process of pattern construction.

Whereas information extraction applications seek to extract every instance of concepts and relations relevant to the domain or application, automatic ontology construction focuses on well-established knowledge, namely, concepts and relations that occur with some frequency in the text collection. Hence, corpus statistics techniques incorporating co-occurrence statistics, machine learning, and data mining can be employed together with pattern matching techniques to extract frequently occurring concept-relation-concept triples from a corpus. These triples can then be used to build a knowledge base of facts or connected together to form a semantic network or an ontology.

Automatic Identification of Semantic Relations Using Pattern Matching

Automatic identification of semantic relations in text involves looking for certain linguistic patterns in the text that indicate the presence of a particular relation. For example, a simple linear pattern for identifying some *cause-effect* information is:

[cause] *is a cause of* [effect]

The tokens in square brackets represent slots to be filled by words/phrases in the text. The slots indicate which part of the sentence represents the *cause* and which part represents the *effect* in the *cause-effect* relation. The following sentence contains a match for this pattern:

Smoking is a cause of *lung cancer*

An extraction pattern is thus a sequence of tokens, each token representing a literal word to be matched in the text, a wildcard that can match any word, or a slot to be filled. The following selectional restrictions can be specified for each token: the syntactic category (e.g., part of speech), type of phrase, syntactic role (e.g., subject, direct object, etc.), and the voice of the verb. Semantic restrictions can also be specified using concept categories from an ontology or type of entity (e.g., organization name, person name, date, amount of money). Pattern-matching is performed to identify the segments of the text that match each pattern.

A major component of any information extraction system is its set of extraction patterns. Construction of patterns can be done manually or automatically by analyzing sample relevant texts and the associated answer keys indicating the information to be extracted. The answer keys are typically constructed by trained human analysts. Pattern construction thus entails constructing patterns that will extract the same information from the text as the human analysts did. The patterns should not be too general lest information be extracted from non-relevant text fragments or incorrect information from relevant text fragments.

Two approaches can be used in the pattern construction: a top-down approach, in which general patterns are first constructed and then gradually specialized to reduce errors, or a bottom-up approach, in which specific patterns are first constructed and then gradually combined to reduce the number of patterns or generalized to cover more situations. Before pattern construction and pattern matching, the text is usually subjected to some amount of preprocessing, which can include tokenizing, stemming or conversion of words to their base forms, syntactic tagging (to identify the part of speech), chunking (to identify particular types of phrases), and semantic tagging (to identify the semantic class, e.g., *inanimate object* and *organization name*, to which the word/phrase belongs). Some information extraction systems make use of a thesaurus or ontology to infer the semantic classes of text tokens and to generalize two or more concepts to a single broader concept.

Automatic Construction of Extraction Patterns

Because manual construction of good extraction patterns is a difficult and time-consuming task, there is a need for automatic or machine-aided pattern construction. Researchers have developed effective techniques for automatic pattern construction. To perform automatic pattern construction, the system needs well-defined heuristics for constructing the initial patterns, for generalizing and specializing the patterns based on positive and negative examples, for selecting which generalization/specialization methods to use in which situation, and for deciding on the order in which the methods are tried. Typically, a variation of the inductive learning algorithm described by Mitchell (1997) is used for pattern learning.

Our survey focuses on information extraction from free text, rather than from structured or semi-structured documents, as our interest is in semantic relations expressed in free text. The learning of patterns for extracting information from structured documents, such as Web pages, is called wrapper induction and it relies on structure identification using HTML tags (Muslea, 1999). An example is Information Extraction based on Pattern Discovery (IEPAD) (Chang & Lui, 2001), a wrapper induction system that generates extraction patterns for Web documents without needing user-labeled examples.

Some well-known systems that learn extraction patterns from free text are AutoSlog (Riloff, 1993), PALKA (Kim, 1996; Kim & Moldovan, 1995), CRYSTAL (Soderland, 1997), WHISK (Soderland, 1999), and

Table 5.1 Pattern templates and examples of instantiated patterns in AutoSlog

Pattern template	Example of pattern constructed
<subj:slot> <passive-verb>	[victim] was murdered
<subj:slot > <active-verb>	[perpetrator] bombed
<subj:slot > <verb> <infinitive-phrase>	[perpetrator] attempted to kill
<subj:slot > <auxiliary-verb> <noun>	[victim] was victim
<passive-verb> <direct-obj:slot>	killed [victim]
<active-verb> <direct-obj:slot>	bombed [target]

RAPIER (Califf & Mooney, 2003). The patterns constructed by these systems generally perform sentence-level extraction, leaving co-reference resolution and merging of extracted information across sentences to later modules, such as discourse parsing modules (Soderland, 1999). A survey of the various types of extraction patterns generated by machine learning algorithms was carried out by Muslea (1999).

AutoSlog (Riloff, 1993) is the earliest system developed to learn text extraction patterns from training examples. It uses partial case frames as linear patterns. Each pattern has only one slot and usually includes a verb and a noun phrase (a subject or direct object). A set of pattern templates define the linear patterns that the system will construct. Each pattern is thus an instantiation of a pattern template. Table 5.1 lists the pattern templates used in AutoSlog and an illustration for each template. The pattern template "<passive-verb> <direct-obj:slot>" was included because a sentence analyzer called CIRCUS (Lehnert, 1991) occasionally confused active and passive constructions.

Before pattern construction, the training corpus is preprocessed by CIRCUS to identify clause boundaries and the major syntactic constituents: subject, verb, direct object, noun phrases, and prepositional phrases. Relevant text segments that contain the semantic relations of interest are identified and answer keys are constructed to indicate which noun phrase should be extracted and what its semantic role is. If the domain of interest is terrorist activities, the semantic roles would include *perpetrators, targets, victims*, and so forth.

During pattern construction, pattern matching is used to match the pattern templates with the training text segments. If a pattern template matches a relevant text segment, then a pattern is constructed by replacing the tokens in the template with the words in the text. If a token in the template indicates a slot, this token is allowed to match a noun phrase in the text only if the noun phrase appears in the answer key (i.e., a human analyst has indicated that this is the information to be extracted). A slot token is placed in the pattern being constructed, and the semantic role for the slot is taken from the answer key. The constructed pattern is thus an instantiation of the pattern template. Finally, a human analyst inspects each pattern and decides which ones should be accepted or rejected. AutoSlog-TS (Riloff, 1996), an extension of AutoSlog, creates dictionaries of extraction patterns using only untagged text. A user needs to provide training texts (relevant and irrelevant texts) and to filter and label the resulting extraction patterns. Generally, extraction patterns occurring in irrelevant texts are filtered out. The accuracy rates come close to those of AutoSlog, in which tagged text is used. However, AutoSlog cannot learn rules that extract values for multiple slots (such as *[victim] was killed by [attacker]*) and does not adjust the patterns by generalizing or specializing them once they are constructed.

In the PALKA system (Kim, 1996; Kim & Moldovan, 1995), the patterns involve the whole clause. Sentences in the training text are first converted to simple clauses. The clauses containing a semantic relation of interest are processed one at a time. If the set of patterns already constructed does not match a clause, then a new pattern is constructed for the clause. This initial pattern covers the main verb, the subject, the object, and the words to be extracted (i.e., the slot). Each of these constituents in the clause is represented by a token in the pattern. Each token is assigned a semantic category from a conceptual hierarchy. Generalizations and specializations are applied only to the semantic constraints. When two similar patterns sharing the same target slots and literal strings are generated, their semantic constraints are generalized by locating a broader concept or ancestor in the conceptual hierarchy that is common to both semantic categories.

The CRYSTAL system (Soderland, Fisher, Aseltine, & Lehnert, 1996) uses a similar approach but is more complex. CRYSTAL learns rules that can extract values for multiple slots. Initially, CRYSTAL constructs a very specific pattern for every sentence in the training text. The sentences are not simplified into simple clauses. The constraints in the initial patterns are gradually relaxed to increase their coverage and to merge similar patterns. CRYSTAL identifies possible generalizations by locating pairs of highly similar patterns. This similarity is measured by counting the number of relaxations required to unify the two patterns. A new pattern is created when constraints are relaxed just enough to merge the two patterns by dropping constraints that the two do not share and finding a common ancestor for their semantic constraints. The new pattern is tested against the training corpus to make sure it does not extract information not specified in the answer keys. If the new pattern is valid, all the patterns subsumed by the new pattern are deleted. This generalization continues until a pattern that exceeds a specified error threshold is generated.

WHISK (Soderland, 1999) induces rules in a top-down manner, first finding the most general rule that covers the seed (i.e., hand-tagged training examples), then constraining the rule by adding terms one at a time. The learned rules are in the form of regular expressions that can extract either single or multiple slots.

RAPIER (Califf & Mooney, 2003) is a bottom-up learning algorithm that incorporates techniques from several inductive logic programming systems (Lavrac & Dzeroski, 1994). Its algorithm starts with initial specific rules created from the input corpus and then incrementally replaces the rules with more general rules using automatic rule evaluation. The rule learning is done separately for each slot; thus, RAPIER cannot learn rules that extract values for multiple slots.

SRV (Freitag, 2000) employs a top-down rule learner that uses a covering algorithm. As each rule is learnt, all positive examples covered by the new rule are removed from consideration for the creation of future rules. Rule learning ends when all positive examples have been covered. SRV utilizes the length of a fragment, the location of a particular token, the relative locations of two tokens, and various user-defined token features, such as capitalization, digits, and word length. SNoW-IE (Roth & Yih, 2001) learns extraction patterns by means of propositional learning mechanisms. Ciravegna (2001) developed a pattern learner that makes use of rule induction and generalization.

Text Mining for Semantic Relations

Text mining for semantic relations is concerned with the extraction of new and implicit relationships between different concept entities from large collections of textual data. Although some semantic relations are clearly expressed through the use of well-defined syntactic structures, other semantic relations are not, and only a multistep sequence of reasoning based on semantic analysis of the text collection can extract them. Most semantic extraction systems take advantage of an existing domain knowledge source (i.e., semantic information) and make use of cue words and syntactic tags provided by a syntactic parser.

Various approaches to automatic semantic extraction from corpus documents have been developed. Girju, Badulescu, and Moldovan (2003) worked on the discovery of semantic relations, especially part-whole relations, from text. They used rich syntactic and semantic features to discover useful and implicit relations from text. The C4.5 decision tree learning algorithm (Quinlan, 1993) was used to learn semantic constraints so as to detect part-whole relations, while WordNet served as the domain knowledge base to identify and disambiguate target concepts (i.e., part and whole components). Girju (2002) also investigated the extraction of causal relations in her dissertation work. The Artequakt system (Alani, Kim, Millard, Weal, Hall, Lewis, et al., 2003) automatically extracts knowledge about artists from the Web, populates a knowledge base, and uses it to generate personalized biographies. Artequakt links a knowledge extraction tool with an ontology to identify entity relationships using ontology relation declarations, such as "[Person] – *place of birth* – [Place]," where [Person] and [Place] are concepts and *"place of birth"* is a semantic relation between them. Dyvik (2004) investigated a method for deriving semantic relations in WordNet from data extracted from the English-Norwegian Parallel Corpus (Johansson, 1997), which comprises around 2.6 million words. The method was based on the hypotheses that semantically closely related words have strongly overlapping sets of translations, and words with a wide range of meanings have a higher number of translations than words with few meanings. The implementation took words with their sets of translations from the corpus as input and returned thesaurus-like entries containing senses, synonyms, hyperonyms, and hyponyms. Calzolari and Picchi (1989; see also Calzolari, 1992) looked into the acquisition of semantic information from machine-readable dictionaries, in which semantic information is implicitly contained. They aimed at reorganizing free-text definitions in natural language form into informationally equivalent structured forms in a lexical knowledge base.

In the medical area, semantic tagging that uses domain knowledge is important for effective text mining. Many studies make use of the Unified Medical Language System (UMLS) (Humphreys, Lindberg, Schoolman, & Barnett, 1998; U.S. National Library of Medicine, 2004) as the domain knowledge base. Blake and Pratt (2001) mined for semantic relationships between medical concepts from medical texts. They mapped the terms in the texts to concepts in UMLS in order to reduce the number of features for data mining. Blake and Pratt focused on *Breast Cancer Treatment* using association rule mining (Borgelt & Kruse, 2002) to find associated concept pairs like *magnesium–migraines*. They were mainly interested in mining the *existence* of relationships between medical concepts (i.e., finding associated concept pairs in breast cancer treatment), not in identifying the specific semantic relations for the associated concept pairs. Lee, Na, and Khoo (2003) carried out a small experiment using a sample of medical abstracts from MEDLINE, a biomedical bibliographic database maintained by the U.S. National Library of Medicine, to identify concept pairs related to *Colon Cancer Treatment*. The semantic relations between the concepts in each pair were then inferred using the UMLS semantic network. They were able to infer semantic relations between concepts automatically from the UMLS semantic network 68 percent of the time, although the method could not distinguish between a few possible relation types.

The Semantic Knowledge Representation (SKR) project at the National Library of Medicine has developed programs that extract usable semantic information from biomedical texts (Rindflesch & Aronson, 2002). Two programs, MetaMap (Aronson, 2001) and SemRep (Rindflesch, Jayant, & Lawrence, 2000), play a major role in semantic information extraction. MetaMap maps noun phrases in free text to concepts in the UMLS Metathesaurus, while SemRep uses the Semantic Network in UMLS to infer possible relationships between those concepts. Consider the input phrase "ablation of pituitary gland." SemRep looks up a semantic rule (i.e., extraction pattern), which declares that the preposition *of* matches the Semantic Network relation *location_of*, and also notes that one of the relationships in the Semantic Network with this predicate is "[Body Part, Organ, or Organ Component] – LOCATION_OF – [Therapeutic or Preventive Procedure]." The

Metathesaurus concept for *ablation* is *Excision, NOS,* found by MetaMap. The semantic type for this concept is *Therapeutic or Preventive Procedure*; the type for *Pituitary Gland* is *Body Part, Organ, or Organ Component.* Because these semantic types match those found in the relationship indicated by the preposition *of* (*location_of*), "Pituitary Gland – *location_of* – Excision, NOS" is extracted as a new semantic relation.

Srinivasan and Rindflesch (2002) have used SemRep in combination with MeSH index terms to find potentially interesting semantic relationships in large sets of MEDLINE abstracts. Rindflesch et al. (2000) built ARBITER (Assess and Retrieve BInding TErminology), which uses UMLS as domain knowledge and relies on syntactic cues (such as the single verb *bind*) provided by a syntactic parser, to identify and extract molecular binding semantic relations from MEDLINE records. Rindflesch, Libbus, Hristovski, Aronson, and Kilicoglu (2003) also built a natural language processing program, called SemGen, to identify and extract causal relations between genetic phenomena and diseases from MEDLINE records. They were able to achieve 76 percent precision with sample sentences.

Automatic Construction of Case Frames

Text mining using co-occurrence statistics is employed in the automatic construction of case frames. The process has three main stages:

- Constructing "subcategorization frames" (Chomsky, 1965) (i.e., identifying the combination of syntactic constituents or arguments that the verb expects).

- Identifying the selectional restriction for each syntactic constituent or slot (e.g., which semantic class of nouns can be the direct object of the verb).

- Assigning a case role to each syntactic constituent or slot in the case frame.

Typically, statistical collocations are mined from the text collection as a first step to finding the words/phrases and types of words/phrases that tend to co-occur with each verb. Some syntactic preprocessing—part-of-speech tagging, chunking to identify types of phrases, or syntactic parsing—is first performed. Associations between verbs and types of co-occurring syntactic constituents can be used to build subcategorization frames (Basili, Pazienza, & Vindigni, 1997; Brent, 1993; Manning, 1993; Nedellec, 2000). The head nouns of the constituent phrases can be generalized to a semantic class so as to identify the selectional restriction for a slot. This semantic generalization is performed with the aid of a thesaurus or ontology (Framis, 1994; Li & Abe, 1998). If a thesaurus is not available, nouns in the text collection can be clustered according to the context in which they tend to appear. For example, clusters of nouns that tend to co-occur as direct objects of the same verbs can be identified. The noun clusters can be accepted as semantic classes, or a similarity measure between the nouns can be used to generalize the selectional restrictions (Grishman & Sterling, 1994).

Automatic assignment of case role labels to case frame slots is more difficult. To some extent it can be determined by examining the semantic classes of nouns filling the roles. Verbs in the text collection can also be clustered to identify sets of verbs that tend to co-occur with the same nouns. This can help to identify clusters of verbs with similar semantics, an aid to identifying the semantic roles assigned by the verbs (Pereira, Tishby, & Lee, 1993). A more promising approach is to use a machine-learning technique to learn the characteristics of verb-noun combinations for each case role. New verb-noun combinations can be assigned a case role label based on their similarity to prototypical verb-noun combinations for each case role. Wanner (2004) used this approach to extract verb-noun collocations from text and categorize them into one or more of 20 lexical functions. A centroid was computed for each lexical function through the use of training verb-noun examples for each lexical function and the utilization of concept classes in EuroWordNet as features.

Finally, dictionary definitions have also been mined to construct case frames (e.g., Calzolari, 1992).

Semantic Relations in Information Retrieval

Overview

To date, research and development in information retrieval have focused on term and concept matching. Some researchers have, however, explored the possibility of using semantic relations to enhance recall and precision. Recall enhancement—increasing the number of relevant documents retrieved—is usually accomplished through query expansion, in other words, adding alternative terms to the query. Typically, paradigmatic relations, especially synonyms and partial synonyms, are used for query expansion, although syntagmatic relations can be used as well. Terms that are semantically related to each query term are added to the search query using the Boolean disjunction operator *OR*.

Precision enhancement—reducing the proportion of nonrelevant documents retrieved—is accomplished through relation matching. This involves specifying additional relational criteria for retrieval, in other words, the documents retrieved must contain not only the terms/concepts specified in the query but must also express the same relations between the concepts as expressed in the query. The relations are in a sense added to the search by means of the Boolean conjunction operator *AND*. Typically, syntagmatic relations are used in relation matching.

A more precise form of information retrieval is question-answering—answering a user's question with facts or text passages extracted from documents. This requires identifying specific semantic relations between document concepts and concepts in the user's question. The appropriate semantic relation to be used for identifying potential answers in documents is determined by the question type (e.g., definition question, list question, and so forth).

Automatic text summarization extracts the most important information from a document or set of documents, then generates an abridged version for a particular user or task (Mani & Maybury, 1999). This helps users skim through a set of retrieved documents to determine their relevance and potential usefulness. Semantic relations are useful for identifying within a document related concepts and statements that can be compressed, as well as for analyzing the document's discourse structure, which can then be used to identify its central concepts. Multidocument summarization can provide an overview of a set of documents, pointing out information that is common to the document set, information unique to each document, and contradictory statements found in the set. Semantic relations between concepts and statements across the documents (cross-document discourse structure) are useful for multidocument summarization.

In this section, we survey research applying semantic relations to query expansion, precision enhancement, question-answering, and automatic text summarization.

Semantic Relations in Query Expansion

Query expansion with related terms is important for improving recall, although it can improve information retrieval precision as well (Wang, Vandendorpe, & Evens, 1985). Related terms can be taken from a knowledge structure, such as a thesaurus, a taxonomy, a semantic network or an ontology, or from a more informal term association list. As explained earlier, knowledge structures such as thesauri and ontologies distinguish between a few types of semantic relations: minimally, the synonymy relation, the hierarchical relations (IS-A and part-whole), and the associative relation (related term). Such knowledge structures are usually manually constructed, although some are constructed semi-automatically. On the other hand, informal term association lists are often constructed using corpus analysis and co-occurrence statistics. (Two terms are associated if they co-occur in the same document or in close proximity in text more often than can be attributed to pure chance alone.) A commonly used term association measure is the mutual information measure (Church & Hanks, 1989).

Query expansion can be performed either automatically without user intervention or manually by a user selecting appropriate related terms from a thesaurus. The usefulness of query expansion depends on many factors: the size and type of the document collection, whether the searching is performed "free text" or on an indexing field using controlled vocabulary, whether the thesaurus is domain-specific or generic, whether the system is a Boolean or best-match search system, and so forth. Most of the large-scale studies have been conducted on TREC (Text REtrieval Conference) corpora (http://trec.nist.gov), using free-text best-match systems and automatic query expansion (see the chapter on TREC in the present volume). However, manual query expansion on a Boolean search system, featuring controlled vocabulary searches conducted with terms drawn from a domain-specific thesaurus, has been performed by generations of librarians and there is perhaps less doubt as to its usefulness.

Query Expansion Using Term Association

Automatic query expansion using term associations derived from a corpus using co-occurrence statistics has not produced promising results. Sparck Jones (1971) even obtained a decrease in retrieval performance. Peat and Willett (1991) demonstrated that the effectiveness of term association is limited because the similar terms identified by co-occurrence data tend to occur very frequently in the database, and frequently occurring terms are poor at discriminating between relevant and nonrelevant documents.

Some researchers managed to obtain positive results with variations of the standard term association method. Qiu and Frei (1993) obtained positive results with their concept-based query expansion method, in which the query is expanded with terms that are strongly related to *all* the query terms. They suggested that the usual term association methods fail because these tend to add terms that are strongly related only to *individual* query terms.

Chen and Lynch (1992) developed a different association measure and "cluster algorithm" for constructing term association lists. Their work was not strictly on automatic query expansion because their term association file was used to display related terms for the user to select. However, they showed that a word co-occurrence algorithm can produce terms that are semantically related. Ruge (1992) introduced a term association method that made use of head/modifier relations (a kind of syntactic relation). She combined linguistic knowledge and co-occurrence in her experiments to produce linguistically based thesaurus relations.

Grefenstette (1992) and Strzalkowski (1995) made use of second-order term association, specifically, they regarded two terms as related if they each tended to co-occur with a third term and bore the same syntactic relation to it. Grefenstette obtained a small improvement in retrieval effectiveness on a collection of medical abstracts.

Information retrieval researchers participating in the TREC series of conferences have carried out large-scale experiments investigating the usefulness of query expansion for full-text searching in large heterogeneous document collections using state-of-the-art, best-match information retrieval systems. From the TREC experiments, researchers have learned that the most effective method of query expansion using associated terms is pseudo-relevance feedback (also called blind, or local, feedback). This involves using the original query to retrieve an initial ranked list of documents. The terms in the top-ranked documents are weighted in some way and added to the original query, and the retrieval process is repeated with this expanded query (Belkin, Head, Jeng, Kelly, Lin, Park, et al., 1999; Buckley, Singhal, Mitra, & Salton, 1996; Hawking, Thistlewaite, & Craswell, 1998; Kwok & Chan, 1998; Xu & Croft, 1996). In this way, the terms added to the query are related to the query as a whole and not just to individual terms within the query.

More recent work has focused on selecting which documents and words to use. Usually, only the most frequently occurring words are used (Buckley et al., 1996). From the top-ranked documents, Buckley, Mitra, Walz, and Cardie (1998) identified clusters of documents corresponding to different query concepts, selected high frequency words from each cluster, and weighted them appropriately. Xu and Croft (1996) retrieved a ranked list of passages instead of whole documents to make pseudo-relevance feedback more precise. In a later study, Xu and Croft (2000) used an additional criterion: The terms selected from the top-ranked passages should co-occur with query terms in those passages. Terms that co-occur with more query terms are preferred. We hypothesize that even better results can be obtained by considering the semantic relations between the associated terms in these top-ranked documents/passages and the query terms found in the documents.

Query Expansion Using Lexical-Semantic Relations

Lexical-semantic relations can be used to distinguish between different kinds of term associations to use for query expansion. Some researchers have investigated what types of semantic relations are useful for query expansion.

Fox (1980) used 73 classes of lexical relations for query expansion. The lexically related words for each query term were identified manually. Some of the relations (e.g., those between *dog* and *bark,* and *lion* and *Africa*) were syntagmatic and associative relations. Using the SMART best-match retrieval system, Fox found that the most effective retrieval was accomplished by using all categories of relations except the antonym relation. In a follow-up study, Wang et al. (1985) used 44 relations, a different weighting scheme, and a different document collection, as well as constructing a relational thesaurus—something not explicitly done by Fox. The results were comparable to Fox's (1980), indicating that the synonym relation and the broader-narrower term relation are not the only relations that can be employed for query expansion. However, these studies involved only very small document collections using single-domain thesauri.

Using the MEDLINE database and MeSH, Rada and Bicknell (1989) found that automatic query expansion using broader-narrower term relations as well as non-hierarchical relations could improve retrieval effectiveness, if the semantic relations were selected carefully. In another study using the Excerpta Medica database and an enriched

EMTREE thesaurus, Rada, Barlow, Potharst, Zanstra, and Bijstra (1991) found that only when the query explicitly mentioned a particular non-hierarchical relation could the retrieval system make use of the specific relation in the thesaurus to improve document ranking.

Wan, Evens, Wan, and Pao (1997) used a relational thesaurus for automatic indexing in a Chinese information retrieval system. They reported that their relational thesaurus, which employed 11 types of semantic relations, did improve retrieval effectiveness in terms of average precision with both manual and automatic indexing. However, the experiment was based on a small database of only 555 Chinese abstracts in computer and information science, with retrieval based on the index field. However, the thesaurus could be used interactively—users could select terms for query expansion. Abu-Salem (1992) also used an interactive relational thesaurus to improve recall in an Arabic retrieval system.

Greenberg (2001) investigated the effect of different thesaural relationships for query expansion using the ProQuest Controlled Vocabulary on the ABI/Inform database, which was searched via a Boolean retrieval system (the Dialog system). She found that synonyms and narrower terms increased relative recall with a nonsignificant decrease in precision, whereas related terms and broader terms increased relative recall with a statistically significant decrease in precision.

Using the TREC-2 test collection and a best-match retrieval system, Voorhees (1994) performed query expansion with various types of semantic relations encoded in WordNet. Even in a best-case scenario with the expanded terms selected by hand, query expansion did not improve retrieval results for long queries that were relatively complete. On the other hand, short queries, consisting of a single sentence describing the topic of interest, obtained significantly better results with the expansion.

Mandala, Tokunaga, and Tanoka (1999), carried out query expansion with a combination of three different types of thesauri—WordNet, a co-occurrence-based thesaurus, and one based on head-modifier relations. Head-modifier relations include four syntactic relations—subject-verb, verb-object, adjective-noun, and noun-noun relations. The expanded terms were also weighted based on their similarity to all the terms in the original query and to those in all three thesauri. Using the TREC-7 test collection, Mandala et al. found that query expansion with a combination of the three thesauri gave better average precision than was the case when no expansion was used or when it involved only one thesaurus.

Working with a Finnish full-text newspaper database and a Boolean information retrieval system, Kristensen and Järvelin (1990) found that expanding a query with synonyms and near-synonyms improved recall substantially with a small loss of precision. Kristensen (1993) experimented with broader-term, narrower-term, related-term, and synonym relations, and concluded that automatic query expansion using all these relations together improved recall twofold with a small reduction in precision. Using a best-match, full-text retrieval system (INQUERY) and the Finnish newspaper database, Kekalainen and Järvelin (1998) showed that the effect of query expansion depended on how the query was structured. Query expansion worked well with strongly structured queries but was detrimental to weakly structured queries in which, for example, the query terms and the expanded terms were treated as one list of weighted terms. The best results were obtained by expanding with all the relations.

It is clear that query expansion with related terms is crucial for improving information retrieval effectiveness and that in addition to the IS-A or broader-narrower term relations, associative relations are useful for query expansion. However, available experimental results do not suggest that it is beneficial to distinguish between specific types of associative relations. It is possible that different types of semantic relations will prove useful for expanding different queries. Rada et al. (1991) suggested that if a particular associative relation is mentioned in the query, then that relation may be useful for expanding the query. More research is needed to investigate whether specific types of semantic relations are useful for expanding specific types of queries.

A literature survey of the use of thesaural relations in information retrieval was carried out by Evens (2002).

Relation Matching for Precision Enhancement

Relation matching in information retrieval can be performed by the use of either syntactic or semantic relations. A *syntactic relation* is the relation between two words derived from the syntactic structure of the sentence; a *semantic relation* is only partly dependent on the syntactic structure of the sentence. As a semantic relation can be expressed in many syntactic forms, semantic relation matching involves matching across different syntactic relations and can yield more matches than syntactic relation matching.

Most studies on relation matching are on syntactic relations. Croft (1986), Croft, Turtle, and Lewis (1991), Dillon and Gray (1983), Hyoudo, Niimi, and Ikeda (1998), Smeaton and van Rijsbergen (1988) all recorded a small improvement in retrieval effectiveness when syntactic relations in documents and queries were taken into account in the retrieval process. Strzalkowski, Carballo, and Marinescu (1995) obtained an improvement of 20 percent, but their system included other enhancements as well. Smeaton, O'Donnell, and Kelledy (1995) obtained worse results from relation matching (using a tree-matching procedure) than from keyword matching. The retrieval results from syntactic relation matching appear to be no better than the results obtainable using index phrases generated through statistical methods, such as those described by Fagan (1989).

Metzler and Haas (1989), Metzler, Haas, Cosic, and Weise (1990), Schwarz (1990), and Ruge, Schwarz, and Warner (1991) performed syntactic processing to produce dependency trees that indicate which terms modify which other terms. Smeaton and van Rijsbergen (1988) found that the premodifier-head noun relation (e.g., adjective-noun) has a greater impact on retrieval than other relations.

In the 1980s and early 1990s, some researchers developed *conceptual information retrieval systems* that made use of complex linguistic processing and knowledge-based inferencing to extract information from text to store in a semantic representation or knowledge representation system. Examples of such systems are RIME (Berrut, 1990), the patent-claim retrieval system described by Nishida and Takamatsu (1982), SCISOR (Rau, 1987; Rau et al., 1989), and FERRET (Mauldin, 1991). Information retrieval was performed by comparing the information in the store with the semantic representation of the user's query. These systems required extensive domain knowledge, much of which was stored in case frames that specified the participant roles in an event, what types of entities could fill those roles, and what syntactic function each participant would have in the sentence (Fillmore, 1968; Somers, 1987). Because the domain knowledge had to be constructed manually, such systems were necessarily restricted to narrow domains.

The DR-LINK project (Liddy & Myaeng, 1993; Myaeng et al., 1994) investigated general methods for extracting semantic relations for information retrieval using machine-readable versions of the *Longman Dictionary of Contemporary English* (2nd ed.) (1987) and *Roget's International Thesaurus* (3rd ed.) (1962). Case frames were constructed semimanually for all verb entries and senses in the *Longman Dictionary*. However, researchers found few relation matches between queries and documents.

Lu (1990) also did not obtain good retrieval results with case relation matching. Case relations exist between words that occur close together within the same clause. Semantic relations between terms occurring in such close proximity can probably be inferred from their co-occurrence; hence, explicit semantic relation identification probably confers no advantage to retrieval effectiveness.

Gay and Croft (1990) focused on the identification of semantic relations between the members of compound nouns. The knowledge base they used included case frames and associations between entities and events. Although their system correctly interpreted compound nouns 76 percent of the time, it was not deemed likely to yield a substantial improvement in retrieval effectiveness.

Liu (1997) investigated partial relation matching. Instead of trying to match the whole concept-relation-concept triple, he sought to match each individual concept together with the semantic role that the concept has in the sentence. Instead of trying to find matches for "word1 → (relation) → word2", his system sought to find matches for "word1 → (relation)" and "(relation) → word2" separately. Liu used case roles and was able to obtain positive results only for long queries (i.e., abstracts used as queries).

Khoo, Myaeng, and Oddy (2001) developed an automatic method to identify causal relations in text and attempted to match causal relations in documents with those in queries. Causal relation matching did not perform better than word proximity matching within the same sentence. Causal relation matching worked best when one member of the causal relation (either the cause or the effect) was represented as a wildcard that could match any word.

In reviewing six years of TREC experiments (1992–1997), Sparck Jones (2000) and Perez-Carballo and Strzalkowski (2000) concluded that sophisticated natural language processing was not helpful for full-text retrieval. They noted that extracting normalized syntactic phrases (e.g., head-modifier pairs) did not give better results than statistical phrases defined by adjacency and proximity. Sparck Jones (2000) commented that there was a lack of clear evidence that a thesaurus helped in manual query construction because many other factors were involved. "It is therefore impossible to determine whether, for example, a good result is attributable to the use of vocabulary aids or just to spending a lot of time on query formation" (Sparck Jones, 2000, p. 65). She further noted that the use of elaborately structured thesauri had not been proven to be better than using a term association database.

Overall, the use of specific semantic relations either for query expansion or relation matching does not appear to be useful for document retrieval. Perhaps document retrieval is too coarse-grained to require the subtlety of semantic relations, which may be more useful for more refined kinds of information retrieval, such as question-answering.

Question-Answering with Full-Text Documents

The technology for question-answering based on full-text documents is still immature. Current approaches in TREC are focused on term matching and passage extraction. Voorhees (2003) has outlined the general approach to question-answering as comprising three steps: a) determining the expected answer type of the question, b) employing information retrieval methods to retrieve documents or passages likely to contain the answers, and c) performing more refined matching to extract the answer or trim nonrelevant text.

Some researchers have applied information extraction techniques such as pattern matching to extract the final answer from the shortlisted document passages. Paranjpe, Ramakrishnan, and Srinivasan (2004) used WordNet to score document passages using Bayesian inferencing and then used different regular expression patterns to select text segments for different kinds of questions. Harabagiu, Moldovan, Clark, Bowden, Williams, and Bensley (2004) also employed WordNet and information extraction using pattern matching. Gaizauskas, Greenwood, Hepple, Roberts, Saggion, et al. (2004) passed the top-ranked passages retrieved by an information retrieval system to an information extraction system, which converted sentences to a predicate-argument logical form.

Different patterns were used to extract answers for different kinds of questions. Litkowski (2001, 2002) extracted concept-relation-concept triples from both documents and questions, and used relational matches as one of the criteria for ranking sentences.

Semantic Relations in Automatic Text Summarization

Mani and Maybury (1999) have provided a good overview of the use of various kinds of relations in text summarization. Summarization includes three kinds of condensation operations: selection of salient or non-redundant information, aggregation of information, and generalization or abstraction. Each of these operations makes use of relations between terms/concepts and between text passages. They further identified three main approaches to text summarization:

- *The surface-level features approach*, which relies on term frequency statistics, location of a sentence within a text, presence of terms from title or user query, cue words indicating summarizing sentences or important concepts.

- *The entity-level approach*, which models the terms/ concepts in the text and their relationships as a semantic network, with relations between concepts based on similarity, proximity in the text, co-occurrence, thesaural relations, co-reference, syntactic relations, and logical relations.

- *The discourse-level approach*, which models the structure of the text.

Some researchers have adapted information extraction systems for text summarization. Others have used sophisticated natural language processing to convert the text to a semantic representation and then performed summarization using knowledge-based inferencing—similar to the approach used in conceptual information retrieval systems. Text summarization can be performed on individual documents, namely, *single document summarization*, or on a set of documents, namely, *multi-document summarization*.

As Radev, Hovy, and McKeown (2002) noted, most summarization systems perform sentence extraction or passage extraction—identifying sentences/passages in the document containing important information based on surface-level features. Paice (1990) provided an overview of this approach, and argued that both the processing of anaphoric and rhetorical relations in the document and analysis of the text structure are necessary for generating high-quality abstracts. Kupiec, Pedersen, and Chen (1999) and Myaeng and Jang (1999) developed statistical models for assigning a probabilistic score to each document sentence based on the presence of surface features. The models were developed on the basis of a collection of training documents, in which sentences had been manually tagged to indicate good summary sentences. Passage extraction methods have also been applied to multidocument summarization (e.g., Goldstein, Mittal, Carbonell, & Callan, 2000).

Entity-level approaches were adopted by Hovy and Lin (1999), who used WordNet as a thesaurus to generalize the terms, and Boguraev and Kennedy (1999), who made use of cohesion relations (including anaphoric references) between terms. Barzilay and Elhadad (1999) linked up the terms in the text into lexical chains, based on cohesion relations of synonymy, repetition, hypernymy, antonymy, and holonymy. Some of the term relations were derived from WordNet. Sentences were then extracted on the basis of "strong" chains with the aid of a number of heuristics.

Entity-level approaches have also been applied to multidocument summarization. Salton, Singhal, Mitra, and Buckley (1999) constructed a network of related paragraphs based on information retrieval similarity measures. Text units that were strongly connected to other units were considered salient and good candidates for extraction. Mani and Bloedorn (1999) constructed a network of terms and text units based on cohesion relations. Spreading activation was used to identify salient nodes on the basis of connectivity and the strengths of the links. Commonalities and differences between documents were then computed on the basis of the salient nodes for each document.

Marcu (1999, 2000) developed a parser to identify rhetorical relations in text to form a rhetorical structure tree, which was then used to identify important clauses. Each rhetorical relation links two text segments—one text segment is considered the nucleus node representing the central information and the other, the satellite node representing secondary information. Nucleus nodes are considered more salient than satellite nodes, with nucleus nodes linked to higher-level nucleus nodes at the top of the tree considered the most salient. Salience scores were computed for the nodes of the rhetorical tree and used to extract corresponding sentences or clauses to form summaries.

Teufel and Moens (1999) made use of macrolevel text structure, focusing on sections of the document that they called the *argumentative structure* of the text. The document sections were also identified with "global rhetorical relations"—relations of the text segment with respect to the content of the whole document. They used the following roles: background, topic, related work, purpose/problem, solution/method, result, and conclusion/claim. The abstract they created also used this argumentative template and sentences were extracted from the corresponding document section to fill the abstract template.

Strzalkowski, Stein, Wang, and Wise (1999) used a discourse structure of news summaries to combine query-relevant information with related but "out-of-context information." They made use of background-main news relations to identify such out-of-context information.

Radev (2000) introduced a theory of cross-document structure, which can be used to describe the rhetorical structure of a set of related documents. Cross-document structure theory makes use of a multidocument graph to represent text simultaneously at different levels of granularity (words, phrases, sentences, paragraphs, and documents). It contains links representing cross-document semantic relationships among text units, such as equivalence, cross-reference, contradiction, and historical background. Different summaries can be generated from the graph according to user needs by preserving some links in the graph while removing others.

Information extraction techniques have also been applied to text summarization. The SUMMONS system (McKeown & Radev, 1999) used information extraction for multidocument summarization. Information was first extracted from each document to fill a template. When the templates for different documents were merged, operations were performed to identify the following logical relations between templates—change of perspective, contradiction, addition, refinement, agreement, superset, trend, and no information.

The RIPTIDES system (White, Korelsky, Cardie, Ng, Pierce, & Wagstaff, 2001) also used an information extraction system to fill templates for summarization in the natural disasters domain. However, additional potentially relevant information not found in the templates were also extracted from selected sentences and added to the summary to round it off.

Knowledge-based approaches to summarization using a semantic representation of the text were adopted in the SUSY (Fum, Guida, & Tasso, 1985), SCISOR (Rau et al., 1989), and TOPIC systems (Hahn & Reimer, 1999; Reimer & Hahn, 1988). The TOPIC system converted the text into a terminological logic representation scheme. From this representation, "salience operators" extracted concepts, relations, and properties, which were then synthesized into a hierarchical text graph incorporating discourse and concept relations.

Lehnert (1999) proposed an inference-based technique for summarizing narratives based on structural relations around plot units. Primitive plot units, including *problem, success, failure, hidden blessing,* and *mixed blessing,* are building blocks for more complex plot units. The method focuses on affect or emotional states, and the relations between events and affect states. Lehnert listed three affect states: *positive event, negative event,* and *mental state* (neutral affect). The relations between events and affect states include *motivation, actualization, termination,* and *equivalence.* These can be used to build primitive plot units, from which more complex plot units can be derived.

Conclusion

Information retrieval in the 20th century focused on terms, especially nouns, and concepts. We seem to be approaching the limit of what term-based and concept-based approaches can accomplish. For example, in the TREC series of conferences, the ad hoc information retrieval track, once considered the main retrieval task, has been discontinued because of the lack of improvement in participating systems.

We believe that natural language processing and semantic relations, in particular, point the way forward for information retrieval in the 21st century. But as we have seen, semantic relations are subtle things. They are difficult for computer programs to identify and process. Yet human minds process semantic relations effortlessly. Our facility with symbolic processing and semantic relations certainly distinguishes us from machines.

Two factors have retarded progress in the effective use of semantic relations in information processing applications. One is the difficulty of automatically identifying semantic relations in text with accuracy. The other is the difficulty of identifying suitable application areas that require the subtlety of semantic relations. Ad-hoc, full-text document retrieval does not appear to require the use of semantic relations. Coarse-grained methods of term matching, appropriate term weighting and document length normalization, and query expansion with term associations based on term co-occurrence statistics, seem to yield as good a retrieval result as we are likely to get. More promising applications for the use of semantic relations are question-answering, document summarization, and information extraction. Effective text processing and text mining tools for identifying semantic relations in text will help to promote more research in their use.

Further studies of relevance relationships between documents and user information needs can also yield deeper insights into how information retrieval effectiveness can be improved. Although several studies have identified different types of relevance relations and factors that affect relevance judgments, we know little about the thought processes, inferencing mechanisms, and domain knowledge used by humans to judge relevance. We need more in-depth studies of the types of relationships between the user's information need, task, situation, and the document content that determine the relevance and usefulness of the document.

It is also not known whether making fine distinctions between the different types of semantic relations and their properties is useful in information processing applications. Because such fine distinctions are found in both language and human information processing, we hypothesize that they are important in information processing, but it is not clear in what way and for what applications they might be used.

Two exciting new areas for research are the manual and automatic construction of ontologies for various applications and the development of methods for exploiting ontologies effectively in different real-life applications. With the availability of vast quantities of textual documents on the World Wide Web, mining the Web for concepts and relations to build relational knowledge bases and ontologies will become increasingly important.

Other promising research areas not covered in this survey are user profiling and personalization (e.g., Jung, Rim, & Lee, 2004), and special types of text categorization and automated content analysis. For example, in the area of automatic sentiment categorization (categorizing documents into those expressing positive or favorable sentiment versus negative or unfavorable sentiment), Nasukawa and Yi (2003) and Na, Sui, Khoo, Chan, and Zhou (2004) found that it was not sufficient to consider just the sentiment-bearing terms in the text; it was important to determine the subject and object to which the sentiments were linked.

Lack of understanding of semantic relations among information science researchers and practitioners has also retarded its use in information science. One purpose of this *ARIST* chapter has been to pull together information about semantic relations from several disciplines to provide a deeper understanding of their nature and types, as well as to suggest some possible applications.

Acknowledgments

We are very grateful to three anonymous reviewers for their close reading of an earlier draft of this chapter. Any errors that remain are ours alone.

We also gratefully acknowledge the excellent editorial work of Ms. Soon-Kah Liew.

Endnotes

1. *Categories* refer to sets of objects, whereas *concepts* refer to the mental representations of the categories. The terms are often used interchangeably when it is not necessary to distinguish between them.

2. A word within square brackets is a label for a concept. A word within round brackets is a label for a relation. Arrows indicate the direction of the relation.

3. Saussure used the term *associative relations* for what is now known as *paradigmatic relations*.

4. We follow the convention that a question mark indicates a sentence is grammatically or semantically odd; an asterisk indicates a sentence is grammatically or semantically abnormal.

References

Abu-Salem, H. (1992). *A microcomputer-based Arabic bibliographic information retrieval system with relational thesauri.* Unpublished doctoral dissertation, Illinois Institute of Technology.

Ahn, W.-K. (1999). Effect of causal structure on category construction. *Memory & Cognition, 27*(6), 1008–1023.

Ahn, W.-K., & Kim, N. S. (2001). The causal status effect in categorization: An overview. In D. L. Medin (Ed.), *The psychology of learning and motivation: Advances in research and theory, Vol. 40* (pp. 23–65). San Diego, CA: Academic Press.

Aitchison, J., Gilchrist, A., & Bawden, D. (1997). *Thesaurus construction and use: A practical manual* (3rd ed.). London: Aslib.

Alani, H., Kim, S., Millard, D. E., Weal, M. J., Hall, W., Lewis, P. H., et al. (2003). Automatic ontology-based knowledge extraction from Web documents. *IEEE Intelligent Systems, 18*(1), 14–21.

Alba, J. W., & Hasher L. (1983). Is memory schematic? *Psychological Bulletin, 93*, 203–231.

Alonge, A., Calzolari, N., Vossen, P., Bloksma, L., Castellon, I., Marti, M. A., et al. (1998). The linguistic design of the EuroWordNet. *Computers and the Humanities, 32*(2–3), 91–115.

Appelt, D. E., & Israel, D. J. (1999). *Introduction to information extraction technology: IJCAI-99 tutorial.* Retrieved December 31, 2004, from http://www.ai.sri.com/~appelt/ie-tutorial

Aristotle. (1961). *Aristotle's metaphysics* (J. Warrington, Ed. & Trans.). London: J. M. Dent.

Aristotle. (1996). *Physics* (R. Waterfield, Trans.). Oxford, UK: Oxford University Press.

Aronson, A. R. (2001). Effective mapping of biomedical text to the UMLS Metathesaurus: The MetaMap program. In S. Bakken (Ed.), *Proceedings of the 2001 AMIA Annual Symposium* (pp. 17–21). Bethesda, MD: American Medical Informatics Association.

Artale, A., Franconi, E., Guarino, N., & Pazzi, L. (1996). Part–whole relations in object-centered systems: An overview. *Data & Knowledge Engineering, 20*, 347–383.

Asher, R. E. (Ed.). (1994). *The encyclopedia of language and linguistics.* Oxford, UK: Pergamon Press.

Baader, F., Calvanese, D., McGuinness, D., Nardi, D., & Patel-Schneider, P. (2003). *The description logic handbook: Theory, implementation and applications.* Cambridge, UK: Cambridge University Press.

Baker, C. F., Fillmore, C. J., & Cronin, B. (2003). The structure of the Framenet database. *International Journal of Lexicography, 16*(3), 281–296.

Baker, C. F., Fillmore, C. J., & Lowe, J. B. (1998). The Berkeley FrameNet project. *Proceedings of the 17th International Conference on Computational Linguistics* (vol. 1), 86–90.

Barriere, C. (1997). *From a children's first dictionary to a lexical knowledge base of conceptual graphs.* Unpublished doctoral dissertation, Simon Fraser University, British Columbia, Canada.

Barriere, C. (2002). Hierarchical refinement and representation of the causal relation. *Terminology, 8*(1), 91–111.

Barry, C. L. (1994). User-defined relevance criteria: An exploratory study. *Journal of the American Society for Information Science, 45*, 149–159.

Barzilay, R., & Elhadad, M. (1999). Using lexical chains for text summarization. In I. Mani & M. T. Maybury (Eds.), *Advances in automatic text summarization* (pp. 111–122). Cambridge, MA: MIT Press.

Basili, R., Pazienza, M. T., & Vindigni, M. (1997). Corpus-driven unsupervised learning of verb subcategorization frames. In *AI*IA 97: Advances in Artificial Intelligence: 5th Congress of the Italian Association for Artificial Intelligence* (Lecture Notes in Artificial Intelligence, Vol. 1321, pp. 159–170). Berlin: Springer-Verlag.

Bateman, J. A., Fabris, G., & Magnini, B. (1995). The Generalized Upper Model knowledge base: Organization and use. In N. Mars (Ed.), *Second International Conference on Building and Sharing of Very Large-Scale Knowledge Bases (KBKS '95)* (pp. 60–72). Amsterdam: IOS Press.

Bean, C. A. (1998). The semantics of hierarchy: Explicit parent-child relationships in MeSH tree structures. *Proceedings of the Fifth International ISKO Conference,* 133–138.

Bean, C. A., & Green, R. (2001). Relevance relationships. In C. A. Bean & R. Green (Eds.), *Relationships in the organization of knowledge* (pp. 115–132). Dordrecht, NL: Kluwer.

Beekman, J., Callow, J., & Kopesec, M. (1981). *The semantic structure of written communication* (5th ed.). Dallas, TX: Summer Institute of Linguistics.

Belkin, N. J., Head, J., Jeng, J., Kelly, D., Lin, S., Park, S. Y., et al. (1999). Relevance feedback versus local context analysis as term suggestion devices: Rutgers' TREC-8 Interactive Track experience. *Eighth Text Retrieval Conference (TREC-8)* (NIST Special Publication 500–246, pp. 565–574). Retrieved September 1, 2004, from http://trec.nist.gov/pubs/trec8/t8_proceedings.html

Bell, A. (1998). The discourse structure of news stories. In A. Bell & P. Garrett (Eds.), *Approaches to media discourse* (pp. 64–104). Oxford, UK: Blackwell.

Bernauer, J. (1996). Analysis of part–whole relation and subsumption in the medical domain. *Data and Knowledge Engineering, 20,* 405–415.

Berners-Lee, T., Hendler, J., & Lassila, O. (2001). The Semantic Web. *Scientific American, 284*(5/May), 34–43. Retrieved March 11, 2004, from http://www.sciam.com/print_version. cfm?articleID=00048144-10D2-1C70-84A9809EC588EF21

Berrut, C. (1990). Indexing medical reports: The RIME approach. *Information Processing & Management, 26*(1), 93–109.

Blair, D. C. (2003). Information retrieval and the philosophy of language. *Annual Review of Information Science and Technology, 37,* 3–50.

Blake, C., & Pratt, W. (2001). Better rules, fewer features: A semantic approach to selecting features from text. *Proceedings of the 2001 IEEE International Conference on Data Mining,* 59–66.

Boguraev, B., & Kennedy, C. (1999). Salience-based content characterization of text documents. In I. Mani & M. T. Maybury (Eds.), *Advances in automatic text summarization* (pp. 99–110). Cambridge, MA: MIT Press.

Borgelt, C., & Kruse, R. (2002). Induction of association rules: Apriori implementation. *15th Conference on Computational Statistics (Compstat 2002).* Retrieved December 31, 2004, from http://fuzzy.cs.uni-magdeburg.de/~borgelt/papers/cstat_02.pdf

Brent, M. R. (1993). From grammar to lexicon: Unsupervised learning of lexical syntax. *Computational Linguistics, 19*(2), 243–262.

Brewer, W. F., & Nakamura, G. V. (1984). The nature and functions of schemas. In R. S. Wyer & T. K. Srull (Eds.), *Handbook of social cognition* (Vol. 1, pp. 119–160). Hillsdale, NJ: Erlbaum.

Buckley, C., Mitra, M., Walz, J., & Cardie, C. (1998). Using clustering and superconcepts within SMART: TREC 6. *The Sixth Text REtrieval Conference (TREC-6)* (NIST Special Publication 500–240, pp. 107–124). Retrieved September 1, 2004, from http://trec.nist.gov/pubs/trec6/t6_proceedings.html

Buckley, C., Singhal, A., Mitra, M., & Salton, G. (1996). New retrieval approaches using SMART: TREC 4. *The Fourth Text Retrieval Conference (TREC-4)* (NIST Special

Publication 500–236, pp. 25–48). Retrieved September 1, 2004, from http://trec.nist.gov/pubs/trec4/t4_proceedings.html

Butcher, K. R., & Kintsch, W. (2003). Text comprehension and discourse processing. In I. B. Weiner (Ed.), *Handbook of psychology* (Vol. 4, pp. 575–595). Hoboken, NJ: Wiley.

Califf, M., & Mooney, R. (2003). Bottom-up relational learning of pattern matching rules for information extraction. *Journal of Machine Learning Research, 4*, 177–210.

Calzolari, N. (1988). The dictionary and the thesaurus can be combined. In M. W. Evens (Ed.), *Relational models of the lexicon: Representing knowledge in semantic networks* (pp. 75–96). Cambridge, UK: Cambridge University Press.

Calzolari, N. (1992). Acquiring and representing semantic information in a lexical knowledge base. In *Lexical Semantics and Knowledge Representation: First SIGLEX Workshop: Proceedings* (pp. 235–243). Berlin: Springer-Verlag.

Calzolari, N., & Picchi, E. (1989). Acquisition of semantic information from an on-line dictionary. *Proceedings of the 12th International Conference on Computational Linguistics* (Vol. 1), 87–92.

Cann, R. (1993). *Formal semantics: An introduction.* Cambridge, UK: Cambridge University Press.

Caplan, L. J., & Barr, R. A. (1991). The effects of feature necessity and extrinsicity on gradedness of category membership and class inclusion. *British Journal of Psychology, 82*, 427–440.

Caplan, L. J., & Herrmann, D. J. (1993). Semantic relations as graded concepts. *Zeitschrift für Psychologie, 201*, 85–97.

Chaffin. R. (1992). The concept of a semantic relation. In A. Lehrer & E. F. Kittay (Eds.), *Frames, fields, and contrasts: New essays in semantic and lexical organization* (pp. 253–288). Hillsdale, NJ: Erlbaum.

Chaffin, R., & Herrmann, D. J. (1984). The similarity and diversity of semantic relations. *Memory & Cognition, 12*(2), 134–141.

Chaffin, R., & Herrmann, D. J. (1987). Relation element theory: A new account of the representation and processing of semantic relations. In D. Gorfein & R. Hoffman (Eds.), *Learning and memory: The Ebbinghaus Centennial Conference* (pp. 221–251). Hillsdale, NJ: Erlbaum.

Chaffin, R., & Herrmann, D. J. (1988a). Effects of relation similarity on part–whole decisions. *Journal of General Psychology, 115*,131–139.

Chaffin, R., & Herrmann, D. J. (1988b). The nature of semantic relations: A comparison of two approaches. In M. W. Evens (Ed.), *Relational models of the lexicon: Representing knowledge in semantic networks* (pp. 289–334). Cambridge, UK: Cambridge University Press.

Chan, S. W. K., & Franklin, J. (2003). Dynamic context generation for natural language understanding: A multifaceted knowledge approach. *IEEE Transactions on Systems, Man, and Cybernetics—Part A: Systems and Humans, 33*(1), 23–41.

Chang, C.-H., & Lui, S.-C. (2001). IEPAD: Information extraction based on pattern discovery. *Proceedings of the 10th International Conference on World Wide Web*, 681–688.

Chen, H., & Lynch, K. J. (1992). Automatic construction of networks of concepts characterizing document databases. *IEEE Transactions on Systems, Man and Cybernetics, 22*(5), 885–902.

Chomsky, N. (1965). Aspects of the theory of syntax. Cambridge, MA: MIT Press.

Chowdhury, G. G. (2003). Natural language processing. *Annual Review of Information Science and Technology, 37,* 51–90.

Church, K. W., Gale, W., Hanks, P., Hindle, D., & Moon, R. (1994). Lexical substitutability. In B. T. S. Atkins & A. Zampolli (Eds.), *Computational approaches to the lexicon* (pp. 153–177). Oxford, UK: Oxford University Press.

Church, K. W., & Hanks, P. (1989). Word association norms, mutual information and lexicography. *Proceedings of the 27th Annual Meeting of the Association for Computational Linguistics,* 76–83.

Ciravegna, F. (2001). Adaptive information extraction from text by rule induction and generalization. *Proceedings of the Seventeenth International Joint Conference on Artificial Intelligence,* 1251–1256.

Clark, E. V. (1973). What's in a word? On the child's acquisition of semantics in his first language. In T. E. Moore (Ed.), *Cognitive development and the acquisition of language* (pp. 65–110). New York: Academic Press.

Clarke, S. G. D. (2001). Thesaural relationships. In C. A. Bean & R. Green (Eds.), *Relationships in the organization of knowledge* (pp. 37–52). Dordrecht, NL: Kluwer.

Collins, A. M., & Loftus, E. F. (1975). A spreading activation theory of semantic processing. *Psychological Review, 82,* 407–428.

Collins, A. M., & Quillian, M. R. (1969). Retrieval time from semantic memory. *Journal of Verbal Learning and Verbal Behavior, 8,* 240–247.

Cook, W. A. (1989). *Case grammar theory.* Washington, DC: Georgetown University Press.

Croft, W. B. (1986). Boolean queries and term dependencies in probabilistic retrieval models. *Journal of the American Society for Information Science, 37(2),* 71–77.

Croft, W. B., Turtle, H. R., & Lewis, D. D. (1991). The use of phrases and structured queries in information retrieval. *Proceedings of the Fourteenth Annual International ACM/SIGIR Conference on Research and Development in Information Retrieval,* 32–45.

Crombie, W. (1985). *Process and relation in discourse and language learning.* London: Oxford University Press.

Cruse, D. A. (1979). On the transitivity of the part–whole relation. *Journal of Linguistics, 15,* 29–38.

Cruse, D. A. (1986). *Lexical semantics.* Cambridge, UK: Cambridge University Press.

Cruse, D. A. (2000). *Meaning in language: An introduction to semantics and pragmatics.* Oxford, UK: Oxford University Press.

Cruse, D. A. (2002). Hyponymy and its varieties. In R. Green, C. Bean, & S. H. Myaeng (Eds.), *The semantics of relationships: An interdisciplinary perspective* (pp. 3–22). Dordrecht, NL: Kluwer.

Cruse, D. A. (2004). *Meaning in language: An introduction to semantics and pragmatics* (2nd ed.). Oxford, UK: Oxford University Press.

Dick, J. P. (1997). Modeling cause and effect in legal text. In D. Lukose, H. Delugach, M. Keeler, L. Searle, & J. Sowa (Eds.), *Conceptual Structures: Fulfilling Peirce's Dream: Fifth International Conference on Conceptual Structures, ICCS'97* (pp. 244–259). Berlin: Springer-Verlag.

Dillon, M., & Gray, A. S. (1983). FASIT: A fully automatic syntactically based indexing system. *Journal of the American Society for Information Science, 34(2),* 99–108.

Dimino, J. A., Taylor, R. M., & Gersten, R. M. (1995). Synthesis of the research on story grammar as a means to increase comprehension. *Reading & Writing Quarterly: Overcoming Learning Difficulties, 11*(1), 53–72.

Dooley, R. A., & Levinsohn, S. H. (2001). *Analyzing discourse: A manual of basic concepts.* Dallas, TX: SIL International.

Dorr, B. J., Levow, G.-A., & Lin, D. (2002). Construction of a Chinese-English verb lexicon for machine translation and embedded multilingual applications. *Machine Translation, 17*, 99–137.

Dowty, D. (1991). Thematic proto-roles and argument selection. *Language, 67*(3), 547–619.

Dyvik, H. (2004). Translations as semantic mirrors: From parallel corpus to WordNet. *Language and Computers, 49*(1), 311–326.

Eells, E. (1991). *Probabilistic causality.* Cambridge, UK: Cambridge University Press.

Efthimiadis, E. N. (1996). Query expansion. *Annual Review of Information Science and Technology, 31*, 121–188.

Eggins, S. (1994). *An introduction to systemic functional linguistics.* London: Pinter Publishers.

Ehring, D. (1997). *Causation and persistence: A theory of causation.* New York: Oxford University Press.

Evens, M. W. (Ed.). (1988). *Relational models of the lexicon: Representing knowledge in semantic networks.* Cambridge, UK: Cambridge University Press.

Evens, M. W. (2002). Thesaural relations in information retrieval. In R. Green, C. Bean, & S. H. Myaeng (Eds.), *The semantics of relationships: An interdisciplinary perspective* (pp. 143–160). Dordrecht, NL: Kluwer.

Evens, M. W., Litowitz, B. E., Markowitz, J. A., Smith, R. N., & Werner, O. (1980). *Lexical-semantic relations: A comparative survey.* Edmonton, AB: Linguistic Research.

Fagan, J. L. (1989). The effectiveness of a nonsyntactic approach to automatic phrase indexing for document retrieval. *Journal of the American Society for Information Science, 40*(2), 115–132.

Farradane, J. E. L. (1950). A scientific theory of classification and indexing and its practical applications. *Journal of Documentation, 6*(2), 83–99.

Farradane, J. E. L. (1952). A scientific theory of classification and indexing: Further considerations. *Journal of Documentation, 8*(2), 73–92.

Farradane, J. E. L. (1967). Concept organization for information retrieval. *Information Storage & Retrieval, 3*(4), 297–314.

Fellbaum, C. (Ed.). (1998). *WordNet: An electronic lexical database.* Cambridge, MA: MIT Press.

Fellbaum, C. (2002). On the semantics of troponymy. In R. Green, C. Bean, & S. H. Myaeng (Eds.), *The semantics of relationships: An interdisciplinary perspective* (pp. 23–34). Dordrecht, NL: Kluwer.

Fellbaum, C., & Chaffin, R. (1990). Some principles of the organization of the verb lexicon. *Proceedings of the 12th Annual Conference of the Cognitive Science Society*, 420–428.

Fellbaum, C., & Miller, G. A. (1991). Semantic networks of English. *Cognition, 41*, 197–229.

Fensel, D. (2001). Ontologies and electronic commerce (editor's introduction). *IEEE Intelligent Systems, 16*(1), 8–14.

Fillmore, C. J. (1968). The case for case. In E. Bach & R. T. Harms (Eds.), *Universals in linguistic theory* (pp. 1–88). New York: Holt, Rinehart and Wilson.

Fillmore, C. J. (1971). Some problems for case grammar. In R. J. O'Brien (Ed.), *22nd Annual Round Table. Linguistics: Developments of the Sixties—Viewpoints of the Seventies* (Monograph Series on Language and Linguistics, 24) (pp. 35–56). Washington, DC: Georgetown University Press.

Firth, J. R. (1957). *Papers in linguistics 1934–1951.* London: Oxford University Press.

Firth, J. R. (1968). *Selected papers of J. R. Firth 1952–1959* (F. R. Palmer, Ed.). London: Longman.

Fox, E. A. (1980). Lexical relations: Enhancing effectiveness of information retrieval systems. *SIGIR Forum, 15*(3), 5–36.

Framis, F. R. (1994). An experiment on learning appropriate selectional restrictions from a parsed corpus. *Proceedings of the Fifteenth International Conference on Computational Linguistics* (Vol. 2), 769–774.

Freitag, D. (2000). Machine learning for information extraction in informal domains. *Machine Learning, 39*, 169–202.

Fum, D., Guida, G., & Tasso, C. (1985). Evaluating importance: A step towards text summarization. *Proceedings of the 9th International Joint Conference on Artificial Intelligence (IJCAI'85)* (Vol. 2), 840–844.

Gaizauskas, R., Greenwood, M. A., Hepple, M., Roberts, I., Saggion, H., et al. (2004). The University of Sheffield's TREC 2003 Q&A experiments. *The Twelfth Text Retrieval Conference (TREC 2003)* (NIST Special Publication 500–255, pp. 782–790). Retrieved December 31, 2004, from http://trec.nist.gov/pubs/trec12/t12_proceedings.html

Ganter, B., & Wille, R. (1997). *Formal concept analysis: Mathematical foundations* (C. Franzke, Trans.). Berlin: Springer.

Gardin, J.-C. (1965). *SYNTOL.* New Brunswick, NJ: Graduate School of Library Service, Rutgers University.

Gay, L. S., & Croft, W. B. (1990). Interpreting nominal compounds for information retrieval. *Information Processing & Management, 26*(1), 21–38.

Gentner, D. (1983). Structure-mapping: A theoretical framework for analogy. *Cognitive Science, 7*, 155–170.

Gentner, D. (1989). The mechanisms of analogical learning. In S. Vosniadou & A. Ortony (Eds.), *Similarity and analogical reasoning* (pp. 199–241). Cambridge, UK: Cambridge University Press.

Gerstl, P., & Pribbenow, S. (1995). Midwinters, end games, and body parts: A classification of part–whole relations. *International Journal of Human–Computer Studies, 43*, 865–889.

Girju, R. (2002). *Text mining for semantic relations.* Unpublished doctoral dissertation, University of Texas at Dallas.

Girju, R., Badulescu, A., & Moldovan, D. (2003). Learning semantic constraints for the automatic discovery of part–whole relations. *Proceedings of the Human Language Technology Conference*, 80–87.

Glass, A. L., Holyoak, K. J., & Kiger, J. I. (1979). Role of antonymy relations in semantic judgments. *Journal of Experimental Psychology: Human Learning & Memory, 5*, 598–606.

Goldstein, J., Mittal, V., Carbonell, J., & Callan, J. (2000). Creating and evaluating multi-document sentence extract summaries. *Proceedings of the 9th International Conference on Information and Knowledge Management*, 165–172.

Gómez-Pérez, A., Fernandez-Lopez, M., & Corcho, O. (2004). *Ontological engineering.* London: Springer-Verlag.

Green, R. (1995a). The expression of conceptual syntagmatic relationships: A comparative survey. *Journal of Documentation, 51*, 315–338.

Green, R. (1995b). Syntagmatic relationships in index languages: A reassessment. *Library Quarterly, 65*, 365–385.

Green, R. (2001). Relationships in the organization of knowledge: An overview. In C. A. Bean & R. Green (Eds.), *Relationships in the organization of knowledge* (pp. 3–18). Dordrecht, NL: Kluwer.

Green, R. (2002). Internally-structured conceptual models in cognitive semantics. In R. Green, C. Bean, & S. H. Myaeng (Eds.), *The semantics of relationships: An interdisciplinary perspective* (pp. 73–90). Dordrecht, NL: Kluwer.

Green, R., & Bean, C. A. (1995). Topical relevance relationships: An exploratory study and preliminary typology. *Journal of the American Society for Information Science, 46*, 654–662.

Green, R., Bean, C., & Myaeng, S. H. (Eds.). (2002). *The semantics of relationships: An interdisciplinary perspective.* Dordrecht, NL: Kluwer.

Greenberg, J. (2001). Automatic query expansion via lexical-semantic relationships. *Journal of the American Society for Information Science and Technology, 52*(5), 402–415.

Grefenstette, G. (1992). Use of syntactic context to produce term association lists for text retrieval. *Proceedings of the Fifteenth Annual International ACM SIGIR Conference on Research and Development in Information Retrieval*, 89–97.

Grishman, R., & Sterling, J. (1994). Generalizing automatically generated selectional patterns. *Proceedings of the Fifteenth International Conference on Computational Linguistics*, 742–747.

Gruber, T. R. (1993). A translation approach to portable ontology specification. *Knowledge Acquisition, 5*(2), 199–220.

Guarino, N. (1997). Understanding, building and using ontologies. *International Journal of Human–Computer Studies, 46*, 293–310.

Guarino, N., & Giaretta, P. (1995). Ontologies and knowledge bases: Towards a terminological clarification. In N. J. I. Mars (Ed.), *Towards very large knowledge bases: Knowledge building and knowledge sharing* (pp. 25–32). Amsterdam: IOS Press.

Hahn, U., & Reimer, U. (1999). Knowledge-based text summarization: Salience and generalization operators for knowledge base abstraction. In I. Mani & M. T. Maybury (Eds.), *Advances in automatic text summarization* (pp. 215–232). Cambridge, MA: MIT Press.

Halliday, M. A. K., & Hasan, R. (1976). *Cohesion in English.* London: Longman.

Harabagiu, S., Moldovan, D., Clark, C., Bowden, M., Williams, J., & Bensley, J. (2004). Answer mining by combining extraction techniques with abductive. *The Twelfth Text Retrieval Conference (TREC 2003)* (NIST Special Publication 500–255, pp. 375–382). Retrieved December 31, 2004, from http://trec.nist.gov/pubs/trec12/t12_proceedings. html

Harris, R. (1987). *Reading Saussure: A critical commentary on the Cours de linguistique generale.* La Salle, IL: Open Court.

Hausmann, F. J. (1985). Kollokationen im deutschen Wörterbuch: Ein Beitrag zur Theorie des lexikographischen Beispiels. In H. Bergenholtz & J. Mugdan (Eds.), *Lexikographie und Grammatik. Akten des Essener Kolloquiums zur Grammatik im Wörterbuch* (pp. 118–129). Tübingen, Germany: Niemeyer.

Hawking, D., Thistlewaite, P., & Craswell, N. (1998). ANU/ACSys TREC-6 experiments. *The Sixth Text Retrieval Conference (TREC-6)* (NIST Special Publication 500–240, pp. 275–290). Retrieved September 1, 2004, from http://trec.nist.gov/pubs/trec6/t6_proceedings. html

Herrmann, D. J. (1987). Representational forms of semantic relations and the modeling of relation comprehension. In E. van der Meer & J. Hoffmann (Eds.), *Knowledge-aided information processing* (pp. 13–29). Amsterdam: North-Holland.

Herrmann, D. J., & Chaffin, R. (1986). Comprehension of semantic relations as a function of the definitions of relations. In F. Klix & H. Hagendorf (Eds.), *Human memory and cognitive capabilities: Mechanisms and performances* (pp. 311–319). Amsterdam: North-Holland.

Herrmann, D. J., Chaffin, R. J. S., Conti, G., Peters, D., & Robbins, P. H. (1979). Comprehension of antonymy and the generality of categorization models. *Journal of Experimental Psychology: Human Learning and Memory, 5*, 585–597.

Herrmann, D. J., & Raybeck, D. (1981). Similarities and differences in meaning in six cultures. *Journal of Cross-Cultural Psychology, 12*, 194–206.

Hobbs, J. (1985). *On the coherence and structure of discourse* (Technical Report CSLI–85–37) Stanford, CA: Center for the Study of Language and Information, Stanford University.

Hoffmann, J., & Trettin, M. (1980). Organizational effects of semantic relations. In F. Klix & J. Hoffmann (Eds.), *Cognition and memory* (pp. 95–102). Amsterdam: North-Holland.

Holyoak, K. J., & Thagard, P. (1995). *Mental leaps: Analogy in creative thought*. Cambridge, MA: MIT Press.

Hovy, E., & Lin, C. (1999). Automated text summarization in SUMMARIST. In I. Mani & M. T. Maybury (Eds.), *Advances in automatic text summarization* (pp. 81–94). Cambridge, MA: MIT Press.

Hudon, M. (2001). Relationships in multilingual thesauri. In C. A. Bean & R. Green (Eds.), *Relationships in the organization of knowledge* (pp. 67–80). Dordrecht, NL: Kluwer.

Hume, D. (1955). *An enquiry concerning human understanding, and selections from a treatise of human nature*. La Salle, IL: The Open Court Publishing Company

Humphreys, B. L., Lindberg, D. A. B., Schoolman, H. M., & Barnett, G. O. (1998). The Unified Medical Language System: An informatics research collaboration. *Journal of the American Medical Informatics Association, 5*(1), 1–13.

Hyoudo, Y., Niimi, K., & Ikeda, T. (1998). Comparison between proximity operation and dependency operation in Japanese full-text retrieval. *Proceedings of the 21st Annual International ACM SIGIR Conference on Research and Development in Information Retrieval*, 341–342.

International Organization for Standardization. (1986). *Guidelines for the establishment and development of monolingual thesauri* (2nd ed.) (ISO 2788–1986(E)). Geneva: International Standards Organization.

Iris, M. A., Litowitz, B. E., & Evens, M. (1988). Problems of the part–whole relation. In M. W. Evens (Ed.), *Relational models of the lexicon: Representing knowledge in semantic networks* (pp. 261–288). Cambridge, UK: Cambridge University Press.

Jaspars, J. (1983). The process of attribution in common-sense. In M. Hewstone (Ed.), *Attribution theory: Social and functional extensions* (pp. 28–44). Oxford, UK: Blackwell.

Jaspars, J., Hewstone, M., & Fincham, F. D. (1983). Attribution theory and research: The state of the art. In J. Jaspars, F. D. Fincham, & M. Hewstone (Eds.), *Attribution theory and research: Conceptual, developmental and social dimensions* (pp. 3–36). London: Academic Press.

Johansson, S. (1997). Using the English-Norwegian Parallel Corpus—a corpus for contrastive analysis and translation studies. In B. Lewandowska-Tomaszczyk & P. J. Melia (Eds.), *Practical applications in language corpora* (pp. 282–296). Lodz, Poland: Lodz University.

Jones, S. (2002). *Antonymy: A corpus-based perspective*. London: Routledge.

Jung, K.-Y., Rim, K.-W., & Lee, J.-H. (2004). Automatic preference mining through learning user profile with extracted information. In *Structural, Syntactic and Statistical Pattern Recognition: Joint IAPR International Workshops SSPR 2004 and SPR 2004* (Lecture Notes in Computer Science 3138, pp. 815–823). Berlin: Springer-Verlag.

Justeson, J. S., & Katz, S. M. (1991). Co-occurrence of antonymous adjectives and their contexts. *Computational Linguistics, 17*, 1–19.

Justeson, J. S., & Katz, S. M. (1992). Redefining antonymy: The textual structure of a semantic relation. *Literary and Linguistic Computing, 7*(3), 176–184.

Katz, J. J. (1972). *Semantic theory*. New York: Harper and Row.

Keil, F. C. (1989). *Concepts, kinds, and cognitive development*. Cambridge, MA: MIT Press.

Keil, F. C. (2003). Categorisation, causation, and the limits of understanding. *Language and Cognitive Processes, 18*(5/6), 663–692.

Kekalainen, J., & Järvelin, K. (1998). The impact of query structure and query expansion on retrieval performance. *Proceedings of the 21st Annual International ACM SIGIR Conference on Research and Development in Information Retrieval*, 130–135.

Khoo, C. S. G. (1995). *Automatic identification of causal relations in text and their use for improving precision in information retrieval*. Unpublished doctoral dissertation, Syracuse University.

Khoo, C., Chan, S., & Niu, Y. (2002). The many facets of the cause–effect relation. In R. Green, C. Bean, & S. H. Myaeng (Eds.), *The semantics of relationships: An interdisciplinary perspective* (pp. 51–70). Dordrecht, NL: Kluwer.

Khoo, C., Myaeng, S. H., & Oddy, R. (2001). Using cause–effect relations in text to improve information retrieval precision. *Information Processing & Management, 37*(1), 119–145.

Kim, J.-T. (1996). Automatic phrasal pattern acquisition for information extraction from natural language texts. *Journal of KISS (B), Software and Applications, 23*(1), 95–105.

Kim, J.-T., & Moldovan, D. I. (1995). Acquisition of linguistic patterns for knowledge-based information extraction. *IEEE Transactions on Knowledge and Data Engineering, 7*(5), 713–724.

Kishore, J. (1986). *Colon classification: Enumerated & expanded schedules along with theoretical formulations*. New Delhi, India: Ess Publications.

Klix, F. (1980). On structure and function of semantic memory. In F. Klix & J. Hoffmann (Eds.), *Cognition and memory* (pp. 11–25). Amsterdam: North-Holland.

Klix, F. (1986). On recognition processes in human memory. In F. Klix & H. Hagendorf (Eds.), *Human memory and cognitive capabilities: Mechanisms and performances* (pp. 321–338). Amsterdam: North-Holland.

Kounios, J., & Holcomb, P. J. (1992). Structure and process in semantic memory: Evidence from event-related brain potentials and reaction times. *Journal of Experimental Psychology: General, 121*, 459–479.

Kounios, J., Montgomery, E. C., & Smith, R. W. (1994). Semantic memory and the granularity of semantic relations: Evidence from speed-accuracy decomposition. *Memory & Cognition, 22*(6), 729–741.

Kristensen, J. (1993). Expanding end-user's query statements for free text searching with a search-aid thesaurus. *Information Processing & Management, 29*(6), 733–744.

Kristensen, J., & Järvelin, K, (1990). The effectiveness of a searching thesaurus in free-text searching in a full-text database. *International Classification, 17*(2), 77–84.

Kukla, F. (1980). Componential analysis of the recognition of semantic relations between concepts. In F. Klix & J. Hoffmann (Eds.), *Cognition and memory* (pp. 169–176). Amsterdam: North-Holland.

Kupiec, J., Pedersen, J., & Chen, F. (1999). Trainable document summarizer. In I. Mani & M. T. Maybury (Eds.), *Advances in automatic text summarization* (pp. 55–60). Cambridge, MA: MIT Press.

Kwok, K. L., & Chan, M. (1998). Improving two-stage ad-hoc retrieval for short queries. *Proceedings of the 21st Annual International ACM SIGIR Conference on Research and Development in Information Retrieval*, 250–256.

Lakoff, G. (1993). The contemporary theory of metaphor. In A. Ortony (Ed.), *Metaphor and thought* (2nd ed.) (pp. 202–251). Cambridge, UK: Cambridge University Press.

Lambrix, P. (2000). *Part–whole reasoning in an object-centered framework* (Lecture Notes in Artificial Intelligence 1771). Berlin: Springer-Verlag.

Lancaster, F. W. (1986). *Vocabulary control for information retrieval*. Arlington, VA: Information Resources Press.

Landis, T. Y., Herrmann, D. J., & Chaffin, R. (1987). Developmental differences in the comprehension of semantic relations. *Zeitschrift für Psychologie, 195*, 129–139.

Lang, R. R. (2003). Story grammars: Return of a theory. In M. Mateas & P. Sengers (Eds.), *Narrative intelligence* (pp. 199–212). Amsterdam: John Benjamins.

Lavrac, N., & Dzeroski, S. (1994). *Inductive logic programming: Techniques and applications*. New York: Ellis Horwood.

Lee, C. H., Na, J. C., & Khoo, C. (2003). Ontology learning for medical digital libraries. *6th International Conference on Asian Digital Libraries, ICADL 2003* (Lecture Notes in Computer Science 2911), 302–305.

Lee, J. W. T., Chan, S. C. F., & Yeung, D. S. (1995). Modelling constraints in part structures. *Computers & Industrial Engineering, 28*(3), 645–657.

Lehmann, F. (Ed.). (1992). *Semantic networks in artificial intelligence*. Oxford, UK: Pergamon Press.

Lehnert, W. (1991). Symbolic/subsymbolic sentence analysis: Exploiting the best of two worlds. In J. Barnden & J. Pollack (Eds.), *Advances in connectionist and neural computation theory* (Vol. 1, pp. 135–164). Norwood, NJ: Ablex.

Lehnert, W. G. (1999). Plot units: A narrative summarization strategy. In I. Mani & M. T. Maybury (Eds.), *Advances in automatic text summarization* (pp. 177–214). Cambridge, MA: MIT Press.

Lehrer, A., & Lehrer, K. (1982). Antonymy. *Linguistics and Philosophy, 5*, 483–501.

Lenat, D., Miller, G., & Yokoi, T. (1995). CYC, WordNet, and EDR: Critiques and responses. *Communications of the ACM, 38*(11), 45–48.

Levin, B. (1993). *English verb classes and alternations: A preliminary investigation.* Chicago: University of Chicago Press.

Levy, F. (1967). On the relative nature of relational factors in classifications. *Information Storage & Retrieval, 3*(4), 315–329.

Li, H., & Abe, N. (1998). Generalizing case frames using a thesaurus and the MDL principle. *Computational Linguistics, 24*(2), 217–244.

Liddy, E. D., & Myaeng, S. H. (1993). DR-LINK's linguistic-conceptual approach to document detection. *The First Text REtrieval Conference (TREC-1)* (NIST Special Publication 500-207, pp. 1–20).

Litkowski, K. C. (2001). Syntactic clues and lexical resources in question-answering. *Ninth Text REtrieval Conference (TREC-9)* (NIST special publication 500-249, pp. 157–166).

Litkowski, K. C. (2002). CL research experiments in TREC-10 question answering. *Tenth Text Retrieval Conference (TREC 2001)* (NIST special publication 500-250, pp. 122–131). Retrieved December 31, 2004, from http://trec.nist.gov/pubs/trec10/t10_proceedings. html

Liu, G. Z. (1997). Semantic vector space model: Implementation and evaluation. *Journal of the American Society for Information Science, 48*(5), 395–417.

Liu, M. (1993). Progress in documentation: The complexities of citation practice: A review of citation studies. *Journal of Documentation, 49,* 370–408.

Longacre, R. E. (1996). *The grammar of discourse* (2nd ed.). New York: Plenum Press.

Longman dictionary of contemporary English. (1987). (2nd ed.) Edinburgh Gate, Harlow: Longman.

Longman dictionary of contemporary English. (1995). (3rd ed.) Edinburgh Gate, Harlow: Pearson Education.

Lu, X. (1990). *An application of case relations to document retrieval.* Unpublished doctoral dissertation, University of Western Ontario.

Lyons, J. (1968). *Introduction to theoretical linguistics.* Cambridge, UK: Cambridge University Press.

Lyons, J. (1977). *Semantics.* Cambridge, UK: Cambridge University Press.

Lyons, J. (1981). *Language and linguistics.* Cambridge, UK: Cambridge University Press.

Lyons, J. (1995). *Linguistic semantics: An introduction.* Cambridge, UK: Cambridge University Press.

Mackie, J. L. (1980). *The cement of the universe: A study of causation.* Oxford, UK: Oxford University Press.

Mandala, R., Tokunaga, T., & Tanoka, H. (1999). Combining multiple evidence from different types of thesaurus for query expansion. *Proceedings of the 22nd Annual International ACM/SIGIR Conference on Research and Development in Information Retrieval,* 191–197.

Mandler, J. M. (1987). On the psychological reality of story structure. *Discourse Processes, 10,* 1–29.

Mandler, J. M., & Johnson, N. S. (1977). Remembrance of things parsed: Story structure and recall. *Cognitive Psychology, 9*(1), 111–151.

Mani, I., & Bloedorn, E. (1999). Summarizing similarities and differences among related documents. In I. Mani, & M. T. Maybury (Eds.), *Advances in automatic text summarization* (pp. 357–380). Cambridge, MA: MIT Press.

Mani, I., & Maybury, M. T. (Eds.). (1999). *Advances in automatic text summarization.* Cambridge, MA: MIT Press.

Mann, W. C., Matthiessen, C. M. I. M., & Thompson, S. A. (1992). Rhetorical structure theory and text analysis. In W. C. Mann & S. A. Thompson (Eds.), *Discourse description: Diverse linguistic analyses of a fund-raising text* (pp. 39–78). Amsterdam: John Benjamins.

Mann, W. C., & Thompson, S. A. (1988). Rhetorical structure theory: Toward a functional theory of text organization. *Text, 8*(3), 243–281.

Mann, W. C., & Thompson, S. A. (1989). Rhetorical structure theory: A theory of text organization. (Information Sciences Institute Research Report 87-190). Marina del Rey, CA: The Institute.

Manning, C. D. (1993). Automatic acquisition of a large subcategorization dictionary from corpora. *31st Annual Meeting of the Association for Computational Linguistics,* 235–242.

Marcu, D. (1999). Discourse trees are good indicators of importance in text. In I. Mani & M. T. Maybury (Eds.), *Advances in automatic text summarization* (pp. 123–136). Cambridge, MA: MIT Press.

Marcu, D. (2000). *The theory and practice of discourse parsing and summarization.* Cambridge, MA: MIT Press.

Markowitz, J. (1988). An exploration into graded set membership. In M. W. Evens (Ed.), *Relational models of the lexicon: Representing knowledge in semantic networks* (pp. 239–260). Cambridge, UK: Cambridge University Press.

Markowitz, J., Nutter, T., & Evens, M. (1992). Beyond IS-A and part–whole: More semantic network links. In F. Lehmann & E. Y. Rodin (Eds.), *Semantic networks in artificial intelligence* (pp. 377–390). Oxford, UK: Pergamon.

Martin, P. (1995). Using the WordNet concept catalog and a relation hierarchy for knowledge acquisition. In *Proceedings of Peirce '95: 4th International Workshop on Peirce, University of California, Santa Cruz.* Retrieved December 31, 2004, from http://www.inria.fr/acacia/Publications/1995/peirce95phm.ps.Z

Martin, P. (1996). *The CGKAT top-level relation ontology.* Retrieved December 31, 2004, from http://www-sop.inria.fr/acacia/personnel/phmartin/ontologies/reCGKAT.html

Mauldin, M. L. (1991). Retrieval performance in FERRET: A conceptual information retrieval system. *Proceedings of the Fourteenth Annual International ACM/SIGIR Conference on Research and Development in Information Retrieval,* 347–355.

McKeon, R. (Ed.). (1941/2001). *The basic works of Aristotle.* New York: Random House.

McKeown, K., & Radev, D. R. (1999). Generating summaries of multiple news articles. In I. Mani & M. T. Maybury (Eds.), *Advances in automatic text summarization* (pp. 381–390). Cambridge, MA: MIT Press.

McNamara, T. P., & Holbrook, J. B. (2003). Semantic memory and priming. In I. B. Weiner (Ed.), *Handbook of psychology* (Vol. 4, pp. 447–474). Hoboken, NJ: Wiley.

Mel'cuk, I. (1996). Lexical functions: A tool for the description of lexical relations in a lexicon. In L. Wanner (Ed.), *Lexical functions in lexicography and natural language processing* (pp. 37–102). Amsterdam: John Benjamins.

Mel'cuk, I. A. (1988). *Dependency syntax: Theory and practice.* Albany: State University of New York Press.

Mellor, D. H. (1995). *The facts of causation*. London: Routledge.

Metzler, D. P., & Haas, S. W. (1989). The Constituent Object Parser: Syntactic structure matching for information retrieval. *ACM Transactions on Information Systems, 7*(3), 292–316.

Metzler, D. P., Haas, S. W., Cosic, C. L., & Weise, C. A. (1990). Conjunction, ellipsis, and other discontinuous constituents in the constituent object parser. *Information Processing & Management, 26*(1), 53–71.

Mill, J. S. (1974). *A system of logic: Ratiocinative and inductive* (books 1V–V1 and appendices). Toronto: University of Toronto Press.

Miller, G. A. (1995). WordNet: A lexical database. *Communications of the ACM, 38*(11), 39–41.

Miller, G. A., & Fellbaum, C. (1991). Semantic networks of English. *Cognition, 41*, 197–229.

Milstead, J. L. (2001). Standards for relationships between subject indexing terms. In C. A. Bean & R. Green (Eds.), *Relationships in the organization of knowledge* (pp. 53–66). Dordrecht, NL: Kluwer.

Minker, W., Bennacef, S., & Gauvain, J.-L. (1996). A stochastic case frame approach for natural language understanding. *Proceedings ICSLP 96: Fourth International Conference on Spoken Language Processing* (Vol. 2), 1013–1016.

Minsky, M. (1975). A framework for representing knowledge. In P. Winston (Ed.), *The psychology of computer vision* (pp. 211–280). New York: McGraw-Hill.

Mitchell, T. M. (1997). *Machine learning*. New York: McGraw-Hill.

Murphy, G. L., & Medin, D. L. (1985). The role of theories in conceptual coherence. *Psychological Review, 92*(3), 289–316.

Murphy, M. L. (2003). *Semantic relations and the lexicon: Antonymy, synonymy, and other paradigms*. Cambridge, UK: Cambridge University Press.

Muslea, I. (1999). Extraction patterns for information extraction tasks: A survey. In *AAAI–99 Workshop on Machine Learning for Information Extraction* (AAAI Workshop Technical Report WS-99-11, pp. 1–6). Menlo Park, CA: AAAI Press. Retrieved December 31, 2004, from http://www.isi.edu/muslea/ RISE/ML4IE

Myaeng, S. H., & Jang, D. (1999). Development and evaluation of a statistically based document summarization system. In I. Mani & M. T. Maybury (Eds.), *Advances in automatic text summarization* (pp. 61–70). Cambridge, MA: MIT Press.

Myaeng, S. H., Khoo, C., & Li, M. (1994). Linguistic processing of text for a large-scale conceptual information retrieval system. *Conceptual Structures: Current Practices: Second International Conference on Conceptual Structures, ICCS '94*, 69–83.

Myaeng, S. H., & McHale, M. L. (1992). Toward a relation hierarchy for information retrieval. *Advances in Classification Research: Proceedings of the 2nd ASIS SIG/CR Classification Research Workshop*, 101–113.

Na, J. C., Sui, H., Khoo, C., Chan, S., & Zhou, Y. (2004). Effectiveness of simple linguistic processing in automatic sentiment classification of product reviews. *Proceedings of the Eighth International ISKO Conference*, 49–54.

Nasukawa, T., & Yi, J. (2003). Sentiment analysis: Capturing favorability using natural language processing. In *Proceedings of the International Conference on Knowledge Capture* (pp. 70–77). New York: ACM.

National Information Standards Organization. (1994). *Guidelines for the construction, format, and management of monolingual thesauri*. Bethesda, MD: NISO Press.

(ANSI/NISO Z39.19–1993). Retrieved September 30, 2004, from http://www.niso.org/standards/resources/Z39-19.html

Nedellec, C. (2000). Corpus-based learning of semantic relations by the ILP system, Asium. In *Learning Language in Logic* (Lecture Notes in Artificial Intelligence, Vol. 1925, pp. 259–278). Berlin: Springer-Verlag.

Neelameghan, A. (1998). Lateral relations and links in multi-cultural, multimedia databases in the spiritual and religious domains: Some observations. *Information Studies, 4,* 221–246.

Neelameghan, A. (2001). Lateral relationships in multicultural, multilingual databases in the spiritual and religious domains: The OM Information Service. In C. A. Bean & R. Green (Eds.), *Relationships in the organization of knowledge* (pp. 185–198). Dordrecht, NL: Kluwer.

Neelameghan, A., & Maitra, R. (1978). *Non-hierarchical associative relationships among concepts: Identification and typology* (part A of FID/CR report no. 18). Bangalore, India: FID/CR Secretariat.

Nishida, F., & Takamatsu, S. (1982). Structured-information extraction from patent-claim sentences. *Information Processing & Management, 18*(1), 1–13.

Noy, N. F., & Hafner, C. D. (1997). The state of the art in ontology design: A survey and comparative review. *AI Magazine, 18*(3), 53–74. Retrieved March 11, 2004 from http://www.aaai.org/Library/Magazine/Vol18/18-03/vol18-03.html

Ogden, C. K. (1967). *Opposition: A linguistic and psychological analysis.* Bloomington: Indiana University Press.

Olson, M. W., & Gee, T. C. (1988). Understanding narratives: A review of story grammar research. *Childhood Education, 64*(5), 302–306.

Owens, D. (1992). *Causes and coincidences.* Cambridge, UK: Cambridge University Press.

Paice, C. D. (1990). Constructing literature abstracts by computer: Techniques and prospects. *Information Processing & Management, 26*(1), 171–186.

Paranjpe, D., Ramakrishnan, G., & Srinivasan, S. (2004). Passage scoring for question answering via Bayesian inference on lexical relations. *Twelfth Text Retrieval Conference (TREC 2003)* (NIST Special Publication 500-255, pp. 305–310). Retrieved December 31, 2004, from http://trec.nist.gov/pubs/trec12/t12_proceedings.html

Park, H. (1997). Relevance of science information: Origins and dimensions of relevance and their implications to information retrieval. *Information Processing & Management, 33*(3), 339–352.

Peat, H. J., & Willett, P. (1991). The limitations of term co-occurrence data for query expansion in document retrieval systems. *Journal of the American Society for Information Science, 42*(5), 378–383.

Pereira, F., Tishby, N., & Lee, L. (1993). Distributional clustering of English words. *31st Annual Meeting of the Association for Computational Linguistics*, 183–190.

Perez-Carballo, J., & Strzalkowski, T. (2000). Natural language information retrieval: Progress report. *Information Processing & Management 36*, 155–178.

Pribbenow, S. (2002). Meronymic relationships: From classical mereology to complex part–whole relations. In R. Green, C. Bean, & S. H. Myaeng (Eds.), *The semantics of relationships: An interdisciplinary perspective* (pp. 35–50). Dordrecht, NL: Kluwer.

Priss, U. (1999). Efficient implementation of semantic relations in lexical databases. *Computational Intelligence, 15*(1), 79–87.

Qiu, Y., & Frei, H. P. (1993). Concept Based Query Expansion. *Proceedings of the Sixteenth Annual International ACM SIGIR Conference on Research and Development in Information Retrieval*, 160–169.

Quillian, M. R. (1967). Word concepts: A theory and simulation of some basic semantic capabilities. *Behavioral Science, 12*, 410–430.

Quillian, M. R. (1968). Semantic memory. In M. L. Minsky (Ed.), *Semantic information processing* (pp. 227–270). Cambridge, MA: MIT Press.

Quine, W. V. (1982). *Methods of logic* (4th ed.). Cambridge, MA: Harvard University Press.

Quinlan, J. R. (1993). *C4.5: Programs for Machine Learning*. San Mateo, CA: Morgan Kaufmann.

Rada, R., Barlow, J., Potharst, J., Zanstra, P., & Bijstra, D. (1991). Document ranking using an enriched thesaurus. *Journal of Documentation, 47*(3), 240–253.

Rada, R., & Bicknell, E. (1989). Ranking documents with a thesaurus. *Journal of the American Society for Information Science, 40*(5), 304–310.

Radev, D. (2000). A common theory of information fusion from multiple text sources step one: Cross-document structure. *Proceedings of the 1st SIGdial Workshop on Discourse and Dialogue*. Retrieved December 31, 2004, from http://www.sigdial.org/workshops/workshop1/proceedings/radev.pdf

Radev, D. R., Hovy, E., & McKeown, K. (2002). Introduction to the special issue on summarization. *Computational Linguistics, 28*(4), 399–408.

Raitt, D. I. (1980). Recall and precision devices in interactive bibliographic search and retrieval systems. *Aslib Proceedings, 37*, 281–301.

Ranganathan, S. R. (1965). *The Colon Classification*. New Brunswick, NJ: Graduate School of Library Service, Rutgers University.

Rau, L. (1987). Knowledge organization and access in a conceptual information system. *Information Processing & Management, 23*(4), 269–283.

Rau, L. F., Jacobs, P. S., & Zernik, U. (1989). Information extraction and text summarization using linguistic knowledge acquisition. *Information Processing & Management, 25*(4), 419–428.

Raybeck, D., & Herrmann, D. (1990). A cross-cultural examination of semantic relations. *Journal of Cross-Cultural Psychology, 21*(4), 452–473.

Rehder, B. (2003). Categorization as causal reasoning. *Cognitive Science, 27*, 709–748.

Rehder, B., & Hastie, R. (2001). Causal knowledge and categories: The effects of causal beliefs on categorization, induction, and similarity. *Journal of Experimental Psychology: General, 130*(3), 323–360.

Reimer, U., & Hahn, U. (1988). Text condensation as knowledge base abstraction. *Proceedings of the 4th Conference on Artificial Intelligence Applications (CAIA'88)*, 338–344.

Riloff, E. (1993). Automatically constructing a dictionary for information extraction tasks. *Proceedings of the Eleventh National Conference on Artificial Intelligence*, 811–816.

Riloff, E. (1996). Automatically generating extraction patterns from untagged text. *Proceedings of the Thirteenth National Conference on Artificial Intelligence (AAAI–96)* (Vol. 2), 1044–1049.

Rindflesch, T. C., & Aronson, A. R. (2002). Semantic processing for enhanced access to biomedical knowledge. In V. Kashyap & L. Shklar (Eds.), *Real world Semantic Web applications* (pp. 157–172). Amsterdam: IOS Press. Retrieved December 31, 2004, from http://nls5.nlm.nih.gov/pubs/semwebapp.5a.pdf

Rindflesch, T. C., Jayant, R., & Lawrence, H. (2000). Extracting molecular binding relationships from biomedical text. *Proceedings of the 6th Applied Natural Language Processing Conference*, 188–195.

Rindflesch, T. C., Libbus, B., Hristovski, D., Aronson, A. R., & Kilicoglu, H. (2003). Semantic relations asserting the etiology of genetic diseases. *Proceedings of 2003 AMIA Annual Symposium*, 554–558. Retrieved December 31, 2004, from http://skr.nlm.nih.gov/papers/references/sgn_amia03.final.pdf

Ringland, G. A., & Duce, D. A. (Eds.). (1988). *Approaches to knowledge representation: An introduction*. Letchworth, Herts.: Research Studies Press.

Roget's international thesaurus (3rd ed.). (1962). New York: Thomas Y. Crowell.

Romney, A. K., Moore, C. C., & Rusch, C. D. (1997). Cultural universals: Measuring the semantic structure of emotion terms in English and Japanese. *Proceedings of the National Academy of Sciences of the United States, 94*, 5489–5494.

Rosner, M., & Somers, H. L. (1980). Case in linguistics and cognitive science. *UEA Papers in Linguistics 13*, 1–29.

Roth, D., & Yih, W. (2001). Relational learning via propositional algorithms: An information extraction case study. *Proceedings of the Seventeenth International Joint Conference on Artificial Intelligence*, 1257–1263.

Ruge, G. (1992). Experiments on linguistically-based term associations. *Information Processing & Management, 28*(3), 317–332.

Ruge, G., Schwarz, C., & Warner, A. J. (1991). Effectiveness and efficiency in natural language processing for large amounts of text. *Journal of the American Society for Information Science, 42*(6), 450–456.

Rumelhart, D. E. (1975). Notes on a schema for stories. In D. G. Bobrow & A. Collins (Eds.), *Representation and understanding: Studies in cognitive science* (pp. 211–236). New York: Academic Press.

Rumelhart, D. E., & Ortony, A. (1977). The representation of knowledge in memory. In R. C. Anderson, R. J. Spiro, & W. E. Montague (Eds.), *Schooling and the acquisition of knowledge* (pp. 99–135). Hillsdale, NJ: Erlbaum.

Salmon, W. (1984). *Scientific explanation and the causal structure of the world*. Princeton, NJ: Princeton University Press.

Salton, G., Singhal, A., Mitra, M., & Buckley, C. (1999). Automatic text structuring and summarization. In I. Mani & M. T. Maybury (Eds.), *Advances in automatic text summarization* (pp. 341–356). Cambridge, MA: MIT Press.

Sattler, U. (1995). A concept language for an engineering application with part–whole relations. In A. Borgida, M. Lenzerini, D. Nardi, & B. Nebel (Eds.), *Proceedings of the 1995 International Workshop on Description Logics* (pp. 119–123). Rome: Universita degli Studi di Roma "La Sapienza", Dipartimento di Informatica e Sistemistica.

Saussure, F. de. (1959). *Course in general linguistics* (C. Bally & A. Sechehaye, Eds.; W. Baskin, Trans). New York: McGraw-Hill.

Schamber, L. (1991). *Users' criteria for evaluation in multimedia information seeking and use situations*. Unpublished doctoral dissertation, Syracuse University.

Schamber, L. (1994). Relevance and information behavior. *Annual Review of Information Science and Technology, 29*, 3–48.

Schank, R. C. (1982). *Dynamic memory*. New York: Cambridge University Press.

Schank, R. C., & Abelson, R. P. (1977). *Scripts, plans, goals, and understanding*. Hillsdale, NJ: Erlbaum.

Schneider, P., & Winship, S. (2002). Judgments of fictional story quality. *Journal of Speech, Language, & Hearing Research, 45*(2), 372–383.

Schwarz, C. (1990). Automatic syntactic analysis of free text. *Journal of the American Society for Information Science, 41*(6), 408–417.

Smadja, F. (1993). Retrieving collocations from text: Xtract. *Computational Linguistics, 19*(1), 143–177.

Smeaton, A. F., O'Donnell, R., & Kelledy, F. (1995). Indexing structures derived from syntax in TREC-3: System description. *Overview of the Third Text REtrieval Conference (TREC-3)* (NIST Special Publication 500-225, pp. 55–67).

Smeaton, A. F., & van Rijsbergen, C. J. (1988). Experiments on incorporating syntactic processing of user queries into a document retrieval strategy. *11th Annual International ACM/SIGIR Conference on Research and Development in Information Retrieval*, 31–51.

Smith, E. E. (1978). Theories of semantic memory. In W. K. Estes (Ed.), *Handbook of learning and cognitive processes. Vol. 4, Linguistic functions in cognitive theory* (pp. 1–56). Potomac, MD: Erlbaum.

Smith, R. N. (1981). On defining adjectives, part III. *Dictionaries: Journal of the Dictionary Society of North America, 3*, 28–38.

Soderland, S. (1997). *Learning text analysis rules for domain-specific natural language processing*. Unpublished doctoral dissertation (Technical report UM–CS–1996–087), University of Massachusetts, Amherst.

Soderland, S. (1999). Learning information extraction rules for semi-structured and free text. *Machine Learning, 34* (1–3), 233–272.

Soderland, S., Fisher, D., Aseltine, J., & Lehnert, W. (1996). Issues in inductive learning of domain-specific text extraction rules. In S. Wermter, E. Riloff, & G. Scheler (Eds.), *Connectionist, Statistical and Symbolic Approaches to Learning for Natural Language Processing* (pp. 290–301). Berlin: Springer-Verlag.

Somers, H. L. (1987). *Valency and case in computational linguistics*. Edinburgh, UK: Edinburgh University Press.

Sosa, E., & Tooley, M. (Eds.). (1993). *Causation*. Oxford, UK: Oxford University Press.

Sowa, J. F. (1984). *Conceptual structures: Information processing in mind and machine*. Reading, MA: Addison-Wesley.

Sowa, J. F. (2000). *Knowledge representation: Logical, philosophical, and computational foundations*. Pacific Grove, CA: Brooks/Cole.

Sowa, J. F. (2001). *Roles and relations*. Retrieved December 31, 2004, from http://www.jfsowa.com/ontology/roles.htm

Sparck Jones, K. (1971). *Automatic keyword classification for information retrieval*. London: Butterworth.

Sparck Jones, K. (2000). Further reflections on TREC. *Information Processing & Management, 36*, 37–85.

Spellman, B. A., Holyoak, K., & Morrison, R. G. (2001). Analogical priming via semantic relations. *Memory & Cognition, 29*(3), 383–393.

Srinivasan, P., & Rindflesch, T. C. (2002). Exploring text mining from MEDLINE. In *Proceedings of the 2002 AMIA Annual Symposium* (pp. 722–726). Bethesda, MD:

American Medical Informatics Association. Retrieved December 31, 2004 from http://nls5.nlm.nih.gov/pubs/amia02-TR2.pdf

Stasio, T., Herrmann, D. J., & Chaffin, R. (1985). Predictions of relation similarity according to relation definition theory. *Bulletin of the Psychonomic Society, 23*, 5–8.

Stephens, L. M., & Chen, Y. F. (1996). Principles for organizing semantic relations in large knowledge bases. *IEEE Transactions on Knowledge and Data Engineering, 8*(3), 492–496.

Strzalkowski, T. (1995). Natural language information retrieval. *Information Processing & Management, 31*(3), 397–417.

Strzalkowski, T., Carballo, J. P., & Marinescu, M. (1995). Natural language information retrieval: TREC-3 report. *Overview of the Third Text REtrieval Conference (TREC-3)* (NIST Special Publication 500–225, pp. 39–53).

Strzalkowski, T., Stein, G., Wang, J., & Wise, B. (1999). A robust practical text summarizer. In I. Mani & M. T. Maybury (Eds.), *Advances in automatic text summarization* (pp. 137–154). Cambridge, MA: MIT Press.

Takemura, T., & Ashida, N. (2002). A study of the medical record interface to natural language processing. *Journal of Medical Systems, 26*(2), 79–87.

Tang, R., & Solomon, P. (2001). Use of relevance criteria across stages of document evaluation: On the complementarity of experimental and naturalistic studies. *Journal of the American Society for Information Science and Technology, 52*(8), 676–685.

Terenziani, P., & Torasso, P. (1995). Time, action-types, and causation: An integrated analysis. *Computational Intelligence, 11*(3), 529–552.

Teufel, S., & Moens, M. (1999). Argumentative classification of extracted sentences as a first step towards flexible abstracting. In I. Mani & M. T. Maybury (Eds.), *Advances in automatic text summarization* (pp. 155–175). Cambridge, MA: MIT Press.

Trybula, W. J. (1999). Text mining. *Annual Review of Information Science and Technology, 34*, 385–419.

Turner, M. (1993). An image-schematic constraint on metaphor. In R. A. Geiger & B. Rudzka-Ostyn (Eds.), *Conceptualizations and mental processing in language* (pp. 291–306). Berlin: Mouton de Gruyter.

Tversky, B. (1990). Where partonomies and taxonomies meet. In S. L. Tsohatzidis (Ed.), *Meanings and prototypes: Studies in linguistic categorization* (pp. 334–344). New York: Routledge.

U.S. National Library of Medicine. (2004). *Unified Medical Language System: Documentation.* Retrieved December 31, 2004 from http://www.nlm.nih.gov/research/umls/documentation.html

Uschold, M. (1996). The use of the typed lambda calculus for guiding naïve users in the representation and acquisition of part–whole knowledge. *Data and Knowledge Engineering, 20*, 385–404.

Van Dijk, T. A. (1972). *Some aspects of text grammars: A study in theoretical linguistics and poetics.* The Hague: Mouton.

Van Dijk, T. A. (1980). *Macrostructures: An interdisciplinary study of global structures in discourse, interaction, and cognition.* Hillsdale, NJ: Erlbaum.

Van Dijk, T. A. (1988). *News as discourse.* Hillsdale, NJ: Erlbaum.

Vickery, B. (1996). Conceptual relations in information systems [letter to the editor]. *Journal of Documentation, 52*, 198–200.

Voorhees, E. M. (1994). Query expansion using lexical-semantic relations. *Proceedings of the Seventeenth Annual International ACM/SIGIR Conference on Research and Development in Information Retrieval*, 61–69.

Voorhees, E. M. (2003). Overview of the TREC 2003 Question Answering Track. *Twelfth Text Retrieval Conference (TREC 2003)* (NIST Special Publication, SP 500-255, pp. 54–68). Retrieved December 31, 2004, from http://trec.nist.gov/pubs/trec12/t12_proceedings. html

Vossen, P. (1998). Introduction to EuroWordNet. *Computers and the Humanities, 32*(2–3), 73–89.

Wan, T. L., Evens, M., Wan, Y. W., & Pao, Y. Y. (1997). Experiments with automatic indexing and a relational thesaurus in a Chinese information retrieval system. *Journal of the American Society for Information Science, 48*(12), 1086–1096.

Wang, Y.-C., Vandendorpe, J. & Evens, M. (1985). Relational thesauri in information retrieval. *Journal of the American Society for Information Science, 36*(1), 15–27.

Wanner, L. (1996). Introduction. In L. Wanner (Ed.), *Lexical functions in lexicography and natural language processing* (pp. 1–36). Amsterdam: John Benjamins.

Wanner, L. (2004). Towards automatic fine-grained semantic classification of verb-noun collocations. *Natural Language Engineering, 10*(2), 95–143.

Warren, H. C. (1921). *A history of the association psychology*. New York: Scribner's.

Warren, W. H., Nicholas, D. W., & Trabasso, T. (1979). Event chains and inferences in understanding narratives. In R.O. Freedle (Ed.), *Advances in discourse processes, Vol. 2: New directions in discourse processing* (pp. 23–52). Norwood, NJ: Ablex.

Wattenmaker, W. D., Nakamura, G. V., & Medin, D. L. (1988). Relationships between similarity-based and explanation-based categorization. In D. J. Hilton (Ed.), *Contemporary science and natural explanation: Commonsense conceptions of causality* (pp. 204–240). Washington Square, NY: New York University Press.

Werner, O. (1988). How to teach a network: Minimal design features for a cultural acquisition device or C-KAD. In M. W. Evens (Ed.), *Relational models of the lexicon: Representing knowledge in semantic networks* (pp. 141–166). Cambridge, UK: Cambridge University Press.

White, M., Korelsky, T., Cardie, C, Ng, V., Pierce, D., & Wagstaff, K. (2001). Multi-document summarization via information extraction. *First International Conference on Human Language Technology Research (HLT-01)*. Retrieved March 25, 2005, from http://www.hlt.utdallas.edu/~vince/papers/hlt01.html

Whitney, P., Budd, D., Bramucci, R. S., & Crane, R. S. (1995). On babies, bathwater, and schemata: A reconsideration of top-down processes in comprehension. *Discourse Processes, 20*, 135–166.

Wille, R. (2003). Truncated distributive lattices: Conceptual structures of simple-implicational theories. *Order, 20*, 229–238.

Winston, M., Chaffin, R., & Herrmann, D. (1987). A taxonomy of part–whole relations. *Cognitive Science, 11*(4), 417–444.

Wittgenstein, L. (1953). *Philosophical investigations* (G. E. M. Anscombe, Trans.). New York: Macmillan.

World Wide Web Consortium (2004a). *OWL Web Ontology Language: Use cases and requirements* (W3C Recommendation 10 February 2004). Retrieved March 11, 2004, from http://www.w3.org/TR/2004/REC-webont-req-20040210

World Wide Web Consortium. (2004b). *OWL Web Ontology Language overview* (W3C Recommendation 10 Feb 2004). Retrieved March 18, 2004, from http://www.w3.org/TR/2004/REC-owl-features-20040210

World Wide Web Consortium (2004c). *Semantic Web.* Retrieved September 1, 2004, from http://www.w3.org/2001/sw

World Wide Web Consortium. (2004d). *Web services architecture requirements* (W3C Working Group Note 11 February 2004). Retrieved December 1, 2004, from http://www.w3.org/TR/wsa-reqs

Xu, J., & Croft, W. B. (1996). Query expansion using local and global document analysis. *Proceedings of the 19th Annual International ACM SIGIR Conference on Research and Development in Information Retrieval,* 4–11.

Xu, J., & Croft, W. B. (2000). Improving the effectiveness of information retrieval with local context analysis. *ACM Transactions on Information Systems, 18*(1), 79–112.

Xu, Y., Araki, M., & Niimi, Y. (2003). A multilingual-supporting dialog system across multiple domains. *Acoustical Science and Technology, 24*(6), 349–357.

CHAPTER 6

Intelligence and Security Informatics

Hsinchun Chen and Jennifer Xu
University of Arizona

ISI: Challenges and Research Framework

The tragic events of September 11, 2001, and the subsequent anthrax scare had profound effects on many aspects of society. Terrorism has become the most significant threat to domestic security because of its potential to bring massive damage to the nation's infrastructure and economy. In response to this challenge, federal authorities are actively implementing comprehensive strategies and measures to achieve the three objectives identified in the "National Strategy for Homeland Security" report (U.S. Office of Homeland Security, 2002): (1) preventing future terrorist attacks, (2) reducing the nation's vulnerability, and (3) minimizing the damage and expediting recovery from attacks that occur. State and local law enforcement agencies, likewise, have become more vigilant about criminal activities that can threaten public safety and national security.

Academics in the natural sciences, computational science, information science, social sciences, engineering, medicine, and many other fields have also been called upon to help enhance the government's capabilities to fight terrorism and other crime. Science and technology have been identified in the "National Strategy for Homeland Security" report as the keys to winning the new counter-terrorism war (U. S. Office of Homeland Security, 2002). In particular, it is believed that information technology and information management will play indispensable roles in making the nation safer (Cronin, 2005; Davies, 2002; National Research Council, 2002) by supporting intelligence and knowledge discovery through collecting, processing, analyzing, and utilizing terrorism- and crime-related data (Badiru, Karasz, & Holloway, 1988; Chen, Miranda, Zeng, Demchak, Schroeder, & Madhusudan, 2003; Chen, Moore, Zeng, & Leavitt, 2004). With access to high-quality intelligence, federal, state, and local authorities can make timely decisions to select effective strategies and tactics and to allocate appropriate resources to detect, prevent, and respond to future attacks.

This chapter addresses issues regarding the development of information technologies in the intelligence and security domain. We propose a research framework with a primary focus on knowledge discovery from databases (KDD). After a comprehensive literature review of existing technologies used in counter-terrorism and crime-fighting applications, we present a set of case studies to demonstrate how KDD and other technologies can contribute to the critical objectives of national security. We also briefly discuss legal, ethical, and social issues related to the use of information technology for national security.

Information Technology and National Security

Six critical mission areas have been identified where information technology can contribute to accomplishing the three strategic national security objectives identified in the "National Strategy for Homeland Security" report (U.S. Office of Homeland Security, 2002):

- *Intelligence and warning.* Although terrorism depends on surprise to bring damage to targets (U.S. Office of Homeland Security, 2002), terrorist activities are neither random nor impossible to track. Terrorists must plan and prepare before the execution of an attack by selecting a target, recruiting and training operatives, acquiring financial support, and traveling to the country where the target is located (Sageman, 2004). To avoid detection, they may hide their true identities and disguise attack-related activities. Similarly, criminals may use falsified identities during police contacts (Wang, Chen, & Atabakhsh, 2004). Although it is difficult, detecting potential terrorist attacks or crimes is possible with the help of information technology. By analyzing communication and activity patterns among terrorists and their contacts, detecting fake identities, and employing surveillance and monitoring techniques, intelligence and warning systems can provide critical alerts and timely warnings to prevent attacks or crimes from occurring.

- *Border and transportation.* Terrorists enter a targeted country by air, land, or sea. Criminals in narcotics rings travel across borders to purchase, transport, distribute, and sell drugs. Information such as travelers' identities, images, fingerprints, and vehicles used is collected from customs, border, and immigration authorities on a daily basis. Such information can greatly improve the capabilities of counter-terrorism and crime-fighting agencies by creating a "smart border," where information from multiple sources is integrated and analyzed to help detect or locate wanted terrorists or criminals. Information sharing and integration, collaboration and communication, biometrics, and image and speech recognition will all be greatly needed in creating smart borders.

- *Domestic counter-terrorism.* As terrorists may be involved in local crimes, state and local law enforcement agencies also contribute by investigating and prosecuting crimes. Terrorism is regarded as a type of organized crime in which multiple actors cooperate to carry out offenses. Information technologies that help unearth cooperative relationships among criminals and reveal their patterns of interaction would also be helpful for analyzing terrorism.

- *Protecting critical infrastructure.* Roads, bridges, water supply, and many other physical service systems are critical infrastructures and key national assets that may become the target of terrorist attacks because of their vulnerabilities (U.S. Office of Homeland Security, 2002). Moreover, virtual infrastructures such as the Internet are also vulnerable to intrusions and insider threats (Lee & Stolfo, 1998). In addition to physical devices such as sensors and detectors, advanced technologies are needed to model the normal usage behaviors of such systems so that abnormalities and exceptions can be identified. Preemptive or reactive measures can be selected on the basis of the results to secure these assets against attacks.

- *Defending against catastrophic terrorism.* Terrorist attacks can cause devastating damage to a society through the use of chemical, biological, or radiological weapons. Biological attacks, for example, may cause contamination, outbreaks of infectious disease, and significant loss of life. Information systems that can efficiently and effectively collect, access, analyze, and report data about potentially catastrophic events can help agencies prevent, detect, respond to, and manage such attacks (Damianos, Ponte, Wohlever, Reeder, Day, Wilson, et al., 2002).

- *Emergency preparedness and response.* In case of a national emergency, prompt and effective responses are critical to damage containment and control. In addition to the systems that are designed to defend against catastrophes, information technologies that help formulate, experiment with, and optimize response plans (Lu, Huang, & Shekhar, 2003); train response professionals; and manage consequences are beneficial in the long run. Moreover, systems that provide social and psychological support to the victims of terrorist attacks can also help society recover from disasters.

Given the importance of information technology to national security, its development for counter-terrorism and crime-fighting applications is of the highest priority, despite the many associated problems and challenges.

Problems and Challenges

Intelligence and security agencies routinely gather large amounts of data from various sources. Processing and analyzing such data, however, have become increasingly difficult. Treating terrorism as a form of organized crime allows us to categorize the challenges into three types:

- *Understanding characteristics of criminals and crimes.* Some crimes may be geographically diffused and temporally dispersed. For instance, transnational narcotics trafficking criminals often live in different countries, states, and cities. Drug distribution and sales occur in different places at different times. This is true of other forms of organized crime (e.g., terrorism, sex trafficking, labor racketeering). As a result, investigations must track and prosecute multiple offenders who commit criminal activities in different places at different times. Given the limited resources at the disposal of intelligence and security agencies, this can be difficult. Moreover, as computer and Internet technologies advance, criminals are committing various types of cybercrime under the guise of ordinary online transactions and communications.

- *Understanding characteristics of crime and intelligence related data.* A significant challenge is the information stovepipe and overload resulting from diverse data sources, multiple data formats, and large data volumes. Unlike other professional disciplines such as marketing, finance, and medicine, in which data can be collected from particular sources (e.g., sales records, companies, patient medical histories), the intelligence and security domain does not have a well-defined set of data sources. Both authoritative information (e.g., crime incident reports, telephone records, financial statements, immigration and custom records) and open source information (e.g., news stories, journal articles, books, Web pages) need to be gathered for investigative purposes. Data collected from these different sources often exist in different formats, ranging from structured database records to unstructured text, image, audio, and video files. Important information such as evidence of criminal associations may be available but buried in unstructured texts and difficult to access and retrieve. Moreover, as data volumes continue to grow, extracting valuable and credible intelligence and knowledge becomes more difficult.

- *Developing crime and intelligence analysis techniques.* Current research on the technologies for counter-terrorism and crime-fighting applications lacks a consistent framework to address the major challenges. Some information technologies, including data integration, data analysis, text mining, image and video processing, and evidence combination, have been identified as particularly helpful (National Research Council, 2002). However, the question of how to employ them in the intelligence and security domain remains unanswered.

We believe that there is a pressing need to develop a science of "Intelligence and Security Informatics" (ISI) (Chen, Miranda, et al., 2003; Chen, Moore, et al., 2004), with its main objective being the "development of advanced information technologies, systems, algorithms, and databases for national security related applications, through an integrated technological, organizational, and policy-based approach" (Chen, Miranda, et al., 2003, p. v).

In comparing ISI with biomedical informatics, a young discipline addressing information management issues in biological and medical applications, we have found important similarities. In terms of data characteristics, they both face the information stovepipe and information overload problems; in terms of technology development, they both are at the exploratory stage of searching for new approaches, methods, and innovative use of existing techniques; in terms of scientific contributions, they both may add new insights and knowledge to fields such as computer science and decision science. Table 6.1 summarizes the similarities and differences between ISI and biomedical informatics. Most importantly, just as a consistent framework has emerged in biomedical informatics (Shortliffe & Blois, 2000), so ISI needs a framework to guide its research agenda. We believe that the knowledge discovery from databases (KDD) methodology, which has achieved significant success in other domains, including business, engineering, biology, and medicine, could be critical in addressing the challenges and problems facing ISI.

Table 6.1 Analogies between ISI and biomedical informatics

		Biomedical Informatics	ISI
Challenges	Domain-Specific	• Complexity and uncertainty associated with organisms and diseases • Critical decisions regarding patient well-being and biomedical discoveries	• Geographically diffused and temporally dispersed organized crimes • Cybercrimes on the Internet • Critical decisions related to public safety and homeland security
	Data	Information stovepipe and overload • HL7 XML standard • PHIN MS messaging • Patient records, diseases data, medical images	Information stovepipe and overload • Justice XML standard • Criminal incident records • Multilingual intelligence open sources
	Technology	• Ontologies and linguistic parsing • Information integration • Data and text mining • Medical decision-support systems and techniques	• Information integration • Criminal network analysis • Data, text, and Web mining • Identity management and deception detection
	Methodology	KDD	KDD
Contributions	Scientific	• Computer and information science, sociology, policy, legal • Clinical medicine and biology	• Computer and information science, sociology, policy, legal • Criminology, terrorism research
	Practical	• Public health • Patient well-being • Biomedical treatment and discovery	• Crime investigation and counter-terrorism • National and homeland security

The ISI Framework

To address the data and technical challenges facing ISI, we present a research framework with a primary focus on KDD technologies. The framework is discussed in the context of types of crime and security implications.

Crime is the commission of an act that is forbidden or the omission of a duty that is commanded by a public law, thus making the offender liable to punishment under that law. The greater the threat that a particular crime poses to public safety, the more likely it is to be viewed as a national security concern. Some crimes, such as traffic violations, theft, and homicide, lie mainly in the jurisdiction of local law enforcement agencies. Other crimes need to be dealt with by both local law enforcement and national security authorities. Identity theft and fraud, for instance, are related to criminal identity management issues at both local and national levels. Criminals may escape arrest by using false identities; drug smugglers may enter the United States by holding counterfeited passports or visas. Organized crime, such as terrorism and narcotics trafficking, often diffuses geographically and temporally, resulting in common security concerns across cities and states. Cybercrime can pose threats to public safety across multiple jurisdictions because of the nature of computer network technology. Table 6.2 summarizes the different types of crimes, sorted by security level (Chen, Chung, Wu, Chau, & Qin, 2004).

Table 6.2 Types of crime and security concerns

Crime Types		
Type	**Local Law Enforcement Level**	**National Security Level**
Traffic Violations	Driving under the influence (DUI), fatal/personal injury/property damage, traffic accident, road rage	
Sex Crime	Sexual offenses, sexual assaults, child molesting	Transnational child pornography
Theft	Robbery, burglary, larceny, motor vehicle theft, stolen property	Theft of national secrets or weapon information
Fraud	Forgery and counterfeiting, fraud, embezzlement, identity deception	Transnational money laundering, identity fraud, transnational financial fraud
Property crime	Property crime (e.g., arson) on buildings, apartments	Intentional destruction of or damage to national infrastructures and assets
Organized Crime	Narcotic drug offenses (sales or possession), gang-related offenses,	Transnational drug trafficking, terrorism (bioterrorism, bombing, hijacking, etc.), organized prostitution
Violent Crime	Criminal homicide, armed robbery, aggravated assault, other assaults	Terrorism
Cybercrime	Internet fraud (e.g., credit card fraud, advance fee fraud, fraudulent Web sites), theft of confidential information	Network intrusion/hacking, illegal trading, virus spreading, cyberpiracy, cyberpornography, cyberterrorism, theft of confidential information

(left margin, vertical:) Increasing public influence →

We believe that KDD techniques can play a central role in improving the counter-terrorism and crime-fighting capabilities of intelligence and security agencies by reducing cognitive and information overload. Knowledge discovery refers to nontrivial extraction of implicit, previously unknown, and potentially useful knowledge from data. Knowledge discovery techniques promise easy, convenient, and practical exploration of very large collections of data for organizations and users, and have been applied in marketing, finance, manufacturing, biology, and many other domains (e.g., predicting consumer behavior, detecting credit card fraud, or clustering genes that have similar biological functions). Knowledge discovery usually consists of multiple stages, including data selection, data preprocessing, data transformation, data mining, and the interpretation and evaluation of patterns (Fayyad, Piatetsk-Shapiro, & Smyth, 1996). Data mining plays a key role in extracting patterns from data. Traditional data mining techniques include association-rule mining, classification and prediction, cluster analysis, and outlier analysis (Han & Kamber, 2001). As natural language processing (NLP) research advances, text mining approaches that automatically extract, summarize, categorize, and translate text documents are also being widely used (Trybula, 1999).

Many of these KDD technologies could be applied in ISI studies (Chen, Miranda, et al., 2003; Chen, Moore, et al., 2004). We categorize existing ISI technologies into six classes: information sharing and collaboration, crime association mining, crime classification and clustering, intelligence text mining, spatial and temporal crime pattern mining, and criminal network mining. These six classes are grounded in traditional knowledge discovery technologies, but include a few new approaches, such as spatial and temporal crime pattern mining and criminal network analysis, that are more relevant to counter-terrorism and crime investigation. Although information sharing and collaboration are not knowledge discovery per se, they help integrate, warehouse, and prepare data for knowledge discovery and thus are included in the framework.

We present in Figure 6.1 our proposed research framework with the horizontal axis representing crime types and vertical axis the six classes of techniques (Chen, Chung, et al., 2004). The shaded regions on the chart show promising research areas, that is, certain classes of techniques are relevant to solving certain types of crime. Note that more serious crimes may require a more complete set of knowledge discovery techniques. For example, the investigation of terrorism may depend on criminal network analysis technology, which requires the use of other knowledge discovery techniques such as association mining and clustering. An important observation about this framework is that the high-frequency occurrences and strong association patterns of severe and organized crime, such as terrorism and narcotics, present a unique opportunity and potentially high rewards for adopting a knowledge discovery framework.

Caveats for ISI

Before we review the technical foundations and approaches, we want to discuss briefly the legal and ethical caveats regarding crime and intelligence research. The potential negative effects of intelligence gathering

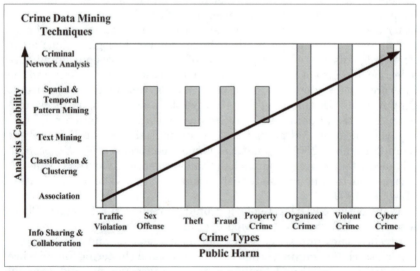

Figure 6.1 A knowledge discovery research framework for intelligence and security informatics.

and analysis on the privacy and civil liberties of the public have been well publicized (Cook & Cook, 2003). Many laws, regulations, and agreements governing data collection, confidentiality, and reporting could influence directly the development and application of ISI technologies. We strongly recommend that intelligence and security agencies and ISI researchers be aware of these laws and regulations in their research efforts (Strickland, 2005). Moreover, we also suggest that a hypothesis-guided, evidence-based approach be used in crime and intelligence analysis research. That is, there should be probable and reasonable causes and evidence for targeting particular individuals or data sets for analysis. Proper investigative and legal procedures need to be strictly followed. It is neither ethical nor legal to "fish" for potential criminals from diverse and mixed crime-, intelligence-, and citizen-related data sources (Strickland, 2005). The well-publicized Defense Advanced Research Program Agency (DARPA) Total Information Awareness (TIA) program and the Multi-State Anti-Terrorism Information Exchange (MATRIX) system, for example, were roundly criticized for their inappropriate use of citizen data and unguided analysis technologies resulting in the potential impairment of Americans' civil liberties (American Civil Liberties Union, 2004). Many new and important privately and publicly funded research projects aim to address these privacy and civil liberties issues in the context of homeland security research. For example, the Electronic Frontier Foundation monitors limits placed on freedom of expression as indicated by Web sites closed for national security reasons by government or Internet service providers (http://www.eff.org/Privacy/Surveillance/Terrorism/antiterrorism_chill.html); the OpenNet Initiative is conducting a three-year study of Internet filtering in Saudi Arabia (http://www.opennetinitiative.net/studies/saudi).

ISI: Technical Foundations and Approaches

In this section, we review the technical foundations of ISI and the six classes of technologies and approaches specified in our research framework. We also summarize relevant past and ongoing research that addresses knowledge discovery in public safety and national security.

Information Sharing and Collaboration

Information sharing across jurisdictional boundaries of intelligence and security agencies has been identified as a key foundation of national security (U.S. Office of Homeland Security, 2002). However, sharing and integrating information from distributed, heterogeneous, and autonomous data sources is a nontrivial task (Hasselbring, 2000; Rahm & Bernstein, 2001). In addition to legal and cultural issues regarding information sharing, it is often difficult to integrate and combine data that are organized in different schemas and stored in different database systems running on different hardware platforms and operating systems

(Hasselbring, 2000). Other data integration problems include: (1) name differences (same entity with different names), (2) mismatched domains (problems with units of measure or reference point), (3) missing data (incomplete data sources or different data available from different sources), and (4) object identification (no global ID values and no inter-database ID tables) (Chen & Rotem, 1998).

Three approaches to data integration have been proposed: *federation, warehousing*, and *mediation* (Garcia-Molina, Ullman, & Widom, 2002). Database federation maintains data in their original, independent sources but provides a uniform data access mechanism (Buccella, Cechich, & Brisaboa, 2003; Haas, 2002). Data warehousing is an integrated system in which copies of data from different data sources are migrated and stored to provide uniform access. Data mediation relies on "wrappers" to translate and pass queries from multiple data sources. The wrappers are "transparent" to an application so that the multiple databases appear to be a single database. These techniques are not mutually exclusive and many hybrid approaches have been proposed (Jhingran, Mattos, & Pirahesh, 2002).

All these techniques are dependent, to a great extent, on the matching between different databases. The task of database matching can be broadly divided into *schema-level* and *instance-level matching* (Lim, Srivastava, Prabhakar, & Richardson, 1996; Rahm & Berhstein, 2001). Schema-level matching is performed by aligning semantically corresponding columns between two sources. Various schema elements such as attribute name, description, data type, and constraints may be used to generate a mapping between the two schemas (Rahm & Bernstein, 2001). For example, prior studies have used linguistic matchers to find similar attribute names based on synonyms, common substrings, pronunciation, and Soundex codes (Newcombe, Kennedy, Axford, & James, 1959) to match attributes from different databases (Bell & Sethi, 2001). Instance-level or entity-level matching connects records describing a particular object in one database to records describing the same object in another. Entity-level matching is frequently performed after schema-level matching is completed. Existing entity matching approaches include (1) key equivalence, (2) user-specified equivalence, (3) probabilistic key equivalence, (4) probabilistic attribute equivalence, and (5) heuristic rules (Lim et al., 1996).

Some of these information integration approaches have been used in law enforcement and intelligence agencies for investigation support. The COPLINK Connect system (Chen, Schroeder, Hauck, Ridgeway, Atabakhsh, Gupta, et al., 2003) employed the database federation approach to achieve schema-level data integration. It provided a common COPLINK schema and a one-stop-shop user interface to facilitate access to different data sources from multiple police departments. Evaluation results showed that COPLINK Connect had out-performed the Record Management System (RMS) of police data in system effectiveness, ease of use, and interface design (Chen, Schroeder, et al., 2003).

Similarly, the Phoenix Police Department Reports (PPDR) is a Web-based, federated intelligence system in which databases share a common schema (Dolotov & Strickler, 2003). The bioterrorism surveillance systems developed at the University of South Florida, on the other hand, used data warehouses to integrate historical and real-time surveillance data and incrementally incorporated data from diverse disease sources (Berndt, Bhat, Fisher, Hevner, & Studnicki, 2004; Berndt, Hevner, & Studnicki, 2003). A transnational information-sharing system developed at the University of Florida employed a data mediation approach (Kasad & Su, 2004). The system accessed different databases via a wrapper query processor, which tailored a user query into database-specific queries. This system was intended to enhance information sharing between immigration and border controls in multiple countries.

Integrating data at the entity level has also been difficult. In addition to existing key equivalence matching and heuristic consolidation approaches (Goldberg & Senator, 1998), use of the National Incident-Based Reporting System (NIBRS) (U.S. Federal Bureau of Investigation, 1992), a crime incident classification standard, has been proposed to enhance data sharing among law enforcement agencies (Faggiani & McLaughlin, 1999; Schroeder, Xu, & Chen, 2003). In the Violent Crime Linkage Analysis System (ViCLAS) (Collins, Johnson, Choy, Davidson, & Mackay, 1998), data collection and encoding standards were used to capture more than 100 behavioral characteristics of offenders in serial violent crimes in order to address the problem of entity-level matching.

Information sharing has also been undertaken in intelligence and security agencies through cross-jurisdictional collaborative systems. The COPLINK Agent ran on top of the COPLINK Connect system (Chen, Schroeder, et al., 2003) and linked crime investigators who were working on related cases at different units to enhance collaborations (Zeng, Qin, Huang, & Chen, 2003). It employed collaborative filtering approaches (Goldberg, Nichols, Oki, & Terry, 1992), which have been widely studied in commercial recommender systems, to identify law enforcement users who had similar search histories. Similar search histories might indicate that these users had similar information needs and thus were working on related crime cases. When one user searched for information about a crime or a suspect, the system would alert other users who worked on related cases so that these users could collaborate and share their information through other communication channels. The FALCON system offered similar monitoring and alerting functionality (M. Brown, 1998). Its collaboration capability, however, was relatively limited. The JNET system (http://www.pajnet.state.pa.us/pajnet/site/default.asp) also provides an alerting capability that immediately notifies relevant agencies via pager or e-mail when a wanted person is found or arrested by other agencies. Research has also been performed to model mathematically collaboration processes across law enforcement and intelligence jurisdictions in order to improve work productivity (Raghu, Ramesh, & Whinston, 2003; Zhao, Bi, & Chen, 2003). Although information sharing and collaboration are not knowledge discovery per se, they prepare data for important subsequent analyses.

Crime Association Mining

Finding associations among data items is an important topic in knowledge discovery research. One of most widely studied approaches is association-rule mining, a process for discovering frequently occurring item sets in a database. Association-rule mining is often used in market basket analysis where the objective is to find which products are bought with which other products (Agrawal, Imielinski, & Swami, 1993; Mannila, Toivonen, & Inkeri, 1994; Silverstein, Brin, & Motwani, 1998). An association is expressed as a rule $X \Rightarrow Y$, indicating that item set X and item set Y occur together in the same transaction (Agrawal et al., 1993). Each rule is evaluated using two probability measures, *support* and *confidence*, where *support* is defined as $prob(\cap Y)$ and *confidence* as $prob(X \cap Y)/prob(X)$. For example, "diaper $\Rightarrow$ milk with 60 percent *support* and 90 percent *confidence*" means that 60 percent of customers buy both diapers and milk in the same transaction and that 90 percent of the customers who buy diapers tend to buy milk at the same time.

In the intelligence and security domain, spatial association-rule mining (Koperski & Han, 1995) has been proposed to extract cause-effect relations among geographically referenced crime data to identify environmental factors that attract crime (Estivill-Castro & Lee, 2001). Moreover, the research on association mining is not limited to association-rule mining but covers the extraction of a wide variety of relationships among crime data items. Crime association mining techniques can include *incident association mining* and *entity association mining* (Lin & Brown, 2003).

The purpose of incident association mining is to find crimes that might have been committed by the same offender; unsolved crimes are linked to solved crimes to identify the suspect. This technique is often used to solve serial sexual offenses and serial homicides. However, finding associated crime incidents can be time-consuming if it is performed manually. It is estimated that pairwise, manual comparisons on just a few hundred crime incidents would take more than one million human hours (Brown & Hagen, 2002). When the number of crime incidents is large, manual identification of associations between crimes is prohibitively expensive. Two approaches, *similarity-based* and *outlier-based*, have been developed for incident association mining. For example, the Violent Criminal Apprehension Program (ViCAP) identifies similar features or traits of violent crimes such as homicides to detect serial offenders (Icove, 1986).

Similarity-based methods detect associations between crime incidents by comparing features such as spatial locations of the incidents and the offender's modus operandi (MO), often regarded as a criminal's "behavioral signature" (O'Hara & O'Hara, 1980). Expert systems relying on decision rules acquired from domain experts used to be a common approach to associating crime incidents (Badiru et al., 1988; Bowen, 1994; Brahan, Lam, Chan, & Leung, 1998). However, as the collection of human decision rules requires considerable knowledge engineering effort and the rules collected are often hard to update, the expert system approach has been replaced by more automated approaches. Brown and Hagen (2002) developed a total similarity measure between two crime records as a weighted sum of similarities of various crime features. For features that take on categorical values (such as an offender's eye color), Brown developed a similarity table based on heuristics that specified the similarity level for each pair of categorical values. Evaluation showed that this approach enhanced both accuracy and efficiency for associating crime records. Similarly, Wang, Lin, Shieh, and Deng (2003) proposed measuring similarity between a new crime incident and existing criminal information stored in police databases by representing the new incident as a query and existing criminal information as vector space. The vector space model is widely employed in information retrieval applications; various similarity measures such as the Jaccard function (Rasmussen, 1992) could be used.

Unlike similarity-based methods, which identify associations based on a number of crime features, the outlier-based method focuses only on the distinctive features of a crime (Lin & Brown, 2003). Imagine a series of robberies in which a Japanese sword was used as the weapon. Because a Japanese sword is a very uncommon weapon, unlike, say, a shotgun, it is probable that this series of robberies was committed by the same offender. Based on this outlier concept, crime investigators need first to cluster incidents into cells and then use an outlier score function to measure the distinctiveness of the incidents in a specific cell. If the outlier score of a cell is larger than a threshold value, the incidents contained in the cell are assumed to be associated and committed by the same offender. Evaluation has shown that the outlier-based method is more effective than the similarity-based method proposed in Brown and Hagen (2002).

The task of finding and charting associations between crime entities such as persons, weapons, and organizations is often referred to as entity association mining (Lin & Brown, 2003) or link analysis (Sparrow, 1991) in law enforcement. The purpose is to find out whether crime entities that appear to be unrelated at the surface are actually linked, and if so, how. Law enforcement officers and criminal investigators throughout the world have long used link analysis to search for and analyze relationships among criminals. For example, the Federal Bureau of Investigation (FBI) used link analysis in the investigation of the Oklahoma City bombing case and the Unabomber case to look for criminal associations and investigative leads (Schroeder et al., 2003). Although link analysis helps trace criminals through chains of relations, manually identifying and detecting criminal relations from large amounts of criminal-justice data is very time-consuming.

Three types of automated link analysis approaches have been suggested: *heuristic-based, statistically-based*, and *template-based*. Heuristic-based approaches rely on decision rules used by domain experts to determine whether two entities in question are related. For example, Goldberg and Senator (1998) suggested that links or associations between individuals in financial transactions be created based on a set of heuristics, such as whether the individuals had shared addresses, shared bank accounts, or related transactions. This technique has been employed by the FinCEN system of the U.S. Department of the Treasury to detect money laundering transactions and activities (Goldberg & Senator, 1998; Goldberg & Wong, 1998). The COPLINK Detect system (Hauck, Atabakhsh, Ongvasith, Gupta, & Chen, 2002) employed a statistically based approach called Concept Space (Chen & Lynch, 1992). This approach measures the weighted co-occurrence associations between records of entities (persons, organizations, vehicles, and locations) stored in crime databases. An association exists between a pair of entities if they appear together in the same criminal incident. The more frequently they occur together, the stronger the association. Zhang, Salerno, and Yu (2003) proposed to use a fuzzy resemblance function to calculate the correlation between two individuals' past financial transactions in order to detect associations between the individuals who might have been involved in a specific money-laundering crime. If the correlation between two individuals is higher than a threshold value, these two individuals are regarded as being related. The template-based approach has been used primarily to identify associations between entities extracted from textual documents, such as police report narratives. Lee (1998) developed a template-based technique using relation-specifying words and phrases. For example, the phrase "member of" indicates an entity–entity association between an individual and an organization. Coady (1985) proposed to use the PROLOG language to derive rules of entity associations automatically from text data and use the rules to detect associations in similar documents. Template-based approaches rely heavily on a fixed set of predefined patterns and rules, and thus may have limited application scope.

Crime Classification and Clustering

Classification is the process of mapping data items into one of several predefined categories based on attribute values of the items (Hand, 1981; Weiss & Kulikowski, 1991). Examples of classification applications include fraud detection (Chan & Stolfo, 1998), computer and network intrusion detection (Lee & Stolfo, 1998), bank failure prediction (Sarkar & Sriram, 2001), and image categorization (Fayyad, Djorgovish, & Weir, 1996). Classification is a type of supervised learning that consists of a training stage and a testing stage. Accordingly, the dataset is divided into a training set and a testing set. The classifier is designed to "learn" from the training set classification models governing the membership of data items. The accuracy of the classifier is assessed using the testing set.

Discriminant analysis (Eisenbeis & Avery, 1972), Bayesian models (Duda & Hart, 1973; Heckerman, 1995), decision trees (Quinlan, 1986, 1993), artificial neural networks (Rummelhart, Hinton, & Williams, 1986), and support vector machines (SVM) (Vapnik, 1995) are widely used classification techniques. In discriminant analysis the class membership of a data item is modeled as a function of the item's attribute values. Through regression analysis a class membership discriminant function can be obtained and used to classify new data items. Bayesian classifiers assume that all data attributes are conditionally independent, given the class membership outcome. The task is to learn the conditional probabilities among the attributes, given the class membership outcome. The learned model is then used to predict the class membership of new data items based on their attribute values. Decision tree classifiers organize decision rules learned from training data in the form of a tree. Algorithms such as ID3 (Quinlan, 1986, 1993) and C4.5 (Quinlan, 1993) are popular decision tree classifiers. An artificial neural network consists of interconnected nodes to imitate the functioning of neurons and synapses of human brains. It usually contains an input layer with nodes taking on the attribute values of data items and the output layer with nodes representing class membership labels. Neural networks learn and encode knowledge through connection weights. SVM is a novel learning classifier based on the Structural Risk Minimization principle from computational learning theory. SVM is capable of handling millions of inputs and does not require feature selection (Cristianini & Shawe-Taylor, 2000). Each of these classification techniques has its advantages and disadvantages in terms of accuracy, efficiency, and interpretability. Researchers have also proposed hybrid approaches to combine these techniques (Kumar & Olmeda, 1999).

Several of these techniques have been applied in the intelligence and security domain to detect financial fraud and computer network intrusion. For example, in order to identify fraudulent financial transactions, Aleskerov, Freisleben, and Rao (1997) employed neural networks to detect anomalies in customers' credit card transactions based on their transaction histories. Hassibi (2000) employed a feed-forward back-propagation neural network to compute the probability that a given transaction was fraudulent. Two types of intrusion detection, *misuse detection* and *anomaly detection*, have been studied in computer network security applications (Lee & Stolfo, 1998). Misuse detection identifies attacks by matching them onto previously known attack patterns or signatures. Anomaly detection, on the other hand, identifies abnormal user behaviors based on historical data. Lee and Stolfo (1998) employed decision rule induction approaches to classify *sendmail* system call traces into normal or abnormal traces. Ryan, Lin, and Mikkulainen (1998) developed a neural network-based intrusion detection system to detect unusual user activity based on the patterns of users' past system command usage.

Stolfo, Hershkop, Wang, Nimeskern, and Hu (2003) applied Bayesian classifiers to distinguish between normal e-mail and spamming e-mail.

Unlike classification, clustering is a type of unsupervised learning. It groups similar data items into clusters without knowing their class membership. The basic principle is to maximize intra-cluster similarity while minimizing intercluster similarity (Jain, Murty, & Flynn, 1999). Clustering has been used in a variety of applications including image segmentation (Jain & Flynn, 1996), gene clustering (Eisen, Spellman, Brown, & Botstein, 1998), and document categorization (Chen, Houston, Sewell, & Schatz, 1998; Chen, Schuffels, & Orwig, 1996). Various clustering methods have been developed, including *hierarchical approaches,* such as complete-link algorithms (Defays, 1977), *partitional approaches,* such as *k*-means (Anderberg, 1973; Kohonen, 1995), and *Self-Organizing Maps* (SOM) (Kohonen, 1995). These clustering methods group data items based on different criteria and may not generate the same clustering results. Hierarchical clustering groups data items into a series of nested clusters and generates a tree-like dendrogram in which each node represents a merging of clusters. Partitional clustering algorithms generate only one partition level rather than nested clusters. Partitional clustering is more efficient and scalable for large datasets than hierarchical clustering, but has difficulty determining the appropriate number of clusters (Jain et al., 1999). In contrast to the hierarchical and partitional clustering that relies on the similarity or proximity measures between data items, SOM is a neural network-based approach that directly projects multivariate data items onto two-dimensional maps. SOM can be used for clustering and visualizing data items and groups (Chen, Schuffels, et al., 1996).

The use of clustering methods in the law enforcement and security domains can be categorized into two types: *crime incident clustering* and *criminal clustering*. The purpose of crime incident clustering is to find a set of similar crime incidents based on an offender's behavioral traits or a geographical area with a high concentration of certain types of crimes. For example, Adderley and Musgrove (2001) employed the SOM approach to cluster sexual attack crimes based on a number of offender MO attributes (e.g., the precaution methods taken and the verbal themes during the crime) in order to identify serial sexual offenders. The clusters found were used to form offender profiles containing MO and other information such as offender motives and racial preferences when choosing victims. Similarly, Kangas, Terrones, Keppel, and La Moria (2003) employed the SOM method to group crime incidents in order to identify serial murderers and sexual offenders. D. Brown (1998) proposed *k*-means and the nearest neighbor approach to clustering spatial data of crimes to find "hot spot" areas in a city. Spatial clustering methods are often used in "hot spot analysis," which will be reviewed in detail in the section on spatial and temporal mining.

Criminal clustering is often used to identify groups of criminals who are closely related. Instead of using similarity measures, this type of clustering relies on relational strength that measures the intensity and frequency of relationships between offenders. Stolfo et al. (2003) proposed grouping e-mail users who frequently communicated with each other into clusters so that unusual e-mail behavior that violated the group communication patterns could be detected. Offender clustering is more often used in criminal network analysis, which will be reviewed in detail in the section with that title.

Intelligence Text Mining

A large amount of intelligence- and security-related data is represented in text form such as police narrative reports, court transcripts, news stories, and Web articles. Valuable information in such texts is often difficult to retrieve, access, and use for the purposes of criminal investigation and counter-terrorism. It is desirable to mine the text data automatically in order to discover valuable knowledge about criminal or terrorism activities.

Text mining has attracted increasing attention in recent years as natural language processing capabilities advance (Chen, 2001). An important task of text mining is information extraction, a process of identifying and extracting from free text select types of information such as entities, relationships, and events (Grishman, 2003). The most widely studied information extraction subfield is named entity extraction. It helps to automatically identify from text documents the names of entities of interest, such as persons (e.g., "John Doe"), locations (e.g., "Washington, DC"), and organizations (e.g., "National Science Foundation"). It has also been extended to identify other text patterns, such as dates, times, number expressions, dollar amounts, e-mail addresses, and Web addresses (URLs). The Message Understanding Conference (MUC) series has served as the major forum for researchers in this area to compare the performance of their entity extraction approaches (Chinchor, 1998).

Four major named-entity extraction approaches have been proposed: lexical lookup, rule-based, statistical models, and machine learning.

- *Lexical lookup.* Most research systems maintain hand-crafted lexicons that contain lists of popular names for entities of interest, such as all registered organizational names in the U.S. and all personal surnames obtained from government census data. These systems work by looking up phrases in texts that match the items specified in their lexicons (e.g., Borthwick, Sterling, Agichtein, & Grishman, 1998).

- *Rule-based.* Rule-based systems rely on hand-crafted rules to identify named entities. The rules may be structural, contextual, or lexical (Krupka & Hausman, 1998). An example rule would look like the following:

  capitalized last name + , + capitalized first name ⇒ *person name*

Although such human-created rules are usually of high quality, this approach may not be easy to apply to other entity types.

- *Statistical models.* Such systems often use statistical models to identify occurrences of certain cues of particular patterns for entities in texts. A training data set is needed for a system to acquire the statistics. The statistical language model reported in Witten, Bray, Mahoui, and Teahan (1999) is an example of such a system. It uses the Prediction by Partial Matching (PPM) model to extract entities from text based on conditional probability distributions of characters. The probability of occurrence of later characters in a word or phrase depends on the occurrence of preceding characters; for example, "12Jan2005" in a newsletter can be correctly identified as a time phrase using this model.

- *Machine learning.* This type of system relies on machine learning algorithms rather than human-created rules to extract knowledge or identify patterns from textual data. Examples of machine learning algorithms used in entity extraction include neural networks, decision trees (Baluja, Mittal, & Sukthankar, 1999), Hidden Markov Models (Miller, Crystal, Fox, Ramshaw, Schwartz, Stone, et al., 1998), and entropy maximization (Borthwick et al., 1998).

Instead of relying on a single method, most existing information extraction systems combine two or more of these approaches. Many systems were evaluated at the MUC-7 conference. The best systems were able to achieve over 90 percent in both precision and recall rates in extracting persons, locations, organizations, dates, times, currencies, and percentages from a collection of *New York Times* news stories.

Recent years have seen research on named-entity extraction for intelligence and security applications (Patman & Thompson, 2003; Wang, Huang, Teng, & Chien, 2004). For example, Chau, Xu, and Chen (2002) developed a neural network-based entity extraction system to identify personal names, addresses, narcotic drugs, and personal property names from police report narratives. Rather than relying entirely on manual rule generation, this system combines lexical lookup, machine learning, and some hand-crafted rules. The system achieved over 70-percent precision and recall rates for personal names and narcotic drug names. However, it was difficult to achieve satisfactory performance for addresses and personal property because of their wide variation. Sun, Naing, Lim, and Lam (2003) converted the entity extraction problem into a classification problem in order to identify relevant entities from the MUC text collection on terrorism. They first identified all noun phrases in a document and then used the support vector machine to classify those entity candidates on the basis of both content and context features. The results showed that for the specific terrorism text collection, the performance of this approach in regards to precision and F measure was comparable to AutoSlog (Riloff, 1996), one of the best entity extraction systems reported earlier.

Several news and event extraction systems have been reported recently, such as Columbia's Newsblaster (McKeown, Barzilay, Chen, Elson, Evans, Klavans, et al., 2003) and CMU's (Carnegie Mellon University) system (Yang, Carbonell, Brown, Pierce, Archibald, & Liu, 1999), which automatically extract, categorize, and summarize events from international online news sources. Some of these systems can also work for multilingual documents and have great potential for automatic detection and tracking of terrorism events for intelligence purposes.

Crime Spatial and Temporal Mining

Most crimes, including terrorism, have significant spatial and temporal characteristics (Brantingham & Brantingham, 1981). Analysis of spatial and temporal patterns of crimes continues to be one of the most important crime investigation techniques. It aims to gather intelligence about environmental factors that prevent or encourage crimes (Brantingham & Brantingham, 1981), identify geographic areas of high crime concentration (Levine, 2000), and detect criminal trends (Schumacher & Leitner, 1999). The discovery of such patterns makes possible the use of effective and proactive control strategies, such as allocating the appropriate amount of policing resources in certain areas at certain times, to prevent crimes.

Spatial pattern analysis and geographical profiling play important roles in solving crimes (Rossmo, 1995). Three approaches for crime spatial pattern mining have been reported: *visual approaches, clustering approaches*, and *statistical approaches* (Murray, McGuffog, Western, & Mullins, 2001). The visual approach is also called crime mapping. It presents a city or regional map annotated with various crime-related information. For example, a map can be color-coded to present the densities of a specific type of crime in different geographical areas. Such an approach can help users visually detect relationships between spatial features and the occurrence of crime. The clustering approach has been used in hot spot analysis, a process of automatically identifying areas with high crime concentration. This type of analysis helps law enforcement effectively allocate policing resources to reduce crime in hot spot areas. Partitional clustering algorithms such as the k-means methods are often used for finding hot spots (Murray & Estivill-Castro, 1998). For example, Schumacher and Leitner (1999) used the k-means algorithm to identify hot spots in the downtown areas of Baltimore. Comparing these for different years, they found evidence of the displacement of crimes following redevelopment of the downtown area. Corresponding proactive strategies were then suggested on the basis of the patterns found. Although efficient and scalable in comparison to hierarchical clustering algorithms, partitional clustering algorithms usually require the user to predefine the number of clusters to be found. This, however, is not always feasible (Grubesic & Murray, 2001). Accordingly, researchers have tried to use statistical approaches to conduct hot spot analysis or to test the significance of hot spots (Craglia, Haining, & Wiles, 2000). The test statistics G_i (Getis & Ord, 1992; Ord & Getis, 1995) and Moran's I (Moran, 1950), which are used to test the significance of spatial autocorrelation, can be used to detect hot spots. If a variable is correlated with itself through space, it is said to be spatially autocorrelated. For example, Ratchliffe and McCullagh (1999) employed G_i and G_i^* statistics to identify the hot spots of residential burglary and motor vehicle crimes in a city. Compared with a domain expert's perception of the hot spots, this approach was shown to be effective (Ratchliffe & McCullagh, 1999). Statistical approaches have also been used in crime prediction. Based on spatial choice theory (McFadden, 1973), Xue and Brown (2003) modeled the probability of a criminal choosing a target location as a function of multiple spatial characteristics of the location such as family density per unit area and distance to highway. Using regression analysis, they predicted the locations of future crimes in a city. Evaluation showed that their models significantly outperformed conventional hot spot models. Similarly, Brown, Dalton, and Hoyle (2004) built a logistic regression model to predict suicide bombing in counter-terrorism applications.

Commercially available geographical information systems (GIS) and crime mapping tools, such as ArcView and MapInfo, have been widely used in law enforcement and intelligence agencies for analyzing and visualizing spatial patterns of crimes. Geographical coordinate information as well as various spatial features, such as the distance between the location of a crime to major roads and police stations, is often used in GIS (Harris, 1990; Weisburd & McEwen, 1997).

Research on temporal patterns of crimes is relatively scarce in comparison to crime mapping. Two major approaches have been reported, namely *visualization* and *statistical modeling* approaches. Visualization approaches present individual or aggregated temporal features of crimes using a periodic or timeline view. Common methods of viewing periodic data include sequence charts, point charts, bar charts, line charts, and spiral graphs displayed in 2-D or 3-D (Tufte, 1983). In a timeline view, a sequence of events is presented based on its temporal order. For example, LifeLines provides the visualization of a patient's medical history using a timeline view. The Spatial Temporal Visualizer (STV) (Buetow, Chaboya, O'Toole, Cushna, Daspit, Peterson, et al., 2003) seamlessly incorporates periodic view, timeline view, and GIS view in the system to support criminal investigations. Visualization approaches rely on human users to interpret data presentations and to find temporal patterns of events. Statistical approaches, on the other hand, build statistical models from observations to capture the temporal patterns of events. For instance, Brown and Oxford (2001) developed several statistical models including a log-normal regression model, a Poisson regression model, and cumulative logistic regression models to predict the number of breaking and entering crimes in Richmond, Virginia. The log-normal regression model was found to fit the data best.

Criminal Network Analysis

Criminals seldom operate in a vacuum but instead interact with one another to carry out various illegal activities. Relationships between individual offenders form the basis for organized crime and are essential for the smooth operation of a criminal enterprise (Cronin, 2005; Strickland, 2002a, 2002b, 2002c, 2002d, 2002e). Unlike bureaucratic organizations, criminal enterprises often operate in networks consisting of nodes (individual offenders) and links (relationships). In criminal networks, there may exist groups or teams, within which members have close relationships. One group may also interact with other groups to obtain or transfer illicit goods, services, or information. Moreover, individuals play different roles in their groups. For example, some key members may act as leaders to control the activities of a group, while others may serve as gatekeepers to ensure the smooth flow of information or illicit goods (Strickland, 2002a, 2002b, 2002c, 2002d, 2002e).

Structural network patterns in terms of subgroups, intergroup interactions, and individual roles thus are important for understanding the organization, structure, and operation of criminal enterprises. Such knowledge can help law enforcement and intelligence agencies disrupt criminal networks and develop effective control strategies to combat organized crime (Cronin, 2005). For example, removal of central members in a network may effectively upset the operational network and put a criminal enterprise out of action (Baker & Faulkner, 1993; McAndrew, 1999; Sparrow, 1991). Subgroups and interaction patterns between groups are helpful for finding a network's overall structure, which often reveals points of vulnerability (Evan, 1972; Ronfeldt & Arquilla, 2001). For a centralized structure such as a star or a wheel, the point of vulnerability lies in its central members. A decentralized network such as a chain or clique, however, does not have a single point of vulnerability and thus may be more difficult to disrupt (Strickland, 2002a, 2002b, 2002c, 2002d, 2002e).

Social Network Analysis (SNA) provides a set of measures and approaches for structural network analysis (Wasserman & Faust, 1994). These techniques were originally designed to discover social structures in social networks (Wasserman & Faust, 1994) and are especially appropriate for studying criminal networks (McAndrew, 1999; Sparrow, 1991). Studies involving evidence mapping in fraud and conspiracy cases have employed SNA measures to identify central members in criminal networks (Baker & Faulkner, 1993; Saether & Canter, 2001). In general,

SNA is capable of detecting subgroups, identifying central individuals, discovering between-group interaction patterns, and uncovering a network's overall structure:

- *Subgroup detection.* With networks represented in a matrix format, the matrix permutation approach and cluster analysis have been employed to detect underlying groups that are not otherwise apparent in data (Wasserman & Faust, 1994). Burt (1976) proposed to apply hierarchical clustering methods based on a structural equivalence measure (Lorrain & White, 1971) to partition a social network into positions in which members have similar structural roles. Xu and Chen (2003) employed hierarchical clustering to detect criminal groups in a narcotics network based on the relational strength between criminals.

- *Central member identification.* Centrality deals with the roles of network members. Several measures, such as degree, betweenness, and closeness, are related to centrality (Freeman, 1979). The degree of a particular node is its number of direct links; its betweenness is the number of geodesics (i.e., the shortest paths between any two nodes) passing through it; and its closeness is the sum of all the geodesics between the particular node and every other node in the network. Although these three measures are all intended to illustrate the importance or centrality of a node, they support interpretation of the roles of network members differently. An individual having a high degree measurement, for instance, may be inferred to have a leadership function, whereas an individual with a high level of betweenness may be seen as a gatekeeper in the network. Baker and Faulkner employed these three measures, especially degree, to find the key individuals in a price-fixing conspiracy network in the electrical equipment industry (Baker & Faulkner, 1993). Krebs found that, in the network consisting of the September 11 hijackers (19 in all), Mohamed Atta scored the highest on degree (Krebs, 2001).

- *Discovery of patterns of interaction.* Patterns of interaction between subgroups can be discovered using an SNA approach called blockmodel analysis (Arabie, Boorman, & Levitt, 1978). Given a partitioned network, blockmodel analysis determines the presence or absence of an association between a pair of subgroups by comparing the density of the links between them at a predefined threshold value. In this way, blockmodeling introduces summarized individual interaction details into interactions between groups so that the overall structure of the network becomes more apparent.

SNA also includes visualization methods that present networks graphically. The Smallest Space Analysis (SSA) approach (Wasserman & Faust, 1994), a branch of Multi-Dimensional Scaling (MDS), is used extensively in SNA to produce two-dimensional representations of social networks. In a graphical portrayal of a network produced by SSA, the stronger the association between two nodes or two groups, the closer they appear on the graph; the weaker the association, the farther apart (McAndrew, 1999). Several network analysis tools, such as Analyst's Notebook (Klerks, 2001), Netmap (Goldberg & Senator, 1998), and Watson (Anderson, Arbetter, Benawides, & Longmore-Etheridge, 1994), can automatically draw a graphical representation of a criminal network. However, these tools do not provide much structural analysis functionality and rely on investigators' manual examinations to extract structural patterns.

The six classes of KDD techniques reviewed here constitute the key components of our proposed ISI research framework. Our focus on the KDD methodology, however, does not exclude other approaches. For example, studies using simulation and multi-agent models have shown promise in the "what-if" analysis of the robustness of terrorist and criminal networks (Carley, Dombroski, Tsvetovat, Reminga, & Kamneva, 2003; Carley, Lee, & Krackhardt, 2002).

In the next section, we present several case studies showing the value and potential of different KDD technologies to accomplish the critical objectives of national security.

ISI in Critical Mission Areas: Case Studies

In response to the challenges of national security, the COPLINK Center at the University of Arizona has developed several research projects to address five of the six critical mission areas identified in the National Strategy for Homeland Security report (U.S. Office of Homeland Security, 2002): intelligence and warning, border and transportation security, domestic counter-terrorism, protecting critical infrastructure and key assets, and emergency preparedness and response. The center's main goal is to develop information and knowledge management technologies appropriate for capturing, accessing, analyzing, visualizing, and sharing law enforcement and intelligence-related information (Chen, Zeng, Atabakhsh, Wyzga, & Schroeder, 2003). Through the following eight case studies, we demonstrate how critical mission issues could be addressed using the knowledge discovery approach. For each case study, we discuss its relevance to national security missions, data characteristics, technology used, and selected evaluation results. Quantitative studies focused primarily on the performance of the techniques in terms of effectiveness, accuracy, efficiency, usefulness, and so forth. In qualitative studies where quantitative results are not yet available, we summarize and report comments and feedback from our domain experts.

Intelligence and Warning

Although terrorism depends on surprise (U.S. Office of Homeland Security, 2002), terrorist attacks are not random but require careful planning, preparation, and cooperation before execution. To avoid being preempted by authorities, terrorists may disguise their true identities or hide their illegal objectives and intents behind legal activities. Similarly, criminals may try to minimize the possibility of being identified and captured by using falsified identities. To detect hidden intent and potential for future attacks or offenses is the main goal of intelligence and warning systems. In this section, we present two case studies addressing intelligence and warning needs. The first helped to detect deceptive identity records in police data (Wang, Chen, et al., 2004), while in the second, we present our design for an intelligence Web portal to help trace and monitor the Web sites of terrorist organizations (Chen, Qin, Reid, Chung, Zhou, Xi, et al., in press; Reid, Qin, Chung, Xu, Zhou, Schumaker, 2004).

Case Study 1: Detecting Deceptive Criminal Identities

It is common practice for criminals to lie about the particulars of their identities, such as name, date of birth, address, and social security number, in order to deceive police investigators. Inability to validate identity can be used as a warning mechanism because the deception signals an intent to commit future offenses. In this case study, we focus on uncovering patterns of criminal identity deception based on actual criminal records and suggest an algorithmic approach to revealing false identities (Wang, Chen, et al., 2004).

Data used in this study were authoritative criminal identity records obtained from the Tucson Police Department (TPD). These records were structured database entries containing criminal identity information, such as name, date of birth (DOB), address, identification number (e.g., social security number), race, weight, and height. The total number of criminal identity records stored in the TPD databases was over 1.5 million. In order to study the patterns of criminal identity deception, we selected from the TPD database 372 records involving 24 criminals, each having one real identity record and several deceptive records. These sets of deceptive records were not randomly sampled from the database, but were manually extracted by a police detective expert who has served in law enforcement for 30 years. The expert used convenience sampling, in which he reviewed the list of all identity records and chose the deceptive identity records that he encountered. Because deceptive identities are sparsely distributed in the criminal database, convenience sampling is more effective than random sampling for experimental purposes. As a result, the conclusions may not be statistically valid.

We carefully examined these 372 records and found that deception occurred most often in specific attributes: name, address, birth date, and Social Security Number (SSN). The identity deception patterns in this dataset are shown in Figure 6.2. Name deception, occurring in most cases, includes giving a false first name and a true last name or vice versa, changing the middle initial, and giving a name pronounced similarly but spelled differently. Deception on DOB can consist of, for example, switching places between the month of birth and the day of birth. Similarly, ID deception is often made by changing a few digits of an SSN or by switching their places. In residency deception, criminals usually change only one portion of the address. For example, we found that, in about 87 percent of cases, criminals provided a false street number along with the true street direction, street name, or street type.

To detect deceptive identity records automatically, we employed a similarity-based association mining method to extract associated (similar) record pairs. Based on the deception patterns found, we selected four attributes (name, DOB, SSN, and address) for our analysis. We compared and calculated the similarity between the values of corresponding attributes of each pair of records. If two records were significantly similar, we assumed that at least one of them was deceptive.

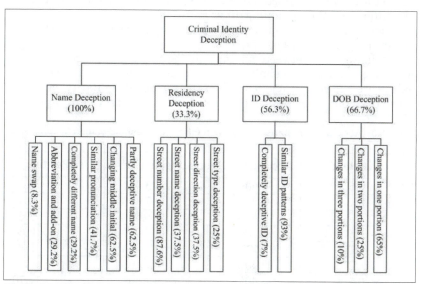

Figure 6.2 Identity deception patterns (each percentage number represents the proportion of records that contain the particular type of deception in the selected dataset).

Because the four selected attributes primarily have string values, we compared two attribute values based on their edit distance (Levenshtein, 1966) and Soundex code (Newcombe et al., 1959). The edit distance between two strings is the minimum number of single character insertions, deletions, and substitutions required to transform one string into the other. Soundex code represents the phonetic pattern of a string. For example, "PEARSE" and "PIERCE" are both coded as "P620." To detect both spelling and phonetic variations between two name strings, edit distance similarity and Soundex similarity were computed separately. In order to capture name exchange deception, similarities were also computed based on different sequences of first name and last name. We took the similarity value from the sequence that had the maximal value between two names. We used only edit distance to compare non-phonetic attributes of DOB, SSN, and address. Each similarity value was normalized between 0 and 1. The similarity value over all four attributes was calculated by means of a normalized Euclidean distance function.

In order to test the performance of our approach, we used convenience sampling again to select another set of 120 records. However in this case, we chose only records with complete information in the name, address, DOB, and SSN fields. The 120 records involved 44 criminals, each of whom had an average of three records in the sample set. Some data were used to train and test our algorithm so that records pointing to the same suspect could be associated with each other. Training and testing were validated by a standard hold-out sampling method. Of the 120 records in the testbed, 80 (66.7 percent) were used for training the algorithm, and the remaining 40 were used for testing.

A similarity matrix was built for all training records. Similarity values in the matrix were used to establish the threshold values appropriate to distinguish between similar and dissimilar pairs. Accuracy rates for correctly recognized similar pairs of records using different threshold values are shown in Table 6.3. When the threshold similarity value was set to 0.52, our algorithm achieved its highest accuracy of 97.4 percent,

Table 6.3 Accuracy comparison based on different threshold values

Threshold	Accuracy	False Negative[*]	False Positive[**]
0.6	76.60%	23.40%	0.00%
0.55	92.20%	7.80%	0.00%
0.54	93.50%	6.50%	2.60%
0.53	96.10%	3.90%	2.60%
0.52	**97.40%**	**2.60%**	**2.60%**
0.51	97.40%	2.60%	6.50%
0.5	97.40%	2.60%	11.70%

*False negative: consider disimilar records as similar ones
**False positive: consider similar records as disimilar ones

with relatively small false negative and false positive rates; both were 2.6 percent.

A similarity matrix was also built for the 40 test records. By application of the optimal threshold value to the testing similarity matrix, records having a similarity value of more than 0.52 were considered to be pointing to the same offender. The accuracy of association in the testing data set is shown in Table 6.4. The result shows that the algorithm is effective (with an accuracy level of 94 percent) in linking deceptive records pointing to the same offender.

Although the case study produced promising results, much more research is needed for deception detection, which we believe is a unique and critical problem for ISI.

Table 6.4   The accuracy of association in the testing data set

Threshold	Accuracy	False Negative	False Positive
0.52	94.0%	6.0%	0.0%

Case Study 2: The "Dark Web" Portal

Because the Internet has become a global platform for information dissemination and communication, terrorists also take advantage of the freedom of cyberspace and construct their own Web sites to propagate terrorist ideology, share information, and recruit new members. Web sites of terrorist organizations may also connect to one another through hyperlinks, forming a "dark Web." We are building an intelligent Web portal, called the Dark Web Portal, to help terrorism researchers collect, access, analyze, and understand terrorist groups (Chen, Qin, et al., in press; Reid et al., 2004). This project consists of three major components: Dark Web testbed building, Dark Web link analysis, and Dark Web Portal building.

- *Dark Web Testbed Building*. Drawing on reliable governmental sources such as the Anti-Defamation League (ADL), FBI, and United States Committee for a Free Lebanon (USCFL), we identified 224 U.S. domestic terrorist groups and 440 international terrorist groups. For U.S. domestic groups, group-generated URLs can be found in FBI reports and the Google Directory. For international groups, we used the group names as queries to search major search engines such as Google and manually identified the group-created URLs from the result lists. To ensure that our testbed covered all the major regions in the world, we sought the assistance of language experts in English, Arabic, Spanish, and Japanese to help us collect URLs in different regions. All URLs collected were manually checked by experts to make sure that they were created by terrorist groups. Once a group's URL was identified, we used the SpidersRUs toolkit, a multilingual Digital Library building tool developed by our own group, to collect all the Web pages under that URL and store them in our testbed. We have collected 500,000 Web pages created by U.S. domestic groups, 400,000 Web pages created by Arabic-speaking groups, 100,000 Web pages created by Spanish-speaking groups, and 2,200 Web pages created by Japanese-speaking groups. This testbed is updated bimonthly.

- *Dark Web Link Analysis and Visualization.* Terrorist groups are not atomized individuals but actors linked to each other through complex networks of direct or mediated exchanges. Identifying how relationships between groups are formed and dissolved in the terrorist group network would enable us to reveal the social milieux and communication channels among terrorist groups across different jurisdictions. Previous studies have shown that the link structure of the Web represents a considerable amount of latent human annotation (Gibson, Kleinberg, & Raghavan, 1998). Thus, by analyzing and visualizing hyperlink structures between terrorist-generated Web sites and their content, we could discover the structure and organization of terrorist group networks, capture network dynamics, and understand their emerging activities (e.g., exploiting formal or informal banking systems, changing identities to take on characteristics more identifiable with Western societies, or creating their own online communities). To test our ideas, we conducted an experiment in which we analyzed and visualized the hyperlink structure between approximately 100,000 Web pages from 46 Web sites in our current testbed. These 46 Web sites were created by four major Arabic-speaking terrorist groups, namely Al-Gama's al-Islamiyya (Islamic Group, IG), Hizballa (Party of God), Al-Jihad (Egyptian Islamic Jihad), and Palestinian Islamic Jihad (PIJ) and their supporters. Hyperlinks between each pair of the 46 Web sites were extracted from the Web pages and a closeness value was calculated for each pair of the 46 Web sites as shown in Figure 6.3. Each node represents a Web site created by one of the 46 groups. A link existing between two nodes means there are hyperlinks between the Web pages of the two sites. We presented this network to several domain experts and confirmed that the structure of the diagram matched the experts' knowledge of how the groups related to each other in the real world. The four clusters represent a logical mapping of the existing relations among the 46 groups. For instance, the Palestinian terrorist group's cluster includes many of these groups' Web sites, as well as their leaders' sites. Examples include the Al-Aqsa Martyrs' Brigade (http://www.katae.baqsa.org), HAMAS (http://www.ezzedeen.net), and PIJ (http://www.abrarway.com).

- *Dark Web Portal Building.* Using the Dark Web Portal, experts are able to locate specific dark Web information in the testbed quickly through keyword search. To address the information overload problem, the Dark Web Portal is designed with post-retrieval components. A modified version of a text summarizer called TXTRACTOR, which uses sentence-selection heuristics to rank and select important text segments (McDonald & Chen, 2002), has been incorporated into the Dark Web Portal. The summarizer can flexibly summarize Web pages so that experts can quickly get the main idea of a page without having to read through it. A categorizer organizes the search results into various folders labeled with the key phrases extracted by the Arizona Noun Phraser (AZNP) (Tolle & Chen, 2000) from the page summaries or titles, thereby facilitating the understanding of different groups of Web pages. A visualizer clusters Web pages into colored regions using the SOM algorithm (Kohonen, 1995), thus reducing information overload when a large number of search results is obtained. Post-retrieval analysis could further reduce information overload, but researchers are limited to data in their native languages and cannot fully utilize the multilingual information in the testbed. To address this problem, we have added a cross-lingual information retrieval (CLIR) component into the portal. On the basis of our previous research, we have developed a dictionary-based CLIR system for use in the Dark Web Portal. It currently accepts English queries and retrieves documents in English, Spanish, Chinese, Japanese, and Arabic. A machine translation

(MT) component will be added to the Dark Web Portal to translate the multilingual information retrieved by the CLIR component back into the experts' native languages.

Because terrorist groups continue to use the Internet as a communication, recruiting, and propaganda tool, a systematic and system-aided approach to studying their presence on the Web is critically needed.

Border and Transportation Security

Terrorists enter a targeted country by air, land, or sea. The government can improve its counter-terrorism and crime-fighting capabilities by creating a "smart border," where information from borders, customs, transportation, and local law enforcement agencies is integrated and analyzed to help locate wanted terrorists or criminals. Our "BorderSafe" project for cross-jurisdictional information integration and sharing (Marshall, Kaza, Xu, Atabakhsh, Petersen, Violette, et al., in press) illustrates how a smart and safe border might be created.

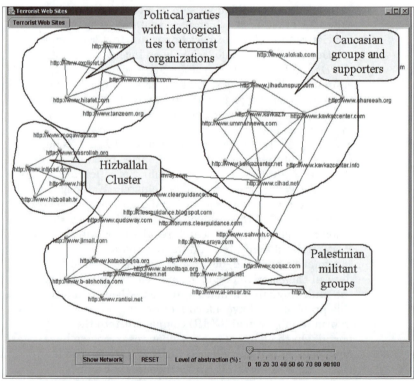

Figure 6.3 **Web site structural relationships between 46 terrorist organizations or affiliated groups.**

Case Study 3: Enhancing BorderSafe

The BorderSafe project is a collaborative research effort involving the University of Arizona's Artificial Intelligence Lab; several law enforcement agencies including the Tucson Police Department (TPD), Phoenix Police Department (PPD), Pima County Sheriff's Office (PCSO), and Tucson Customs and Border Protection (CBP); the San Diego ARJIS (Automated Regional Justice Systems, a regional consortium of more than 50 public safety agencies); the San Diego Supercomputer Center (SDSC); and the Corporation for National Research Initiatives (CNRI).

In this study our objective was to integrate structured, authoritative data from TPD, PCSO, and a limited dataset from CBP containing license plate data of border crossing vehicles. Tables 6.5 and 6.6 present the statistics from the three datasets. TPD's and PCSO's jurisdictions represent a shared community of citizens in Tucson and southern Arizona. They also share intertwined communities of criminals. We found a substantial amount of data overlap among these datasets. Around seven percent of vehicles involved in gang-related, violent, and narcotics crimes were registered outside of Arizona. More than 483,000 people appeared in both the TPD and PCSO datasets, representing 36 percent of the TPD records and 37 percent of the PCSO records. These statistics strongly suggest that sharing information across jurisdictions could help catch criminals.

The federation approach to data integration was employed. We adopted the COPLINK schema as the global schema and developed a transformation mechanism to reconcile the database structure and semantics from a particular database into the global schema. Data were then mapped or transformed to allow shared query processing. In our

Table 6.5   Statistics regarding the TPD and PCSO datasets

	TPD	PCSO
Number of recorded incidents	2.84 million	2.18 million
Number of persons	1.35 million	1.31 million
Number of vehicles	62,656	520,539

Table 6.6   CBP border crossing dataset

Number of records	1,125,155
Number of distinct vehicles	226,207
Number of plates issued in AZ	130,195
Number of plates issued in CA	5,546
Number of plates issued in Mexico	90,466

datasets, establishing automated transformation procedures for legacy PCSO and TPD records into COPLINK format resolved most of the structural and semantic difference issues.

At the instance level, each dataset had a unique key assigned to each person or vehicle, but these unique keys did not match across datasets. To address this problem, vehicles were matched between datasets on the basis of their license plate numbers. We based people matching on input from domain experts and assumed that all records with the same first name, last name, and DOB represented the same person. These heuristics were not perfect; a few incorrect matches resulted and certainly many correct matches may have been missed. We plan to employ our new identity deception detection approach (Wang, Chen, et al., 2004) in the future to improve instance-level matching.

We generated and visualized several criminal networks based on integrated data. We extracted associations between a set of criminals and vehicles from crime incident records. A link was created when two or more criminals or vehicles were listed in the same incident record. In network visualization we differentiated entity types by shape, key attributes by node color, level of activity (measured by number of crimes committed) by node size, data source by link color, and some details in link text or roll-over tool tips. Figure 6.4 shows a network connecting a known narcotics dealer to a border crossing plate.

A qualitative field study provided positive feedback regarding the potential of our data integration approach. Currently, the crime analysts from both TPD and PCSO are using the triangulated, integrated criminal networks generated by our system to monitor vehicles and criminals crossing the border.

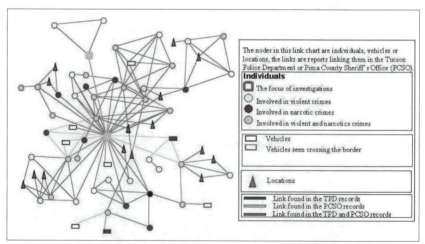

Figure 6.4 A sample criminal network based on integrated data from multiple sources. Nodes and links are color coded in the actual system.

Domestic Counter-Terrorism

As terrorists may be involved in local crimes, state and local law enforcement agencies contribute to national security by investigating and prosecuting these crimes. Terrorism, like gangs and narcotics trafficking, is treated as a type of organized crime in which multiple offenders cooperate to carry out criminal activities. Information technologies that aid in the discovery of cooperative relationships among criminals and reveal the patterns of their interaction would also be helpful in analyzing terrorism. Through three case studies in this section, we show how criminal association information can be extracted from large volumes of data (Hauck et al., 2002) and how structural patterns in criminal or terrorist organizations can be discovered (Xu & Chen, 2003, in press).

Case Study 4: COPLINK Detect

Crime analysts and detectives search for criminal associations to develop investigative leads. However, because association information is not directly available in most existing law enforcement and because intelligence databases and manual searching are extremely time consuming, automatic identification of relationships among criminal entities may significantly speed up investigations. COPLINK Detect is a link analysis system that automatically extracts relationship information from large volumes of crime incident data (Hauck et al., 2002).

Our data were structured crime incident records stored in TPD databases. The TPD's current record management system (RMS) consists of more than 1.5 million crime incident records that contain details of criminal events spanning from 1986 to 2004. Although investigators can access the RMS to tie information together, they must manually search the RMS for connections or existing relationships.

We used the concept space approach (Chen & Lynch, 1992) to identify relationships between entities of interest. Concept space analysis is a type of co-occurrence analysis used in information retrieval. The resulting network-like concept space holds all possible associations between terms—that is, the system retains and ranks every existing link between every pair of concepts. In COPLINK Detect, detailed incident records serve as the underlying space, and concepts are derived from the meaningful terms that occur in each incident. Concept space analysis easily identifies relevant terms and their degree of relationship to the search term. The system output includes relevant terms ranked in the order of their degree of association, thereby distinguishing the most relevant terms from inconsequential ones. From a crime investigation standpoint, concept space analysis can help investigators link known entities to other related entities that might contain useful information for further investigation, such as people and vehicles related to a given suspect. It is considered an example of entity association mining (Lin & Brown, 2003).

Information related to a suspect can move an investigation in the right direction, but revealing relationships among data in one particular incident might fail to capture other relationships from the entire database. In effect, investigators need to review all incident reports related to a suspect and this can be tedious work. The COPLINK Detect system introduces concept space as an alternative method that captures the relationships between four types of entities (person, organization, location, and vehicle) across the entire database. COPLINK Detect also offers an easy-to-use interface and allows searching for relationships among the four types of entities. Figure 6.5 presents the COPLINK Detect interface, showing sample search results for vehicles, relations, and crime case details (Hauck et al., 2002).

We conducted user studies to evaluate the performance and usefulness of COPLINK Detect. Eleven crime analysts and one homicide detective from TPD participated in the longitudinal field study over a four-week period. Crime analysts were experienced in investigating high-profile cases as well as creating statistical reports on criminal activities. They were accustomed to link analysis and are the target user group of COPLINK Detect. Although detectives were not specialized in crime analysis in general, the participating homicide detective was experienced in searching for criminal associations using record management systems. In this study, three major areas were identified where COPLINK Detect provided improved support for crime investigation: link analysis, interface design, and operating efficiency.

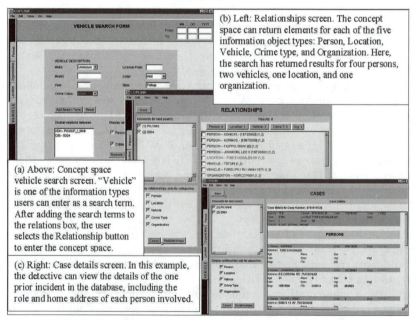

(a) Above: Concept space vehicle search screen. "Vehicle" is one of the information types users can enter as a search term. After adding the search terms to the relations box, the user selects the Relationship button to enter the concept space.

(b) Left: Relationships screen. The concept space can return elements for each of the five information object types: Person, Location, Vehicle, Crime type, and Organization. Here, the search has returned results for four persons, two vehicles, one location, and one organization.

(c) Right: Case details screen. In this example, the detective can view the details of the one prior incident in the database, including the role and home address of each person involved.

Figure 6.5   COPLINK Detect interface showing sample research results.

Case Study 5: Criminal Network Mining

Because organized crime is carried out by networked offenders, investigation naturally depends on network analysis approaches. Grounded in social network analysis methodology, our criminal network-structure mining research aims at helping intelligence and security agencies extract valuable knowledge regarding criminal or terrorist organizations by identifying the central members, subgroups, and overall network structure (Xu & Chen, 2003, in press).

Two datasets from TPD were used in the study. (1) A gang network: The list of gang members consisted of 16 offenders who had been under investigation during the first quarter of 2002. These gang members had been involved in 72 crime incidents of various types (e.g., theft, burglary, aggravated assault, drug offenses) since 1985. We used the concept space approach and generated links between criminals who had committed crimes together, ending with a network of 164 members. (2) A narcotics network: The list for the narcotics network consisted of 71 criminal names. A sergeant from the Gang Unit had been studying the activities of these criminals since 1995. Because most of them had committed crimes related to methamphetamines, the sergeant called this network the "Meth World." These offenders had been involved in 1,206 incidents since 1983. A network of 744 members was generated.

We made use of SNA approaches to extract structural patterns in the criminal networks:

- *Network partition.* We employed hierarchical clustering, namely the complete-link algorithm, to partition a network into subgroups based on relational strength. Clusters obtained represent subgroups. To employ the algorithm, we first transformed co-occurrence weights generated in the previous phrase into distances/dissimilarities. The distance between two clusters was defined as the distance between the pair of nodes drawn from each cluster that was farthest apart. The algorithm worked by merging the two nearest clusters into one cluster at each step and eventually formed a cluster hierarchy. The resulting cluster hierarchy specified groupings of network members at different granularity levels. At lower levels of the hierarchy, clusters (subgroups) tended to be smaller and group members were more closely related. At higher levels of the hierarchy, subgroups are large and group members may be loosely related.

- *Centrality measures.* We used all three centrality measures to identify central members in a given subgroup. The degree of a node could be obtained by counting the total number of links it had to all the other group members. A node's score of betweenness and closeness required the computation of shortest paths (geodesics) using Dijkstra's (1959) algorithm.

- *Blockmodeling.* At a given level of a cluster hierarchy, we compared intergroup link densities with the network's overall link density to determine the presence or absence of intergroup relationships.

- *Visualization.* To map a criminal network onto a two-dimensional display, we employed multi-dimensional scaling (MDS) to generate *x-y* coordinates for each member in a network. We chose Torgerson's (1952) classical metric MDS algorithm because

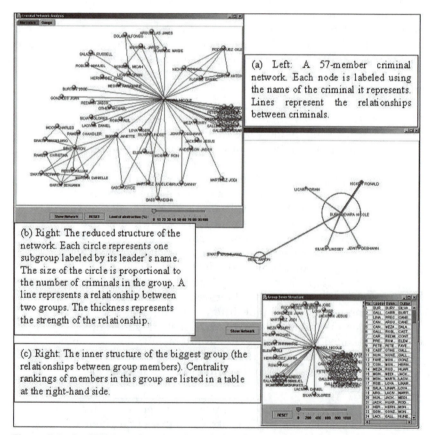

(a) Left: A 57-member criminal network. Each node is labeled using the name of the criminal it represents. Lines represent the relationships between criminals.

(b) Right: The reduced structure of the network. Each circle represents one subgroup labeled by its leader's name. The size of the circle is proportional to the number of criminals in the group. A line represents a relationship between two groups. The thickness represents the strength of the relationship.

(c) Right: The inner structure of the biggest group (the relationships between group members). Centrality rankings of members in this group are listed in a table at the right-hand side.

Figure 6.6   An SNA-based system for criminal network analysis and visualization.

distances transformed from co-occurrence weights were quantitative data.

A graphical user interface was provided to visualize criminal networks. Figure 6.6 shows the screenshot of our prototype system. In this example, each node was labeled with the name of the criminal it represented. Criminal names were scrubbed for data confidentiality. A straight line connecting two nodes indicated that two corresponding criminals committed crimes together and thus were related. To find subgroups and interaction patterns between groups, a user could adjust the "level of abstraction" slider at the bottom of the panel. A high level of abstraction corresponded with a high distance level in the cluster hierarchy. Group members' rankings in centrality are listed in a table.

A qualitative study was conducted to evaluate the prototype system. We presented the two testing networks to domain experts at TPD and received encouraging feedback (Xu & Chen, 2003):

- *Subgroups detected were mostly correct.* The domain experts checked and validated the members in each group. These groups had different characteristics with different specialties or crime preferences. We also found that although relationships in our network were extracted based on crime incidents, they reflected relationships between criminals based on friendship, kinship, and even conflicts.

- *Centrality measures provided ways of identifying key members in a network.* According to our domain experts, betweenness was a reliable measure to identify gatekeepers between subgroups. However, degree sometimes misidentified leaders because the criminals with the most connections to others may not always be the leaders. Leaders may be smart enough to hide behind other criminals to avoid police contact.

- *Interaction patterns identified could help reveal relationships that previously had been overlooked.* Our system could generate the "big picture" for a complex network. As a result some relationships between criminal groups that had been overlooked before the system were made easier to identify.

- *Saving investigation time.* Our domain experts had obtained knowledge about the gang and narcotics organizations based on several years of work. Using information gathered from a large number of arrests and interviews, they had built the networks incrementally by linking new criminals to known gangs in the network and then studying the organization of these networks. Because there was no structural analysis tool available, they did all of this by hand. With the help of our system, they expected that substantial time would be saved in network creation and structural analysis.

- *Saving training time for new investigators.* New investigators who did not have sufficient knowledge of criminal organizations and individuals could use the system to grasp the essence of the network and related crime history quickly. They would not have to spend as much time studying hundreds of incident reports.

- *Helping prove guilt of criminals in court.* The relationships discovered between individual criminals and criminal groups would be helpful for proving guilt when presented at court for prosecution.

Case Study 6: Analyzing Terrorist Networks

As part of the worldwide Islamic Jihadist movement, a number of terrorist organizations have targeted the West. Terrorism and terrorist attacks pose severe threats and have caused significant damage worldwide. Only with an in-depth understanding of terrorism and terrorist organizations can societies defend themselves against the threats. Because terrorist organizations often operate in networks through which individual terrorists collaborate to carry out attacks (Klerks, 2001; Krebs, 2001), network analysis can help uncover valuable information by studying the networks' structural properties (Xu & Chen, in press). We have employed techniques and methods from SNA and Web mining to address the problem of structural analysis of terrorist networks.

The objective of this case study was to examine the potential of network analysis tools for terrorist analysis. By comparing our findings with experts' input we sought to ascertain whether automatic analysis of structural properties of a terrorist network would generate information consistent with expert knowledge.

In this study, we focused on the structural properties of a set of Islamic terrorist networks, including Osama bin Laden's Al Qaeda. In a recently published book, Sageman (2004) documented the history and evolution of these terrorist organizations, which he terms Global Salafi Jihad (GSJ). Sageman is a social psychologist and formerly served as a foreign service officer. During the Afghan-Soviet war, from 1986 to 1989, he dealt with Islamic fundamentalists on a daily basis and developed substantial expertise in terrorism and terrorist organizations. Drawing upon various open sources, such as news articles and court transcripts, he collected data on

364 terrorists in the GSJ network regarding their background, religious beliefs, social relations, and the terrorist attacks in which they participated. There are three types of social relations among these terrorists: personal links (e.g., acquaintance, friendship, and kinship), operational links (e.g., collaborators in the same attack), and relations formed after attacks (Sageman, 2004). Sageman identified four major terrorist groups on the basis of their geographical locations: Central Staff, Core Arab, Maghreb Arab, and Southeast Asian. Each group has its own leaders. For example, Osama bin Laden is the leader of the Central Staff group, which connects to the other three groups through several lieutenants.

We analyzed the GSJ network based on the social relation data contained in a spreadsheet provided by Sageman. Using the SNA visualization approach, we depicted the GSJ network graphically as shown in Figure 6.7.

- *Centrality analysis.* Considering all three types of social relations, we found that the four group leaders were among the 11 most popular members, where popularity was represented by degree measure. For

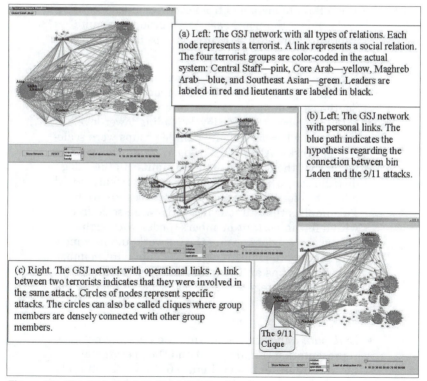

(a) Left: The GSJ network with all types of relations. Each node represents a terrorist. A link represents a social relation. The four terrorist groups are color-coded in the actual system: Central Staff—pink, Core Arab—yellow, Maghreb Arab—blue, and Southeast Asian—green. Leaders are labeled in red and lieutenants are labeled in black.

(b) Left: The GSJ network with personal links. The blue path indicates the hypothesis regarding the connection between bin Laden and the 9/11 attacks.

(c) Right. The GSJ network with operational links. A link between two terrorists indicates that they were involved in the same attack. Circles of nodes represent specific attacks. The circles can also be called cliques where group members are densely connected with other group members.

The 9/11 Clique

Figure 6.7 The Global Salafi Jihad (GSJ) network.

example, Osama bin Laden had 72 links to other terrorists and ranked second in degree. Although he was not a leader, Hambali had the highest degree score and played an important role in connecting different terrorist groups (see Figure 6.7a). Moreover, the lieutenants tended to have high scores in betweenness and served as gatekeepers between groups. The analysis implies that centrality measures could be useful for identifying important members of a terrorist network.

- *Subgroup analysis*. The four terrorist groups depicted in Figure 6.7 were color coded in the actual GSJ network system using Sageman's advice. To find out whether these geographically based groups were also structurally cohesive, we calculated the cohesion score (Wasserman & Faust, 1994) of each group. We found that all these groups had high cohesion scores. The Southeast Asian group scored the highest in cohesion. This may suggest that members in this group tended to be more closely related to members of their own group than to members from other groups. According to Sageman, the Southeast Asian group was quite different from the other three groups in terms of their religious beliefs and missions.

- *Network structure analysis*. Sageman had reported that these groups had different structures: The Southeast Asian group's structure was hierarchical with members at higher levels leading lower-level members, whereas the other three groups were scale-free networks (Albert & Barabási, 2002). However, we found that the four groups were similar in their degree distribution, which was a power-law distribution with a long tail for large values of degree (see Figure 6.8). This implies that all four networks were scale-free, with a few important members (nodes with high degree scores) dominating the network and new members tending to join through these dominant members. This finding has an important policy implication: Disruptive strategies should potentially be focused on central members in a terrorist network (Strickland, 2002a, 2002b, 2002c, 2002d, 2002e).

- *Link path analysis*. Comparing the personal network representation (Figure 6.7b) and the operational network representation (Figure 6.7c), we found that some important members did not have direct personal links to an attack prior to execution. For example, neither Osama bin Laden, Khalid Sheikh Mohammed, nor Hambali had direct personal links to terrorists in the 9/11 attack clique. We performed link path analysis to find out the shortest paths of personal links leading to the 9/11 terrorists. One of our hypotheses was that Osama bin Laden connected to the 9/11 clique through a four-hop path: bin Laden—Nashiri—ZaMihd—Mihdhar—Shibh (the dark path in Figure 6.7b). Although this hypothesis turned out to be wrong according to Sageman's feedback (other information was needed to establish the link), the analysis showed the potential of using link path analysis to generate hypotheses about the motives and planning processes behind terrorist attacks.

Protecting Critical Infrastructure and Key Assets

The Internet is a critical infrastructure and asset in the information age. Cybercriminals have been using various Web-based channels (e.g., e-mail, Web sites, Internet newsgroups, chat rooms) to distribute illegal materials. One common characteristic of these channels is anonymity. People usually do not need to provide information about their real identity, such as name, age, gender, and address, in order to participate in cyberactivities. Compared with conventional crimes, cybercrime conducted through such anonymous channels creates novel challenges for

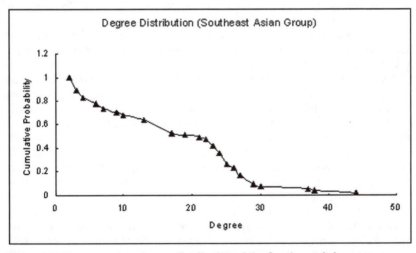

Figure 6.8 The power-law degree distribution of the Southeast Asian group.

researchers and law enforcement agencies engaged in criminal identity tracing. The situation is further complicated by the enormous number of cyberusers and activities, making the manual approach to criminal identity tracing impossible. Law enforcement agencies urgently need approaches that automate criminal identity tracing in cyberspace and allow investigators to prioritize their tasks and focus on major criminals. This case study demonstrates the potential of using authorship analysis with carefully selected feature sets and effective classification techniques for criminal identity tracing in cyberspace (Zheng et al., 2003).

Case Study 7: Identity Tracing in Cyberspace

Data used in this study were from open sources. Three datasets, two in English and one in Chinese, were collected. One of the English datasets consisted of 153 Usenet newsgroup illegal sales of pirated CDs and software messages. We manually identified the nine most active users (represented by a unique ID and e-mail address) who posted messages in these newsgroups. The Chinese dataset contained 70 Bulletin Board System (BBS) illegal CD and software for-sale messages downloaded from a popular Chinese BBS.

The two key techniques used in this study were feature selection and classification. The objective was to classify text messages into different classes with each class representing one author. Based on a review of previous studies on text and e-mail authorship analysis, along with the specific characteristics of the messages in our datasets, we selected a large number of features that were potentially useful for identifying message authors. Three types of features were used: *style markers* (content-free features such as frequency of function word, total number of punctuation marks, and average sentence length), *structural features* (such as use of a greeting statement, position of requoted text, use of farewell statement), and *content-specific features* (such as frequency of keywords, special character of content).

For classification analysis, three popular classifiers were selected including the C4.5 decision tree algorithm (Quinlan, 1986), backpropagation neural networks (Lippmann, 1987), and support vector machines (SVM) (Cristianini & Shawe-Taylor, 2000; Hsu & Lin, 2002). Each individual classifier had been employed in previous authorship analysis research (Diederich, Kindermann, Leopold, & Paass, 2003). In general, SVM and neural networks had exhibited better performance than decision trees (Diederich et al., 2003). However, most previous authorship studies had been based on corpora of newspaper articles such as *The Federalist Papers*. Because online messages are quite different from formal articles in style, we needed to test the performances of these three algorithms on our datasets.

The procedure of the experiment was as follows: Three experiments were conducted on the newsgroup dataset with one classifier at a time. First, 205 style markers (67 for the Chinese BBS dataset) were used, nine structural features were added in the second run, and nine content-specific features were added in the third run. A 30-fold cross-validation testing method was used in all experiments.

We used *accuracy, recall,* and *precision* to evaluate the prediction performance of the three classifiers. Accuracy represents the overall prediction performance of a classifier. For each author, we used precision and recall to measure the effectiveness of a classifier. The three measures are defined in equations (1)–(3).

(1) $\quad$ Accuracy= $\dfrac{\text{Number of messages with author correctly identified}}{\text{Total number of messages}}$

(2) $\quad$ Precision= $\dfrac{\text{Number of messages correctly assigned to the author}}{\text{Total number of messages assigned to the author}}$

(3) $\quad$ Recall= $\dfrac{\text{Number of messages correctly assigned to the author}}{\text{Total number of messages written by the author}}$

We summarize the results as follows:

- *SVM and neural networks outperformed the C4.5 decision tree algorithm.* For example, in regards to the application of style markers to the e-mail dataset, the C4.5, neural networks, and SVM achieved accuracies of 74.29 percent, 81.11 percent, and 82.86 percent, respectively. SVM also consistently achieved higher accuracy, precision, and recall than the neural networks. However, the performance differences between SVM and neural networks were relatively small. Our results were generally consistent with previous studies, in that neural networks and SVM typically achieve better performance than decision tree algorithms (Diederich et al., 2003).

- *Use of style markers and structural features outperformed use of style markers only.* We achieved significantly higher accuracy levels for all three datasets (p-values were below 0.05) by adopting the structural features. This possibly resulted from an author's consistent writing patterns being evident in the message's structural features.

- *Use of style markers, structural features, and content-specific features did not achieve better performance than use of style markers and structural features.* The results indicated that using content-specific features as additional features did not improve the authorship prediction performance significantly (with p-value of 0.3086). We thought this was because authors of illegal messages typically included diverse content in their messages and little additional information could be derived from the message content to determine authorship. We also observed that high levels of accuracy were obtained when style markers alone were used as input features for the English datasets. The accuracy level ranged from 71 to 89 percent. The results indicated that style markers alone contain a large amount of information about people's online message writing styles and are surprisingly robust in predicting the authorship.

- *There was a significant drop in prediction performance measures for the Chinese BBS dataset in comparison to the English datasets.* For example, when using style markers only, C4.5 achieved average accuracies of 86.28 and 74.29 percent for the English newsgroup and e-mail datasets, whereas for the Chinese dataset, it achieved an average accuracy of only 54.83 percent. A possible reason was that only 67 Chinese style markers were used in the experiments, significantly fewer than the 205 style markers used with the English dataset. We expect to achieve higher prediction performances if additional Chinese style markers are identified and included. We also observed that when structural features were added, all three algorithms achieved relatively high precision, recall, and accuracy (from 71 to 83 percent) for the Chinese dataset. Considering the significant language differences, our proposed approach to the problem of online message identity tracing appears promising in a multilingual context.

Similar to "finger-print" and "voice-print" that could help identify a person, we believe that there is a need and potential for developing a robust multilingual "write-print" model based on an individual's unique writing style. Such a model, possibly building on research in stylometrics (Williams, 1975) would have strong value for cybercrime investigation.

Emergency Preparedness and Responses

Terrorist attacks can cause devastating damage to a society through the use of chemical, biological, or radiological weapons. Currently, a large amount of infectious disease data is being collected by various laboratories, health care providers, and government agencies at local, state, national, and international levels (Pinner, Rebmann, Schuchat, & Hughes, 2003). However, access to some of these data sources and related search and reporting functionalities may be limited to the agencies that have developed such systems (Kay, Timperi, Morse, Forslund, McGowan, & O'Brien, 1998), reducing the effective use of infectious disease data in national and global contexts. In addition, real-time data sharing, especially of databases across species and jurisdictions, could enhance expert scientific review and rapid response using input and action triggers provided by multiple government and public health partners. In this case study we discuss our ongoing research and system development efforts designed to address some of these challenges. We aim to develop scalable technologies and related standards and protocols needed for a national infectious disease information infrastructure (Zeng, Chen, Tseng, Larson, Eidson, Gotham, et al., 2004).

Case Study 8: The WNV-BOT Portal

Our research focuses on two prominent infectious diseases: *West Nile Virus* (WNV) and *Botulism*. These two diseases were chosen because of their significant public health and national security implications and the availability of related datasets for the states of New York and California. We developed a research prototype called the WNV-BOT Portal system, which provides integrated, Web-enabled access to a variety of distributed data sources including the New York State Department of Health (NYSDH), the California Department of Health Services (CADHS), and other federal sources (e.g., the United States Geological Survey [USGS]). It also provides advanced information visualization capabilities as well as predictive modeling support.

Architecturally, the WNV-BOT Portal consists of three major components: a *Web portal*, a *data store*, and a *communication backbone*. The Web portal implements the user interface and provides the following main functionalities: (1) searching and querying available WNV/BOT datasets, (2) visualizing WNV/BOT datasets using spatial-temporal visualization, (3) accessing analysis and prediction functions, and (4) accessing the alerting mechanism.

To enable data interoperability, we use Health Level Seven (HL7) standards (http://www.hl7.org) as the main storage format. In our data warehousing approach, contributing data providers transmit data to WNV-BOT Portal as HL7-compliant XML messages (through a secure network connection if necessary). After receiving these XML messages, the WNV-BOT Portal adds them directly to its data store. To alleviate potential computational performance problems associated with this HL7 XML-based approach, we have identified a core set of data fields, on which searches could be performed efficiently.

An important function of the data store layer is data ingest and access control. The data ingest control module is responsible for checking the integrity and authenticity of data feeds from the underlying information sources. The access control module is responsible for granting and restricting user access to sensitive data.

The communication backbone component enables data exchanges between the WNV-BOT Portal and the underlying WNV/BOT sources based upon the CDC's (Centers for Disease Control and Prevention) Electronic Disease Surveillance System (NEDSS) and HL7 standards. It uses a collection of source-specific "connectors" to communicate with underlying sources. We use the connector linking NYSDOH's Health Information Network (HIN) system and WNV-BOT Portal to illustrate a typical design of such connectors. The data sent from HIN to the portal system are transmitted in a "push" manner. HIN sends secure Public Health Information Network Messaging System (PHIN MS) messages to the portal at prespecified time intervals. The connector at the portal side runs a data receiver daemon listening for incoming messages. After a message is received, the connector checks for data integrity syntactically and invokes the data normalization subroutine. Then the connector stores the verified message in the portal's internal data store through its data ingest control module. Other data sources (e.g., those from USGS) may have "pull-" type connectors, which periodically download information from the source Web sites and examine and store data in the portal's internal data store. In general, the communication backbone component provides data receiving and sending functionalities, source-specific data normalization, as well as data encryption capabilities.

The WNV-BOT Portal makes available the Spatial Temporal Visualizer (STV) (Buetow et al., 2003) to facilitate exploration of infectious disease case data and to summarize query results. STV has three integrated and synchronized views: periodic, timeline, and GIS. Figure

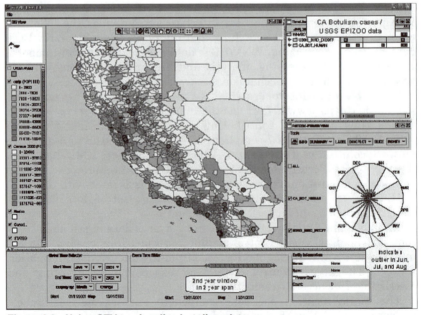

Figure 6.9   Using STV to visualize botulism data.

6.9 illustrates how these three views can be used to explore the infectious disease dataset. The top-left panel shows the GIS view. The user can select multiple datasets to be shown on the map in a layered manner using the checkboxes. The top-right panel corresponds to the timeline view displaying the occurrences of various cases using a Gantt chart-like display. The user can also access case details easily by using the tree display located left of the timeline display. Below the timeline view is the periodic view through which the user can identify periodic temporal patterns (e.g., which months have an unusually high number of cases). The bottom portion of the interface allows the user to specify subsets of data to be displayed and analyzed.

Our project has supported exploration of, and experimentation with, technological infrastructures needed for a full-fledged implementation of a national infectious disease information infrastructure and has helped foster information sharing and collaboration among related government agencies at state and federal levels. In addition, we have obtained important insights into, and hands-on experience with, various important policy-related challenges faced by developing a national infrastructure. For example, a nontrivial part of our project activity has been devoted to developing data-sharing agreements between project partners from different states.

Our ongoing technical research is focusing on two aspects of infectious disease informatics: hotspot analysis and efficient alerting and dissemination. For WNV, localized clusters of dead birds typically identify high-risk disease areas. Automatic detection of dead bird clusters using hotspot analysis can help predict disease outbreaks and allocate prevention/control resources effectively. Initial experimental results indicate that these techniques are promising for disease informatics analysis. We are planning to augment existing predictive models by considering additional environmental factors (e.g., weather information, bird migration patterns), and tailoring data mining techniques for infectious disease datasets that have prominent temporal features.

Case Study Summary

We summarize in Table 6.7 the eight case studies in terms of their data characteristics, technologies employed, and the national security missions they addressed using our proposed ISI research framework.

The ISI Partnership Framework

In order to accomplish the six critical mission areas of national security, the Department of Homeland Security has proposed establishing a network of laboratories consisting of satellite research centers across the nation (U.S. Office of Homeland Security, 2002). The purpose is to create a multidisciplinary environment for developing technologies to counter various threats to homeland security. However, information sharing and

Table 6.7 Summary of ISI case studies

Case Study	Project	Data Characteristics	Technologies Used	Critical Mission Area Addressed
1	Identity deception detection	• Authoritative source • Structured criminal identity records	• Association mining • Similarity-based	Intelligence and warning
2	Dark Web Portal	• Open source • Web hyperlink data	• Cluster analysis • Visualization	Intelligence and warning
3	BorderSafe	• Authoritative source • Structured data	• Information sharing and integration • Database federation	Border and transportation security
4	COPLINK Detect	• Authoritative source • Structured data	• Association mining • Statistical-based	Domestic counter-terrorism
5	Criminal network analysis	• Authoritative source • Structured data	• Social network analysis • Cluster analysis • Visualization	Domestic counter-terrorism
6	Terrorist network analysis	• Open source • Text data, structured data	• Intelligence text mining • Social network analysis	Domestic counter-terrorism
7	Identity tracing in cyberspace	• Open source • Structured data	• Intelligence text mining • Classification	Protecting critical infrastructure and key assets
8	WNV-BOT Portal	• Authoritative source • Structured data	• Information sharing and integration • Spatial and temporal visualization	Emergency preparedness and responses

collaboration across different jurisdictions, agencies, and research institutes is not merely a technical issue. A variety of social, organizational, and political barriers needs to addressed, including:

- *Security and confidentiality.* In the intelligence and law enforcement domain, security is of great concern. Data regarding crimes, criminals, terrorist organizations, and potential terrorist attacks may be highly sensitive and confidential in nature. Improper use of data could lead to fatal consequences.

- *Trust and willingness to share information.* Different agencies may not be motivated to share information and collaborate if there is no immediate gain. They may also fear that information being shared will be misused, resulting in legal liabilities.

- *Data ownership and access control.* The questions that need to be addressed are: Who owns a particular data set? Who is allowed to access, aggregate, or input data? Who owns the derivative data (knowledge)? For both original and derivative data, who is allowed to distribute them to whom?

The COPLINK Center at the Artificial Intelligence Lab of the University of Arizona, as a leading research center for law enforcement and intelligence information and knowledge management, intends to become a part of the national network of laboratories. During its development over the past decade the COPLINK Center has encountered many of these non-technical challenges in its partnerships with various law enforcement and federal agencies. In this section, we summarize some of our experiences and lessons learned.

Ensuring Data Security and Confidentiality

In any data sharing initiative, it is essential to make sure that the data shared between agencies are secure and that the privacy of individuals is respected. In our research we have taken the necessary measures to ensure data privacy, security, and confidentiality. Data shared among law enforcement agencies, such as TPD, PPD, and CBP, contained only law enforcement data and were available only to individuals screened by these agencies using a combination of TPD Background Check, Employee Non-Disclosure Agreement (NDA), and the Terminal Operator Certificate (TOC) test.

All personnel who have access to law enforcement data fill out background forms provided by TPD and have their fingerprints taken at TPD. They also sign a nondisclosure agreement provided by TPD. In addition, they take the TOC test every year. The background information and fingerprints are then checked by TPD investigators to ensure lack of involvement in criminal activity and to verify identity.

In addition to these forms and test, all law enforcement data in the University of Arizona COPLINK Center reside behind a firewall and in a secure room accessible only by activated cards to those who have met the security criteria. As soon as an employee stops working on projects related to law enforcement data, his or her card is deactivated. However, the NDA is perpetual and remains in effect even after a researcher or employee leaves. These requirements are similar to those imposed upon noncommissioned civilian personnel in a police department.

Reaching Agreements Among Partners

Federal, state, and local regulations require that agreements between agencies within their respective jurisdictions receive advanced approval from their governing hierarchy. This precludes informal information sharing agreements between those agencies. We found that requirements varied from agency to agency according to the statutes by which they were governed.

For instance, the ordinances governing information sharing by the city of Tucson differed somewhat from those governing the city of Phoenix. This necessitated numerous attempts and passes at proposed documents by each city's law enforcement and legal staffs before a final draft could be settled upon for approval by the city councils. We found that similar language existed in the ordinances and statutes governing this exchange, but that the processes varied significantly. It appears that the level of bureaucracy is proportional to the size of the jurisdiction.

TPD has recently developed a generic Inter-Governmental Agreement (IGA) that could be adopted between different law enforcement agencies. This IGA was condensed from memoranda of understanding (MOUs), policies, and agreements that previously existed in various forms between numerous agencies. The IGA was drafted to be generic, including language from those laws but excluding reference to any particular chapter or section. This allowed the required verbiage to exist in the document without being specific to any jurisdiction.

Sharing information between agencies with disparate information systems has also led to the bridging of boundaries between software vendors and agencies (their customers). We took care not to violate licensing terms by ensuring that nondisclosure agreements existed and that contract language assured compliance with the vendors' licensing policies.

We believe MOUs and IGAs can be used as templates of information sharing agreements and contracts, and can serve as components of an ISI partnership framework. We plan to provide free access to these legal agreement templates to help facilitate the process of information sharing and collaboration across agencies and research institutions in the future.

Conclusions and Future Directions

In this chapter we have discussed the technical issues related to intelligence and security informatics research, which supports accomplishment of the critical missions of national security. We have proposed a research framework addressing the technical challenges facing counterterrorism and crime-fighting applications, with a primary focus on knowledge discovery from databases (KDD). We have identified and incorporated into the framework six classes of ISI technologies: information sharing and collaboration, crime association mining, crime classification and clustering, intelligence text mining, spatial and temporal analysis of crime patterns, and criminal network analysis. We have also presented a set of COPLINK case studies, ranging from the detection of criminal identity deception to an intelligent Web portal for monitoring terrorist Web sites, thus demonstrating the potential of ISI technologies for contributing to the critical missions of national security.

As this new ISI domain continues to evolve, several important directions need to be pursued, including technology development; testbed creation; and social, organizational, and policy studies:

- New technologies need to be developed and many existing information technologies should be re-examined and adapted for national security applications. The knowledge discovery perspective provides a promising direction. However, new technologies should be developed in a legal and ethical framework that does not compromise the privacy or civil liberties of private citizens.

- Large scale, nonsensitive data testbeds that incorporate data from diverse, authoritative, and open sources and in different formats should be created and made available to the ISI research community. Lack of real data has been a long-standing problem in intelligence- and security-related research. Many researchers are forced to use simulated or synthetic data that may not resemble actual crime data characteristics. Furthermore, comparing competing technical approaches has been difficult because of the lack of standard test collections. A comprehensive and non-sensitive open source data collection, analogous to the Message Understanding Conference collection, would be of great value for ISI researchers to experiment, test, and evaluate various technologies and to compare and share findings, insights, and knowledge. Advanced methods may need to be employed to scrub data contained in the non-open source testbed to ensure data confidentiality while preserving its characteristics and underlying structures.

The ultimate goal of ISI research is to enhance national security. However, the question of how this type of research has and will have an impact on society, organizations, and the general public remains unanswered. Researchers from sociology, political science, organizational and management sciences, psychology, and education can contribute substantially to this task.

We hope that active ISI research will help improve knowledge discovery and dissemination; enhance information sharing and collaboration among academics, industry, and local, state, and federal agencies; and thereby promote positive societal outcomes.

Acknowledgments

The projects reported in the case studies have been funded mainly by the following grants:

National Institute of Justice (NIJ), COPLINK: Database Integration and Access for a Law Enforcement Intranet, 1997–2000.

National Science Foundation (NSF), Digital Government Program, COPLINK Center: Information and Knowledge Management for Law Enforcement, 2000–2003.

National Science Foundation (NSF) and Central Intelligence Agency (CIA), Knowledge Discovery and Dissemination Program, Creating an Intelligence Research Testbed, 2002–2003.

National Science Foundation (NSF), Information Technology Research Program COPLINK Center for Intelligence and Security Informatics Research, 2003–2005.

Department of Homeland Security (DHS), BorderSafe Program, Criminal Activity Network Analysis and Visualization, 2002–2005.

National Science Foundation (NSF), Disease Informatics Program, NV/BOT Portal: Developing a National Infectious Disease Information Infrastructure, 2003–2004.

References

Adderley, R., & Musgrove, P. B. (2001). Data mining case study: Modeling the behavior of offenders who commit serious sexual assaults. *Proceedings of the 7th ACM SIGKDD International Conference on Knowledge Discovery and Data Mining*, 215–220.

Agrawal, R., Imielinski, T., & Swami, A. (1993). Mining association rules between sets of items in large databases. *Proceedings of the ACM SIGMOD International Conference on Management of Data*, 207–216.

Albert, R., & Barabási, A.-L. (2002). Statistical mechanics of complex networks. *Reviews of Modern Physics, 74*(1), 47–97.

Aleskerov, E., Freisleben, B., & Rao, B. (1997). CARDWATCH: A neural network based database mining system for credit card fraud detection. *Proceedings of Computational Intelligence for Financial Engineering (CIFEr)*, 220–226.

American Civil Liberties Union. (2004). *MATRIX: Myths and reality*. Retrieved July 27, 2004, from http://www.aclu.org/Privacy/Privacy.cfm?ID=14894&c=130

Anderberg, M. R. (1973). *Cluster analysis for applications*. New York: Academic Press.

Anderson, T., Arbetter, L., Benawides, A., & Longmore-Etheridge, A. (1994). Security works. *Security Management, 38*(17), 17–20.

Arabie, P., Boorman, S. A., & Levitt, P. R. (1978). Constructing blockmodels: How and why. *Journal of Mathematical Psychology, 17*, 21–63.

Badiru, A. B., Karasz, J. M., & Holloway, B. T. (1988). AREST: Armed Robbery Eidetic Suspect Typing expert system. *Journal of Police Science and Administration, 16*, 210–216.

Baker, W. E., & Faulkner, R. R. (1993). The social organization of conspiracy: Illegal networks in the heavy electrical equipment industry. *American Sociological Review, 58*(12), 837–860.

Baluja, S., Mittal, V., & Sukthankar, R. (1999). Applying machine learning for high performance named-entity extraction. In N. Cercone, K. Naruedomkul, & K. Kogure (Eds.), *PACLING '99: Proceedings of the Conference (Pacific Association of Computational Linguistics)* (pp. 1–14). Waterloo, Ont.: Department of Computer Science, University of Waterloo.

Bell, G. S., & Sethi, A. (2001). Matching records in a national medical patient index. *Communications of the ACM, 44*(9), 83–88.

Berndt, D. J., Bhat, S., Fisher, J. W., Hevner, A. R., & Studnicki, J. (2004). Data analytics for bioterrorism surveillance. *Proceedings of the Second Symposium on Intelligence and Security Informatics (ISI'04)*, 17–28.

Berndt, D. J., Hevner, A. R., & Studnicki, J. (2003). Bioterrorism surveillance with real-time data warehousing. *Proceedings of the First NSF/NIJ Symposium on Intelligence and Security Informatics (ISI'03)*, 322–335.

Borthwick, A., Sterling, J., Agichtein, E., & Grishman, R. (1998). NYU: Description of the MENE named entity system as used in MUC-7. *Proceedings of the 7th Message Understanding Conference (MUC-7)*. Retrieved August 19, 2004, from http://www.itl.nist.gov/iaui/894.02/related_projects/muc/proceedings/muc_7_proceedings/nyu_english_named_entity.pdf

Bowen, J. E. (1994). An expert system for police investigators of economic crimes. *Expert Systems with Applications, 7*(2), 235–248.

Brahan, J. W., Lam, K. P., Chan, H., & Leung, W. (1998). AICAMS: Artificial Intelligence Crime Analysis and Management System. *Knowledge-Based Systems, 11*, 355–361.

Brantingham, P., & Brantingham, P. (1981). *Environmental criminology*. Beverly Hills, CA: Sage.

Brown, D. E. (1998). The Regional Crime Analysis Program (RECAP): A framework for mining data to catch criminals. *Proceedings of the 1998 International Conference on Systems, Man, and Cybernetics* (Vol. 3, pp. 2848–2853). Piscataway, NJ: IEEE.

Brown, D. E., Dalton, J., & Hoyle, H. (2004). Spatial forecast methods for terrorism events in urban environments. *Proceedings of the Second Symposium on Intelligence and Security Informatics (ISI'04)*, 426–435.

Brown, D. E., & Hagen, S. (2002). Data association methods with applications to law enforcement. *Decision Support Systems, 34*(4), 369–378.

Brown, D. E., & Oxford, R. B. (2001). Data mining time series with applications to crime analysis. *Proceedings of the 2001 IEEE International Conference on Systems, Man & Cybernetics Conference* (Vol. 3, pp. 1453–1458). Piscataway, NJ: IEEE.

Brown, M. (1998). *Future Alert Contact Network: Reducing crime via early notification*. Retrieved July 27, 2004, from http://pti.nw.dc.us/solutions/solutions98/public_safety/charlotte.html

Buccella, A., Cechich, A., & Brisaboa, N. R. (2003). An ontology approach to data integration. *Journal of Computer Science and Technology, 3*(2), 62–68.

Buetow, T., Chaboya, L., O'Toole, C., Cushna, T., Daspit, D., Peterson, T., et al. (2003). A spatial temporal visualizer for law enforcement. *Proceedings of the First NSF/NIJ Symposium on Intelligence and Security Informatics (ISI'03)*, 181–193.

Burt, R. S. (1976). Positions in networks. *Social Forces, 55*, 93–122.

Carley, K. M., Dombroski, M., Tsvetovat, M., Reminga, J., & Kamneva, N. (2003). Destabilizing dynamic covert networks. *Proceedings of the 8th International Command*

and *Control Research and Technology Symposium*. Retrieved August 19, 2004, from http://www.dodccrp.org/events/2003/8th_ICCRTS/pdf/021.pdf

Carley, K. M., Lee, J., & Krackhardt, D. (2002). Destabilizing networks. *Connections, 24*(3), 79–92.

Chan, P. K., & Stolfo, S. J. (1998). Toward scalable learning with non-uniform class and cost distributions: A case study in credit card fraud detection. *Proceedings of the 4th International Conference on Knowledge Discovery and Data Mining (KDD'98)*, 164–168.

Chau, M., Xu, J., & Chen, H. (2002). Extracting meaningful entities from police narrative reports. *Proceedings of the National Conference on Digital Government Research*. Retrieved August 19, 2004, from http://www.digitalgovernment.org/library/library/pdf/chau2.pdf

Chen, H. (2001). *Knowledge management systems: A text mining perspective*. Tucson: The University of Arizona.

Chen, H., Chung, W., Xu, J., Wang, G., Chau, M., & Qin, Y. (2004). Crime data mining: A general framework and some examples. *IEEE Computer, 37*(4), 50–56.

Chen, H., Houston, A. L., Sewell, R. R., & Schatz, B. R. (1998). Internet browsing and searching: User evaluation of category map and concept space techniques. *Journal of the American Society for Information Science, 49*(7), 582–603.

Chen, H., & Lynch, K. J. (1992). Automatic construction of networks of concepts characterizing document databases. *IEEE Transactions on Systems, Man and Cybernetics, 22*(5), 885–902.

Chen, H., Miranda, R., Zeng, D. D., Demchak, C., Schroeder, J., & Madhusudan, T. (Eds.). (2003). *Intelligence and security informatics: Proceedings of the First NSF/NIJ Symposium on Intelligence and Security Informatics*. Berlin: Springer.

Chen, H., Moore, R., Zeng, D., & Leavitt, J. (Eds.). (2004). *Intelligence and security informatics: Proceedings of the Second Symposium on Intelligence and Security Informatics*. Berlin: Springer.

Chen, H., Qin, J., Reid, E., Chung, W., Zhou, Y., Xi, W., et al. (in press). The Dark Web Portal: Collecting and analyzing the presence of domestic and international terrorist groups on the Web. *Proceedings of the 7th Annual IEEE Conference on Intelligent Transportation Systems (ITSC 2004)*.

Chen, H., Schroeder, J., Hauck, R., Ridgeway, L., Atabakhsh, H., Gupta, H., et al. (2003). COPLINK Connect: Information and knowledge management for law enforcement. *Decision Support Systems, 34*(3), 271–285.

Chen, H., Schuffels, C., & Orwig, R. (1996). Internet categorization and search: A self-organizing approach. *Journal of Visual Communication and Image Representation, 7*(1), 88–102.

Chen, H., Zeng, D., Atabakhsh, H., Wyzga, W., & Schroeder, J. (2003). COPLINK: Managing law enforcement data and knowledge. *Communications of the ACM, 46*(1), 28–34.

Chen, I.-M. A., & Rotem, D. (1998). Integrating information from multiple independently developed data sources. *Proceedings of the 7th International Conference on Information and Knowledge Management*, 242–250.

Chinchor, N. A. (1998). Overview of MUC-7/MET-2. *Proceedings of the 7th Message Understanding Conference (MUC-7)*. Retrieved August 19, 2004, from http://www.itl.nist.gov/iaui/894.02/related_projects/muc/proceedings/muc_7_proceedings/overview.html

Coady, W. F. (1985). Automated link analysis: Artificial intelligence-based tool for investigators. *Police Chief, 52*(9), 22–23.

Collins, P. I., Johnson, G. F., Choy, A., Davidson, K. T., & Mackay, R. E. (1998). Advances in violent crime analysis and law enforcement: The Canadian Violent Crime Linkage Analysis System. *Journal of Government Information, 25*(3), 277–284.

Cook, J. S., & Cook, L. L. (2003). Social, ethical and legal issues of data mining. In J. Wang (Ed.), *Data mining: Opportunities and challenges* (pp. 395–420). Hershey, PA: Idea Group Publishing.

Craglia, M., Haining, R., & Wiles, P. (2000). A comparative evaluation of approaches to urban crime pattern analysis. *Urban Studies, 37*(4), 711–729.

Cristianini, N., & Shawe-Taylor, J. (2000). *An introduction to support vector machines and other kernel-based learning methods*. New York: Cambridge University Press.

Cronin, B. (2005). Intelligence, terrorism, and national security. *Annual Review of Information Science and Technology, 39*, 395–432.

Damianos, L., Ponte, J., Wohlever, S., Reeder, F., Day, D., Wilson, G., et al. (2002). MiTAP for bio-security: A case study. *AI Magazine, 23*(4), 13–29.

Davies, P. H. J. (2002). Intelligence, information technology, and information warfare. *Annual Review of Information Science and Technology, 36*, 313–352.

Defays, D. (1977). An efficient algorithm for a complete link method. *Computer Journal, 20*(4), 364–366.

Diederich, J., Kindermann, J., Leopold, E., & Paass, G. (2003). Authorship attribution with support vector machines. *Applied Intelligence, 19*(1–2), 109–123.

Dijkstra, E. (1959). A note on two problems in connection with graphs. *Numerische Mathematik, 1*, 269–271.

Dolotov, A., & Strickler, M. (2003). Web-Based Intelligence Reports System. *Proceedings of the First NSF/NIJ Symposium on Intelligence and Security Informatics (ISI'03)*, 39–58.

Duda, R. O., & Hart, P. E. (1973). *Pattern recognition and scene analysis*. New York: Wiley.

Eisen, M. B., Spellman, P. T., Brown, P. O., & Botstein, D. (1998). Cluster analysis and display of genome-wide expression patterns. *Proceedings of the National Academy of Sciences, 95*(25), 14863–14868.

Eisenbeis, R., & Avery, R. (1972). *Discrimination analysis and classification procedures*. Lanham, MA: Lexington Books.

Estivill-Castro, V., & Lee, I. (2001). Data mining techniques for autonomous exploration of large volumes of geo-referenced crime data. *Proceedings of the 6th International Conference on GeoComputation*. Retrieved August 19, 2004, from http://www.geocompu tation.org/2001/papers/estivillcastro.pdf

Evan, W. M. (1972). An organization-set model of interorganizational relations. In M. Tuite, R. Chisholm, & M. Radnor (Eds.), *Interorganizational decision-making* (pp. 181–200). Chicago: Aldine.

Faggiani, D., & McLaughlin, C. (1999). Using nation incident-based reporting system data for strategic crime analysis. *Journal of Quantitative Criminology, 15*(2), 181–191.

Fayyad, U. M., Djorgovski, S. G., & Weir, N. (1996). Automating the analysis and cataloging of sky surveys. In U. Fayyad, G. Piatetsky-Shapiro, P. Smyth, & R. Uthurusamy (Eds.), *Advances in knowledge discovery and data mining* (pp. 471–493). Menlo Park, CA: AAAI Press.

Fayyad, U., Piatetsk-Shapiro, G., & Smyth, P. (1996). The KDD process for extracting useful knowledge from volumes of data. *Communications of the ACM, 39*(11), 27–34.

Freeman, L. C. (1979). Centrality in social networks: Conceptual clarification. *Social Networks, 1*, 215–240.

Garcia-Molina, H., Ullman, J. D., & Widom, J. (2002). *Database systems: The complete book.* Upper Saddle River, NJ: Prentice-Hall.

Getis, A., & Ord, J. K. (1992). The analysis of spatial association by use of distance statistics. *Geographical Analysis, 24*, 189–199.

Gibson, D., Kleinberg, J., & Raghavan, P. (1998). Inferring Web communities from link topology. *Proceedings of the 9th ACM Conference on Hypertext and Hypermedia*, 225–234.

Goldberg, D., Nichols, D., Oki, B., & Terry, D. (1992). Using collaborative filtering to weave an information tapestry. *Communications of the ACM, 35*(12), 61–69.

Goldberg, H. G., & Senator, T. E. (1998). Restructuring databases for knowledge discovery by consolidation and link formation. *Proceedings of the 1998 AAAI Fall Symposium on Artificial Intelligence and Link Analysis*, 47–52.

Goldberg, H. G., & Wong, R. W. H. (1998). Restructuring transactional data for link analysis in the FinCen AI System. *Proceedings of the 1998 AAAI Fall Symposium on Artificial Intelligence and Link Analysis*, 38–46.

Grishman, R. (2003). Information extraction. In R. Mitkov (Ed.), *The Oxford handbook of computational linguistics* (pp. 545–559). New York: Oxford University Press.

Grubesic, T. H., & Murray, A. T. (2001). Detecting hot spots using cluster analysis and GIS. *Proceedings of 2001 Crime Mapping Research Conference.* Retrieved August 19, 2004, from http://www.ojp.usdoj.gov/nij/maps/Conferences/01conf/Grubesic.doc

Haas, L. M. (2002). Data integration through database federation. *IBM Systems Journal, 41*(4), 578–596.

Han, J., & Kamber, M. (2001). *Data mining: Concepts and techniques.* San Francisco, CA: Morgan Kaufmann.

Hand, D. J. (1981). *Discrimination and classification.* Chichester, UK: Wiley.

Harris, K. D. (1990). *Geographic factors in policing.* New York: McGraw-Hill.

Hasselbring, W. (2000). Information system integration. *Communications of the ACM, 43*(6), 33–38.

Hassibi, K. (2000). Detecting payment card fraud with neural networks. In P. J. G. Lisboa, A. Vellido, & B. Edisbury (Eds.), *Business applications of neural networks* (pp. 141–158). Singapore: World Scientific.

Hauck, R. V., Atabakhsh, H., Ongvasith, P., Gupta, H., & Chen, H. (2002). Using COPLINK to analyze criminal justice data. *IEEE Computer, 35*(3), 30–37.

Heckerman, D. (1995). A tutorial on learning with Bayesian networks. In M. Jordan (Ed.), *Learning in Graphical Models* (pp. 301–354). Cambridge, MA: MIT Press. (Also available as Research Report No. MSR–TR–95–06 from Microsoft.)

Hsu, C. W., & Lin, C. J. (2002). A comparison of methods for multi-class support vector machines. *IEEE Transactions on Neural Networks, 13*, 415–425.

Icove, D. J. (1986). Automated crime profiling. *Law Enforcement Bulletin, 55*, 27–30.

Jain, A. K., & Flynn, P. J. (1996). Image segmentation using clustering. In N. Ahuja & K. Bowyer (Eds.), *Advances in image understanding* (pp. 65–83). Piscataway, NJ: IEEE Press.

Jain, A. K., Murty, M. N., & Flynn, P. J. (1999). Data clustering: A review. *ACM Computing Surveys, 31*(3), 264–323.

Jhingran, A. D., Mattos, N., & Pirahesh, H. (2002). Information integration: A research agenda. *IBM Systems Journal, 41*(4), 555–562.

Kangas, L. J., Terrones, K. M., Keppel, R. D., & La Moria, R. D. (2003). Computer Aided Tracking and Characterization of Homicides and sexual assaults (CATCH). In J. Mena (Ed.), *Investigative data mining for security and criminal detection* (pp. 364–375). Amsterdam: Butterworth Heinemann.

Kasad, T., & Su, S. (2004). Transnational information sharing and event notification. In *Proceedings of the International Association for Development of the Information Society, International e-Society (IADIS e-Society '04)*, 52–62.

Kay, B. A., Timperi, R. J., Morse, S. S., Forslund, D., McGowan, J. J., & O'Brien, T. (1998). Innovative information-sharing strategies. *Emerging Infectious Diseases, 4*(3). Retrieved August 19, 2004, from http://www.cdc.gov/ncidod/eid/vol4no3/kay.htm

Klerks, P. (2001). The network paradigm applied to criminal organizations: Theoretical nit-picking or a relevant doctrine for investigators? Recent developments in the Netherlands. *Connections, 24*(3), 53–65.

Kohonen, T. (1995). *Self-organizing maps*. Berlin: Springer-Verlag.

Koperski, K., & Han, J. (1995). Discovery of spatial association rules in geographic information databases. *Proceedings of the 4th International Symposium on Large Spatial Databases (Advances in Spatial Databases)*, 47–66.

Krebs, V. E. (2001). Mapping networks of terrorist cells. *Connections, 24*(3), 43–52.

Krupka, G. R., & Hausman, K. (1998). IsoQuest Inc.: Description of the NetOwl text extractor system as used for MUC-7. *Proceedings of the 7th Message Understanding Conference (MUC-7)*. Retrieved August 19, 2004, from http://www.itl.nist.gov/iaui/894.02/related_projects/muc/proceedings/muc_7_proceedings/isoquest.pdf

Kumar, A., & Olmeda, I. (1999). A study of composite or hybrid classifiers for knowledge discovery. *INFORMS Journal on Computing, 11*(3), 267–277.

Lee, R. (1998). Automatic information extraction from documents: A tool for intelligence and law enforcement analysts. *Proceedings of the 1998 AAAI Fall Symposium on Artificial Intelligence and Link Analysis*, 63–67.

Lee, W., & Stolfo, S. (1998). Data mining approaches for intrusion detection. *Proceedings of the 7th USENIX Security Symposium*. Retrieved August 20, 2004, from http://citeseer.ist.psu.edu/cache/papers/cs/3327/http:zSzzSzwww.cs.columbia.eduzSz~wenkezSzpaperszSzusenix.pdf/lee98data.pdf

Levenshtein, V. L. (1966). Binary codes capable of correcting deletions, insertions, and reversals. *Soviet Physics Doklady, 10*, 707–710.

Levine, N. (2000). CrimeStat: A spatial statistics program for the analysis of crime incident locations. *Crime Mapping News, 2*(1), 8–9.

Lim, E.-P., Srivastava, J., Prabhakar, S., & Richardson, J. (1996). Entity identification in database integration. *Information Sciences, 89*, 1–38.

Lin, S., & Brown, D. E. (2003). Criminal incident data association using the OLAP technology. *Proceedings of the First NSF/NIJ Symposium on Intelligence and Security Informatics (ISI'03)*, 13–26.

Lippmann, R. P. (1987). An introduction to computing with neural networks. *IEEE Acoustics Speech and Signal Processing Magazine, 4*(2), 4–22.

Lorrain, F. P., & White, H. C. (1971). Structural equivalence of individuals in social networks. *Journal of Mathematical Sociology, 1*, 49–80.

Lu, Q., Huang, Y., & Shekhar, S. (2003). Evacuation planning: A capacity constrained routing approach. *Proceedings of the First NSF/NIJ Symposium on Intelligence and Security Informatics (ISI'03)*, 111–125.

Mannila, H., Toivonen, H., & Inkeri, V. A. (1994). Efficient algorithms for discovering association rules. *Proceedings of Knowledge Discovery in Databases (KDD'94)*, 181–192.

Marshall, B., Kaza, S., Xu, J., Atabakhsh, H., Petersen, T., Violette, C., et al. (in press). Cross-jurisdictional criminal activity networks to support border and transportation security. *Proceedings of the 7th Annual IEEE Conference on Intelligent Transportation Systems (ITSC 2004)*.

McAndrew, D. (1999). The structural analysis of criminal networks. In D. Canter & L. Alison (Eds.), *The social psychology of crime: Groups, teams, and networks* (pp. 53–94). Dartmouth, UK: Aldershot.

McDonald, D., & Chen, H. (2002). Using sentence-selection heuristics to rank text segments in TXTRACTOR. *Proceedings of the Second ACM/IEEE-CS Joint Conference on Digital Libraries (JCDL'02)*, 28–35.

McFadden, D. (1973). Conditional logit analysis of qualitative choice behavior. In P. Zarembka, (Ed.), *Frontiers of econometrics* (pp. 105–142). New York: Academic Press.

McKeown, K., Barzilay, R., Chen, J., Elson, D., Evans, D., Klavans, J., et al. (2003). Columbia's Newsblaster: New features and future directions. *Proceedings of Human Language Technology Conference (HLT-NAACL 2003)*, 15–16.

Miller, S., Crystal, M., Fox, H., Ramshaw, L., Schwartz, R., Stone, R., et al. (1998). BBN: Description of the SIFT system as used for MUC-7. *Proceedings of the 7th Message Understanding Conference (MUC-7)*. Retrieved August 19, 2004, from http://www.itl.nist.gov/iaui/894.02/related_projects/muc/proceedings/muc_7_proceedings/bbn_muc7.pdf

Moran, P. A. P. (1950). Notes on continuous stochastic phenomena. *Biometrika, 37*, 17–23.

Murray, A. T., & Estivill-Castro, V. (1998). Cluster discovery techniques for exploratory spatial data analysis. *International Journal of Geographical Information Science, 12*, 431–443.

Murray, A. T., McGuffog, I., Western, J. S., & Mullins, P. (2001). Exploratory spatial data analysis techniques for examining urban crime. *British Journal of Criminology, 41*, 309–329.

National Research Council. (2002). *Making the nation safer: The role of science and technology in countering terrorism.* Washington, DC: National Academy Press.

Newcombe, H. B., Kennedy, J. M., Axford S. J., & James, A. P. (1959). Automatic linkage of vital records. *Science, 130*(3381), 954–959.

O'Hara, C. E., & O'Hara, G. L. (1980). *Fundamentals of criminal investigation* (5th ed.). Springfield, IL: Charles C. Thomas.

Ord, J. K., & Getis, A. (1995). Local spatial autocorrelation statistics: Distributional issues and an application. *Geographical Analysis, 27*, 286–296.

Patman, F., & Thompson, P. (2003). Names: A new frontier in text mining. *Proceedings of the First NSF/NIJ Symposium on Intelligence and Security Informatics (ISI'03)*, 27–38.

Pinner, R. W., Rebmann, C. A., Schuchat, A., & Hughes, J. M. (2003). Disease surveillance and the academic, clinical, and public health communities. *Emerging Infectious Diseases, 9*(7). Retrieved August 19, 2004, from http://www.cdc.gov/ncidod/eid/vol9no7/03-0083.htm

Quinlan, J. R. (1986). Induction of decision trees. *Machine Learning, 1*, 86–106.

Quinlan, J. R. (1993). *C4.5: Programs for machine learning*: San Francisco: Morgan Kaufmann.

Raghu, T. S., Ramesh, R., & Whinston, A. B. (2003). Addressing the homeland security problem: A collaborative decision-making framework. *Proceedings of the First NSF/NIJ Symposium on Intelligence and Security Informatics (ISI'03)*, 249–265.

Rahm, E., & Bernstein, P. A. (2001). A survey of approaches to automatic schema matching. *The VLDB Journal, 10*, 334–350.

Rasmussen, E. (1992). Clustering algorithms. In W. B. Frakes & R. Baeza-Yates (Eds.), *Information retrieval: Data structures and algorithms* (pp. 419–442). Englewood Cliffs, NJ: Prentice Hall.

Ratchliffe, J. H., & McCullagh, M. J. (1999). Hotbeds of crime and the search for spatial accuracy. *Journal of Geographical Systems, 1*(4), 385–398.

Reid, E., Qin, J., Chung, W., Xu, J., Zhou, Y., Schumaker, R., et al. (2004). Terrorism Knowledge Discovery Project: A knowledge discovery approach to address the threats of terrorism. *Proceedings of the Second Symposium on Intelligence and Security Informatics (ISI'04)*, 125–145.

Riloff, E. (1996). Automatically generating extraction patterns from untagged text. *Proceedings of the 13th National Conference on Artificial Intelligence (AAAI '96)*, 1044–1049.

Ronfeldt, D., & Arquilla, J. (2001). What next for networks and netwars? In J. Arquilla & D. Ronfeldt (Eds.), *Networks and netwars: The future of terror, crime, and militancy* (pp. 311–362). Santa Monica, CA: Rand Press.

Rossmo, D. K. (1995). Overview: Multivariate spatial profiles as a tool in crime investigation. In C. R. Block, M. Dabdoub, & S. Fregly (Eds.), *Crime analysis through computer mapping* (pp. 65–97). Washington, DC: Police Executive Research Forum.

Rumelhart, D. E., Hinton, G. E., & Williams, R. J. (1986). Learning internal representations by error propagation. In D. E. Rumelhart & J. L. McLelland (Eds.), *Parallel distributed processing: Explorations in the microstructure of cognition* (pp. 318–362). Cambridge, MA: MIT Press.

Ryan, J., Lin, M., & Mikkulainen, R. (1998). Intrusion detection with neural networks. In M. I. Jordan, M. J. Kearns, & S. A. Solla (Eds.), *Advances in neural information processing systems* (pp. 943–949). Cambridge, MA: MIT Press.

Saether, M., & Canter, D. V. (2001). A structural analysis of fraud and armed robbery networks in Norway. *Proceedings of the 6th International Investigative Psychology Conference*. Retrieved August 20, 2004, from http://www.i-psy.com/conferences/sixth/multimedia/powerPoint/saether/saether.htm

Sageman, M. (2004). *Understanding terror networks*. Philadelphia: University of Pennsylvania Press.

Sarkar, S., & Sriram, R. S. (2001). Bayesian models for early warning of bank failures. *Management Science, 47*(11), 1457–1475.

Schroeder, J., Xu, J., & Chen, H. (2003). CrimeLink Explorer: Using domain knowledge to facilitate automated crime association analysis. *Proceedings of the First NSF/NIJ Symposium on Intelligence and Security Informatics (ISI'03)*, 168–180.

Schumacher, B. J., & Leitner, M. (1999). Spatial crime displacement resulting from large-scale urban renewal programs in the city of Baltimore, MD: A GIS modeling approach.

Proceedings of the 4th International Conference on GeoComputation. Retrieved August 19, 2004, from http://www.geocomputation.org/1999/047/gc_047.htm

Shortliffe, E. H., & Blois, M. S. (2000). The computer meets medicine and biology: Emergence of a discipline. In K. J. Hannah & M. J. Ball (Eds.), *Health informatics* (pp. 1–40). New York: Springer-Verlag.

Silverstein, C., Brin, S., & Motwani, R. (1998). Beyond market baskets: Generalizing association rules to dependence rules. *Data Mining and Knowledge Discovery, 2,* 39–68.

Sparrow, M. K. (1991). The application of network analysis to criminal intelligence: An assessment of the prospects. *Social Networks, 13,* 251–274.

Stolfo, S. J., Hershkop, S., Wang, K., Nimeskern, O., & Hu, C.-W. (2003). Behavior profiling and email. *Proceedings of the First NSF/NIJ Symposium on Intelligence and Security Informatics (ISI'03),* 74–90.

Strickland, L. S. (2002a). Information and the war against terrorism (Part I). *Bulletin of the American Society for Information Science and Technology, 28*(2), 12–17.

Strickland, L. S. (2002b). Information and the war against terrorism (Part II): Were American intelligence and law enforcement effectively positioned to protect the public? *Bulletin of the American Society for Information Science and Technology, 28*(3), 18–22.

Strickland, L. S. (2002c). Information and the war against terrorism (Part III): New information-related laws and the impact on civil liberties. *Bulletin of the American Society for Information Science and Technology, 29*(3), 23–27.

Strickland, L. S. (2002d). Information and the war against terrorism (Part IV): Civil liberties vs. security in the age of terrorism. *Bulletin of the American Society for Information Science and Technology, 28*(4), 9–13.

Strickland, L. S. (2002e). Information and the war against terrorism (Part V): The business implications. *Bulletin of the American Society for Information Science and Technology, 28*(6), 18–21.

Strickland, L. S. (2005). Domestic security surveillance and civil liberties. *Annual Review of Information Science and Technology, 39,* 433–513.

Sun, A., Naing, M.-M., Lim, E.-P., & Lam, W. (2003). Using support vector machines for terrorism information extraction. *Proceedings of the First NSF/NIJ Symposium on Intelligence and Security Informatics (ISI'03),* 1–12.

Tolle, K. M., & Chen, H. (2000). Comparing noun phrasing techniques for use with medical digital library tools. *Journal of the American Society for Information Science, 51*(4), 352–370.

Torgerson, W. S. (1952). Multidimensional scaling: Theory and method. *Psychometrika, 17,* 401–419.

Trybula, W. J. (1999). Text mining. *Annual Review of Information Science and Technology, 34,* 385–419.

Tufte, E. (1983). *The visual display of quantitative information.* Cheshire, CT: Graphics Press.

U.S. Federal Bureau of Investigation. (1992). *Uniform crime reporting handbook: National incident-based reporting system (NIBRS).* Washington, DC: The Bureau.

U.S. Office of Homeland Security. (2002). *National Strategy for Homeland Security.* Washington DC: Office of Homeland Security. Retrieved August 19, 2004, from http://www.whitehouse.gov/homeland/book/nat_strat_hls.pdf

Vapnik, V. (1995). *The nature of statistical learning theory.* New York: Springer-Verlag.

Wang, G., Chen, H., & Atabakhsh, H. (2004). Automatically detecting deceptive criminal identities. *Communications of the ACM, 47*(3), 71–76.

Wang, J.-H., Huang, C.-C., Teng, J.-W., & Chien, L.-F. (2004). Generating concept hierarchies from text for intelligence analysis. *Proceedings of the Second Symposium on Intelligence and Security Informatics (ISI'04)*, 100–113.

Wang, J.-H., Lin, B. T., Shieh, C.-C., & Deng, P. S. (2003). Criminal record matching based on the vector space model. *Proceedings of the First NSF/NIJ Symposium on Intelligence and Security Informatics (ISI'03)*, 386.

Wasserman, S., & Faust, K. (1994). *Social network analysis: Methods and applications.* Cambridge, UK: Cambridge University Press.

Weisburd, D., & McEwen, T. (Eds.). (1997). *Crime mapping and crime prevention.* Monsey, NY: Criminal Justice Press.

Weiss, S. I., & Kulikowski, C. A. (1991). *Computer systems that learn: Classification and prediction methods from statistics, neural networks, machine learning, and expert systems.* San Francisco: Morgan Kaufmann.

Williams, C. (1975). Mendenhall's studies of word-length distribution in the works of Shakespeare and Bacon. *Biometrika, 62*, 207–212.

Witten, I. H., Bray, Z., Mahoui, M., & Teahan, W. J. (1999). Using language models for generic entity extraction. *Proceedings of the ICML Workshop on Text Mining.* Retrieved August 20, 2004, from http://www-ai.ijs.si/DunjaMladenic/ICML99/WittenFinal.ps

Xu, J., & Chen, H. (2003). Untangling criminal networks: A case study. *Proceedings of the First NSF/NIJ Symposium on Intelligence and Security Informatics (ISI'03)*, 232–248.

Xu, J., & Chen, H. (in press). Criminal network analysis and visualization: A data mining perspective. *Communications of the ACM.*

Xue, Y., & Brown, D. E. (2003). Decision based spatial analysis of crime. *Proceedings of the First NSF/NIJ Symposium on Intelligence and Security Informatics (ISI'03)*, 153–167.

Yang, Y., Carbonell, J., Brown, R., Pierce, T., Archibald, B. T., & Liu, X. (1999). Learning approaches for detecting and tracking news events. *IEEE Intelligent Systems, 14*(4), 32–43.

Zeng, D., Chen, H., Daspit, D., Shan, F., Nandiraju, S., Chau, M., et al. (2003). COPLINK Agent: An architecture for information monitoring and sharing in law enforcement. *Proceedings of the First NSF/NIJ Symposium on Intelligence and Security Informatics (ISI'03)*, 281–295.

Zeng, D., Chen, H., Tseng, C., Larson, C., Eidson, M., Gotham, I., et al. (2004). West Nile virus and botulism portal: A case study in infectious disease informatics. *Proceedings of the Second Symposium on Intelligence and Security Informatics (ISI'04)*, 28–41.

Zhang, Z., Salerno, J. J., & Yu, P. S. (2003). Applying data mining in investigating money laundering crimes. *Proceedings of the 9th ACM SIGKDD International Conference on Knowledge Discovery and Data Mining*, 747–752.

Zhao, J. L., Bi, H. H., & Chen, H. (2003). Collaborative workflow management for interagency crime analysis. *Proceedings of the First NSF/NIJ Symposium on Intelligence and Security Informatics (ISI'03)*, 266–280.

Zheng, R., Qin, Y., Huang, Z., & Chen, H. (2003). Authorship analysis in cybercrime investigation. *Proceedings of the First NSF/NIJ Symposium on Intelligence and Security Informatics (ISI'03)*, 59–73.

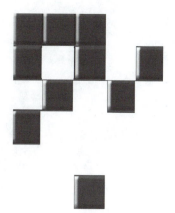

Information
Needs and Use

CHAPTER 7

Information Behavior

Donald O. Case
University of Kentucky

Introduction

Wilson (2000, p. 49) defined information behavior as "the totality of human behavior in relation to sources and channels of information, including both active and passive information seeking and information use." Seen this way, information behavior includes purposive information seeking; serendipitous encountering of information; and the giving, sharing, and use of information. Pettigrew, Fidel, and Bruce (2001) described recent debates over the scope and terminology of this topic, including difficulties with the label "information behavior"; they concluded that information behavior was the most appropriate term for this area of research.

This chapter reviews recent literature on information behavior, covering publications appearing during the four-year period 2001 to 2004. The year 2001 has been chosen as the starting point because the most recent *ARIST* chapters relevant to information behavior (see the citations in the next section) were published in 2001 and 2002, and their bibliographies cover works from 2001 and earlier. In addition, I published a comprehensive overview of the information behavior literature that includes items published through late 2001.

The citations were gathered through electronic searches of bibliographic databases (such as Wilson's *Library Literature and Information Science Full Text*) and print and electronic journals on the topic (such as *The New Review of Information Behaviour Research* and *Information Research*), along with manual scanning of general publications in library and information science. Searches coupled the term "information" with "behavior," "seeking," "needs," and "uses."

More than 2,000 potentially relevant documents were identified, appearing between January, 2001 and December, 2004. With hundreds of relevant items being published every year, it is impossible to be comprehensive, even when restricting the literature to a four-year period. Consequently, several restrictions were applied.

The chapter assumes that a central component of "information behavior" is the notion of interacting with an array of potential sources that might address one's interests and information needs. Following the logic of the last comprehensive review, I have excluded items that are "site-specific, system-specific, or service-specific" (Hewins, 1990, p. 145). That is, this review will not include studies of an isolated source (e.g., use of a particular electronic journal), solitary site (use of an individual library, for example), or a single service (e.g., Web access from home), unless the investigator attempted to situate these in the context of other sources, sites, or services; investigations of Internet usage are included if studied as one element of an omnibus mix of mass media (e.g., Web pages, journals) and interpersonal channels (e.g., e-mail, discussion groups) in the context of other sources (e.g., verbal exchanges). A portion of such excluded material—that related to information retrieval, searching behaviors, use of electronic journals, and search engines—has been recently reviewed in *ARIST* chapters by Vakkari (2003), Kling and Callahan (2003), and Bar-Ilan (2004).

History of *ARIST* Information Seeking Reviews

This chapter is the latest in a long—but interrupted—series of reviews. *ARIST* chapters on "information needs and uses" first appeared in 1966 (Menzel), 1967 (Herner & Herner), 1968 (Paisley), 1969 (Allen), 1970 (Lipetz), 1971 (Crane), 1972 (Lin & Garvey), 1974 (Martyn), and 1978 (Crawford). Perhaps because of the increasing availability of bibliographies on the topic, there was a pause in reviewing information needs and uses publications following Crawford's 1978 chapter. Later, comprehensive *ARIST* chapters reappeared in 1986 (Dervin & Nilan) and 1990 (Hewins). These chapters reviewed between 26 and 136 items, reflecting accumulations of literature over periods of one to eight years; all of them imposed restrictions of some kind, e.g., reviewing only science and engineering studies or not reviewing investigations that employed questionnaires or were specific to one site.

Gradually, reviews of the information behavior literature grew more specialized. Since 1990, *ARIST* has not published any general "information needs and uses" reviews, although several related reviews have appeared. A 1991 survey by Tibbo of "information systems, services, and technology for the humanities" included some information seeking studies. Three 1993 *ARIST* chapters, respectively by Choo and Auster ("Environmental Scanning"), Chang and Rice ("Browsing"), and Metoyer-Duran ("Information Gatekeepers"), covered contrasting behaviors that traditionally have been considered in the information behavior literature. Since then Pettigrew, Fidel, and Bruce (2001) have written on the conceptual models used in information behavior research; King and Tenopir (2001) on the use of scholarly literature; Wang (2001) on methods for studying user behavior; Cool (2001) on the concept of "situation" in information science; and Solomon (2002) on "discovering information in context." Together, these reviews cover many topics typically found in the information behavior literature; additional works from the 1990s are reviewed in Case (2002).

A Framework for Reviewing the Literature

The information behavior literature presents a bewildering array of topics, populations, samples, sites, theories, and methods. Case's (2002) review of information behavior literature presents an argument for categorizing the literature into one or more of the following areas:

- Information seekers by occupation (e.g., scientists, managers)

- Information seekers by role (e.g., patient or student)

- Information seekers by demographics (e.g., by age or ethnic group)

- Theories, models, and methods used to study information seekers

This last category—theories, models, and methods—suggests how and why to study information behavior. Occupations have constituted the most popular framework for investigating human information behavior, as when a researcher studies a group of engineers or managers (Julien & Duggan, 2000). Non-employment roles, such as student or patient, are the next most prevalent. Finally, demographic characterizations, such as age, gender, or ethnicity are less common and typically distinct from the other approaches.

The review will follow the same order, ending with some generalizations about the growth and scope of the literature on information behavior.

Information Seekers by Occupation

Research by Julien and Duggan (2000) and by McKechnie, Baker, Greenwood, and Julien (2002) makes it obvious that occupations are the most common entry point for investigations of information behavior. McKechnie et al. found that 32 percent of a large sample of recent investigations studied some kind of "worker," typically a professional, and another 17 percent of studies focused on academics or other researchers. Past *ARIST* chapters also have exhibited a fondness for studies of engineers, scientists, scholars, and managers.

Scientists, Engineers, and Scholars

Investigations of the information sources and habits of engineers and scientists continue, although the latter group has been of less interest in recent years. In fact, recent studies of scientists tend to replicate the conventional research questions and methods of the past, typically employing questionnaires and interviews in studying the reading and information-gathering habits of small samples within a single discipline. Flaxbart's (2001) interviews with six university chemistry faculty and

Hallmark's (2001) interviews with 43 academic meteorologists are examples of this approach, each covering a range of sources but emphasizing the impact of electronic journals on the habits of these groups. Murphy's (2003) Internet survey of 149 toxicologists, biochemists, and other scientists working at the U.S. Environmental Protection Agency is also of this type. It is perhaps notable that Flaxbart's is the solitary empirical study in a journal issue devoted to "information and the professional scientist and engineer"; even though "information needs" and "information seeking" are the subjects of several other articles, the authors' comments are mostly based on older studies of scientists and engineers. Scientists are no longer the frontier of information seeking research, as they were 30 years ago; perhaps the lack of novel findings in these studies means that we already know enough about scientists.

Recent investigations of engineers show more depth in their research questions. Fidel and Green (2004) chose to study the accessibility of information sources as perceived by engineers. Numerous studies have found accessibility to be the factor that most influences engineers' selection of information. However, Fidel and Green found some variation among their respondents in how they interpreted "source accessibility." Saving time was the chief criterion for selecting among documents, but familiarity was the guiding factor in selecting human sources of information. Bruce, Fidel, Pejtersen, Dumais, Grudin, and Poltrock (2003) and Fidel, Pejtersen, Cleal, and Bruce (2004) used a variety of techniques to investigate collaborative information gathering and sharing among members of design teams at Microsoft and Boeing; the investigators illustrate the use of "cognitive work analysis" to explore seven dimensions of the tasks they studied. Yitzhaki and Hammershlag (2004) contrasted academic computer scientists' with industrial software engineers' use of information and their perceptions of the accessibility of sources. Their mail survey of 233 respondents demonstrated differences among the two groups in age, education, seniority, type of research, and the use of most sources. Within both groups the accessibility of information was only partly correlated with its use; this relationship was stronger among the academics than those working in industry. Kwasitsu (2003) focused on engineers involved in microprocessor design and manufacturing, finding that the higher the level of education, the less likely the engineers were to rely on memory and the more likely to use libraries.

Academic scholars of all types have traditionally been a subject of information behavior investigations. This continued to be true during the period 2001 to 2004. Such studies formerly restricted themselves to one discipline but now sample broader populations. An example is Talja's (2002) study of information sources, peer influence, and information sharing among 44 faculty members at two Finnish universities. Talja conducted lengthy interviews with samples of 10 to 12 nursing specialists, historians, literature scholars, and environmental scientists. Her findings demonstrate that scholars define their research areas and disciplines through social interaction and that collaboration and information sharing are essential aspects of scholarship.

Belefant-Miller and King (2001, 2003) examined a sample of faculty at a single university to chart their reading habits and use of e-mail. They documented the changing nature of scholarship as many sources became electronically available and emphasized the continuing value of browsing in searches for information. Herman (2004) used a critical incident technique to determine 11 aspects of information needs common to academic researchers.

Recent investigations of the information seeking habits of social scientists and humanists include those by Meho and Haas (2001), Meho and Tibbo (2003), and Brown (2001, 2002). Meho and Haas's (2001) survey of Kurdish Studies scholars demonstrated that interdisciplinary scholars often need to work harder and employ more elaborate methods of information seeking in order to locate and use relevant research. That theme is further explored by Meho and Tibbo (2003), who partially replicate earlier work by Ellis (see Ellis, Wilson, Ford, Foster, Lam, Burton, et al., 2002, and Wilson, Ford, Ellis, Foster, & Spink, 2002, for elaboration of the Ellis model). Theirs is a multinational study of 60 social scientists who study "stateless nations." Meho and Tibbo use Ellis's characterization of information seeking as a sequence of different stages and actions: starting, chaining, browsing, differentiating, monitoring, and extracting. They develop a model that adds other types of actions—accessing, networking, verifying, and managing—to Ellis's earlier model.

Humanities scholars have received continuing attention from information behavior researchers. Brown (2001) examined how music scholars in the U.S. and Canada communicated via e-mail and electronic discussion groups to facilitate their research. Using Diffusion Theory, interviews, and a survey, she found that music scholars rated e-mail as more helpful than discussion groups. Overall, both modes of communication played marginal roles in the research of these scholars. In her subsequent work, Brown (2002) proposed a six-stage model of the research process of music scholars, based on interviews with 30 respondents who described recent research projects. Dalton and Charnigo (2004) and Duff and Johnson (2002) focused on information acquired by historians. Caidi (2001), Palmer and Neumann (2002), and Westbrook (2003) discussed the challenges of interdisciplinarity for scholars in the humanities and social sciences. And as noted earlier, an *ARIST* chapter by King and Tenopir (2001) reviewed the personal and situational factors affecting the use of both print and electronic scholarly literature.

Managers

For several decades, managerial decision making has received a great deal of attention from scholars of organizational behavior. Increasingly, these scholars use the terms "scanning," "sense-making," and "information seeking" to describe the behaviors of interest. For example, Farhoomand and Drury (2002) asked 124 managers across various companies and government agencies in four English-speaking countries to define "information overload"; identify its frequency, sources, and effects; and report the actions they took in response. Most described overload in terms of excessive volume, irrelevant content, or an inability to manage or understand information. Over half of the respondents experienced the feeling often and most said they "filtered" information to combat overload. Allen and Wilson (2003) and Eppler and Mengis (2004) also address the definition and extent of information overload.

Choo (2001a, 2001b) described four modes of environmental scanning frequently observed within organizations, claiming that each reflected typical needs, habitual information seeking, and standard uses. Choo's model correlated needs, information seeking, and information uses with managerial traits, organizational strategies, and external situations. It also suggested future research approaches and applications.

Correia and Wilson (2001) interviewed 47 individuals in 19 Portuguese firms of differing size to discover factors that influenced environmental scanning. Using a case-study approach, coupled with grounded theory, they discovered factors that were partly individual in character—information consciousness (attitude toward information-related activities) and exposure to information (frequency of opportunities of contact with well-informed people and information-rich contexts)—and also partly related to both organizational information climate (conditions that determine access to and use of information in an organization) and "outwardness" (links to other organizations). They concluded that the more open the organization is to its environment, the more likely it is that individuals in the organization will be exposed to relevant information; correspondingly, to the extent that openness occurs, the organization is more likely to develop an information climate that supports the individual.

Houtari and Wilson (2001) focused on "Critical Success Factors" (CSF) in their case studies of the information needs of managers at U.K. and Finnish universities and business firms; CSF are linked to objectives that, if not achieved, may result in organizational failure. Qualitative interviews and social network analysis were combined with a grounded theory approach to identify main themes across the two universities and two companies, confirming the validity of CSF in differing kinds of organizations. Houtari and Chatman (2001) have explained the theories (Social Networks, Small Worlds) underlying such studies.

Mackenzie (2003a) surveyed 50 business managers and 50 non-managers, finding significant differences between the two groups in terms of their information behaviors and motivations. The results demonstrate that managers tend to gather information they do not need, in a quest to simplify their environment and make faster decisions. The respondents believed that gathering information gave them the reputation of being well connected and knowledgeable. In addition, her interviews with 22 line managers (Mackenzie, 2003b) reveal that, in some cases, they were drawn to a source that represented the best (e.g., most trusted or liked) relationship rather than the best information. Other studies by Mackenzie (2002, 2004) suggested that managers consciously cultivated other individuals as information sources. Hall (2003) took on similar themes to those of Mackenzie, exploring the motives behind the sharing of knowledge in information-intensive organizations.

Widén-Wulff (2003) examined how 15 Finnish insurance companies (virtually a census of the industry in that country) built their respective knowledge bases. After conducting 40 interviews, she outlined three categories of companies on the basis of the characteristics of their internal environments: "closed businesses" (in which tradition and safety were emphasized), "open companies" (those that were innovative and integrated social capital and individual employees in their planning process), and firms "in the middle," perhaps transitioning from closed to open.

Hirsh and Dinkelacker (2004) followed the information-seeking behaviors of 180 researchers from Hewlett Packard Labs and Compaq Computers during the merger of those two companies. They found heavy use of Web sources, the corporate library, information from standards bodies, and information from colleagues outside the firm. Their results suggested that the factors most influencing selection of sources were time-saving, authoritativeness, and convenience; currency, reliability, and familiarity were less important.

Journalists are another occupational group that has received attention of late, particularly as the Internet changes both the way they gather information and how they publish their work. Attfield and Dowell (2003) and Attfield, Blandford, and Dowell (2003) based their conclusions on interviews with reporters for the (London) *Times*. They examined the role of uncertainty in the work of newspaper reporters in Britain, looking at how they perceived newsworthiness, generated "angles" for stories, exercised creativity, and gathered information in the process of writing. The ideas of Kuhlthau (see Kuhlthau & Tama, 2001) and Dervin (see Dervin, 2003) figured prominently in this research. Although more focused on Web usage, Garrison (2001) used Diffusion Theory to consider other jobs and roles in the newsroom. The results of Garrison's large-scale survey suggested that daily newspapers typically have three types of roles (news researchers, specialists, and reporters/editors) that differ in the sophistication of their searching skills. The efforts of news librarians to train reporters and editors in online searching have resulted in less dependence on librarians and other news researchers.

Other Occupations

Attorneys are among the other occupational groups often encountered in past information behavior literature. Wilkinson (2001) conducted over 150 interviews with lawyers about how they solved problems in their practice. She concluded that "legal research" was not synonymous with "information seeking." Wilkinson's respondents named other tasks, such as administration, that entailed both problem-solving and information-seeking activities. In general, they preferred informal and internal sources of information, especially those who were working in larger firms. Haruna and Mabawonku (2001) took a more conventional approach in studying the needs and seeking behaviors of lawyers in Nigeria. They concluded that the most pressing information needs of Nigerian lawyers related to recent decisions of superior courts, new legislation, and advice on bettering their knowledge and skills. They concluded that law libraries in their country were not fully meeting lawyers' needs. In another study relevant to law libraries, Kuhlthau and Tama (2001) concluded that lawyers desire information services that are highly customized to their needs.

Investigations of health care providers appear to be increasing. For example, Sundin (2002) conducted multiple interviews with 20 Swedish nurses to explore the distinctions made between practical and theoretical knowledge and the relationship of that knowledge to the nurse's professional identity (that is, as a legitimate specialty distinct from that of physicians). Sundin argued persuasively for a sociocultural approach to studying information behavior as one aspect of professionalization. Cogdill (2003) studied the information-seeking practices of 300 primary-care nurses through a questionnaire and interviews with 20 of the respondents following episodes with patients. He found that nurse practitioners most frequently needed information related to drug therapy and diagnosis and that they most frequently consulted colleagues, drug reference manuals, textbooks, and protocol manuals. Gorman, Lavelle, Delcambre, and Maier (2002) used individual and group interviews coupled with participant observation of the information behaviors of a sample of physicians, nurses, and pharmacists in order to design better digital libraries for them. They found that an overabundance of records, coupled with severe time constraints, forced their informants to focus tightly on data related to the problem of patient care. MacIntosh-Murray (2001) offered a framework for studying the monitoring of "adverse clinical events." She argued that the incidence and seriousness of medical errors made the scanning behavior of health care workers an important topic for research and suggested variables that could influence the incidence of adverse events. Ocheibi and Buba (2003) described a conventional survey of the information needs of Nigerian doctors. Urquhart (2001) reviewed some earlier studies of health care professionals in the course of explaining the use of vignettes in information research. Vignettes, it may be noted, are clinical case histories used (in this context) to elicit from physicians and nurses the information sources or actions they would most likely employ in response to a situation presented to them.

Donat and Pettigrew (2002) reviewed literature on both doctors and patients in describing typical information behavior surrounding the dying patient. Harrison, Hepworth, and de Chazal (2004) studied the information behavior of hospital social workers by means of questionnaires, focus groups, and interviews. Their results suggested that these social workers were relatively "information poor," given their needs and their lack of access to the Internet and other useful sources; consequently, information tended to be gathered in face-to-face exchanges with other people. Baker, Case, and Policicchio (2003) raised the issue of how information professionals might help sex workers cope with health problems. They carried out nonparticipant observation of 75 sex workers, using a social services van in a Midwestern U.S. city as a research platform; a similar study was carried out in South Africa by Stilwell (2002). Hepworth (2004) interviewed 60 non-professionals who provided substantial health care for a relative or other person and suggested a model of information service based on his findings.

Two publications in this category by Ikoja-Odongo and Ocholla (2003, 2004) concern unusual occupations; both employed the critical incident technique to gather information. For the first study they interviewed members of the "artisan fisher folk of Uganda," a group that includes a range of occupations associated with the fishing industry: fish and equipment sales, processing, boat building, net making, fisheries research, government extension, and so forth. Ikoja-Odongo and Ocholla (2004) interviewed 602 entrepreneurs in various businesses in Uganda, including fishermen, metal fabricators, blacksmiths, quarry workers, brick makers, carpenters, builders, mechanics, and craftsmen. Observation of the entrepreneurs' work environments and historical methods were also employed. Their results demonstrated the importance of oral traditions and local knowledge in the trades they examined. Information behavior research, Ikoja-Odongo and Ocholla stated, must be sensitive to the circumstances of poverty, illiteracy, and lack of infrastructure often found in developing areas. In doing so, it could suggest ways of "repackaging" information for use by such entrepreneurs. These findings were echoed by Serema (2002) in an investigation of communities in Botswana, by Meyer (2003) in a study of maize farmers in South Africa, and by Ekoja (2004), who focused on Nigerian farmers.

Information Seekers by Role

Julien and Duggan's (2000) work indicated that the second most common approach to studying information behavior is the investigation of roles such as citizen, consumer, patient, student, or gatekeeper. Investigations of "citizens," "voters," or "consumers" may have practical outcomes (e.g., improving social services or marketing efforts) yet also cover many other areas of interest to the average person. As McKechnie,

Baker, Greenwood, and Julien (2002) pointed out, reports on "ordinary people" make up about 22 percent of the information seeking literature; investigations of "students" (a role we take on for the majority of our childhood and often part of our adult lives as well) make up another 19 percent of such studies.

The General Public

Studies of the information behavior of the public have been rare since the large-scale studies of the 1970s and 1980s (e.g., Chen & Hernon, 1982). An exception is Marcella and Baxter's (2001) random sample of almost 900 British citizens. They and their colleagues conducted "doorstep" interviews using a careful sampling plan, preceded by a questionnaire survey of almost 1,300 residents. They advocated door-to-door interviews as a method that could probe deeply and reach individuals who might be missed by other approaches.

A broad range of methods—surveys, observations, interviews, focus groups, and case studies—was used by Pettigrew, Durrance, and Unruh (2002) to assess the use of community information by the general public. Libraries in the states of Illinois, Pennsylvania, and Oregon were used as entry points to see how the Internet and libraries disseminated local information, answered questions, provided access to governmental services, and connected citizens to one another. Pettigrew, Durrance, and Unruh concluded that such networks were highly beneficial even when deficient in terms of interface design, organization, authority, currency, security, and other factors.

Beer (2004) conducted interviews with representatives of over 100 community groups, businesses, and information providers in eight remote communities in Shetland and the Western Isles of Scotland. She found that strong personal ties within the community enabled residents to find answers from other people. Complaints were made about the lack of relevance of some information from outside (e.g., "urban solutions") and the withholding of some information by local parties (sometimes due to journalistic sensitivity within such small communities). Difficulty of travel—even within the islands themselves—was judged to be a key barrier to finding information.

One aspect of everyday information seeking has dramatically changed since Hewin's (1990) review: the emergence of the Internet as an omnibus channel that complements (and, in part, replicates) the usual array of interpersonal and mass media sources of information. The diffusion of access to the World Wide Web is frequently discussed in reports of information seeking research. Case, Johnson, Andrews, Allard, and Kelly (2004), for example, argued that patterns of source preferences common 30 years ago (e.g., information gained in face-to-face or telephone exchanges with friends and family members) have shifted in light of the widespread availability of e-mail and Web pages. They based their findings on data from a 2002 telephone survey of 882 adults regarding information seeking about the genetic basis of disease. In the context of voting-related behavior, Kaye and Johnson (2003) used the results of an online survey of 442 respondents to demonstrate that the Internet is gradually substituting for other media usage—particu- larly television, radio, and magazines.

Although usage of Web pages (in isolation) falls outside the scope of this review, those investigations that consider Web searching in the con- text of other sources are deemed relevant. Hektor (2003) conducted an investigation of this type among 10 Swedish citizens. His study consid- ered the place of Web sources among others available in the respondents' environment, including other people, the television, and the telephone. Based on interviews and diaries, Hektor noted that the Internet is used broadly for both seeking and giving information, yet is most often a com- plement to or substitute for other sources, not a unique source (a point also made by Flanagin & Metzger, 2001). The Web is but one channel among many that may be monitored habitually.

A series of articles by Savolainen, co-authored in three cases with Kari, advanced similar claims concerning the role of the Web among other sources and channels available in daily life. Savolainen (2001a) carried out an empirical study of a Finnish newsgroup on consumer issues, exploring the interaction between information needs, sources, and the social network of newsgroup users; a related article (Savolainen, 2001b) considered the relationship of Bandura's Social Cognitive Theory to the finding of information. Kari and Savolainen (2003) made the case that Web searching needs to be considered within the larger contexts of other sources and the person's "life-world," or everyday reality. Savolainen and Kari (2004a) extended the consideration of larger con- texts by studying the "information source horizon" of the Internet in the context of self-development. Source horizons place information sources and channels in order of preference, based on attributes such as accessi- bility and quality. Savolainen and Kari's study drew on interviews with 18 Internet users who positioned information sources within three cate- gories that were defined by degree of relevance to the respondents' inter- ests and goals. Human sources such as friends and colleagues were preferred, followed by print media such as newspapers and books; net- worked sources were ranked third. In interviews with 18 Finnish citi- zens, Savolainen and Kari (2004b) found that they conceptualized the Internet as a space or place and that they assessed the quality of what they found there by comparing it with other information sources.

A study that barely fits in this category is that of Julien and Michels (2003). They documented "intra-individual information behavior," by which they meant patterns of need, information seeking, context, and source selection across one individual's various daily life situations. Through participant diaries and interviews, Julien and Michels found that time constraints, motivations, context, type of initiating event, loca- tion, intended application of the information found, and source type were the most common influences on the information behavior of their single respondent.

Patients

Perhaps because of the steady growth in the complexity and importance of health care, studies of health-related information needs in general, and patients in particular, have been popular.

The information-seeking behaviors of spinal surgery patients, both before and after the procedure, were examined by Holmes and Lenz (2002). Lion and Meertens (2001) considered cases in which patients sought information about a potentially risky medicine. Rees and Bath (2001) conducted a study of the information needs and seeking behaviors of women with breast cancer, utilizing the Monitoring/Blunting Scale (MBS). (The eight-item MBS is the most widely used instrument for measuring how people react to threatening information; a "monitoring" response is to scan the environment for potential threats, while a "blunting" behavior [e.g., going to the movies] ignores threatening information or distracts the person from it.) A later investigation by Williamson and Manaszewicz (2002) reviewed additional literature on the use of the MBS and conducted interviews with 34 women to aid in the design of a Web portal; its findings cast doubt on the utility and validity of the MBS in this particular context.

Warner and Procaccino (2004) surveyed a broad sample (by age and education) of 119 women regarding their seeking of health-related information. They found that physicians, medical or health books, people with similar medical conditions, family or friends, nurses or pharmacists, Web sites, and public libraries were the most common sources of health information, in that order. A similar investigation was conducted among Somali women living in the U.K. by Davies and Bath (2003). McKenzie (Carey, McKechnie, & McKenzie, 2001; McKenzie, 2002a, 2002b, 2003a, 2003b) studied the information behavior of pregnant women, in some cases along with those of their midwives (McKenzie, 2004).

In a review of the health care literature, Donat and Pettigrew (2002) addressed the topic of the terminally ill patient, pointing out that the patient, as well as her or his caregivers, may have information needs at this difficult time. Baker (2004) carried out empirical research on this topic through a content analysis of a book of conversations between a husband who was dying and his wife (a grief counselor). Baker concluded that a person near death may need a variety of information to help her or him cope with dying and death. These needs reflected physical, emotional, spiritual, and financial dimensions of the person's situation.

Studies by Hepworth, Harrison, and James (2003a, 2003b); Box, Hepworth, and Harrison (2003); and Hepworth and Harrison (2004) used a variety of methods (focus group interviews, audio diaries, and questionnaires) to investigate the information needs of people with multiple sclerosis. The thousands of responses they gathered indicated a need for information regarding the disease, its symptoms, and treatment, to be tailored to various audiences (including the patients, the public, and health care providers). Matthews, Sellergren, Manfredi, and Williams (2002) employed focus group interviews to explore factors affecting medical information seeking among African-American cancer patients. They identified several cultural and socioeconomic barriers, including limited knowledge and misinformation about cancer, mistrust of the medical community, privacy concerns, religious beliefs, fear, and stigma associated with seeking help.

Johnson, Andrews, and Allard (2001) offered a model for studying cancer genetics information seeking, drawing upon research on cancer patients. Johnson, Andrews, Case, and Allard (in press) argue that issues surrounding genomics make the topic a "perfect information seeking research problem." Case, Johnson, Andrews, Allard, and Kelly (2004) reported on a telephone survey of 882 adults regarding the public's sources of information about genetic screening and the genetic bases for cancer. The respondents said they would be more likely to access the Internet before turning to health care providers or relatives—both of whom are better sources of information about a person's genetic basis for disease. Johnson, Case, Andrews, Allard, and Johnson (in press) presented contrasting ways of considering survey data about health information sources as either "fields" or "pathways." The former approach is the traditional view of individuals choosing among one or more information sources, whereas the latter sees the search for an answer as a serial chain of sources that is followed until the seeker is satisfied or exhausted. Taylor, Alman, David, and Manchester (2001) also focused on genetics-related information available through the Internet.

Marton's (2003) examination of health information seeking by 265 women ranked Web information high on relevance but only moderately on perceived reliability; in contrast, health care providers, books, and pamphlets received high ratings on both of those attributes. Yet other studies (Wikgren, 2001, 2003) of Internet health discussion groups emphasized this channel as a source of interpersonal communication and emotional support. Wikgren (2003) found that 80 percent of references for supporting information were to Web pages and that 60 percent of all references were to sources with scientific medical content.

Students

As has been noted, studies of students constitute 19 percent of the literature on information seeking. Indeed, almost any publication on "learning" is relevant to information behavior (Kuhlthau, 2004b), making this section necessarily more selective in what it covers.

Toms and Duff (2002) studied 11 history students, mostly at the doctoral level. Respondents were interviewed and also kept diaries describing their visits to six different archives. Toms and Duff noted that the diaries provided strong evidence complementing the evidence derived from interviews, yet the work depended on the commitment of respondents to maintaining the diary.

Gross (2001) and Gross and Saxton (2001) reported two investigations of "imposed" information seeking—queries developed by one person but given to someone else to resolve—in public and school libraries. The first study took place in three elementary school libraries serving children ages 4 through 12. The investigators found that between 32 and 43 percent of all circulation transactions in the school libraries involved imposed queries. The second survey, undertaken in 13 public libraries and involving 1,107 users, did not include minors, but it was clear that instructors' assignments were still a major source of imposed queries, along with those of spouses and, especially, the children of library users. Gross (2004a, 2004b) reports another study of imposed queries and information seeking in schools. In a similar vein, Hultgren and Limberg's (2003) review of the learning and information seeking literature suggested a strong relationship between the nature of school assignments and the ways in which students seek and use information.

Whitmire (2003) looked at the information-seeking behavior of 20 senior undergraduates as they researched a major paper. She employed Kuhlthau's Information Search Process model and four other research models from educational psychology to create the theoretical foundation of her investigation. Whitmire found that students' epistemological beliefs (e.g., the belief that "right and wrong answers exist for everything," as opposed to the belief that "all knowledge is contextual") affected their choice of topic, the ways they looked for information, how they evaluated information, and their ability to recognize cognitive authority.

Foster and Ford (2003) examined the role of serendipity in the information-seeking behavior of 45 university students and faculty, particularly how they accidentally or incidentally acquired information of interest to them. Foster (2004) identified three core processes and three levels of interaction with the context of the information—likening the resulting behavioral patterns to an artist's palette.

Given (2002a, 2002b) reported qualitative interviews with 25 "mature" university undergraduates. Taking Savolainen's (see Kari & Savolainen, 2003) framework for the study of everyday-life information seeking, Given's investigation explored how the academic and non-academic information needs of these students were related to one another, including the role of social and cultural capital. Jeong's (2004) interviews with, and observations of, Korean graduate students in the United States revealed gaps in their knowledge about American culture and described language and financial barriers that inhibited them from learning about their surroundings. Seldén (2001) used interviews, observation, and textual analysis to investigate the information-seeking behaviors, career, identity, and independence of 10 doctoral students in business administration.

Heinström (2003) tested the personality attributes of 305 master's degree students in a variety of disciplines. Her quantitative analysis found that five personality dimensions—neuroticism, extraversion, openness to experience, competitiveness, and conscientiousness—interacted with contextual factors to affect students' information behavior.

Other Roles: Hobbyists

Beyond students, other studies of "roles" tend to focus on narrowly defined groups. An example is Hartel's (2003) research on "hobbyist cooks"—people who do not cook for a living and yet are devoted to collecting and using information about food and its preparation. Hartel used this population as an example of the potential for the study of hobbies as a serious lifetime pursuit that features concentrated episodes of information seeking. In the same vein, Yakel's (2004) interviews with 29 genealogists and family historians explored their motivations and use of sources. Yakel concluded that being a family historian involves seeking meaning and self-identity as well as collecting facts; another study of genealogists by Duff and Johnson (2003) focused more on the latter aspect.

Information Seekers by Demographic or Social Group

Compared to occupational or role-based investigations, relatively few studies have involved the information-seeking behavior of demographic groups. However, demographic variables still form a common schema for analyzing the results of these other investigations.

Children and Youth

Children and adolescents are under-studied groups (considering their numbers and importance) according to Todd (2003). Those studies that have appeared tend to look at aspects of library or Web usage, but with the broader intent of understanding the child's thinking, learning, and social interactions.

Cooper (2002) observed seven-year-olds browsing in a library in order to understand how people who are still learning to read are able to search through printed material. She found that children often picked books on the basis of their covers rather than a closer examination of the contents. Using much different approaches and research questions, Alexandersson and Limberg (2003) performed an ethnographic study in which eleven-year-olds were observed, interviewed, and surveyed regarding how they constructed meaning from books, films, CD-ROMs, and the Internet.

The information-seeking behavior of young people was the subject of a series of articles by Shenton and Dixon (2003a, 2003b, 2004a, 2004b). Rather than studying sources or habits, their qualitative investigation of English children examined characteristics that correlated with the use of a range of information sources. Shenton and Dixon (2003a) offered a model of the information behavior of the young. Specifically, they focused on how youngsters used other people, particularly friends, as information sources (Shenton & Dixon, 2003b). They also emphasized that some information-seeking patterns reflected personal problems that the youngsters were facing, whereas others were attempts to simplify the process (Shenton & Dixon, 2004a). Shenton and Dixon (2004b) have discussed general approaches to studying the information behavior of the young. Carey et al. (2001) also included some preschool children in their sample of informants.

Chelton and Cool's (2004) edited book is mostly about how children and youth have used electronic information systems in schools. However, two chapters discuss broader, what Chelton and Cool (2004, p. xii) call "personal, as opposed to school-based," information behaviors. One is Todd and Edwards's (2004) review of investigations of how teenage girls find out about drug usage, drawing on the work of Chatman (1996), Dervin (1989), and Kuhlthau (2004a). Not surprisingly, the main sources of the informants were other teenagers. Also in this volume, Julien (2004) described the results of interviews with 30 adolescent men and women regarding how they made decisions about future careers. Julien categorized the informants into five types of decision-making styles, based on locus of control and degree of active information seeking.

Agosto and Hughes-Hassell (in press) investigated the everyday-life information-seeking behavior of urban young adults through group interviews, surveys, audio journals, "photo tours," and activity logs. The informants were 27 Philadelphia teenagers, aged 14 to 17, nearly all belonging to racial minorities. Agosto and Hughes-Hassell found that friends and family members were preferred sources; cell phones served as the favorite medium; and schoolwork, social life, and the time or date of events were the most common topics of interest.

Hamer (2003) adopted a social constructivist perspective in a study of young men's information needs related to coming out and forming a gay identity. Their information seeking most often took place through online forums. Hamer relates his findings to Chatman's (1996) Theory of Information Poverty. A related investigation concerned the information needs of gay, lesbian, bisexual, and transgendered health care professionals (Fikar & Keith, 2004); although restricted to a sample of medical workers, the results of this Internet survey clearly have implications for the broader public that visits clinics and hospitals.

Other Groups: Immigrants, the Poor, the Homeless, Women, and the Elderly

Fisher, Durrance, and Hinton (2004) expanded the concept of "information grounds"—temporary environments in which information flows abundantly as a by-product of other activities—in their investigation of immigrant users of Queens, New York, public library programs. They observed and interviewed 45 patrons, staff, and other stakeholders to ascertain what immigrants gained from such services and how the programs related to information literacy, concluding that the programs resulted in many benefits to newcomers.

Spink and Cole (2001) investigated the information-seeking channels used in poor African-American households in Dallas, Texas. Their interviews with 300 heads of households revealed that what their respondents most wanted to know about were local events, followed by information relevant to personal security and health. They ranked family and school as the most important sources of news events, followed by television, newspapers, and radio. Although friends and neighbors were the least important source for general news, they were the second- and third-ranked sources (preceded by newspapers—by far the best source) for information on employment.

Hersberger (2001) examined the information needs and sources of homeless parents. Hersberger spent a year as a participant observer in six homeless shelters in Indianapolis, Indiana. In addition to her other observations, she conducted interviews with 28 informants, generating over 800 pages of transcripts. Financial needs were the most pressing issue among the members of this sample, followed by child care, housing, health, employment, education, transportation, public assistance, and problems associated with living in the shelter. Altogether, Hersberger identified 16 major problem categories and 145 specific needs within those categories. Social service staff was the most frequently mentioned information source in nearly all major categories of need, with friends and family, personal experience, and other shelter residents also serving as common sources of help. A second study by Hersberger (2003) was a social network analysis based on interviews with 21 homeless parents in shelters in North Carolina and Washington State. Hersberger found the social networks of these informants to be small and sometimes unconnected. At times, the informants used secrecy and deception to protect themselves.

Dunne (2002) examined the information-seeking behavior of battered women. She advanced a "person-in-progressive-situations" model to chart stages in information seeking and identified three types of barriers that battered women face in finding information. Ikoja-Odongo (2002), Jiyane and Ocholla (2004), and Mooko (2002) each used interviews and questionnaires to study samples of South African women; all three studies stressed reliance on personal experience and the dominance of information obtained by word of mouth.

Finally, Wicks (2004) undertook a qualitative study of 29 older adults, showing how their information sources varied by role, retirement status, and living situation.

Metatheory, Theory, and Models

Aspects of both theory and metatheory have received considerable attention in the recent literature. To the long-standing and continuing influence of Dervin (e.g., Dervin, 2003), Wilson (e.g., Wilson, 2002), Kuhlthau (2004a), and Chatman (e.g., Dawson & Chatman, 2001) on information behavior theory are added the influential voices of Hjørland (2002a, 2002b, 2004), Savolainen (2001a, 2001b), and other mid-career scholars.

Empirical research on the use of theory and metatheory in information seeking, carried out by Fisher (née Pettigrew; e.g., Pettigrew, Fidel, & Bruce, 2001; Pettigrew & McKechnie, 2001; and McKechnie & Pettigrew, 2002), Julien and Duggan (2000), and McKechnie (McKechnie et al., 2002), has made us more aware of both the importance and the changing nature of theories, metatheories, and paradigms in the domain.

Bates (2002) has argued persuasively in favor of metatheoretical diversity in the study of information behavior. She sees in the ongoing epistemological debates a tendency for proponents of the three main metatheories (information transfer, constructivism, and constructionism—in the terms of Tuominen, Talja, & Savolainen, 2002) to celebrate the "triumph" of their metatheory over the others. Bates believes that there is room for multiple approaches and that more effort should be spent on mutual understanding and less on attacking other epistemological stances. Hjørland (2002a, 2004) and Tuominen, Talja, and Savolainen (2002) offer contrasting points of view to those of Bates. Of special importance is Hjørland's (2004) thoughtful essay on the tendency of information scientists to avoid fundamental philosophical issues underlying research. Hjørland points out that some investigators appear ambivalent about whether there exists a reality independent of human minds; in information behavior research this leads to a neglect of the objective possibilities of information resources, an overemphasis on users' mental states, and a corresponding lack of explanatory power.

The degree to which information studies are—or should be—subjective or objective in nature is well addressed in articles by Ford (2004b) and Abbott (2004). Ford was concerned with the implications of subjectivity for research; Abbott considered its relationship to problematic issues like classification and retrieval. The aforementioned work by Hjørland (2004) also contains an analysis of the relationship of objectivity and subjectivity to one another and to information behavior research.

A concept central to many theories—that of "context" or "situation"—has received a great deal of discussion in recent years. Both Johnson (2003) and Cool (2001) have written lengthy reviews on how context and situation have been defined and operationalized. Cool's (2001) *ARIST* chapter is a comprehensive review of the literature up to the year 2000. Since then, an essay by Johnson (2003) has appeared. Johnson maintained that context is commonly used in three progressively more complex senses: as equivalent to the situation in which a process is immersed (a "positivist" orientation that specifies factors that moderate relationships); as contingent aspects of situations that have specific effects (a "post-positivist" view that emphasizes the prediction of outcomes); and as frameworks of meaning (a "post-positivist" sense in which the individual is inseparable from the context). He illustrated his essay with examples from two different contexts: studies of organizational communication, compared with cancer-related information seeking. Johnson argued that these two contexts offer useful contrasts in levels of analysis, rationality, and predictability.

Pettigrew, Fidel, and Bruce (2001) provided a comprehensive overview of the many models and theories used in studying information behavior since 1978, dividing the literature into approaches that are either cognitive, social, or multifaceted. They explored definitions of "information behavior" and noted an apparent disconnect between research on that topic and its application to information system design.

Theories of Information Behavior (Fisher, Erdelez & McKechnie, 2005), a book produced by members of the American Society for Information Science and Technology (ASIST) Special Interest Group on Information Needs, Seeking, and Use, describes over 70 theories used in information-behavior research, many of them developed by information studies faculty (rather than borrowed from other disciplines); a total of 85 authors from 10 countries have contributed to the volume. Examination of the table of contents reveals a wide array of topics. Few are really "theories" in the most formal sense used in the social sciences (i.e., an articulated set of constructs, definitions, and propositions); rather, most are concepts, hypotheses, or models that have been developed to explain information-related phenomena. Chapters in the volume explore, for example, propositions about what knowledge and skills are needed for a researcher to use an archive, models showing how library searches typically proceed, and explanations of factors that influence relevance judgments; also included are decades-old constructs and theories (e.g., Diffusion Theory) adopted from other disciplines. Several of the entries show the influence of Bandura's (2001) Social Cognitive Theory, especially his central concept of self-efficacy. Whatever the granularity of the entries, the book promises be a useful addition to a literature frequently criticized for its lack of theory. Many of the concepts discussed are closely related, so it is reasonable to expect converging definitions and future collaborations among the authors, leading to further development of theory specifically for information behavior.

Several theories have been proposed recently in regard to information behavior. Dawson and Chatman (2001), for example, have suggested Reference Group Theory, as used by Merton and other sociologists. Burnett, Besant, and Chatman (2001) and Houtari and Chatman (2001) are two studies that applied Chatman's Small-World Theory. McKenzie (2004) discussed Positioning Theory in the context of her study of pregnant women. And Budd (2001) has argued for the importance of phenomenology, particularly the work of Bakhtin on dialogic communication.

Hall (2003) discussed the issue of borrowing theory from other disciplines to use in information research. Hall employed Social Exchange Theory, a framework used in sociology, psychology, and anthropology, as an example of borrowed theory. Wilson (2002) addressed the relevance of phenomenology for information behavior studies. Ford (2004a) suggested a wide variety of different theories and models that could be applied in the study of information seeking, including most notably the Conversation Theory of Gordon Pask.

Järvelin and Wilson (2003, online) differentiated "summary-types" from those that are more "analytical," in their review of conceptual models for information seeking and retrieval (IS&R). They discussed the functions of conceptual models in research and explored the attributes of models that facilitate the formulating of research questions and hypotheses.

A number of the items in this review have presented new models. A model of the information behavior of adolescents was advanced by Shenton and Dixon (2003a). Ford (2003) presented a model of learning based on constructs used in both information studies and education, including information processing types and approaches, learning objectives, needs, and relevance. Choo (2001a, 2001b) developed a model of environmental scanning. Niedwiedzka (2003) proposed modifications to Wilson's general model of information behavior, in order to apply it to the information-seeking behaviors of Polish managers. Wilson updated his own model in a chapter in *Theories of Information Behavior* by Fisher, Erdelez & McKechnie (2005). An empirical test of Taylor's value-added model was conducted by Miwa (2003), based on 62 callers to the AskERIC service. McKenzie (2003a) elaborated a model of quotidian life information practices contrasting direct with indirect ways of finding information on the basis of her interviews with 19 pregnant women. Brown (2002), Foster (2004), Hepworth (2004), Kari and Savolainen (2003), Meho and Tibbo (2003), Savolainen (2001b), Warner and Procaccino (2004), and Whitmire (2003) suggested yet other models, several of them based on Kuhlthau's Information Search Process model.

Methods

Wilson (2002) offered a novel typology of research methods. He considers observation to be the "root" method of data collection, dividing it into direct and indirect variants and further subdividing it into more familiar types, such as ethnographic observation, survey questionnaires, interviews. In a subsequent and related article co-authored with Järvelin, Wilson discussed conceptual models for information behavior research (Järvelin & Wilson, 2003).

A series of studies over the past decade has given us an overview of specific methods used in investigations on information seeking. The latest of these, by McKechnie, Baker, et al. (2002), examined 1,739 articles published during the period 1993 to 2000 in seven major journals and proceedings; of these, 247 (14 percent) were classed as concerning human information behavior. The 247 articles were content-analyzed to determine the affiliations of the authors, the populations they studied, and the methods they used—the last of which were not well described in some cases. The study found that 35 percent consisted of interviews; 20 percent, of other kinds of surveys; 14 percent, of observation studies; and about 12 percent, of content or document analysis; the remainder used a variety of measurement designs, including (in order of frequency) diaries, transaction logs, focus groups, "think aloud" protocols, secondary analysis, experiments, tests, bibliometric analysis, and discourse analysis.

Numerous methods have been advocated by individual researchers. Carey et al. (2001), for example, discussed both theoretical stances and interviewing methods that can tease out information behavior in everyday life. They argued for a shift in focus away from the individual as a unit of analysis toward a more general understanding of cultural conditions, illustrating their points with examples from studies of three different populations: pregnant women, members of a self-help support group, and preschool children.

In a similar vein, Urquhart (2001), Urquhart, Light, Thomas, Barker, Yeoman, Cooper, et al. (2003), and Bates (2004) described various interviewing approaches to studying information behavior. Methods including the critical incident technique, vignettes, scenarios, storytelling, narrative interviewing, and focus group interviews have been used in studies of hospital staff, doctors, nurses, midwives, and patients, among other respondents. Penzhorn (2002) reported on the use of a qualitative, participatory research approach for studying information needs.

Gorman, Lavelle, Delcambre, and Maier (2002) described participant observation and "think aloud" techniques that enabled them to understand the "information spaces" of medical clinicians. And Wildemuth (2002) compared the methods of Gorman et al. with those used in some other information-seeking studies.

Stefl-Mabry (2003) described the measurement technique of Social Judgment Analysis (SJA), which can be used to interpret information source preferences and can aid the understanding of how and why users may "satisfice," meaning that they decide to terminate their information seeking before all sources have been consulted or all available information has been found. A tool widely used in the social sciences, SJA relies on scenarios (or vignettes—see Urquhart, 2001) to represent patterns of information from different media, with the sources varying by the degree to which they offer supporting or conflicting information. Using data from 90 human subjects, Stefl-Mabry's multiple regressions illustrated the utility of the technique for studying source preferences.

Finally, Shim (2003) described how a researcher can use handheld computers (e.g., PDAs) in research on information seeking. These devices can function as diaries for the collection of information-seeking actions and episodes.

Conclusions

A topic with a long history, information behavior is more popular than ever. Both the individual and society have come into focus, resulting in more attention to context and social influence, more effort to "get inside the head" of the seeker, more time spent with individual informants, and greater depth of description overall.

The research community is increasingly international. Thirty years ago the majority of information-seeking research was conducted in the U.K. and North America. Now the research community has become global, with leading investigators found in other parts of Europe (especially Scandinavia), along with Africa and Asia. The field has many talented researchers, some of them highly influential and productive even at relatively early stages of their careers. This development is partly due to the popularity and effectiveness of the Information Seeking in Context (ISIC) conferences, of which there have been five to date; these meetings have provided fertile ground for the exchange of research ideas regarding information behavior.

The ways in which information-seeking behavior has been conceptualized and studied have changed profoundly over the last three decades. Perhaps the most obvious influences have been various strains of the "sense-making" paradigm as well as constructivist and constructionist models of thought. The shift in these new directions started about 30 years ago, when Brenda Dervin questioned the static ways in which "needs and uses" had been characterized. Now the dynamic, personal, and context-laden nature of information behavior seems to be taken as a given by all.

This paradigmatic shift has resulted in more attention being paid to, and more diversity in, both theory and methods. Researchers continue to embrace concepts and theories from many other disciplines, including sociology, psychology, communication, organizational behavior, and computer science. They are also developing their own concepts and theories. It would be heartening to see more agreement emerge from the current confusion—or at least a few clearly articulated camps within which everyone would agree on the nature of reality (see Bates, 2002, and Hjørland, 2004).

Examining the topics addressed in the Fisher, Erdelez, and McKechnie volume (2005), along with other work reviewed here, I am struck by how often spatial metaphors are employed in information behavior research; about a quarter of the entries in the theory book use some kind of quasi-geographical notion such as field, network, horizon, ground, boundary, domain, environment, browsing, or foraging. Of course, we live in a physical world populated with information-laden objects such as books and people, so spatial perception and movement is an unavoidable aspect of information behavior. Yet, as in other disciplines (e.g., Silber, 1995, regarding sociological theory), spatial metaphors have come to be used in the study of nonspatial aspects of social and mental life (see also Zook's chapter in the present volume). Indeed, Silber's work points out the strong influence of such metaphors in the work of particular theorists who are currently popular among information behavior researchers: Bourdieu, Foucault, Giddens, Goffman, Granovetter, Habermas, Lin, Luckmann, Schutz, and Zerubavel, to name a few. The popularity of spatial metaphors in information behavior research is due not only to the influence of these theorists, but also to the creative potential of metaphors for suggesting new meanings and relationships; there is nothing wrong with using them—except that we must keep in mind that any metaphor has its limits.

In terms of topical focus, traditional occupations (especially engineers, managers, university faculty, and health care providers) continue to be the subject of many investigations. A vigorous research agenda on everyday-life information seeking has made ordinary people the target of an ever-expanding number of investigations. The widespread influence of the Internet and World Wide Web on human information behavior has spawned a large number of studies in itself; a future *ARIST* volume could benefit from a chapter focusing on just those investigations.

This trend—the plethora of "Web searching" studies—blurs the identity of the traditional information behavior literature. In previous decades, investigations that focused on searching electronic resources were not typically called "information seeking" studies; they were, rather, a subtopic within other research areas: information retrieval, online searching, system evaluation, or human–computer interaction. Use of the Web is increasingly characterized as a kind of "information seeking." Has information behavior, then, subsumed all of what used to be called "online searching" investigations? It would appear so.

A problem with this broadening of scope is that the importance of "information behavior" as a concept is weakened. In a world in which everything is considered to be "short," height ceases to be a useful construct. Is there any topic in information studies that has nothing to do with "information behavior"? This question brings to mind Fairthorne's (1969, p. 26) comments, now nearly four decades old, regarding the scope of "information science":

> [There is] a dangerous tendency to bring in every and any science or technique or phenomenon under the "information" heading. Certainly hitherto distinct activities and interests should be unified, if indeed they have common principles. However, one does not create common principles by giving different things the same name.

If information behavior includes all aspects of searching, seeking, and use (as Wilson, 2000, implies), then it is even more important for authors to exercise precision in their titles and abstracts. Too many evaluations of searching skills or system features are now labeled "information seeking" or "information behavior"; these terms have simply become too popular.

Will the growth in the scope and size of the information behavior literature continue? Given the cyclical nature of academic research, it would not be surprising to see the number of information behavior investigations eventually subside. Yet, aside from a suggestion by one editorial board member that *Library & Information Science Research* may publish too much information behavior research (see Schwartz, 2003), enthusiasm for the topic appears to be growing. In particular, the increasing attention paid to theory is a sign of maturity in the investigation of information behavior.

References

Abbott, R. (2004). Subjectivity as a concern for information science: A Popperian perspective. *Journal of Information Science, 30,* 95–106.

Agosto, D. E., & Hughes-Hassell, S. (in press). People, places, and questions: An investigation of the everyday life information-seeking behaviors of urban young adults. *Library & Information Science Research, 27.*

Alexandersson, M., & Limberg, L. (2003). Constructing meaning through information artefacts. *The New Review of Information Behaviour Research, 4,* 17–30.

Allen, D., & Wilson, T. D. (2003). Information overload: Context and causes. *The New Review of Information Behaviour Research, 4,* 31–44.

Allen, T. J. (1969). Information needs and uses. *Annual Review of Information Science and Technology, 4,* 3–29.

Attfield, S., Blandford, A., & Dowell, J. (2003). Information seeking in the context of writing: A design psychology interpretation of the 'problematic situation.' *Journal of Documentation, 59,* 430–453.

Attfield, S., & Dowell, J. (2003). Information seeking and use by newspaper journalists. *Journal of Documentation, 59,* 187–204.

Baker, L. M. (2004). Information needs at the end of life: A content analysis of one person's story. *Journal of the Medical Libraries Association, 92,* 78–82.

Baker, L. M., Case, P., & Policicchio, D. L. (2003). General health problems of inner-city sex workers: A pilot study. *Journal of the Medical Library Association, 91,* 67–71.

Bandura, A. (2001). Social cognitive theory: An agentic perspective. *Annual Review of Psychology, 52,* 1–26.

Bar-Ilan, J. (2004). The use of Web search engines in information science research. *Annual Review of Information Science and Technology, 38,* 231–288.

Bates, J. A. (2004). Use of narrative interviewing in everyday information behavior research. *Library & Information Science Research, 26,* 15–28.

Bates, M. J. (2002). Toward an integrated model of information seeking and searching. *The New Review of Information Behaviour Research, 3,* 1–16.

Beer, S. (2004). Information flow and peripherality in remote island areas of Scotland. *Libri, 54*(3), 148–157.

Belefant-Miller, H., & King, D. W. (2001). How, what and why science faculty read. *Science and Technology Libraries, 19*, 91–112.

Belefant-Miller, H., & King, D. W. (2003). A profile of faculty reading and information-use behaviors on the cusp of the electronic age. *Journal of the American Society for Information Science and Technology, 54*, 179–181.

Box, V., Hepworth, M., & Harrison J. (2003). Identifying information needs of people with multiple sclerosis. *Nursing Times, 99*(49), 32–36.

Brown, C. D. (2001). The role of computer-mediated communication in the research process of music scholars: An exploratory investigation. *Information Research, 6*. Retrieved July 3, 2004, from http://InformationR.net/ir/6-2/infres62.html

Brown, C. D. (2002). Straddling the humanities and social sciences: The research process of music scholars. *Library & Information Science Research, 24*, 73–94.

Bruce, H., Fidel, R., Pejtersen, A. M., Dumais, S., Grudin, J., & Poltrock, S. (2003). A comparison of the collaborative information retrieval behaviour of two design teams. *The New Review of Information Behaviour Research, 4*, 139–153.

Budd, J. (2001). Information seeking in theory and practice: Rethinking public services in libraries. *Reference & User Services Quarterly, 40*, 256–263.

Burnett, G., Besant, M., & Chatman, E. A. (2001). Small worlds: Normative behavior in virtual communities and feminist bookselling. *Journal of the American Society for Information Science and Technology, 52*, 536–547.

Caidi, N. (2001). Interdisciplinarity: What is it and what are its implications for information seeking? *Humanities Collections, 1*(4), 35–46.

Carey, R. F., McKechnie, L., & McKenzie, P. (2001). Gaining access to everyday life information seeking. *Library & Information Science Research, 23*, 319–334.

Case, D. O. (2002). *Looking for information: A survey of research on information seeking, needs, and behavior.* San Diego, CA: Academic Press.

Case, D. O., Johnson, J. D., Andrews, J. E., Allard, S., & Kelly, K. M. (2004). From two-step flow to the Internet: The changing array of sources for genetics information seeking. *Journal of the American Society for Information Science and Technology, 55*, 660–669.

Chang, S., & Rice, R. (1993). Browsing: A multidimensional framework. *Annual Review of Information Science and Technology, 28*, 231–276.

Chatman, E. A. (1996). The impoverished life-world of outsiders. *Journal of the American Society for Information Science, 47*, 193–206.

Chelton, M. K., & Cool, C. (Eds.). (2004). *Youth information-seeking behavior: Theories, models and approaches.* Metuchen, NJ: Scarecrow Press.

Chen, C., & Hernon, P. (1982). *Information-seeking: Assessing and anticipating user needs.* New York: Neal-Schuman.

Choo, C. W. (2001a). Environmental scanning as information seeking and organizational learning. *Information Research, 7*. Retrieved December 3, 2003, from http://InformationR.net/ir/7-1/infres71.html

Choo, C. W. (2001b). *Information management for the intelligent organization: The art of scanning the environment.* Medford, NJ: Information Today.

Choo, C. W., & Auster, E. (1993). Environmental scanning: Acquisition and use of information by managers. *Annual Review of Information Science and Technology, 28*, 279–314.

Cogdill, K. W. (2003). Information needs and information seeking in primary care: A study of nurse practitioners. *Journal of the Medical Libraries Association, 91*, 203–215.

Cool, C. (2001). The concept of situation in information science. *Annual Review of Information Science and Technology, 35*, 5–42.

Cooper, L. Z. (2002). A case study of information-seeking behavior in 7-year-old children in a semistructured situation. *Journal of the American Society for Information Science and Technology, 53*, 904–922.

Correia, Z., & Wilson, T. D. (2001). Factors influencing environmental scanning in the organizational context. *Information Research, 7*. Retrieved December 3, 2003, from http://InformationR.net/ir/7-1/infres71.htm

Crane, D. (1971). Information needs and uses. *Annual Review of Information Science and Technology, 6*, 3–39.

Crawford, S. (1978). Information needs and uses. *Annual Review of Information Science and Technology, 13*, 61–81.

Dalton, M. S., & Charnigo, L. (2004). Historians and their information sources. *College & Research Libraries, 65*, 400–425.

Davies, M. M., & Bath, P. A. (2003). Interpersonal sources of health and maternity information for Somali women living in the UK: Information seeking and evaluation. *Journal of Documentation 58*, 302–318.

Dawson, M., & Chatman, E. A. (2001). Reference group theory with implications for information studies: A theoretical essay. *Information Research, 6*. Retrieved December 3, 2003, from http://InformationR.net/ir/6-3/infres63.htm

Dervin, B. (2003). Human studies and user studies: A call for methodological interdisciplinarity. *Information Research, 9*. Retrieved July 3, 2004, from http://InformationR.net/ir/9-1/infres91.htm

Dervin, B. (1989). Users as research inventions: How research categories perpetuate inequities. *Journal of Communication, 39*, 216–232.

Dervin, B., & Nilan, M. (1986). Information needs and uses. *Annual Review of Information Science and Technology, 21*, 1–25.

Donat, J. F., & Pettigrew, K. E. (2002). The final context: Information behaviour surrounding the dying patient. *The New Review of Information Behaviour Research, 3*, 175–186.

Duff, W. M., & Johnson, C. A. (2002). Accidentally found on purpose: Information seeking behavior of historians in archives. *Library Quarterly, 72*, 475–499.

Duff, W. M., & Johnson, C. A. (2003). Where is the list with all the names? Information-seeking behavior of genealogists. *American Archivist, 66*(1), 79–95.

Dunne, J. E. (2002). Information seeking and use by battered women: A "person-in-progressive-situations" approach. *Library & Information Science Research, 24*, 343–355.

Ekoja, I. I. (2004). Sensitising users for increased information use: The case of Nigerian farmers. *African Journal of Library, Archives & Information Science, 14*, 193–204.

Ellis, D., Wilson, T. D., Ford, N., Foster, A., Lam, H. M., Burton, R., et al. (2002). Information seeking and mediated searching. Part 5: Intermediary interaction. *Journal of the American Society for Information Science and Technology, 53*, 883–893.

Eppler, M. J., & Mengis, J. (2004). The concept of information overload: A review of literature from organization science, accounting, marketing, MIS, and related disciplines. *The Information Society, 30*, 325–344.

Fairthorne, R. A. (1969). The scope and aims of the information sciences and technologies. In International Federation for Documentation, Committee on Research on the Theoretical Basis of Information, *On theoretical problems of informatics* (FID 435) (pp. 25–31). Moscow: All-Union Institute for Scientific and Technological Information.

Farhoomand, A. F., & Drury, D. H. (2002). Managerial information overload. *Communications of the ACM, 45*(10), 127–131.

Fidel, R., & Green, M. (2004). The many faces of accessibility: Engineers' perception of information sources. *Information Processing & Management, 40,* 563–581.

Fidel, R., Pejtersen, A. M., Cleal, B., & Bruce, H. (2004). A multidimensional approach to the study of human–information interaction: A case study of collaborative information retrieval. *Journal of the American Society for Information Science and Technology, 55,* 939–953.

Fikar, C. R., & Keith, L. (2004). Information needs of gay, lesbian, bisexual, and transgendered health care professionals: Results of an Internet survey. *Journal of the Medical Library Association, 92,* 56–65.

Fisher, K. E., Durrance, J. C., & Hinton, M. B. (2004). Information grounds and the use of need-based services by immigrants in Queens, New York: A context-based, outcome evaluation approach. *Journal of the American Society for Information Science and Technology, 55,* 754–766.

Fisher, K. E., Erdelez, S., & McKechnie, E. F. (Eds.). (2005). *Theories of information behavior*. Medford, NJ: Information Today.

Flanagin, A. J., & Metzger, M. J. (2001). Internet use in the contemporary media environment. *Human Communication Research, 27,* 153–181.

Flaxbart, D. (2001). Conversations with chemists: Information-seeking behavior of chemistry faculty in the electronic age. *Science & Technology Libraries, 21*(3/4), 5–26.

Ford, N. (2003). Towards a model of learning for educational informatics. *Journal of Documentation, 60,* 183–225.

Ford, N. (2004a). Modeling cognitive processes in information seeking: From Popper to Pask. *Journal of the American Society for Information Science and Technology, 55,* 769–782.

Ford, N. (2004b). Creativity and convergence in information science research: The roles of objectivity and subjectivity, constraint, and control. *Journal of the American Society for Information Science and Technology, 55,* 1169–1182.

Foster, A. (2004). A nonlinear model of information-seeking behavior. *Journal of the American Society for Information Science and Technology, 55,* 228–237.

Foster, A., & Ford, N. (2003). Serendipity and information seeking: An empirical study. *Journal of Documentation, 59,* 321–343.

Garrison, B. (2001). Journalists' newsroom roles and their World Wide Web search habits. *Newspaper Research Journal.* Retrieved June 29, 2004 from http://www.miami.edu/com/car/phoenix2.htm

Given, L. (2002a). Discursive constructions in the university context: Social positioning theory and mature undergraduates' information behaviours. *The New Review of Information Behaviour Research, 3,* 127–142.

Given, L. (2002b). The academic and the everyday: Investigating the overlap in mature undergraduates' information-seeking behaviors. *Library & Information Science Research, 24,* 17–29.

Gorman, P., Lavelle, M., Delcambre, L., & Maier, D. (2002). Following experts at work in their own information spaces: Using observational methods to develop tolls for the digital library. *Journal of the American Society for Information Science and Technology, 53*, 1245–1250.

Gross, M. (2001). Imposed information seeking in public libraries and school library media centers: A common behaviour? *Information Research, 6.* Retrieved December 3, 2003, from http://InformationR.net/ir/6-2/infres62.html

Gross, M. (2004a). Children's information seeking at school: Findings from a qualitative study. In M. K. Chelton & C. Cool (Eds.), *Youth information-seeking behavior: Theories, models and issues* (pp. 211–240). Lanham, MD: Scarecrow Press.

Gross, M. (2004b). *Children's questions: Information seeking behavior in school.* Lanham, MD: Scarecrow Press.

Gross, M., & Saxton, M. (2001). Who wants to know? Imposed queries in the public library. *Public Libraries, 40*(3), 170–176.

Hall, H. (2003). Borrowed theory: Applying exchange theories in information science research. *Library & Information Science Research, 25*, 287–305.

Hallmark, J. (2001). Information-seeking behavior of academic meteorologists and the role of information specialists. *Science & Technology Libraries, 21*(1/2), 53–64.

Hamer, J. S. (2003). Coming-out: Gay males' information seeking. *School Libraries Worldwide, 9*(2), 73–89.

Hartel, J. (2003). The serious leisure frontier in library and information studies: Hobby domains. *Knowledge Organization, 3*, 228–238.

Harrison, J., Hepworth, M., & de Chazal, P. (2004). NHS and social care interface: A study of social workers' library and information needs. *Journal of Librarianship and Information Science, 36*(1): 27–35.

Haruna, I., & Mabawonku, I. (2001). Information needs and seeking behaviour of legal practitioners and the challenges to law libraries in Lagos, Nigeria. *International Information and Library Review, 33*(1): 69–87.

Heinström, J. (2003). Five personality dimensions and their influence on information behaviour. *Information Research, 9.* Retrieved December 3, 2003, from http://InformationR.net/ir/91/infres91.html

Hektor, A. (2003). Information activities on the Internet in everyday life. *The New Review of Information Behaviour Research, 4*, 127–138.

Hepworth, M. (2004). A framework for understanding user requirements for an information service: Defining the needs of informal carriers. *Journal of the American Society for Information Science and Technology, 55*, 695–708.

Hepworth, M., & Harrison, J. (2004). A survey of the information needs of people with Multiple Sclerosis. *Health Informatics Journal, 10*(1), 49–69.

Hepworth, M., Harrison, J., & James, N. (2003a). Information needs of people with MS. *Library & Information Update, 2*(3), 38–39.

Hepworth, M., Harrison, J., & James, N. (2003b). Information needs of people with Multiple Sclerosis and the implications for information provision based on a national UK survey. *Aslib Proceedings, 55*, 290–303.

Herman, E. (2004). Research in progress: Some preliminary and key insights into the information needs of the contemporary academic researcher. Part 1. *Aslib Proceedings, 56*, 34–47.

Herner, S., & Herner, M. (1967). Information needs and uses in science and technology. *Annual Review of Information Science and Technology, 2*, 1–34.

Hersberger, J. (2001). Everyday information needs and information sources of homeless parents. *The New Review of Information Behaviour Research, 2*, 119–134.

Hersberger, J. (2003). A qualitative approach to examining information transfer via social networks among homeless populations. *The New Review of Information Behaviour Research, 4*, 95–108.

Hewins, E. T. (1990). Information need and use studies. *Annual Review of Information Science and Technology, 25*, 145–172.

Hirsh, S., & Dinkelacker, J. (2004). Seeking information in order to produce information: An empirical study at Hewlett Packard Labs. *Journal of the American Society for Information Science and Technology, 55*(9), 807–817.

Hjørland, B. (2002a). Domain analysis in information science. Eleven approaches—traditional as well as innovative. *Journal of Documentation, 58*, 422–462.

Hjørland, B. (2002b). Epistemology and the socio-cognitive perspective in information science. *Journal of the American Society for Information Science and Technology, 53*, 257–270.

Hjørland, B. (2004). Arguments for philosophical realism in library and information science. *Library Trends, 52*, 488–506.

Holmes, K. L., & Lenz, E. R. (2002). Perceived self-care information needs and information-seeking behaviors before and after elective spinal procedures. *Journal of Neuroscience Nursing, 29*, 79–85.

Houtari, M.-L., & Chatman, E. A. (2001). Using everyday life information seeking to explain organizational behavior. *Library & Information Science Research, 23*, 351–366.

Houtari, M.-L., & Wilson, T. D. (2001). Determining organizational information needs: The Critical Success Factors approach. *Information Research, 6*. Retrieved July 1, 2004, from http://InformationR.net/ir/6-3/infres63.html

Hultgren, F., & Limberg, L. (2003). A study of research on children's information behaviour in a school context. *The New Review of Information Behaviour Research, 4*, 1–15.

Ikoja-Odongo, R. (2002). Insights into the information needs of women in the informal sector of Uganda. *South African Journal of Library & Information Science, 68*(1), 39–52.

Ikoja-Odongo, R., & Ocholla, D. N. (2003). Information needs and information-seeking behavior of artisan fisher folk of Uganda. *Library & Information Science Research, 25*, 89–105.

Ikoja-Odongo, R., & Ocholla, D. N. (2004). Information seeking behaviour of the informal sector entrepreneurs: The Uganda experience. *Libri, 11*(1), 54–66.

Järvelin, K., & Wilson, T. D. (2003). On conceptual models for information seeking and retrieval research. *Information Research, 9*. Retrieved July 1, 2004, from http://InformationR.net/ir/9-1/-infres91.html

Jeong, W. (2004). Unbreakable ethnic bonds: Information-seeking behavior of Korean graduate students in the United States. *Library & Information Science Research, 26*, 384–400.

Jiyane, V., & Ocholla, D. N. (2004). An exploratory study of information availability and exploitation by the rural women of Melmoth, KwaZulu-Natal. *South African Journal of Library & Information Science, 70*(1), 1–8.

Johnson, J. D. (2003). On contexts of information seeking. *Information Processing & Management, 39*, 735–760.

Johnson, J. D., Andrews, J. E., & Allard, S. (2001). A model of understanding and affecting cancer genetics information seeking. *Library and Information Science Research, 23*, 335–349.

Johnson, J. D., Andrews, J. E., Case, D. O., & Allard, S. (in press). Genomics: The perfect information seeking research problem. *Journal of Health Communication, 10*.

Johnson, J. D., Case, D., Andrews, J. E., Allard, S. L., & Johnson, N. E. (in press). Fields and pathways: Contrasting or complementary views of information seeking. *Information Processing & Management, 41*.

Julien, H. (2004). Adolescent decision-making for careers: An exploration of information behavior. In M. K. Chelton & C. Cool (Eds.), *Youth information-seeking behavior: Theories, models and issues* (pp. 321–352). Lanham, MD: Scarecrow Press.

Julien, H., & Duggan, L. (2000). A longitudinal analysis of the information needs and uses literature. *Library & Information Science Research, 22*, 291–309.

Julien, H., & Michels, D. (2003). Intra-individual information behaviour. *Information Processing & Management, 40*(3), 547–562.

Kari, J., & Savolainen, R. (2003). Towards a contextual model of information seeking on the Web. *The New Review of Information Behaviour Research, 4*, 155–175.

Kaye, B. K., & Johnson, T. J. (2003). From here to obscurity? Media substitution theory and traditional media in an on-line world. *Journal of the American Society for Information Science and Technology, 54*, 260–273.

King, D. W., & Tenopir, C. (2001). Using and reading scholarly literature. *Annual Review of Information Science and Technology, 34*, 423–477.

Kling, R., & Callahan, E. (2003). The use of Web search engines in information science research. *Annual Review of Information Science and Technology, 37*, 127–177.

Kuhlthau, C. (2004a). *Seeking meaning: A process approach to library and information services* (2nd ed.). Westport, CT: Libraries Unlimited.

Kuhlthau, C. C. (2004b). Student learning in the library: What Library Power librarians say. In M. Chelton & C. Cool (Eds.), *Youth information-seeking behavior: Theories, models and issues* (pp. 37–64). Lanham, MD: Scarecrow Press.

Kuhlthau, C., & Tama, S. L. (2001). Information search process of lawyers: A call for 'just for me' information services. *Journal of Documentation, 57*, 25–43.

Kwasitsu, L. (2003). Information-seeking behavior of design, process, and manufacturing engineers. *Library & Information Science Research, 25*, 459–476.

Lin, N., & Garvey, W. D. (1972). Information needs and uses. *Annual Review of Information Science and Technology, 7*, 5–37.

Lion, R., & Meertens, R. M. (2001). Seeking information about a risky medicine: Effects of risk-taking tendency and accountability. *Journal of Applied Psychology, 31*, 778–795.

Lipetz, B.-A. (1970). Information needs and uses. *Annual Review of Information Science and Technology, 5*, 3–32.

Macintosh-Murray, A. (2001). Scanning and vicarious learning from adverse events in health care. *Information Research, 7*. Retrieved July 3, 2004, from http://InformationR.net/ir/7-1/infres71.html

Mackenzie, M. L. (2002). Information gathering: The information behaviors of line-managers within a business environment. *Proceedings of the 65th Annual Meeting of the American Society for Information Science and Technology*, 164–170.

Mackenzie, M. L. (2003a). An exploratory study investigating the information behaviour of line managers within a business environment. *The New Review of Information Behaviour Research, 4*, 63–78.

Mackenzie, M. L. (2003b). Information gathering revealed within the social network of line-managers. *Proceedings of the 66th Annual Meeting of the American Society for Information Science and Technology*, 85–94.

Mackenzie, M. L. (2004). The cultural influences of information flow at work: Manager information behavior documented. *Proceedings of the 67th Annual Meeting of the American Society for Information Science and Technology*, 184–190.

Marcella, R., & Baxter, G. (2001). A random walk around Britain: A critical assessment of the random walk sample as a method of collecting data on the public's citizenship information needs. *The New Review of Information Behaviour Research, 2*, 87–103.

Marton, C. (2001). Environmental scan on women's health information resources in Ontario, Canada. *Information Research, 7*. Retrieved July 3, 2004, from http://InformationR.net/ir/7-1/infres71.html

Marton, C. (2003). Quality of health information on the Web: User perceptions of relevance and reliability. *The New Review of Information Behaviour Research, 4*, 195–206.

Martyn, J. (1974). Information needs and uses. *Annual Review of Information Science and Technology, 9*, 3–23.

Matthews, A. K., Sellergren, S. A., Manfredi, C., & Williams, M. (2002). Factors influencing medical information seeking among African American cancer patients. *Journal of Health Communication, 7*, 205–219.

McKechnie, L. M., Baker, L., Greenwood, M., & Julien, H. (2002). Research method trends in human information behaviour literature. *The New Review of Information Behaviour Research, 3*, 113–126.

McKechnie, L. M., & Pettigrew, K. E. (2002). Surveying the use of theory in library and information science research: A disciplinary perspective. *Library Trends, 50*, 406–417.

McKenzie, P. J. (2002a). Communication barriers and information-seeking counterstrategies in accounts of practitioner-patient encounters. *Library & Information Science Research, 24*, 31–47.

McKenzie, P. J. (2002b). Connecting with information sources: How accounts of information seeking take discursive action. *The New Review of Information Behaviour Research, 3*, 161–174.

McKenzie, P. J. (2003a). A model of information practices in accounts of everyday-life information seeking. *Journal of Documentation, 59*, 19–40.

McKenzie, P. J. (2003b). Justifying cognitive authority decisions: Discursive strategies of information seekers. *Library Quarterly, 73*, 261–288.

McKenzie, P. J. (2004). Positioning theory and the negotiation of information needs in a clinical midwifery setting. *Journal of the American Society for Information Science and Technology, 55*, 685–694.

Meho, L. I., & Haas, S. W. (2001). Information-seeking behavior and use of social science faculty studying stateless nations: A case study. *Library & Information Science Research, 23*, 5–25.

Meho, L. I., & Tibbo, H. R. (2003). Modeling the information-seeking behavior of social scientists: Ellis's study revisited. *Journal of the American Society for Information Science and Technology, 54*, 570–587.

Menzel, H. (1966). Information needs and uses in science and technology. *Annual Review of Information Science and Technology, 1*, 41–69.

Metoyer-Duran, C. (1993). Information gatekeepers. *Annual Review of Information Science and Technology, 28*, 111–150.

Meyer, H. W. J. (2003). Information use in rural development. *The New Review of Information Behaviour Research, 4*, 109–125.

Miwa, M. (2003). Situatedness in users' evaluation of information and information services. *The New Review of Information Behaviour Research, 4*, 207–224.

Mooko, N. P. (2002). The use and awareness of women's groups as sources of information in three small villages in Botswana. *South African Journal of Library & Information Science, 68*, 104–111.

Murphy, J. (2003). Information-seeking habits of environmental scientists: A study of interdisciplinary scientists at the Environmental Protection Agency in Research Triangle Park, North Carolina. *Issues in Science and Technology Librarianship*. Retrieved December 1, 2004, from http://www.istl.org/03-summer/refereed.html

Niedwiedzka, B. (2003). A proposed general model of information behaviour. *Information Research, 9*. Retrieved December 3, 2003, from http://InformationR.net/ir/9-1/infres91.html

Ocheibi, J. A., & Buba, A. (2003). Information needs and information gathering behaviour of medical doctors in Maiduguri, Nigeria. *Journal of Educational Media & Library Sciences, 40*, 417–427.

Paisley, W. J. (1968). Information needs and uses. *Annual Review of Information Science and Technology, 3*, 1–30.

Palmer, C. L., & Neumann, L. J. (2002). The information work of interdisciplinary humanities scholars: Exploration and translation. *Library Quarterly, 72*, 85–117.

Penzhorn, C. (2002). The use of participatory research as an alternative approach for information needs research. *Aslib Proceedings, 54*, 240–250.

Pettigrew, K. E., Durrance, J. C., & Unruh, K. T. (2002). Facilitating community information seeking using the Internet: Findings from three public library–community network systems. *Journal of the American Society for Information Science and Technology, 53*, 894–903.

Pettigrew, K. E., Fidel, R., & Bruce, H. (2001). Conceptual frameworks in information behavior. *Annual Review of Information Science and Technology, 35*, 43–78.

Pettigrew, K. E., & McKechnie, L. (2001). The use of theory in information science research. *Journal of the American Society for Information Science and Technology, 52*, 62–73.

Rees, C., & Bath, P. (2001). Information seeking behaviors of women with breast cancer. *Oncology Nursing Forum, 28*(5), 899–907.

Savolainen, R. (2001a). 'Living encyclopedia' or idle talk? Seeking and providing consumer information in an Internet newsgroup. *Library & Information Science Research, 23*, 67–90.

Savolainen, R. (2001b). Network competence and information seeking on the Internet: From definitions towards a social cognitive model. *Journal of Documentation, 58*, 211–226.

Savolainen, R., & Kari, J. (2004a). Placing the Internet in information source horizons: A study of information seeking by Internet users in the context of self-development. *Library & Information Science Research, 26*, 415–433.

Savolainen, R., & Kari, J. (2004b). Conceptions of the Internet in everyday life information seeking. *Journal of Information Science, 30*, 219–226.

Schwartz, C. (2003). Where are we, where might we go? *Library & Information Science Research, 25*, 233–237.

Seldén, L. (2001). Academic information seeking: Careers and capital types. *The New Review of Information Behaviour Research, 2*, 195–216.

Serema, B. C. (2002). *Community information structures in Botswana: A challenge for librarians.* The Hague, NL: International Federation of Library Associations and Institutions. Retrieved July 3, 2004, from http://www.ifla.org/IV/ifla68/papers/029–114e.pdf

Shenton, A. K., & Dixon, P. (2003a). Models of young people's information seeking. *Journal of Librarianship and Information Science, 35*, 5–22.

Shenton, A. K., & Dixon, P. (2003b). Youngsters' use of other people as an information-seeking method. *Journal of Librarianship and Information Science, 35*, 219–233.

Shenton, A. K., & Dixon, P. (2004a). Issues arising from youngsters' information-seeking behavior. *Library & Information Science Research, 26*, 177–200.

Shenton, A. K., & Dixon, P. (2004b). The nature of information needs and strategies for their investigation in youngsters. *Library & Information Science Research, 26*, 296–310.

Shim, W. (2003). Using handheld computers in information seeking research. *Journal of Education for Library and Information Science, 44*, 258–265.

Silber, I. F. (1995). Space, fields, boundaries: The rise of spatial metaphors in contemporary sociological theory. *Social Research, 62*, 323–355.

Solomon, P. (2002). Discovering information in context. *Annual Review of Information Science and Technology, 36*, 229–264.

Spink, A., & Cole, C. (2001). Information and poverty: Information-seeking channels used by African American low-income households. *Library & Information Science Research, 23*, 1–22.

Stefl-Mabry, J. (2003). A social judgment analysis of information source preference profiles: An exploratory study to empirically represent media selection patterns. *Journal of the American Society for Information Science & Technology, 54*, 879–904.

Stilwell, C. (2002). The case for informationally based social inclusion for sex workers: A South African exploratory study. *Libri, 52*(2), 67–77.

Sundin, O. (2002). Nurses' information seeking and use as participation in occupational communities. *The New Review of Information Behaviour Research, 3*, 187–202.

Talja, S. (2002). Information sharing in academic communities: Types and levels of collaboration in information seeking and use. *The New Review of Information Behaviour Research, 3*, 143–160.

Taylor, M., Alman, A., David, A., & Manchester, D. (2001). Use of the Internet by patients and their families to obtain genetics-related information. *Mayo Clinic Proceedings, 76*, 772–776.

Tibbo, H. (1991). Information systems, services, and technology for the humanities. *Annual Review of Information Science and Technology, 26*, 287–346.

Todd, R. (2003). Adolescents of the information age: Patterns of information seeking and use, and implications for information professionals. *School Libraries Worldwide, 9*(2), 27–46.

Todd, R. J., & Edwards, S. (2004). Adolescents' information seeking and utilization in relation to drugs. In M. K. Chelton and C. Cool (Eds.), *Youth information-seeking behavior: Theories, models and issues* (pp. 353–386). Lanham, MD: Scarecrow Press.

Toms, E. G., & Duff, W. (2002). "I spent 1 1/2 hours sifting through one large box ..." Diaries as information behavior of the archives user: Lessons learned. *Journal of the American Society for Information Science and Technology, 53*, 1232–1238.

Tuominen, K., Talja, S., & Savolainen, R. (2002). Discourse, cognition, and reality: Toward a social constructionist metatheory for library and information science. *Proceedings of the Fourth International Conference on Conceptions of Library and Information Science (CoLIS 4)*, 271–283.

Urquhart, C. (2001). Bridging information requirements and information needs assessment: Do scenarios and vignettes provide a link? *Information Research, 6.* Retrieved July 3, 2004, from http://InformationR.net/ir/6-2/infres62.html

Urquhart, C., Light, A., Thomas, R., Barker, A., Yeoman, A., Cooper, J., et al. (2003). Critical incident technique and explication interviewing in studies of information behavior. *Library & Information Science Research, 25*, 63–88.

Vakkari, P. (2003). Task-based information searching. *Annual Review of Information Science and Technology, 37*, 413–464.

Wang, P. (2001). Methodologies and methods for user behavioral research. *Annual Review of Information Science and Technology, 34*, 53–100.

Warner, D., & Procaccino, J. D. (2004). Toward wellness: Women seeking health information. *Journal of the American Society for Information Science and Technology, 55*, 709–730.

Westbrook, L. (2003). Information needs and experiences of scholars in women's studies: Problems and solutions. *College & Research Libraries, 64*, 192–209.

Whitmire, E. (2003). Epistemological beliefs and the information-seeking behavior of undergraduates. *Library & Information Science Research, 25*, 127–142.

Wicks, D. A. (2004). Older adults and their information seeking. *Behavioral & Social Sciences Librarian, 22*(2), 1–26.

Widén-Wulff, G. (2003). Information as a resource in the insurance business: The impact of structures and processes on organization information behaviour. *The New Review of Information Behaviour Research, 4*, 79–94.

Wikgren, M. (2001). Health discussions on the Internet: A study of knowledge communication through citations. *Library & Information Science Research, 23*, 305–317.

Wikgren, M. (2003). Everyday health information exchange and citation behaviour in Internet discussion groups. *The New Review of Information Behaviour Research, 4*, 225–239.

Wildemuth, B. M. (2002). Introduction and overview: Effective methods for studying information seeking and use. *Journal of the American Society for Information Science and Technology, 53*, 1218–1222.

Wilkinson, M. A. (2001). Information sources used by lawyers in problem solving: An empirical exploration. *Library & Information Science Research, 23*, 257–276.

Williamson, K., & Manaszewicz, R. (2002). Breast cancer information needs and seeking: Towards an intelligent, user sensitive portal to breast cancer knowledge online. *The New Review of Information Behaviour Research, 3*, 203–219.

Wilson, T. D. (2000). Human information behavior. *Informing Science, 3*(2), 49–55.

Wilson, T. D. (2002). Alfred Schutz, phenomenology and research methodology for information behaviour research. *The New Review of Information Behaviour Research, 3*, 71–82.

Wilson, T. D., Ford, N., Ellis, D., Foster, A., & Spink, A. (2002). Information seeking and mediated searching. Part 2. Uncertainty and its correlates. *Journal of the American Society for Information Science and Technology, 53*, 704–715.

Yakel, E. (2004). Seeking information, seeking connections, seeking meaning: Genealogists and family historians. *Information Research, 10*. Retrieved December 12, 2004, from http://informationr.net/ir/10-1/paper205.html

Yitzhaki, M., & Hammershlag, G. (2004). Accessibility and use of information sources among computer scientists and software engineers in Israel: Academy versus industry. *Journal of the American Society for Information Science and Technology, 55*, 832–842.

CHAPTER 8

Collaborative Information Seeking and Retrieval

Jonathan Foster
University of Sheffield

Introduction

One of the aims of information science is to develop systems and practices that facilitate information seeking, searching, and retrieval. These processes are typically performed by an individual, sometimes in tandem with a professional intermediary. Settings have emerged, however, in which either information is gathered specifically for collaborative purposes, or individuals collaborate directly or indirectly with other users engage in seeking, searching, and retrieving information. This chapter reviews current research in developing systems and practices that enable individuals to collaborate on information seeking, searching, and retrieval tasks.

Scope

Research into collaborative information seeking and retrieval draws from work in information science, information retrieval, human–computer interaction, and computer-supported collaborative work. The focus of this review is on research published from 2000 through 2004 that investigates social and collaborative approaches to information seeking, querying, filtering, and navigating. Each section begins with a brief note on the structure and scope of the section and ends with a summary of current problems and issues. The chapter concludes by identifying future research challenges in collaborative information seeking and retrieval. Prior *ARIST* reviews relevant to this chapter include: Vakkari's (2003) "Task-based information searching," Finholt's (2002) "Collaboratories," Sawyer and Eschenfelder's (2002) "Social informatics," and Twidale and Nichols's (1998) "Computer supported cooperative work in information search and retrieval." Churchill, Sullivan, and Snowdon (1998) and Karamuftuoglu (1998) also act as reference points for this chapter.

Definition

Collaborative information seeking and retrieval is an interdisciplinary phenomenon, research into which has been conducted by a range of academic communities within a variety of settings. Relevant work can be found within information science, information retrieval, human–computer interaction, and computer–supported cooperative work. Practitioner settings that provide a context for such research include the academy, industry, medicine, and the military.

Definitions of collaborative information seeking and retrieval vary in relation to the assumptions of the academic discipline within which the research is conducted. Depending on the discipline, a definition may emphasize information handling, search and retrieval, interaction, or the seeking and retrieving of information in support of collaborative work tasks. The following inclusive definition is adopted here: the study of the systems and practices that enable individuals to collaborate during the seeking, searching, and retrieval of information.

Information Task: Seeking

This section on social and collaborative approaches to information seeking tasks is organized according to the specific communities within which such research has been conducted. They are: the academy, industry, medicine, and the military.

Collaborative Information Seeking in Academe

Several studies have investigated systems and practices for collaborative information seeking within academic communities, including the types of information needed to support collaborative research, academics' information-sharing behavior, and collaborative information seeking as a group learning method. Studies of information seeking to support collaborative research have focused on: developments in collaboratories (e.g., Albrechtsen, 2004), the data intensity of such research (e.g., Karasti, Baker, & Bowker, 2003; Sonnenwald, Maglaughlin, & Whitton, 2004), and requirements for awareness information (e.g., Sonnenwald et al., 2004; Spring & Vathanophas, 2003).

In a combined ethnographic and experimental study of physicists, researchers discovered that successful scientific collaboration requires the collection and use of a range of awareness information that updates team members on the current state of their teams' activities (Sonnenwald et al., 2004). The study investigated the types of information and knowledge that need to be shared to support situation awareness and the ways in which technology can be used to facilitate such information sharing. Data for the ethnographic study were collected via semi-structured interviews, critical incident interviews, and participant observation. Sonnewald et al. conducted interviews with faculty, post-doctorates, and graduate student scientists and observed experiments conducted independently and collaboratively by their subjects. They collected data for the experimental study via a series of controlled experiments in order to compare the processes and outcomes of work completed using the collaboratory system in both face-to-face and remote modes. Data were collected on the performance of authentic scientific tasks such as operation of scientific equipment and the capture and recording of data in electronic notebooks. Data analysis consisted of open and axial coding in order to discover work patterns and practices. The study's main finding was the identification of three types of information that each team member needed to support his or her situation awareness: contextual, task-and-process, and socio-emotional. Additionally, Sonnewald et al. identified a role for technology in supporting access to the awareness information.

Studies of collaborative information seeking in the academy have also focused on academics' information-sharing behavior (Talja, 2002). The term *information sharing* is used to denote "an umbrella concept that covers a wide range of collaboration behaviours, from sharing accidentally encountered information to collaborative query formulation and retrieval" (Talja, 2002, p. 145). Qualitative case studies of scholars across a range of humanistic, social-scientific, and scientific disciplines were conducted. Talja collected data via semi-structured interviews, focusing both on scholars' personal information seeking and on group and department-level collaboration. Data were analyzed inductively with a view to developing testable concepts and hypotheses relating to information sharing.

Talja identified four types of information-sharing practices: *Strategic sharing* occurs when information sharing serves a broader purpose or goal, for example, as part of a "conscious strategy of maximizing efficiency in a research group (Talja, 2002, p. 147). Research groups engage in *paradigmatic sharing* to establish "a novel and distinguishable research approach or area within a discipline or across disciplines" (Talja, 2002, p. 147). *Directive sharing* involves two-way exchange of information, (in contrast with one-way information giving), such as "when senior and junior scholars both benefit from the results of each other's searches (Talja, 2002, p. 150). And *social sharing* contributes to "relationship- and community-building activity" (Talja, 2002, p. 147). Talja also explored nonsharing and concluded that further research is needed to ascertain the extent of collaborative document retrieval practices and the ways in which situational and other factors influence such retrieval. She also identified a role for digital libraries in supporting collaborative information seeking (e.g., collaborative keyword formulation, document retrieval, filtering, and problem solving).

Two further studies of collaborative information seeking in the academy illustrate its use as a group learning method. One study focused on understanding the characteristics of collaborative information behavior in a group-based educational setting and took the form of a case study of two groups of information science students (Hyldegård, in press).

Longitudinal data on the students' completion of the same task over a period of seven weeks were collected via diaries and interviews. Elements of Kuhlthau's information-seeking process model—activities, cognitive experiences, and emotional experiences—were used to analyze the information behaviors exhibited during completion of the tasks. The primary findings of Hyldegård's study were that, over time, the subjects read and wrote more and searched less; that the perceived usefulness of information sources varied within the same group; and that group members' formulations of the focus of the project were generally weak at the beginning but strengthened over time—a process facilitated by increasing intra-group understanding; variation among group members as to their perceived degrees of understanding highlighted the need for good communication and discussion. Hyldegård concluded that both social sources of support and intragroup problems affected individuals' cognitive experiences on the project. Data from the diaries and interviews also yielded information on levels of motivation and states of certainty and/or uncertainty.

A further study focused on understanding the impact of extrinsic motivation on students' information behaviors in relation to their use of a collaborative information-finding system (Shapira, Kantor, & Melamed, 2001). AntWorld (http://aplab.rutgers.edu/ant) is a collaborative information finding system where items predicted to be of interest to a student are recommended to them by calculating the similarity between a student's current "quest" (i.e., an information need represented in the form of a description similar to a query) and the past quests of other students that have been stored in the system. In order for the AntWorld system to perform most effectively, students are expected to evaluate the items that they find, in this case through a rating system. A weakness of collaborative information-finding systems is that they depend, for optimal effectiveness, on the contributions of all students. Although all students may benefit from the evaluations of other students, some use the system without contributing their own evaluations. The study explored extrinsic motivations and their impact on students' contributing evaluations to overcome the "free-riding" phenomenon.

The experimental study of AntWorld consisted of three trials. The first, involving MBA students, influenced the two subsequent trials in adjusting how suggestions were recorded, reducing the frequency of the online surveys of users' satisfaction, and improving the user interface. Trials 2 and 3 were conducted with groups of undergraduate students. In Trial 2, undergraduate groups were provided with no extrinsic incentive for providing evaluations. In Trial 3, undergraduate groups were provided with a reward (a pizza party) for the group that provided the highest number of judgments. The results of these trials demonstrated that contributing activity was greater in the motivated group (Trial 3) than in the non-motivated group (Trial 2). The authors are careful to point out the existence of two motivational factors during the conduct of

Trial 3: In addition to the pizza party, something that was not, in itself, of great value was the effect that the provision of an extrinsic incentive had on "intragroup competitive motivation, because only one group was expected to 'win'" (Shapira et al., 2001, p. 885). Shapira et al. discuss the implications of the experimental study; for example, they explore the combination of extrinsic and intrinsic forms of motivation in the design of collaborative information finding systems and the development of an organization's "shared knowledge base."

Collaborative Information Seeking in Industry

Examples of social and collaborative approaches to information-seeking tasks in industrial settings include two studies of the role of collaborative information seeking in research and development and a study examining the role of collaborative information seeking during patent application approval processing.

A study of two teams engaged in the design of computer-related products focused on how team members collectively sought and shared external information acquired within the team (Poltrock, Grudin, Dumais, Fidel, Bruce, & Pejtersen, 2003). Informed by a work analysis framework, the study observed the processes of collaborative information retrieval defined as "the activities that a group or team of people undertakes to identify and resolve a shared information need" (Poltrock et al., 2003, p. 237). The teams serving as subjects were a software design team at Microsoft and a hardware design team at Boeing.

Poltrock et al. (2003) conducted two rounds of interviews with the design team leaders, team members, and others working with the team. The first round of interviews focused on general questions about their work, its organizational context, the decisions they made, the information they sought, and their work with other people. Team members also received structured notepads in order to take notes about information-seeking events and activities. A second round of interviews elicited information from participants on their information-seeking activities. Additional data were collected via observations and recordings of team meetings, the monitoring of group e-mail communication, observations of members at work, and work shadowing. In their analysis, Poltrock et al. isolated five collaborative information retrieval strategies: (1) identifying needs collaboratively, (2) formulating queries collaboratively, (3) retrieving information collaboratively, (4) communicating about information needs and sharing retrieved information, and (5) coordinating information retrieval activities. They also identified a role for a common information space specifically designed to support these collaborative information retrieval strategies.

The investigation of a complex phenomenon such as collaborative information seeking may require the development of a multidimensional approach (Fidel, Pejtersen, Cleal, & Bruce, 2004) that overcomes the limitations of the current unidimensional approaches that have been applied to the analysis of human information behavior (e.g., psychological, social). A multidimensional approach aims to "account for the complexity that exists in human–information interaction" (Fidel et al., 2004, p. 941). Dimensions considered relevant to an understanding of collaborative information retrieval are the actor's dimension, the task situation dimension, and the organizational analysis dimension. These can be further divided into: the cognitive dimension, the specific task and decision, the nature of the information sources, the nature of the information needed, the organization of the team's work, and the organizational culture. Based on their analysis of a single case, Fidel et al. (2004) make a number of methodological points relevant to the adoption of a multidimensional approach. First, they introduce the notion of a "collaborative information retrieval event" (Fidel et al., 2004, p. 944); second, they suggest that collaborative information retrieval serves not only informational purposes but also social and organizational ones; and third, they recommend that systems designed to support collaborative retrieval should incorporate features that support interactions between actors and "enhance their access to one another's knowledge, ideas, and opinions or help them keep on track" (Fidel et al., 2004, p. 951).

Another study reports on collaborative information seeking during the patent application approval process (Hansen & Järvelin, 2004, 2005). Three interrelated questions motivated this study, which was conducted at the Swedish Patent and Registration Office: How do collaborative activities manifest themselves? How often do they occur? And when do collaborative activities take place in an information storage and retrieval process? Hansen and Järvelin collected data on the collaborative information seeking and retrieval activities of nine patent engineers through semi-structured and open-ended interviews, electronic diaries, and focused observations. They developed a model of the patent approval process, hierarchically decomposing it into a series of tasks and subtasks, and identified the information-seeking and information-retrieval tasks related to each step of the patent approval process. In total, 155 collaborative information-seeking and retrieval events (100 document-related and 55 human-related) were identified as occurring across 12 patent tasks. This represented an average of 8.33 document-related and 4.58 human-related events per patent task. The study concluded by noting the importance of collaborative information seeking and retrieval for the completion of a range of tasks specific to patent application approval processing (e.g., planning tasks, problem definition, search topic selection, and query formulation, search paths and query construction, and relevance assessments). The study also called for frameworks and models that move beyond single user interaction to collaborative interaction during information seeking and retrieval, and recommended further research on the evaluation of collaborative information retrieval systems.

Collaborative Information Seeking in Medicine

Investigators have also examined social and collaborative approaches to information seeking in medicine and have identified a number of factors that affect the efficacy of information seeking in intensive care units (ICUs) (Gorman, Lavelle, Delcambre, & Maier, 2002; Reddy & Dourish, 2002). Such factors include coordination across a range of personnel (e.g., physicians, nurses, and pharmacists); the satisfaction of a range of information needs relating to the familiarization, stabilization, and monitoring of a patient's condition; and the dynamic nature of information seeking and use, as personnel update themselves on a changing clinical situation.

Faced with a fresh clinical problem, a physician's first task is to become familiar with the details of the patient's case. This process requires navigating through "a complex collection of documents, identifying information that is relevant to the problem at hand, while ignoring the rest" (Gorman et al., 2002, p. 1246). During the first phase of their study, Gorman et al. collected data on the selection process through use of think-aloud sessions, in which physicians were asked to carry out simulated work tasks. The researchers noted that the physicians seemed to be "attending only to information that had a direct bearing on the problem presented in the scenario" (Gorman et al., 2002, p. 1246). They also observed that the physicians invested a considerable amount of effort in the process of information selection, deciding what information to examine and what to ignore; that they frequently resorted to informal annotation and note-taking during the process of information selection; and that they were interested in knowing how a previous physician had navigated the documents in relation to the same patient. This last observation led the investigators to focus on "the trace left by an expert as she explores the collection of documents" (Gorman et al., 2002, p. 1246) in the second phase of the project.

In phase two, Gorman et al. (2002) employed ethnographic methods to collect data on how physicians utilized information within the intensive care unit. Field observations confirmed the assumption that the same information selections were being used and re-used by different personnel: "We observed many examples of experts selecting, organizing, and sharing subsets of relevant information, usually drawn from diverse, physically separate information sources" (Gorman et al., 2002, p. 1248). These information selections, designated "information bundles," were of different types: a kardex, a resident's worksheet, an ICU flow sheet, as well as bundles of bundles. Different types of information bundles shared the following properties: Each contained highly selective information, was systematically organized, and was used on a collaborative basis. Collaborative use and reuse of such information bundles led researchers to propose a new user model, based not on the individual user but on the collaborative user:

Collaboration: the "user model" for this work is most often not a single person but a group. Members of the group had roles, including information management roles that although professionally and formally defined, are also dynamic and socially negotiated. Bundles were used in the ICU in a manner that fostered multiauthor, multiuser collaboration. (Gorman et al., 2002, p. 1248)

Gorman et al. (2002) developed SLIMpad, which uses technology to exploit access to traces left by other physicians in information selection and use of information bundles. SLIMpad functions included: (1) for each document, a précis of information about its origin, its content, and the history of its use by experts; (2) for each expert problem solver, a trace that describes the path taken through the collection; (3) and navigation tools that could assist subsequent problem solvers using the collection by exploiting the knowledge inherent in existing traces.

A further study enhances our understanding of collaborative information seeking in medical settings (Reddy & Dourish, 2002). Conducted within the Surgical Intensive Care Unit (SICU) of a large metropolitan teaching hospital, it investigated how the work of the SICU, revolving around patient stabilization and bed management, motivated collaboration in information seeking not only among physicians but also between physicians, nurses, and pharmacists. Reddy and Dourish noted that this collaboration in information seeking, along with the constraints created by staff's work rhythms, created the need for the "continual assembly and collective interpretation of a heterogeneous collection of information" (Reddy & Dourish, 2002, p. 347). They observed the work of the SICU over a period of seven months, collecting data by means of formal and informal interviews, observations, and internal communications. Their main finding was that staff's work rhythms acted both as a resource for and a constraint on collaborative information seeking. This led Reddy and Dourish to propose a new concept of information retrieval, one that moves away from the database query model to one that embeds information retrieval within work activities. Roles for technology in supporting such an approach include the utilization of awareness technologies that integrate information spaces with representations of activity, the development of information displays that incorporate temporal and cyclical information, visualizations, and social navigation approaches that create information spaces enhanced by the inclusion of temporal information on the scheduled work of the unit. In so doing, information retrieval systems act not as locations where information is stored and retrieved but as "places where work gets done" (Reddy & Dourish, 2002, p. 352).

Collaborative Information Seeking in Military Settings

Social and collaborative approaches to information seeking have also been studied in military settings, more specifically in the conduct and organization of simulated command and control exercises (Prekop, 2002; Sonnenwald & Pierce, 2000).

Successful performance of a command and control exercise requires a shared understanding of the mission, the battle plan, and of course the battle itself. This, in turn, requires the communication and sharing of information from multiple domains across groups of people. Sonnenwald and Pierce (2000, p. 468) posited that each person involved in the exercise can be deemed an "information handler." They garnered data on collaborative information behavior through analysis of command and control exercise documentation, observation of a simulated command and control exercise, and semi-structured and critical incident interviews. Data analysis involved open and axial coding and the findings were reported at both organizational and informational levels. At the organizational level, Sonnenwald and Pierce found that the commander played an important role in identifying critical information needs, as did a plans and operations group "responsible for creating situation awareness of the battlefield among the battalion staff and the companies that report to the battalion" (Sonnenwald & Pierce, 2000, p. 469). At the informational level, they identified three characteristics of information sharing and their ultimate impact on team performance. Two characteristics supported achievement of team goals: "interwoven situation awareness" involved shared understandings of the situation and "dense social networks" supported "frequent, bi-directional information flow among team members"; "contested collaboration," however, where "team members challenge[d] the contributions of others" hindered achievement of goals (Sonnenwald & Pierce, 2000, p. 463).

In another study, Prekop (2002) examined the collaborative information-seeking behaviors of a working group created to perform a command and control exercise (Prekop, 2002). He collected data by analyzing working group meeting minutes and conducting semi-structured interviews. For data analysis, Prekop adopted a grounded theory approach and, on the basis of axial and selective coding, identified three components of the working group's collaborative information behavior: information-seeking roles (gatherer, instigator, referrer, verifier, indexer/abstracter, group administrator, and group manager), information-seeking patterns (by recommendation, direct questioning, or advertising information paths), and the contexts (e.g., working group or organization) within which these roles and patterns were performed. He concluded by noting that "collaborative information seeking has emerged as an important field of study. ... However, to be able to support collaborative information seeking with technology, a richer understanding of collaborative information seeking, based on studies of real world collaborative information seeking activities, needs to be developed" (Prekop, 2002, p. 545).

Summary

Studies of collaborative information seeking have been conducted within a variety of settings, such as the academy, industry, medicine, and the military. They have focused on a range of aspects relevant to an understanding of the phenomenon, including awareness of the different types of information that are required by teams to support collaborative work (Sonnenwald et al., 2004), the different types of collaborative information behavior displayed (Talja, 2002), and the collaborative information retrieval strategies used (Poltrock et al., 2003), the distributed nature of collaborative information seeking as information sought for one subtask needing to be coordinated over space and time with information sought for another subtask as part of a larger common work task (Gorman et al., 2002; Gorman, Ash, Lavelle, Lyman, Delcambre, Maier, et al., 2001), problems of and support for collaboration during information seeking (Hyldegård, in press; Shapira et al., 2001), the design of appropriate technology (Sonnenwald et al., 2004), and the nature of the community or domain itself (Prekop, 2002). The majority of studies conducted to date has followed an inductive approach, with studies adopting a deductive approach drawing on theories developed outside the field of collaborative information seeking. A range of research approaches has been used including ethnographic field studies, case studies, and controlled experiments; both qualitative and quantitative analyses have been conducted. In some cases, research approaches have been combined as part of a multimethodological approach to a complex phenomenon (Fidel et al., 2004; Sonnenwald, et al., 2004).

Problems of conducting research into collaborative information seeking include methodological issues specific to the complex nature of the phenomenon and the relative novelty of the research. The finding that collaborative information behavior is often embedded in the conduct of broader work tasks suggests that contextual methods are required to complement existing experimental methods. Current contextual methods tend to model collaborative information seeking in relation to repetitive works tasks that occur in the domain under study (Hansen & Järvelin, 2005). The observability of such seeking processes and the range of levels relevant to the information-intensive nature of such work (e.g., task, domain, organization, group, technology) also continue to constitute challenges. In terms of research gaps, much work has been done on designing features for technology geared toward collaborative information seeking; however, research evaluating the impact of such designs on collaborative information seeking, as a whole or on specific processes within it (such as, e.g., relevance judgments), remains rare.

Information Task: Searching

This section addresses the topic of collaborative information search and retrieval. Two subtopics are selected for review: collaborative querying and collaborative filtering.

Collaborative Querying

Collaborative querying is a technology that enables users of an information retrieval system to draw on the past query preferences of other users at the query formulation and reformulation stages of an information search. The benefits of collaborative querying include the generation of recommendations based on other users' selection of search terms and hence a reduction in the costs associated with conducting an information search.

Research into collaborative querying has addressed a number of issues. These include the automation of collaborative querying and the nature of the algorithms used to compute similarity between queries and identify query clusters (Fu, Goh, & Foo, 2004; Wen, Nie, & Zhang, 2002); interfaces for collaborative querying, for example, the use of collaborative querying as a means of query expansion for information retrieval (Hust, 2004); the implementation of discussion facilities to enable human–human interaction during query refinement (Blackwell, Stringer, Toye, & Rode, 2004); the design and evaluation of prototypical collaborative querying environments that provide a seamless environment for collaborative querying (Chau, Zeng, Chen, Huang, & Hendriawan, 2003; Walkerdine & Rodden, 2001).

Collaborative querying technology works by mining a search engine's query logs, which contain the queries used by prior users of the system. Clusters of queries are identified by means of a similarity measure and are recommended to future users. From a computational perspective, the crucial factor in collaborative querying is the nature of the similarity measure used to identify query clusters, and thereby to transform the raw logs of queries into usable recommendations. Recently, Fu et al. (2004) examined three different approaches to identifying clusters of similar queries: content-based, feedback-based, and results-based. Content-based approaches determine similarity on the basis of a similarity between the terms used in different queries; feedback-based approaches determine similarity on the basis of an overlap between different users' selections of documents from the set generated by querying; results-based approaches determine similarity on the basis of an overlap between the documents that result from the queries. Fu et al. evaluated each of these approaches, together with its associated similarity measure, before presenting their own similarity measure for identifying clusters of similar queries. They designed a set of query-clustering experiments and discussed some of the tradeoffs of adopting different similarity measures. Their study concluded with the suggestion that future experiments should evaluate a number of different measures in combination.

Arguing that a plateau has been reached in the accuracy that can be achieved from ad hoc document retrieval systems, Hust (2004) has suggested that greater use needs to be made of contextual information in order to improve retrieval effectiveness. He defines the collaborative information retrieval approach as one based on "learning to improve retrieval effectiveness from the interaction of different users with the retrieval engine" (Hurst, 2004, p. 254). The archived contextual information on which collaborative information retrieval draws in attempting to improve retrieval effectiveness includes information from previous search processes, specifically, individual queries and complete search processes, and the relevance information gathered during previous searches (Hurst, 2004, p. 254). This last may incorporate human relevance judgments in the form of ratings. Thus, a collaborative information retrieval system is intended to act as an automated "advisor" to users posing new queries, fulfilling this function by finding previous queries similar to the current query together with the documents judged relevant to those previous queries. The current query can then be expanded by adding terms gathered from relevant documents yielded by previous queries (Hurst, 2004, p. 255). Hust describes the method used to set new collaborative querying algorithms and discusses the different types of algorithm used to compute similarity both between queries and between queries and documents, as well as identifying query expansion methods for collaborative information.

Blackwell et al. (2004) have presented another example of a collaborative querying environment, one that also places emphasis on human–human interaction. Their approach, termed Query by Argument, provides not only collaborative querying facilities, but also discussion facilities during query refinement and expansion. Taking the "cooperative management of queries" as their starting point, Walkderine and Rodden (2001, p. 140) have described the design and evaluation of a prototype environment that supports community use of query recommendations, query versions, annotation of queries, query exchange, and user views of query categories, and features the incorporation of awareness mechanisms for updating users on the current state of the system (e.g., the addition of a new query or a new category) and providing a usage history for each query. They describe an evaluation of a prototypical cooperative query management system and present some initial findings.

Collaborative querying systems predict relevant queries on the basis of some form of similarity search. User-driven collaborative judgments as to the relevance of particular queries can be accommodated through the use of discussion facilities. The utilization of collective user judgments in the relevance process has also been studied by Zhang (2002, p. 221), who proposes that the determination of relevance judgments through a "consensus of peers or group members" can enhance the personal, "subjective," relevance judgments of the individual. He describes the steps of a group consensus method and presents an experimental study that evaluates this approach.

Collaborative querying may also form one component in an integrated collaborative retrieval environment. Romano, Roussinov, Nunamaker, and Chen (1999) defined collaborative information retrieval as *multiuser information retrieval (IR)*, a field of investigation that they characterize as lying at the intersection of research and development into information retrieval and group support systems (GSS). Focused on the facilitation of collaboration during directed information search, this approach gives rise to system support for a collaborative information retrieval environment that not only provides shared access to information resources but also provides facilities for shared searching, query formulation, and browsing. Basing themselves on a background in group decision support and its evaluation, Romano et al. maintained that the adoption of collaborative mechanisms during information search and retrieval can enhance both process and productivity. They suggested that the potential benefits to be derived from a collaborative information retrieval environment (CIRE) include (1) automatic creation of an information retrieval memory; (2) the ability to allow multiple users to share both queries and search results; (3) "elimination of *'same-time same-place same-technology'* constraints, thus allowing team members to search together even if they are distributed physically, temporally and technologically" (Romano et al., 1999, p. 5); and (4) significant reduction of "[r]edundancy in query and search results ... when users share a *'structured social awareness'* of the collaborative search process" (Romano et al., 1999, p. 5). They built a prototype collaborative information retrieval environment that included (1) an integrated page/comments/evaluation screen and (2) facilities for electronic polling as a support mechanism for voting (and thus achieving consensus) on the relevance of search results. Diamadis and Polyzos (2004) have further investigated the design of systems for facilitating structured social awareness in collaborative information searching.

Collaborative Filtering

Recommender systems (Furner, 2002; Konstan, 2004; Resnick & Varian, 1997) are technology-based systems designed to relieve information seekers of the task of evaluating the vast amount of information available. Once considered equivalent to recommender systems in general, collaborative filtering (CF) is now usually treated as one among several mechanisms (e.g., content-based filtering, social data mining) by means of which information items can be recommended to users. CF is a specific technology (Konstan & Riedl, 2003) that recommends information items to users by computing the similarity between the past preferences of one user and another. CF systems can be categorized as active (further divisible into push-active and pull-active) and automated

(Konstan & Riedl, 2003). In push-active CF systems, responsibility for filtering rests with a professional human intermediary who actively intervenes in mediating between users' interests and the information resources; in pull-active CF systems, it rests with the users who actively filter the information for themselves; and in automated CF systems, it rests with the system. This section examines, in turn, active and automated CF and concludes with a summary and identification of current problems and issues in CF.

Active Collaborative Filtering

Konstan and Riedl (2003, p. 45) describe pull-active systems, "in which the users 'pull' the information they want" by formulating queries based on characteristics of the items, such as keywords, or the reaction of previous users to those items. On the other hand, in push-active systems "users 'push' items to other users based on an understanding of the other users' information needs" (Konstan & Riedl, 2003, p. 45). Goldberg, Nichols, Oki, and Terry's (1992) Tapestry is an example of a push-active system; Maltz and Ehrlich (1995) present an example of a pull-active system.

Ehrlich and Cash (1999) critically examined the claim that the services of information intermediaries are not required in an age of automation. They reported on three case studies as examples of domain- and search-related expertise in mediated settings. Two of the case studies explored the work of customer support personnel in a technical setting; the third examined the work of corporate libraries. On the basis of their observations and interviews, they assessed the benefits of such work and its invisibility to customer and manager alike. They concluded their study by identifying new roles for intermediaries in a range of settings, considering both their value-adding potential and their possible drawbacks.

Automated Collaborative Filtering

The design of an automated recommender system can be divided into three components: input, transformation, and output (Resnick & Varian, 1997). The input to a CF system typically consists of users' past preferences or evaluations in relation to a specific domain of interest (e.g., citations, music, movies, news). A transformation process operates on the input and the system then recommends other items predicted to be of interest to a user on the basis of the calculation of similarities between his or her own preferences and those stored in the system. How the transformation process occurs is the key component in differentiating CF systems. The output from a CF system takes the form of a set of recommendations (e.g., a ranked list). Enhancements to the output may include annotations explaining the recommendations, summaries, and visualizations. Two examples of collaborative filtering approaches are provided here; one is drawn from academe, the other from industry.

Collaborative Filtering of Citation Information

One way to filter information within the scientific domain is to provide secondary tools such as citation indexes that provide bibliographical access to the primary literature. However, a problem intrinsic to CF systems is that input to such systems requires users to explicitly provide evaluations (e.g., ratings) about items. This can lead both to initial startup problems when the system lacks a critical mass of evaluations and, once a critical mass of evaluations has been reached, to free-riding, whereby users are able to benefit from the evaluations of others without themselves contributing. One way of dealing with this is to develop hybrid systems that combine collaborative and content-based filtering. McNee, Albert, Cosley, Gopalkrishnan, Lam, Rashid, et al. (2002) reported on an attempt to augment automatic citation indexing based on author linkages with semantic information about the papers themselves. In order to circumvent both the startup and free-riding problems, they mapped their CF framework onto a citation web, namely, an existing network of research papers. Offline and online experiments were designed to evaluate the accuracy of four CF algorithms for selecting citations and providing recommendations: co-citation matching, user–item collaborative filtering, item–item collaborative filtering, and naïve Bayesian classifier. Two further algorithms acted as baselines for the information searches, which comprised a citation graph search and a keyword search. McNee et al. demonstrated that the type of algorithm chosen affected the type of recommendations generated, with either very relevant recommendations or very novel recommendations being generated, but not simultaneously. And although there was some variation in how the algorithms behaved, they were able to conclude that "recommenders based on the social web of citations can be valuable aids to researchers" (McNee et al., 2002, pp. 124–125)

A further study relevant to the collaborative filtering of citations is by Torres, McNee, Abel, Konstan, and Riedl (2004). Research aimed at extending our understanding and modeling of user preferences, which constitute the input to such CF systems, is reported by Middleton, Shadbolt, and DeRoure (2004).

Collaborative Filtering of Music Information

Current growth in MP3 technology for the manufacture of music and the use of the Internet as a distribution platform for direct delivery of digital information content to consumers has made both music and evaluations of music content readily available. Traditionally an area in which content-based filtering was difficult to apply, collaborative filtering has emerged as a technique that can aid in the recommendation of music. One approach is to use others' evaluations in selecting music; searching for music can also be collaborative (Cunningham, Reeves, & Britland, 2003). An example of pioneering work in music recommenders is Ringo (Shardanand, 1994; Shardanand & Maes, 1995). Current commercial examples include Amazon, CDNow, SongExplorer, and MediaUnbound (Swearingen & Sinha, n.d.).

The input to a music recommender system will consist of users' preferences in relation to music items. Typically, explicit preferences will be expressed using a rating system option with options for a full-text description. In common with other automated collaborative filtering approaches to recommendation, particularly during their development, music recommenders encounter problems vis-à-vis startup and the achievement of a critical mass of evaluations. Adamczyk (2004) and Crossen, Budzik, and Hammond (2002) provide examples of solving the startup problem in the collaborative filtering of music information by drawing on existing information contained in online music directories. An additional weakness in current automated collaborative filtering systems can be insensitivity to context. This context may be task-based or, in more localized and contingent terms, "representation of the user's interests at a particular moment" (Hayes & Cunningham, 2004, p. 132). Rather than relying on individual music items as input to the recommender, Hayes and Cunningham (2004, p. 132) draw information from a user's playlist, "a compilation of music tracks built by one listener and recommended to other like-minded listeners;" arbitration among group preferences is explored by McCarthy and Anagnost (1998, 2000). Algorithmic methods for transforming these input data into music recommendations include combining content-based filtering with user profiling (Grimaldi & Cunningham, 2004; Hoashi, Matsumoto, & Inoue, 2003), combining content-based filtering with collaborative and statistical methods (Chen & Chen, 2001), and clustering (Li, Kim, Guan, & Oh, 2004). Output for a music recommender may typically be presented as a ranked list of items. In common with other recommender systems (irrespective of the methods used to calculate similarities between items and between items and users), output is typically presented in a ranked list, albeit one amenable to enhancements such as explanations of the reasoning behind evaluations (Herlocker, Konstan, & Riedl, 2000).

Summary

Two areas of research within the field of collaborative information searching have been reviewed: collaborative querying and collaborative filtering. Research in collaborative querying has investigated the design of optimal algorithms for computing similarity between queries, human–computer and human–human interaction for collaborative querying, and the design of integrated collaborative querying and search environments. User-based evaluations of the recommendations resulting from different collaborative querying algorithms or of the impact of collaborative querying environments on the conduct of information tasks are rare and further research is needed here to base the utility of such tools on a user perspective.

Research into the development of CF systems has focused almost exclusively on the specification and evaluation of optimal algorithms. Current problems and issues associated with CF systems can be divided into those associated with the input into a CF system, the transformation process, and output mechanisms.

Reliance on user evaluations is a weakness of such systems that can lead to a cold start and lack of a critical mass, free-riding, and unreasonable bias. Attempts to circumvent these problems have included jump-starting CF systems with input from existing tools that contain evaluative information in one form or another (e.g., citation indexes, music directories), some form of extrinsic motivation; and supplementing recommendations with explanations of the reasoning behind them.

Collaborative filtering has been developed to supplement content-based filtering methods, particularly in domains such as music, images, and movies, where implementing content-based filtering is more costly. Thus, it is common to find CF methods being used in combination with other methods (e.g., content- and statistics-based) as part of a hybrid approach to filtering. A range of enhancements for the output from CF systems is currently being developed, including the use of implicit interest measures, situational recommendations, recommending for groups, manual override, confidence and explanations, and the inclusion of non-human agents in the formation of recommendations (Konstan & Riedl, 2003). In sum, current issues and problems with CF systems can be identified in terms of the domain space, the set of recommendations (input, transformation, output), and in relation to social issues (free-riding, incentives, privacy).

Research has been conducted into the design and evaluation of collaborative tools for use at each stage of information search and retrieval. For the most part, investigators have adopted experimental methods involving statistical analysis of algorithmic performance, with some qualitative analysis emerging either from usability evaluations or user evaluations of the recommendations generated. However, in spite of the domain specificity of the tools developed to date (e.g., citations, music), these have largely been developed in experimental settings rather than their situation of use.

Information Task: Navigation

As users navigate an information space, they leave behind traces of their activity. These traces provide evidence of the use of digital objects, e.g., page visits, links, or paths taken through the data. Individuals' awareness of others' information-seeking activities serves as a source of social information that can provide guidance for their own information-seeking activities, whether in the form of recommendations or establishment of contact with other people who are navigating the same or similar documents or information spaces. Designers' use of such information aims to bring the social aspects into information provision (Höök, Benyon, & Munro, 2003; Munro, Höök, & Benyon, 1999) and to lead users to an understanding of information spaces not as static spaces, but as active, populated, information "places." Indeed, social navigation can be considered a form of interaction rather than a technology (Dourish, 2003), with social navigation systems aiming at not only more efficient information seeking but also enhanced social experience during information seeking (Dieberger, Dourish, Höök, Resnick, & Wexelblat, 2000). Tools to support social navigation can be broadly categorized into those that provide support for asynchronous interaction and those that facilitate synchronous interaction. A final section summarizes the use of social navigation techniques for the support of information tasks.

Asynchronous Social Navigation

This section describes examples of two asynchronous approaches to the social navigation of information: social data mining and history-enriched data objects.

Social Data Mining

Web technologies now afford the possibility of recording the paths or trails that users take as they traverse an information space (Reich, Carr, DeRoure, & Hall, 1999). Thus, unlike collaborative filtering systems that rely, for the most part, on explicit user activity such as adding evaluations to the information items found, social data mining systems do not require users to engage in any new activity; rather, they seek to exploit the user preference information implicit in records of existing activity (Terveen & Hill, 2002). A different approach to gathering social information about people's activities (and, thus, social navigation) is to examine traces not from the point of view of interaction with digital objects, but on the basis of the *paths* that users take while navigating the information space within which the objects are embedded and to make this path information available to other users. Thus, unlike recommendations provided by CL systems, the input to which is users' explicit preferences (e.g., ratings), recommendations provided by social data mining systems are formed on the basis of users' implicit preferences (e.g., trail data) by computing similarity between pathways (Amento, Terveen, Hill, Hix, & Schulman, 2003, p. 3).

The TOPICshop system (Amento, Terveen, & Hill, 2003; Amento, Terveen, Hill, Hix, et al., 2003) is a recommender system, the design goal of which is to make it simpler for users to explore and organize information resources found on the Web. TOPICshop's architecture is designed to carry out two tasks of relevance to this information management problem: information extraction and display of the information in such a way that users can easily manipulate and organize it. The TopicShop Webcrawler collects information about the most visited sites by analyzing the links followed for a particular topic. The Webcrawler selects a set of pages on a specific topic, extracts the links from these pages, and then selects a subset of the linked-to pages to fetch. Once the pages have been collected, the TopicShop Explorer interface provides support to users in evaluating, managing, and sharing their information resources. The Explorer consists of: (1) a site profiles view that makes available to the user Web-based information normally left implicit and not used for site evaluation purposes (e.g., number of in-links, out-links, number of pages, media type) and (2) a work area that enables personalized organization of information, featuring a thumbnail image of the currently featured site, and providing a standard window for navigating hierarchically-organized folder space.

Amento, Terveen, Hill, Hix, et al. (2003) reported on an evaluation of the performance of two aspects of the TopicShop system that deal with the question of enhanced interaction. The first pertains to human-computer interaction: Is user task performance enhanced by the TopicShop interface? The second relates to the value of retrieving normally hidden information: Is the information that TopicShop extracts about Web sites valuable? Amento and colleagues evaluated the performance of the system by means of an evaluation task and an organization task. Drawing on an analysis of the results of the information evaluation task, the authors reported that TopicShop subjects selected significantly higher quality sites than did Yahoo! subjects and suggested that the implicit site profile data made available to users via the TopicShop interface may provide the explanation for this finding: "In summary, TopicShop subjects selected higher quality sites, in less time and with less effort. We believe these benefits are due to TopicShop's site profile data" (Amento et al., 2003, p. 67). As regards the organization task, subjects were asked to arrange their selected sites into groups and to name the groups. The researchers report that TopicShop subjects appeared to do a better job of organizing the items they selected: They created more groups, annotated more sites, and agreed more often about how to group items than their Yahoo! counterparts; moreover, they achieved these results in half the time the Yahoo! subjects devoted to the task (Amento, Terveen, Hill, Hix, et al., 2003). The investigators suggested that the reason for these results lies in the ease with which subjects were able to group and annotate the information in the TopicShop work area and the fact that the rich information contained in site profiles remains visible while users organize sites. They also reported that several comments by subjects showed that they appreciated the ability to integrate tasks and the fact that their task state remained visible.

History-Enriched Digital Objects

The physical world is full of material objects marked by the history of their use. A book, for example, displays evidence of wear that indicates past use (e.g., well-thumbed pages, underlinings). In this instance, a subsequent reader may derive useful information from underlinings to guide his or her own decisions as to how to navigate the text.

Just as material objects in the physical world carry evidence of their use, so also digital objects can be designed in such a way that they carry with them a history of their use, thus enabling later users to draw on this history in making their own choices within the information space. History-rich digital objects bear the imprint of an "interaction history" (Wexelblat, 2003, p. 223), or record of the interactions that past users have had with them.

Wexelblat has recently described the results of the Footprints project (Wexelblat, 2003; Wexelblat & Maes, 1999), which applies the idea of an interaction history to objects in a digital world. Such interaction histories can variously represent "sequences of actions, relationships of elements on which people have acted and the resulting structures," a "temporal collage," and "recurrent opposed states" (Wexelblat, 2003, p. 224). Wexelblat discussed a theoretical framework informing the design of a range of features that transform trail data into user-based navigation tools. Such tools include a map, a path view, annotations, and signposts or comments. On the basis of user studies, Wexelblat (2003) reports that the tools were successful not only in facilitating more efficient information seeking but in creating a higher level of satisfaction among users.

Synchronous Social Navigation

This section discusses examples of two synchronous approaches to the social navigation of information: co-browsing and the use of chat tools and other live awareness information during information seeking and document use tasks.

The physical world provides many opportunities for impromptu, opportunistic communication. Budzik, Fu, and Hammond (2000) have attempted to translate such opportunities to electronic spaces through the use of an awareness application that enables users accessing a common repository of documents or a Web site to communicate opportunistically. They describe a prototype system, I2I, that manages the early stages of initiating informal collaboration by providing its users with opportunities to become aware of the activities of others who share common interests, as represented by the documents with which they interact (Budzik et al., 2000).

The primary objects in the system, and around which people communicate, are conceptually similar documents represented by and located in conceptual space. Secondary objects associated with these documents can also be represented, including people, chat rooms, and calling cards. The I2I interface yields information about system activity and a range of awareness information that provides opportunities for information seekers to establish contact with others sharing similar interests via tabs such as "Who Is Online," "Related Documents," "Active Chat Information," and "Calling Cards" (Budzik, Bradshaw, Fu, & Hammond, 2002a, pp. 729–730). Its developers (Budzik et al. 2002a, 2002b) have focused on making users aware of the activities of other information users by tracking use and computing a similarity between different users' work contexts represented by the content of the document that is being manipulated. Users are then clustered together in a conceptual neighborhood. Budzik et al. present evaluations of the algorithms used to compute document similarity. They also discuss the nature of opportunistic collaborations, proposing that "good" collaboration "brings together people whose knowledge, skills, perspectives, and interests compliment [sic] each other in ways that are mutually beneficial" (Budzik, et al., 2002a, p. 734). To this end, giving the system better models of groups and individuals could allow it to automatically build this kind of complementary collection of people.

A further example of a chat application designed to make information seekers aware of other information seekers is the "chat in context" approach of Livemaps (Cohen, Jacovi, Maarek, & Soroka, 2002, p. 9); an approach that combines elements of document awareness and people awareness to promote "collection awareness ... the knowledge of people who are at the same instant visiting the same collection of virtual places, with a single place granularity" (Cohen et al., 2002, p. 12). In other words, a set of related pages within a site or across sites forms the basis for collaborative information seeking rather than the site itself. Maglio, Barrett, and Farrell (2003) develop a different approach to the same design problem of turning the Web from a space into a place through the notion of Web intermediaries. Such Web intermediaries can perform a range of functions such as personalizing, annotating, transcoding, filtering, and aggregating (Maglio et al., 2003).

Summary

Social navigation tools are designed to enable users to be aware of, and be guided by, the activities of others during information seeking. This guidance may be indirect when drawing on information from interaction histories and trail-based information, or more direct when using collaborative browsing and chat tools. Both synchronous and asynchronous applications technology provide individuals with the opportunity to be aware of people and information that have hitherto remained hidden during information seeking. The benefits of social navigation include not only more efficient information seeking and handling but also the promise of a more socially translucent experience as graphical and communication tools help transform information spaces into information places.

The design of social navigation tools raises a number of issues and problems. The first of these has to do with the nature of the information on which such navigation systems are based. Although these tools help guide users in making their own information-seeking decisions, there is no guarantee that by doing so users will find the optimal as opposed to the well-trodden path—indeed they may be taken down the wrong path.

Social navigation tools develop novel representations of both aggregations of individual activity and of social activity (e.g., Chen, Cribbin, Kuljis, & Macredie, 2002; Erickson & Kellogg, 2000, 2003). Further comparative evaluations of the utility of such tools for specific information-seeking tasks are needed. More generally, there is the need for additional research into the measurable benefits of such systems for information seeking.

Conclusion

This chapter has reviewed examples of current research to develop social and collaborative approaches to a range of information tasks. The research is multidisciplinary, being conducted within information science, information retrieval, human–computer interaction, and computer-supported collaborative work. Each of the studies reviewed shares an interest in developing social and collaborative approaches that enhance the view that information seeking and retrieval are individually motivated activities. The range of studies reviewed also teaches us that the development of systems and practices to support social and collaborative information seeking, search, and retrieval requires attention to a range of contexts, including information tasks, domains (e.g., academe, industry), organizations (e.g., universities, manufacturers), work groups, human–human and human–computer interaction (e.g., information visualization), access to heterogeneous sources of data and knowledge, and algorithmic performance. The main findings of the review and their implications for future research are summarized here.

All information tasks (e.g., seeking, querying, filtering, and navigating) and their subtasks (e.g., evaluation) can be performed in collaboration with others. Collaboration on an information task can be direct (e.g., when members of a team share a common work task) or indirect (e.g., when individuals draw on the explicit or implicit informational activity of others while performing their own information tasks). Collaboration on an information task may be invisible. For example, it may be subsumed within a more clearly identifiable work task (e.g., Hansen & Järvelin, 2005; Poltrock et al., 2003) or be performed by a human intermediary (e.g., Ehrlich & Cash, 1999).

Methods for observing collaborative information seeking and retrieval are dominated by the use of experimental methods to evaluate the performance of collaborative information tools in a laboratory setting, although the use of contextual methods (e.g., case study, ethnography) that accommodate the analysis of qualitative data is beginning to emerge (e.g., Gorman et al., 2002; Hansen & Järvelin, 2005; Hyldegård, in press; Talja, 2002; Talja & Hansen, in press). Mixed-method approaches combining experimental and contextual methods are also being adopted (e.g., Fidel et al., 2004; Sonnenwald et al., 2004).

Research into collaborative information seeking has led to a call for collaborative user models (Gorman et al., 2001; Hansen & Järvelin,

2005). The need for systems to enable multiple collaborating actors (e.g., Gorman et al., 2001) to access heterogeneous sources of information (e.g., Karasti et al., 2003; Sonnenwald & Pierce, 2000) has given rise to collaboration at all stages of the information-seeking, search, and retrieval processes; for example, the use of annotated documents in the specification of an information need (e.g., Gorman et al., 2001), collaborative querying practices (e.g., Hust, 2004; Walkerdine & Rodden, 2001), and the making of collaborative relevance judgments (e.g., Zhang, 2002).

Although systems to support both direct and indirect collaboration on information tasks are in use, it is in the design, implementation, and evaluation of systems to support indirect collaboration that most progress has been made. These systems include automated CF systems that conduct a similarity search of users' explicit preferences (e.g., an evaluation) and social navigation systems that often exploit users' implicit preferences; for example, trails (e.g., Amento, Terveen, Hill, Hix, et al., 2003; Gorman et al., 2002) or interaction histories (e.g., Wexelblat, 2003). Research is currently extending the benefits and usefulness of these systems by capturing information on aspects of the user context. This contextual information may take the form of explanatory information to supplement the provision of recommendations (Herlocker, Konstan, & Riedl, 2004); information on users' short-term, dynamic preferences (e.g., Hayes & Cunningham, 2004); or information on a user's work context (e.g., Budzik et al., 2002b).

There currently exist many models and designs derived from more qualitative research that has been conducted to support direct collaboration (e.g., Fidel, Bruce, Pejtersen, Dumais, Grudin, & Poltrock, 2000; Hansen & Järvelin, 2005; Reddy & Dourish, 2002; Sonnenwald, et al., 2004; Talja, 2002: Talja & Hansen, in press), but such work is comparatively recent and studies of implementations and evaluations of such systems are rare. Within the area of support for direct collaboration during information tasks, the best-developed applications are those that provide awareness information both for purposes of collaborative work (e.g., Reddy & Dourish, 2002; Sonnenwald et al., 2004) and for more opportunistic collaboration (e.g., Budzik et al., 2000; Cohen et al., 2002).

Research in the field needs to continue addressing and evaluating the conditions that influence the deployment and sustainability of systems to support direct and indirect collaboration during information tasks, such as startup, critical mass, and use of incentives. This entails a multidisciplinary approach to the study of the intermediary mechanisms, both human and automated, that bring individuals into contact with others as they conduct their information tasks.

References

Adamczyk, P. D. (2004). Seeing sounds: Exploring musical social networks. *Proceedings of the Twelfth ACM International Conference on Multimedia*, 512–515.

Albrechtsen, H. (Ed.). (2004). Web-based collaboratories: From centres without walls to collaboratories in use. *Journal of Digital Information Management, 2*, 1–39.

Amento, B., Terveen, L., & Hill, W. (2003). From PHOAKS to TopicShop: Experiments in social data mining. In C. Lueg & D. Fisher (Eds.), *From Usenet to CoWebs* (pp. 167–205). London: Springer.

Amento, B., Terveen, L., Hill, W., Hix, D., & Schulman, R. (2003). Experiments in social data mining: The TopicShop system. *ACM Transactions on Computer–Human Interaction, 10*, 54–85.

Blackwell, A. F., Stringer, M., Toye, E. F., & Rode, J. A. (2004). Tangible interface for collaborative information retrieval. *Extended Abstracts of the 2004 Conference on Human Factors and Computing Systems*, 1473–1476.

Budzik, J., Bradshaw, S., Fu, X., & Hammond, K. (2002a). Clustering for opportunistic communication. *Proceedings of the Eleventh International World Wide Web Conference*, 726–735.

Budzik, J., Bradshaw, S., Fu, X., & Hammond, K. (2002b). Supporting online resource discovery in the context of ongoing tasks with proactive assistants. *International Journal of Human–Computer Studies, 56*, 47–74.

Budzik, J., Fu, X., & Hammond, K. J. (2000). Facilitating opportunistic communication by tracking the documents people use. *CSCW 2000 Workshop on Awareness and the WWW*. Retrieved January 30, 2005, from http://infolab.northwestern.edu/infolab/downloads/papers/paper10104.pdf

Chau, M., Zeng, D., Chen, H., Huang, M., & Hendriawan, D. (2003). Design and evaluation of a multi-agent collaborative Web mining system. *Decision Support Systems, 35*, 167–183.

Chen, C., Cribbin, T., Kuljis, J., & Macredie, R. (2002). Footprints of information foragers: Behaviour semantics of visual exploration. *International Journal of Human–Computer Studies, 57*, 139–163.

Chen, H.-C., & Chen, A. L.-P. (2001). A music recommendation system based on music data grouping and user interests. *Proceedings of the Tenth International Conference on Information and Knowledge Management*, 231–238.

Churchill, E. F., Sullivan, J. W., & Snowdon, D. (1998). Collaborative and co-operative information seeking: CSCW '98 workshop report. *ACM SIGGROUP Bulletin, 20*, 56–59.

Cohen, D., Jacovi, M., Maarek, Y. S., & Soroka, V. (2002). Livemaps for collection awareness. *International Journal of Human–Computer Studies, 56*, 7–23.

Crossen, A., Budzik, J., & Hammond, K. J. (2002). Flytrap: Intelligent group music recommendation. *Proceedings of the Seventh International Conference on Intelligent User Interfaces*, 184–185.

Cunningham, S., Reeves, N., & Britland, M. (2003). An ethnographic study of music information seeking: Implications for the design of a music digital library. *Proceedings of the Third ACM/IEEE-CS Joint Conference on Digital Libraries*, 5–16.

Diamadis, E. T., & Polyzos, G. C. (2004). Efficient cooperative searching on the Web: System design and evaluation. *International Journal of Human–Computer Studies, 61*, 699–724.

Dieberger, A., Dourish, P., Höök, K., Resnick, P., & Wexelblat, A. (2000). Social navigation: Techniques for building more usable systems. *Interactions, 7*, 36–45.

Dourish, P. (2003). Where the footprints lead: Tracking down other roles for social navigation. In K. Höök, D. Benyon, & A. J. Munro (Eds.), *Designing information spaces: The social navigation approach* (pp. 273–291). London: Springer.

Ehrlich, K., & Cash, D. (1999). The invisible world of intermediaries: A cautionary tale. *Computer Supported Cooperative Work, 8,* 147–167.

Erickson, T., & Kellogg, W. (2000). Social translucence: An approach to designing systems that support social processes *ACM Transactions on Computer–Human Interaction, 7,* 59–83.

Erickson, T., & Kellogg, W. (2003). Social translucence: Using minimalist visualizations of social activity to support collective interaction. In K. Höök, D. Benyon, & A. J. Munro (Eds.), *Designing information spaces: The social navigation approach* (pp.17–41). London: Springer.

Fidel, R., Bruce, H., Pejtersen, A. M., Dumais, S., Grudin, J., & Poltrock, S. (2000). Collaborative information retrieval (CIR). *New Review of Information Behaviour Research, 1,* 235–247.

Fidel, R., Pejtersen, A. M., Cleal, B., & Bruce, H. (2004). A multidimensional approach to the study of human-information interaction: A case study of collaborative information retrieval. *Journal of the American Society for Information Science and Technology, 55,* 939–953.

Finholt, T. A. (2002). Collaboratories. *Annual Review of Information Science and Technology, 36,* 73–107.

Fu, L., Goh, D. H.-L., & Foo, S. S.-B. (2004). The effect of similarity measures on the quality of query clusters. *Journal of Information Science, 30,* 396–407.

Furner, J. (2002). On recommending. *Journal of the American Society for Information Science and Technology, 53,* 747–763.

Goldberg, D., Nichols, D., Oki, B., & Terry, D. (1992). Using collaborative filtering to weave an information tapestry. *Communications of the ACM, 35*(12), 61–70.

Gorman, P., Ash, J., Lavelle, M., Lyman, J., Delcambre, L., Maier, D., et al. (2001). Bundles in the wild: Managing information to solve problems and maintain situation awareness. *Library Trends, 49,* 266–289.

Gorman, P., Lavelle, M., Delcambre, L., & Maier, D. (2002). Following experts at work in their own information spaces: Using observational methods to develop tools for the digital library. *Journal of the American Society for Information Science and Technology, 53,* 1245–1250.

Grimaldi, M., & Cunningham, P. (2004). Experimenting with music taste prediction by user profiling. *Proceedings of the Sixth ACM SIGMM International Workshop on Multimedia Information Retrieval,* 173–180.

Hansen, P., & Järvelin, K. (2004). Collaborative information searching in an information-intensive work domain: Preliminary results *Journal of Digital Information Management, 2,* 26–29.

Hansen, P., & Järvelin, K. (2005). Collaborative information retrieval in an information-intensive domain. *Information Processing & Management, 41,* 1101–1119.

Hayes, C., & Cunningham, P. (2004). Context boosting collaborative recommendations. *Knowledge-Based Systems, 17,* 131–138.

Herlocker, J. L., Konstan, J. A., & Riedl, J. (2004). Explaining collaborative filtering recommendations. *Proceedings of the 2000 ACM Conference on Computer Supported Cooperative Work,* 241–250.

Herlocker, J. L., Konstan, J. A., Terveen, L. G., & Riedl, J. T. (2004). Evaluating collaborative filtering recommender systems. *ACM Transactions on Information Systems, 22,* 5–53.

Hoashi, K., Matsumoto, K., & Inoue, N. (2003). Personalization of user profiles for content-based music retrieval based on relevance feedback. *Proceedings of the Eleventh ACM International Conference on Multimedia,* 110–119.

Höök, K., Benyon, D., & Munro, A. J. (Eds.). (2003). *Designing information spaces: The social navigation approach.* London: Springer.

Hust, A. (2004). Introducing query expansion methods for collaborative information retrieval. In A. Dengel, M. Junker, & A. Weisbecker (Eds.), *Reading and learning: Adaptive content recognition* (pp. 252–280). London: Springer.

Hyldegård, J. (in press). Collaborative information behaviour: Exploring Kuhlthau's Information Search Process model in a group-based educational setting. *Information Processing & Management.*

Karamuftuoglu, M. (1998). Collaborative information retrieval: Toward a social informatics view of IR interaction. *Journal of the American Society for Information Science, 49,* 1070–1080.

Karasti, H., Baker, K. S., & Bowker, G. C. (2003). SCW2003 Computer Supported Scientific Collaboration (CSSC) workshop report. *ACM SIGGROUP Bulletin, 24,* 6–13.

Konstan, J. A. (2004). Introduction to recommender systems: Algorithms and evaluation. *ACM Transactions on Information Systems, 22,* 1–4.

Konstan, J. A., & Riedl, J. (2003). Collaborative filtering: Supporting social navigation in large, crowded infospaces. In K. Höök, D. Benyon, & A. J. Munro (Eds.), *Designing information spaces: The social navigation approach* (pp. 43–82). London: Springer.

Li, Q., Kim, B., Guan, D. H., & Oh, D. W. (2004). A music recommender based on audio features. *Proceedings of the Twenty-Seventh SIGIR Annual International Conference on Research and Development in Information Retrieval,* 532–533.

Maglio, P. P., Barrett, R., & Farrell, S. (2003). WebPlaces: Using intermediaries to add people to the Web. In K. Höök, D. Benyon, & A. J. Munro (Eds.), *Designing information spaces: The social navigation approach* (pp. 249–269). London: Springer.

Maltz, D., & Ehrlich, E. (1995). Pointing the way: Active collaborative filtering. *Proceedings of the ACM Conference on Human Factors in Computing Systems,* 202–209.

McCarthy, J. F., & Anagnost, T. D. (1998). MusicFX: An arbiter of group preferences for computer supported collaborative workouts. *Proceedings of the 1998 ACM Conference on Computer Supported Cooperative Work,* 363–372.

McCarthy, J. F., & Anagnost, T. D. (2000). MusicFX: An arbiter of group preferences for computer supported collaborative workouts. *Proceedings of the 2000 ACM Conference on Computer Supported Cooperative Work,* 348.

McNee, S. M., Albert, I., Cosley, D., Gopalkrishnan, P., Lam, S. K., Rashid, A. L., et al. (2002). On the recommending of citations for research papers. *Proceedings of the 2002 ACM Conference on Computer Supported Cooperative Work,* 116–125.

Middleton, S. E., Shadbolt, N. R., & DeRoure, D. C. (2004). Ontological user profiling in recommender systems. *ACM Transactions on Information Systems, 22,* 54–88.

Munro, A. J., Höök, K., & Benyon, D. (1999). *Social navigation of information space.* London: Springer.

Poltrock, S., Grudin, J., Dumais, S., Fidel, R., Bruce, H., & Pejtersen, A. M. (2003). Information seeking and sharing in design teams. *Proceedings of the 2003 International ACM SIGGROUP Conference on Supporting Group Work*, 239–247.

Prekop, P. (2002). A qualitative study of collaborative information seeking. *Journal of Documentation, 58*, 533–547.

Reddy, M., & Dourish, P. (2002). A finger on the pulse: Temporal rhythms and information seeking in medical work. *Proceedings of the 2002 ACM Conference on Computer Supported Cooperative Work*, 344–353.

Reich, S., Carr, L. A., DeRoure, D. C., & Hall, W. (1999). Where have you been from here? Trails in hypertext systems. *ACM Computing Surveys, 31*(4es), article no.11.

Resnick, P., & Varian, H. R. (1997). Recommender systems. *Communications of the ACM, 40*(3), 56–58.

Romano, N. C., Jr., Roussinov, D., Nunamaker, J. F., Jr., & Chen, H. (1999). Collaborative information retrieval environment: Integration of information retrieval with group support systems. *Proceedings of the Thirty-Second Hawaii International Conference on System Sciences*, 10 pp.

Sawyer, S., & Eschenfelder, K. (2002). Social informatics: Perspectives, examples, and trends. *Annual Review of Information Science and Technology, 36*, 427–466.

Shapira, B., Kantor, P. B., & Melamed, B. (2001). The effect of extrinsic motivation on user behavior in a collaborative information finding system. *Journal of the American Society for Information Science and Technology, 52*, 879–887.

Shardanand, U. (1994). *Social information filtering for music recommendation.* Unpublished master's thesis, Massachusetts Institute of Technology.

Shardanand, U., & Maes, P. (1995). Social information filtering: Algorithms for automating word-of-mouth. *Proceedings of the SIGCHI Conference on Human Factors in Computing Systems*, 210–217.

Sonnenwald, D. H., Maglaughlin, K. L., & Whitton, M.C. (2004). Designing to support situation awareness across distances: An example from a scientific collaboratory. *Information Processing & Management, 40*, 989–1011.

Sonnenwald, D. H., & Pierce, L. G. (2000). Information behaviour in dynamic group work contexts: Interwoven situational awareness, dense social networks and contested collaboration in command and control. *Information Processing & Management, 36*, 461–479.

Spring, M. B., & Vathanophas, V. (2003). Peripheral social awareness information in collaborative work. *Journal of the American Society for Information Science and Technology, 54*, 1006–1013.

Swearingen, K., & Sinha, P. (n.d.). *Interaction design for recommender systems.* Retrieved January 26, 2005, from http://www.rashmisinha.com/articles.html

Talja, S. (2002). Information sharing in academic communities: Types and levels of collaboration in information seeking and use. *New Review of Information Behaviour Research, 3*, 143–159.

Talja, S., & Hansen, P. (in press). Information sharing. In A. Spink & C. Cole (Eds.), *New directions in information behaviour*. Dordrecht, NL: Kluwer.

Terveen, L., & Hill, W. (2002). Beyond recommender systems: Helping people help each other. In J. M. Carroll (Ed.), *Human–computer interaction in the new millennium* (pp. 487–509). London: Addison-Wesley.

Torres, R., McNee, S. M., Abel, M., Konstan, J. A., & Riedl, J. (2004). Enhancing digital libraries with TechLens. *Proceedings of the ACM/IEEE Joint Conference on Digital Libraries*, 228–236.

Twidale, M. B., & Nichols, D. M. (1998). Computer supported cooperative work in information search and retrieval. *Annual Review of Information Science and Technology, 33*, 259–319.

Vakkari, P. (2003). Task-based information searching. *Annual Review of Information Science and Technology, 37*, 413–464.

Walkerdine, J., & Rodden, T. (2001). Sharing searches: Developing open support for collaborative searching. *Proceedings of Interact 2001*, 140–147.

Wen, J., Nie, J. Y., & Zhang, H. J. (2002). Query clustering using user logs. *ACM Transactions on Information Systems, 20*, 59–81.

Wexelblat, A. (2003). Results from the Footprints project. In K. Höök, D. Benyon, & A. J. Munro (Eds.), *Designing information spaces: The social navigation approach* (pp. 223–248). London: Springer.

Wexelblat, A., & Maes, P. (1999). Footprints: History-rich tools for information foraging. *Proceedings of the SIGCHI Conference on Human Factors in Computing Systems*, 270–277.

Zhang, X. (2002). Collaborative relevance judgment: A group consensus method for evaluating user research performance. *Journal of the American Society for Information Science and Technology, 53*, 220–231.

CHAPTER 9

Information Failures in Health Care

Anu MacIntosh-Murray and Chun Wei Choo
University of Toronto

Introduction

Health care failures, or clinical adverse events, have become highly topical both in the popular media and in professional and clinical journals. On the basis of several large-scale studies, researchers have estimated that 3 percent to 10 percent of inpatient admissions resulted in some form of medically related injury, one-third to one-half of which were preventable (Baker, Norton, Flintoft, Blais, Brown, Cox, et al., 2004; Brennan, Leape, Laird, Hébert, Localio, Lawthers, et al., 1991; Leape, Brennan, Laird, Lawthers, Localio, & Barnes, 1991; Thomas, Studdert, Burstin, Orav, Zeena, Williams, et al., 2000; Vincent, Neale, & Woloshynowych, 2001). Kohn, Corrigan, and Donaldson (1999, p. 24) define "adverse events" as injuries caused by medical management rather than the underlying condition of the patient. For example, major public inquiries in the U.K. (Bristol) and Canada (Winnipeg) have provided detailed accounts of how failures in several health care organizations contributed to significant adverse events—in these instances, the deaths of children undergoing cardiac surgery (Kennedy, 2001; Sinclair, 2000). The issues identified in both of these inquiries may be illustrative of the more widespread problems signaled by the large studies. The testimony given in both inquiries made it painfully clear that adverse events have a devastating impact on all those involved, including not only patients and family members, but also health care providers and health care organizations.

It is also evident from the testimony and final reports from Winnipeg and Bristol that there were failures in the way that the organizations handled information that might have prevented or at least minimized the problems. The aim of this chapter is to highlight the multifaceted role information failures can play in clinical adverse events and patient safety. We use themes based on the Winnipeg and Bristol reports to structure an overview of the interdisciplinary concepts that researchers have used to study health care failures and issues having to do with information handling and management. The themes include culture (organizational, professional, safety, and information); incident reporting and safety monitoring; human factors analysis and systems thinking; and resilience and learning from adverse events in complex environments.

Patient safety can be defined as "the reduction and mitigation of unsafe acts within the health care system, as well as through the use of best practices shown to lead to optimal patient outcomes" (Davies, Hébert, & Hoffman, 2003, p. 12). The emphasis on learning and prevention has prompted calls for improved organizational and safety cultures (Battles & Lilford, 2003; Walshe & Shortell, 2004). Safety culture has been equated with an "informed culture" (Hudson, 2003; Reason, 1997, 1998; Toft & Reynolds, 1994). Both the Sinclair (2000) and Kennedy (2001) reports point to the cultures of the respective health care organizations as factors contributing significantly to the tragic outcomes.

Researchers interested in preventing patient safety failures in hospitals have turned to studies of accidents and disasters in complex environments for insights (Gaba, 2000; Hudson, 2003; Rosenthal & Sutcliffe, 2002; Schulman, 2002; Weick, 2002). Information failures have been cited as a major contributing factor to, and preconditions of, organizational disasters and accidents (Horton & Lewis, 1991; Pidgeon & O'Leary, 2000; Reason, 1997; Toft & Reynolds, 1994; Turner & Pidgeon, 1997; Vaughan, 1996). In these studies, examples abound of missed or ignored warning signals and failure to handle information in ways that could have prevented adverse outcomes. This research, in particular the work of Turner (1976), Turner and Pidgeon (1997), Westrum (1987, 1992, 2004), and Vaughan (1999), raises the interesting possibility that underlying ways of shared thinking, or culture, and related information practices may make it more difficult for an organization to handle information about errors and failures effectively. Horton and Lewis (1991, p. 204) label these phenomena "dysfunctional information attitudes and behaviors." Examples of these types of information breakdown are found in the two inquiry reports and will receive further discussion.

As Sophar (1991, p. 151) notes, "[n]ot all disasters are spectacular. Many, such as environmental and information disasters, are usually the accumulation of many smaller ones." Common concepts and patterns emerge from the study of the diverse mishaps and events covered in the literature (Horton & Lewis, 1991; Turner & Pidgeon, 1997). The events involve people in organizations engaged in activities potentially linked to risks or hazards that could cause injury or damage. The common thread of these elements is present in Turner and Pidgeon's (1997, p. 70) definition of disaster as "an event, concentrated in time and space, which threatens a society or a relatively self-sufficient subdivision of society with major unwanted consequences as a result of the collapse of precautions which had hitherto been culturally accepted as adequate." Studies of how such precautions fail or succeed in organizations have spawned several theoretical approaches (Rijpma, 1997, 2003), including Turner's (1976) disaster incubation theory, Normal Accident Theory (Clarke & Perrow, 1996; Perrow, 1999a, 1999b), and High Reliability

Theory (LaPorte & Consolini, 1991; Roberts, 1990, 1993; Rochlin, 1999; Weick & Roberts, 1993). We will highlight these approaches and consider how they contribute to understanding the role of information failures in patient safety failures.

Given the breadth of this subject area and the potential for linkages to many topics, it should be noted that we offer just one of many possible paths through the literature. This chapter is not intended to be an all-encompassing review, but rather to illustrate possibilities and connections. For useful background reading on organizations as information environments and organizational information processing, see Sutcliffe (2001) and Choo (1998, 2002).

By way of background, the next section will provide a brief summary of the situations that gave rise to the Manitoba inquest and Bristol inquiry.

The Inquiries into Pediatric Cardiac Surgery Deaths in Winnipeg and Bristol

The inquest headed by Mr. Justice Sinclair (2000) looked into circumstances surrounding the deaths of 12 children who had undergone cardiac surgery at the Winnipeg Health Sciences Centre in 1994. The hearings began in 1995 and ended in 1998, resulting in 50,000 pages of transcripts and 10,000 pages of exhibits. The inquest found that five of the deaths were preventable and several more possibly could have been prevented (Sinclair, 2000, p. vi). It also found that the parents of the children were not adequately informed about the inexperience of the surgeon or the risks of the surgery (Sinclair, 2000, p. 480). The procedures had been carried out by a new junior surgeon, who had been recruited to restart the pediatric cardiac surgery program after it had been suspended when the previous surgeon had left. The report indicated that some of the issues related to the skills and abilities of particular individuals, but "other problems were largely systemic in nature" (Sinclair, 2000, p. 465). The surgeries took place in a context beset by problems, including a shortage of cardiologists in the program; inadequate supervision and lack of a phased start-up plan; poor case selection; confusion over lines of authority; and poor leadership, team relations, and communication. The report identified failures in monitoring the program and in both internal and external quality assurance mechanisms. It also pointed out that insufficient attention was paid to individuals (nurses and anesthetists) who raised concerns about the surgical outcomes and especially condemned the treatment of the nurses in this regard (Sinclair, 2000, p. 477). The report noted that poor outcomes were rationalized as part of the "learning curve" to be expected as the new surgeon and surgical team gained experience (Sinclair, 2000, p. 473). Due to the systemic failures, there were delays in dealing with problems related to the team's performance. The report recommended that the program resume only as part of a regional program because of the concern that the number of patients in the province of Manitoba alone would be insufficient to allow it to develop fully, a situation that could increase the risk of deaths and complications (Sinclair, 2000, p. viii).

Shortly after the Sinclair report was published in 2000, *Learning from Bristol*, the report of the inquiry into children's heart surgery at the Bristol Royal Infirmary (BRI), was released (Kennedy, 2001). The time frame and scope were significantly larger, in that the review covered the care given to complex pediatric cardiac patients—over 1,800 children in all—between 1984 and 1995. The review dealt with over 900,000 pages of documents. The Bristol Inquiry was not charged with investigating causes of individual deaths but looked at the adequacy of services and whether appropriate action had been taken in response to concerns about the surgeries (Kennedy, 2001, p. 1). The inquiry found that one-third of the children who had undergone open-heart surgery received "less than adequate care" and, further, that between 1991 and 1995, the mortality rate was higher than expected for comparable units at the time, resulting in 30 to 35 more deaths in children under the age of one (Kennedy, 2001, p. 241). Given that BRI did not have a full-time pediatric surgeon on its staff, the children's procedures were carried out by two surgeons who operated primarily on adults. The pediatric patients received care in the adult intensive care unit, there were inadequate numbers of pediatric nurses, and the service was split between two sites. Because of problems with the physical plant and concerns about inadequate numbers of patients, it had been debated whether BRI should at all have been designated as a pediatric cardiac surgery center. Once again, there were findings of individual failings, but systemic issues were dominant. These related to hierarchical culture and lack of teamwork; poor organization, communication, and leadership; and inadequate resources and staffing. Parents of the children were not informed adequately about the risks nor were they given enough time to consider what they were told, prompting the observation that "sharing of information should be a process" (Kennedy, 2001, p. 220). Over the course of several years, an anesthetist who joined the hospital in 1988 raised concerns about the length of procedures and their outcomes. The report chronicled his efforts to bring the data he had compiled to the attention of various individuals, but these efforts did not result in effective action for a considerable time (Kennedy, 2001, pp. 134–151). The lines of accountability and responsibility for monitoring were confused, both internally and externally. The culture was described as one in which data about bad results were variously explained by the learning curve (Kennedy, 2001, p. 247), the complicated case mix (Kennedy, 2001, p. 161), a run of bad luck, or the small numbers skewing the percentages. The broader context also contributed to the "wishing away" of the problems; "the tradition in the NHS of overcoming the odds drowned out any messages that things were worse than they should be" (Kennedy, 2001, p. 247).

There are striking parallels between the Winnipeg and Bristol inquiries and the two resulting reports. It is a compelling coincidence that two such similar inquiries with overlap in circumstances, mandates, and time frames occurred in separate countries. The fact that they arrived at similar recommendations supports the wider applicability of the lessons learned from these cases.

Each of the institutions involved was staffed by well-intentioned, hard-working but in some instances misguided health care professionals. Both reports emphasized the importance of taking systems and human factors approaches to identify issues rather than blaming individuals and instilling fear (Kennedy, 2001, pp. 4, 258; Sinclair, 2000, p. 488). They described flawed systems, with lack of leadership and teamwork, and confusion over lines of authority and responsibility for monitoring. Although Bristol "was awash with data," these had been handled in a fragmented way and thus had been open to varying interpretations and challenges (Kennedy, 2001, pp. 240–241). By contrast, Winnipeg had inadequate data and "no tracking of common indicators that might point to matters of concern" (Sinclair, 2000, p. 484). However, in both cases the reports found that information and concerns about the problems had been explained away, not fully understood, discounted, ignored, or had fallen through cracks within the organizations (e.g., Kennedy, 2001, p. 247; Sinclair, 2000, p. 233). Consequently, among the many recommendations made in each report, particular emphasis was placed upon the need to change the cultures of health care organizations so as to promote open sharing and learning from errors, near-misses, and incident reporting.

In the sections that follow, we take a closer look at themes reflected in the Winnipeg and Bristol reports: culture (organizational, professional, safety, and information), human factors analysis and systems thinking, and incident reporting and safety monitoring. Connections also are made to the notion of resilience and the ability to recover from errors (Kennedy, 2001, p. 359; Sinclair, 2000, p. 497), topics that have been a focus in learning from adverse events in complex environments. We highlight the literature relevant to each topic, with illustrations from the two inquiry reports.

Cultures—Organizational, Professional, Safety, and Information

Culture is a recurrent theme in both the Winnipeg and Bristol reports, although variations on the term show up much more frequently in Bristol (over 180 instances in the 530-page final report). By sheer weight of emphasis, the Bristol Inquiry clearly accorded the concept a great deal of importance. The report defines culture as "the attitudes, assumptions and values of the NHS and its many professional groups" (Kennedy, 2001, p. 264), "which condition the way in which individuals and organizations work" (Kennedy, 2001, p. 266). It is "the way things are done around here" (Kennedy, 2001, p. 264). The definitions are in keeping with those given by Schein (1992) and Denison (1996) in the context of organizational studies literature. Denison (1996, p. 624) describes culture as

> the deep structure of organizations, which is rooted in the values, beliefs, and assumptions held by organizational members. Meaning is established through socialization to a variety of identity groups that converge in the workplace. Interaction produces a symbolic world that gives culture both a great stability and a certain precarious and fragile nature rooted in the dependence of the system on individual cognition and action.

Denison (1996) points out that researchers have described three levels of cultural phenomena: a surface level, which includes artifacts, symbols, and practices; an intermediate level, which includes values and traits; and a deep level, composed of assumptions. This reflects Schein's (1992) categorization of levels. The notions of "identity groups," "socialization," and "assumptions" present in Denison's definition share roots with Schein's (1992, p. 12) much-quoted definition of the culture of a group:

> A pattern of shared basic assumptions that the group learned as it solved its problems of external adaptation and internal integration that has worked well enough to be considered valid and, therefore, to be taught to new members as the correct way to perceive, think, and feel in relation to those problems.

Schein's emphasis on the role of leaders in creating and managing culture has been criticized as perhaps too narrow and overly functionalist in outlook (Alvesson, 1993; Martin, 1992; Schultz, 1994). However, the emphasis on problem solving has been taken up by other organizational researchers, such as Westrum (2004).

Assumptions are part of the third, deeper level of culture, and tend to be the unquestioned beliefs that unconsciously guide actions. As such, they are very difficult to uncover and change, due to the defensive routines that members invoke when challenged or threatened (Argyris & Schön, 1996).

The complexity of culture as a concept is underscored by the variety of ways in which the term is used in the Bristol report. The term serves to call attention to values and attitudes to which the organization and health care system should aspire—for example, a culture of quality, safety, flexibility, openness, accountability, and public service (Kennedy, 2001, p. 13); culture of teamwork (Kennedy, 2001, p. 276); and culture of high performance (Kennedy, 2001, p. 276). The report depicts the inadequacies of the organization and system (as they were prior to the inquiry) by presenting a distressing litany of values and attitudes, including culture of blame (Kennedy, 2001, p. 16); club culture (Kennedy, 2001, p. 2); over-reliance on an oral culture (Kennedy, 2001, p. 37); culture of medicine (territorial) (Kennedy, 2001, p. 161); management culture of fear (Kennedy, 2001, pp. 171, 201); a culture that excluded nurses (Kennedy, 2001, p. 176); culture of the NHS (chapter 22), with its prevailing culture of blame and stigma (Kennedy, 2001, p. 259); culture of defensiveness (Kennedy, 2001, p. 272); culture of uncertainty (in contrast to accountability) (Kennedy, 2001, p. 273). Even the ostensibly positive mend-and-make-do culture and culture of pragmatism (Kennedy, 2001, p. 274) were found to have contributed to the problems.

In the text of the Winnipeg final report, on the other hand, the word "culture" does not appear until chapter 10 (*Findings and Recommendations*). Yet, although he makes only sparing use of the term, Sinclair (2000, p. 492) forcefully states that

> [t]he [Health Sciences Centre] must develop an institutional culture in which information about safety hazards is actively sought, messengers are trained to gather and transmit such information, and responsibility for dealing with that information is shared by all. This will require new approaches to quality assurance, risk management and team performance.

Echoing these comments from Winnipeg, Kennedy's (2001, p. 16) recommendations call for the development of a culture of safety:

> A culture of safety in which safety is everyone's concern must be created. Safety requires constant vigilance. Given that errors happen, they must be analyzed with a view to anticipate and avoid them. A culture of safety crucially requires the creation of an open, free, non-punitive environment in which health care professionals can feel safe to report adverse events and near misses (sentinel events).

The quoted recommendations weave together aspects of professional, safety, and information cultures. In the following sections we give an overview of some of the related literature that explains how traditional characteristics of professional cultures have made it difficult to achieve the informed safety cultures advocated by the Winnipeg and Bristol reports.

Professional Cultures and Subcultures

Denison (1996, p. 635) states that the social constructivist perspective of culture emphasizes the "recursive dynamics between the individual and the system." As individuals are socialized to various identity groups, multiple subcultures may develop (the differentiation perspective of cultures) rather than a single unified or homogenous organizational culture (the integration perspective) (Martin, 1992). Martin describes a third possibility, the fragmentation perspective, which emphasizes ambiguity as the dominant aspect of a culture. Health care organizations are the workplace for many occupational communities; as a result, they harbor a kaleidoscope of distinct and overlapping work cultures.

Van Maanen and Barley (1984, p. 287) refer to occupational communities as groups of people who consider themselves to be engaged in the same sort of work; whose identity is drawn from the work; who share with one another a set of values, norms, and perspectives that applies to but extends beyond work-related matters; and whose social relationships meld work and leisure. In their analysis of work culture, they consider task rituals, standards for proper and improper behavior, and work codes for routine practices, as well as the occupational group's "compelling accounts attesting to the logic and value of these rituals, standards, and codes" (Van Maanen & Barley, 1984, p. 287). They note that occupational communities strive for control over the way their work is done, how it is evaluated, and who may enter the community.

Schein (1996) identifies and describes three different cultures operating silently within an organization: the operators, the engineers, and the executives. In health care, strong professional subcultures and markedly different worldviews influence decision making and information practices (Davies, Nutley, & Mannion, 2000; Walshe & Rundall, 2001). Bloor and Dawson (1994) suggest that their diverse values and practices help professionals make sense of and manipulate events, possibly as a way to maintain or improve their status or position relative to other groups in the organization. According to Alvesson (1993, p. 117), "[g]iven cultural differentiation, values and ideals will be implemented to different degrees depending on the issue and the amount of influence a particular group has. Compromise, tension, and even conflict can be expected." Traditionally, the medical profession has been dominant in health care, although physicians' position of power and authority as a "sovereign profession" has been eroded somewhat by the rise of corporate medicine, as chronicled by Starr (1982, p. 1).

The culture of a profession, including its long-held beliefs and practices, can be at odds with broader organizational goals and environmental changes. For example, learning and improvement require team skills and understanding of patient care as multidisciplinary processes embedded in complex systems (Feldman & Roblin, 1997; Leape & Berwick, 2000; Nolan, 2000). West (2000) suggests that the increasing specialization of health care professionals over time has contributed to compartmentalization of knowledge and information, which Vaughan (1996, p. 62) refers to as "structural secrecy." This view is further reinforced by West's (2000, p. 123) finding that "nurses and doctors rarely discuss important professional matters informally with each other. ... These boundaries, around medicine in particular, could be a barrier to communication with, and monitoring by, other professional groups" (see also West, Barron, Dowsett, & Newton, 1999). In her thoughtful analysis of the neglect of the nurses' concerns in Winnipeg, Ceci (2004) draws on insights from Foucault for one explanation of why this happened. She suggests that social norms and rules constitute and privilege some knowledge claims as more credible than others: "[n]urses, it seems, before they even spoke, were confined within already existing relations of power and knowledge that determined them to be, that positioned them as, the sorts of persons whose concerns need not be taken seriously" (Ceci, 2004, p. 1884).

The norm of hierarchical organization among health care professionals can result in reporting relationships impaired by too great an adherence to an authority gradient. Nurses and junior medical staff are not in a position to challenge physicians' erroneous judgment calls; as a consequence, communication and collaboration may be undermined (Sexton, Thomas, & Helmreich, 2000; Thomas, Sexton, & Helmreich, 2003). Davies, Nutley, and Manion (2000, p. 113) observe that "health care is notoriously tribal," as can be seen in the rivalry, competition, and discordant subcultures found in some organizations. Given these factors, teamwork in health care may sometimes seem like an oxymoron. This is in keeping with the observation from Bristol that

> complexity lies in the coexistence of competing cultures. This is very much the case within the NHS, where the cultures, for example, of nursing, medicine, and management are so distinct and internally closely-knit that the words 'tribe' and 'tribalism' were commonly used by contributors to the Inquiry seminars on this subject. (Kennedy, 2001, p. 266)

Both inquiries condemned the traditional disciplinary hierarchy and its impact on communication. "The continued existence of a hierarchical approach within and between the healthcare professions is a significant cultural weakness. ... This sense of hierarchy also influences who gets listened to within the organization when questions are raised" (Kennedy, 2001, pp. 268–269).

Physicians have traditionally been seen as independent contractors and the "captain of the ship," an image that has encouraged perpetuation of a myth of medical infallibility (Helmreich & Merritt, 1998; Weick, 2002). In the Winnipeg inquest, Sinclair (2000, p. 485) bluntly criticized this aspect of medical culture, which, in his words,

> reflected the concept of the surgeon as the supreme and infallible captain of the ship. This meant that what should have been the collective concern about the team's ability to handle certain cases turned into highly charged conflicts centering on the surgeon.

Sharpe (1998, p. 17) traces the historical roots of this view of the medical profession in North America, stating that

> [T]he dominant view of medical error and ways in which it is institutionalized presupposes, expresses, and reinforces the assumption that medical quality itself is essentially a function of the competence and integrity of individuals and that error prevention, therefore, is largely about their technical and moral improvement.

As a result, Sharpe notes, the litigating public has been quite willing to embrace this view and hold individual physicians accountable through lawsuits for adverse outcomes that they have suffered. The combined effect of professional culture and societal culture appears to be circular and self-amplifying. Physicians have generated a culture of independence with a belief in individual, not system, causes of human error (Bosk, 1979). Patients sue to hold them accountable for adverse outcomes, even if these may be the result of multiple systemic causes. The physicians become more wary of litigation and less likely to engage in the open reflection required for learning, for fear of producing data that may be used as evidence against them. Consequently, analyses of root causes are not pursued, learning does not take place, adverse outcomes continue, and the cycle goes on. To help break this vicious cycle, the Bristol report recommended that clinical negligence be abolished and replaced with an alternate system to provide compensation to patients injured by clinical adverse events (Kennedy, 2001, p. 16).

Safety Culture

Safety implies preventing adverse events, and the occurrence of adverse events is used as a safety indicator, the one being the converse of the other (Flin, Mearns, O'Connor, & Bryden, 2000; Mearns & Flin, 1999). However, Hale (2000, p. 10) voices a note of caution about applying lessons learned from the retrospective identification of cultural factors implicated in adverse events, a process that involves "tracing causal chains into the past." It is still difficult to know which specific cultural factors to measure in order to assess safety in organizations, as there may be widely varying combinations that constitute lethal time bombs. Despite this caveat, Pidgeon (1991, p. 131) suggests that "the notion of safety culture may provide at least heuristic normative guidance for ongoing management and control of risk."

Turner (1991, p. 241) has defined safety culture as "the specific set of norms, beliefs, roles, attitudes and practices within an organization which is concerned with minimizing exposure of employees, managers, customers, suppliers, and members of the general public to conditions considered to be dangerous or injurious." This definition is comparable to many other definitions of culture, such as that of Hale (2000, p. 7), who refers to "the attitudes, beliefs, and perceptions shared by natural groups as defining norms and values, which determine how they act and react in relation to risks and risk control systems." Gherardi and Nicolini (2000) made use of Lave and Wenger's (1991) notion of communities of practice to study how members learn about safety practices in the workplace and how the learning process shapes safety cultures. According to Gherardi and Nicolini (2000, p. 13), "knowing is a contested and negotiated phenomenon," a characterization that echoes Turner and Pidgeon's (1997, p.188) notion of a "distinctive organizational discourse about 'the way safety is handled around here.'" As noted earlier, Ceci (2004) would agree that knowing is contested, but her study indicated that the power structures in health care organizations do not allow some of the members much latitude to negotiate. Gherardi, Nicolini, and Odella (1998, p. 211) found that safety cultures vary by different groups or communities of practice; as suggested by the two inquest reports, this may be the case in health care organizations as well:

> Dispersed communities have diverse and non-overlapping organizational information, world-views, professional codes, organizational self-interests, and different interpretations of what is happening, why it is happening, and what its implications are. ... The limits on safety derive from the isolation of potentially dissenting points of view, but simultaneously learning about safety occurs because learning is mediated by differences of perspective among co-participating communities.

For an extensive discussion of the literature relating to communities of practice, see Davenport and Hall (2002).

Mearns and Flin (1999) also emphasized the need to understand how a shared view of safety is constructed through the interaction among the members of an organization. Richter and Koch's (2004) case study of safety cultures in Danish companies has built on Martin's (1992, 2002) and Alvesson's (1993, 2002) conceptual approaches to organizational culture. Richter and Koch use the metaphors of "production," "welfare," and "master" to characterize the multiple safety cultures they observed in one company. The production metaphor emphasizes the view that risks are an acceptable part of work and can be minimized by the workers. What counts is productivity, and safety measures get in the way. The welfare metaphor stresses that risks are unacceptable and can jeopardize workers' long-term participation as productive members of society. Safe technology and social practices can prevent accidents (Richter & Koch, 2004, p. 713). The master metaphor stresses that risk taking is unacceptable. It emphasizes the safe mastery of the trade through learning from good examples modeling habits of risk avoidance (Richter & Koch, 2004, p. 714).

Flin et al. (2000) state that discussions of safety culture and climate in published studies generally focus on such dimensions as members' perceptions of management and supervision, safety systems and arrangements, risk, work pressures, competence, and procedures. A subset of this cluster of factors is the way that responsibility and blame are handled. Safety researchers point out that the most common reaction in an organization is to focus on the actual event itself and the immediate response is to find responsible culprits to blame (Berwick, 1998b; Cook, Woods, & Miller, 1998; Reason, 1997, 1998; Sharpe, 1998). This is in keeping with anthropologist Mary Douglas's (1992, p. 19) wry observation that the culture of the organization will govern what will count as information and that "blaming is a way of manning the gates through which all information has to pass." If prevailing hospital values are perceived to favor learning and prevention of future mistakes through human factors analysis of error, this may influence how staff expect to be treated when adverse events happen and how they react in the future to information about adverse events. If the prevailing values and practice lean toward holding individuals accountable and placing blame, mistakes may be seen as an occasion for fear and less open communication (Hofmann & Stetzer, 1998; Nieva & Sorra, 2003): "in the politics of blaming, information is tailored to be ammunition" (Hart, Heyse, & Boin, 2001, p. 184). The Kennedy and Sinclair reports explicitly emphasized the need for human factors and systems approaches both for their own analyses of events, as well as for ongoing learning in the health care organizations (Kennedy, 2001, pp. 182, 256; Sinclair, 2000, p. 488). By avoiding inappropriate fault-finding and blame, organizations can encourage staff to report the incidents and near-misses that are crucial for review and learning.

Information Culture and Safety Culture

In the Winnipeg report, Sinclair (2000, p. 492) emphasized the need for a culture that actively supports the seeking and use of information about safety hazards. Likewise, Kennedy (2001, p. 366) wrote that improvement of safety requires creation of "an environment of openness so as to give rise to a systematic flow of information." The importance of information has also been underscored in the research literature linking safety and culture: "Failures in information flow figure prominently in many major accidents, but information flow is also a type marker for organizational culture" (Westrum, 2004, p. ii23). Toft and Reynolds (1994) considered a safety culture as the appropriate environment for facilitating the information flow necessary to learn from adverse events. Reason (1998, p. 294) has characterized the intertwined nature of information and safety cultures:

> In the absence of frequent bad events, the best way to induce and then sustain a state of intelligent and respectful wariness is to gather the right kinds of data. This means creating a safety information system that collects, analyses and

disseminates information from accidents and near misses, as well as from regular proactive checks on the system's vital signs. All of these activities can be said to make up an informed culture—one in which those who manage and operate the system have current knowledge about the human, technical, organizational, and environmental factors that determine the safety of the system as a whole. In most important respects an informed culture is a safety culture.

The "information flow" to which these authors refer includes measurement of quantifiable safety and risk indicators as well as descriptive reports of both near misses and actual incidents involving harm or damage. In addition to the Bristol and Winnipeg inquiries, other studies and government-sponsored reports have emphasized the importance of such information for improving patient safety, at the same time pointing out the limitations of many current data-gathering approaches used in health care organizations (Baker & Norton, 2002; Institute of Medicine, 2003; Karson & Bates, 1999; Kohn et al., 1999; National Steering Committee on Patient Safety, 2002; Thomas & Petersen, 2003; Wald & Shojania, 2001). One of the major deficiencies explored in studies is the substantial underreporting of both adverse events and near misses (Stanhope, Crowley-Murphy, Vincent, O'Connor, & Taylor-Adams, 1999; Weingart, Ship, & Aronson, 2000). Researchers have linked underreporting to many of the cultural issues highlighted in previous sections, including fear of punishment or litigation (Leape, 1999; Wild & Bradley, 2005), as well as workload, lack of knowledge of how to report, disagreement about the necessity or utility of reporting (Vincent, Stanhope, & Crowley-Murphy, 1999), unwillingness to report unless a written protocol has been violated (Lawton & Parker, 2002), and variability in definition and interpretation of what constitutes a reportable "event" (Kaplan & Fastman, 2003; Sutcliffe, 2004; Tamuz, Thomas, & Franchois, 2004). In the Winnipeg inquest, the report notes that "[n]o member of the HSC [Health Sciences Centre] staff made use of the incident reporting system to flag any of the issues. Indeed, it is distressing that many staff members did not even believe that the reporting system applied to them" (Sinclair, 2000, p. 199). The Bristol report discusses many of the same barriers to reporting (Kennedy, 2001, p. 362).

Various recommendations have been put forward to improve the situation, including modeling reporting systems after those used successfully in other industries, such as aviation (Barach & Small, 2000; Billings, 1998; Thomas & Helmreich, 2002). To counter the cultural and institutional barriers, some argue that reporting should be voluntary, confidential, nonpunitive, and protective of those reporting, with an emphasis on capturing near misses (Barach & Small, 2000; Cohen 2000). These suggestions reflect both the tone and substance of the recommendations in the Bristol report (Kennedy, 2001, p. 370). However, Johnson (2003) warned that the expected benefits may be based on overly optimistic assessments of experience in other industries and underestimation of the limitations of incident reporting, as well as the complexity of proper analysis of incidents. Leape (1999, 2000) expressed concern about the high costs of such systems and suggested that focused data collection and analysis methods may be more productive. On a more optimistic note, Kaplan and Barach (2002) proposed that staff participation in incident reporting, if properly supported, could contribute to the development of safety culture and mindfulness.

In addition to risk reports, traditional technical safety and risk management has also relied on codified knowledge such as policies and procedures to promote understanding of safety practice requirements. However, as discussed earlier, according to Gherardi and Nicolini (2000), it is not enough to have concrete policies, procedures, and indicator reports. Sustaining an informed safety culture also depends on understanding how members of an organization become part of a community, how they actually perform their work, and how they communicate information and knowledge. Weick (2002, p. 186) echoes this thought, referring to the danger of what Westrum, quoted in Weick (1995, p. 2), calls the "fallacy of centrality," or

> the belief that one is at the center of an information network, rather than just one interdependent player among many in a complex system. The reality is that systems have lots of centers, each with its own unique expertise, simplifications, and blind spots. It is the exchanging and coordinating of the information distributed among these centers that separates more from less intelligent systems. Systems that fail to coordinate effectively, and systems in which people assume that things they don't know about are not material, tend toward higher error rates. (Weick, 2002, p. 186)

An organization's cultures shape assumptions about what constitutes valid information and how it should be interpreted and transmitted (Choo, 2002; Turner & Pidgeon, 1997; Weick, 1995). Westrum (1992, p. 402) has put forth the argument that the very safety of an organization is dependent on a culture of "conscious inquiry," which supports the early warning system alluded to in connection with effective information flows (see also Westrum, 2004). This is another way of stating Sinclair's prescription for a culture that actively seeks and uses hazard information. A culture of conscious inquiry may be characterized as one in which "the organization is able to make use of information, observations or ideas wherever they exist within the system, without regard for the location or the status or the person or group having such information, observations or ideas" (Westrum, 1992, p. 402). This formulation brings to the fore the issue of information politics and the power that may be wielded by sharing or withholding information (Davenport, 1997). Individuals may be disenfranchised in a politicized information environment and lack the influence to persuade those in power of the validity of their hazard information, with the result that their warning signals are not taken seriously (Ceci, 2004; Turner & Pidgeon, 1997).

Westrum (1992) characterizes organizations as pathological, bureaucratic, or generative, according to how well they "notice" information. In a more recent study, Westrum (2004, p. ii24) has elaborated on this typology by explaining that "the processes associated with fixing the hidden problems that Reason has called latent pathogens would seem strongly connected with information flow, detection, reporting, problem solving, and implementation." He describes six types of responses to "anomalies" or indicators of problems (Westrum, 2004, p. ii25):

- Suppression—Harming or stopping the person bringing the anomaly to light; "shooting the messenger."

- Encapsulation—Isolating the messenger, with the result that the message is not heard.

- Public relations—Putting the message "in context" to minimize its impact.

- Local fix—Responding to the present case, but ignoring the possibility of others elsewhere.

- Global fix—An attempt to respond to the problem wherever it exists. Common in aviation, when a single problem will direct attention to similar ones elsewhere.

- Inquiry—Attempting to get at the "root causes" of the problem.

Applying these categories to the Bristol and Winnipeg situations shows that both health care organizations responded to signs of problems through encapsulation and public relations, and possibly suppression. The concerns of the nurses in Winnipeg were at best encapsulated and more likely suppressed, largely through being ignored. Although he was not discouraged from collecting data on problems, the anesthetist in Bristol was told repeatedly to go away and verify his data. The poor outcomes cited in both cases were explained away and "put in context" by invocation of the learning curve, severity of cases, and low volumes, all of which precluded more generative responses such as a "global fix."

Hudson (2003, p. i9) has adapted and expanded Westrum's categories to describe the "evolution of safety cultures" from pathological through reactive, calculative, and proactive, to generative, driven by increasing levels of "informedness" and trust. Citing Reason (1998), he suggests that a safety culture is one that is based on learning and is informed, wary (vigilant), just, and flexible (Hudson, 2003, p. i9). These characteristics are reminiscent of the characteristics attributed to reliability-seeking organizations, as will be discussed in a later section. Generative organizations are active in scanning, sensing, and interpreting and are more successful than pathological organizations at using information about adverse events. Bureaucratic (or calculative) information cultures may be as prone to information failures as pathological cultures. Although the behaviors may not be as overtly toxic to constructive sense making, information failures may nonetheless occur due to not-so-benign neglect and passivity, which may be inadvertently nurtured in a bureaucratic information culture. It appears that elements of pathological and bureaucratic information cultures were at work in Bristol. Those in leadership positions made it clear that problems were not welcome and that only solutions should be brought forward, thus taking a stance that "failed to encourage staff and patients to share their problems and speak openly" (Kennedy, 2001, p. 202).

Other researchers working outside the health care context have considered the interaction of culture and information handling. Brown and Starkey (1994, p. 808) stated that organizational culture "is an important factor affecting attitudes to, and systems and processes pertaining to, the management of information and communication." Ginman (1987, p. 103) defined CEO (chief executive officer) information culture as "the degree of interest in information and the attitude to factors in the external company environment" and suggested a connection with business performance. In a similar vein, Marchand, Kettinger, and Rollins (2000, 2001) have described information orientation as a composite of a company's capabilities to manage and use information effectively. According to them, information orientation is comprised of three categories of practices: information technology, information management, and information behaviors and values. The set of information behaviors and values includes *integrity*, or the absence of manipulation of information for personal gain (which relates to the issue of information politics noted earlier); *formality*, or the degree of use of and trust in formal information sources; *control and sharing*, or the degree of exchange and disclosure of information; *proactiveness*, or the degree to which members actively seek out information about changes in the environment; *transparency*, or the degree to which there is enough trust to be open about errors and failures (Marchand et al., 2000, p. 71). The last three information behaviors and values are clearly reflected in Westrum's information culture characteristics.

Davenport (1997, p. 5) has identified information culture and behaviors as one of the elements of an organization's information ecology, which "puts how people create, distribute, understand, and use information at its center." He suggests that sharing, handling overload, and dealing with multiple meanings are three behaviors associated with successful information ecologies. Once again taking the obverse view, a pathological information organization may show evidence of inadequate sharing, overwhelming information overload, and inability to reconcile the multiple meanings of ambiguous hazard signals constructively, a situation consonant with Turner and Pidgeon's (1997, p. 40) notion of the "variable disjunction of information." On the basis of the inquiry reports, it can be argued that many of these symptoms were in evidence in both Winnipeg and Bristol.

The Role of Human Error Vs. Systems Thinking

Because of the emphasis given to systems thinking and human factors analysis in both the Winnipeg and Bristol reports, it is important to show that these concepts relate to the health care context. Researchers have found that the gradual erosion of margins of safety in a system is attributable to various causes, for example, the design of the work environment and the pressure managers and staff may feel to take short cuts (Rasmussen, 1997; Reason, 1998; Sagan, 1993; Snook, 1996, 2000; van Vuuren, 1999, 2000; Vicente, 2004). Because safety tends to be equated with the absence of negative outcomes, "the associated information is indirect and discontinuous" (Reason, 1998, p. 4) so that the erosion is not evident until a catastrophic event occurs. The same pattern may well be occurring in cash-strapped hospitals, as the number of support staff is cut and more work is expected from fewer people, with nurses being expected to carry more responsibilities. Ironically, this is happening in the context of a serious nursing shortage, with the result that experienced nurses are in great demand and short supply. Novices have less practical experience and may have less access to adequate orientation and mentoring, and so may be in a vulnerable position. If learning, knowing, and collective mindfulness are the products of social construction and interaction, it is possible that cutbacks may disrupt occupational social networks and erode knowledge of safe practice (Fisher & White, 2000). Reason's (1995, p. 80) systems and human factors approach to the role of human error suggests that these frontline staff at the "sharp end" of the systems may commit errors and violations, which he calls active failures, with immediately visible adverse effects or outcomes. However, Reason emphasizes that these sharp-end human failures or unsafe acts occur in the context of the conditions latent within the systems. The latent conditions result from, for example, managerial decisions concerning resource allocation and can include "poor design, gaps in supervision, undetected manufacturing defects or maintenance failures, unworkable procedures, clumsy automation, shortfalls in training, less than adequate tools and equipment" (Reason, 1997, p. 10). Such latent conditions build up over time, becoming part and parcel of the organizational context.

Vicente (2004) has applied human factors engineering—engineering that tailors the design of technology to people—to a broader set of problems that arises out of the interactions and relationships between people and technology. He developed a conceptual framework based on a systematic analysis of the principles that govern human behavior, which can be organized into five levels: physical, psychological, team, organizational, and political. The *physical* level refers to how individuals differ in their physiology, strength, dexterity, and other capabilities. The *psychological* level includes our knowledge of how human memory works, how we make sense of situations, and how we seek and use information. The *team* level focuses on the communication and coordination activities of the group, comprising both the advantages and drawbacks of teamwork. The *organizational* level covers a range of factors, including organizational culture, leadership, reward structures, information flows, and staffing levels. Decisions made at the organizational level can have important effects at the lower levels: for example, when the number of nurses assigned to a hospital ward is too low, the workload of the individual nurse may push his or her psychological ability to cope to a breaking point (Vicente, 2004). The topmost level is the *political*, where basic considerations include public opinion, social values, policy agendas, budget allocations, laws, and regulations. This hierarchy of levels forms what Vicente (2004, p. 52) calls "the Human-Tech ladder." In this model, design should begin by understanding a specific human or societal need (e.g., public health, transportation, counterterrorism), identifying the levels that need to be considered, and then tailoring the technology or system to reflect, and fit with, the human factor principles at each of these levels. At lower levels (for example, when designing a toothbrush), design is concerned with achieving a good fit between physical and psychological factors. However, when we are designing large-scale social systems within which people and technology interact, it becomes necessary to consider higher-level factors such as authority relationships, staffing policies, laws, and regulations. Vicente used this model to analyze a number of cases and systems in health care, nuclear power, aviation, the environment, and other safety-critical sectors.

Rasmussen (1997, p. 189) presented a behavioral model of accident causation focusing on "a natural migration of work activities towards the boundary of acceptable performance." In any work system, human behavior is shaped by multiple objectives and constraints. Within these targets, however, many degrees of freedom are left open, allowing groups and individuals to search for work practices guided by criteria such as workload, cost-effectiveness, risk of failure, and joy of exploration. This search space is defined by four important boundaries: (1) a boundary to economic failure (beyond which work is not being done cost-effectively), (2) a boundary to unacceptable workload, (3) a boundary specified by official procedures, and (4) a safety boundary beyond which accidents can occur. Over time, work practices drift or migrate under the influence of two sets of forces resulting in two gradients. The first gradient moves work practices toward least effort, so that the work can be completed with a minimum of mental and physical effort. The second gradient is management pressure that moves work practices toward cost-effectiveness. The combined effect is that work practices drift toward the boundary of safety. In order to improve the safety of skilled activities, Rasmussen (1997, p. 192) suggested that rather than attempting to control behavior and stop the natural migration of work practices, the most promising general approach to improved risk management appears to be an explicit identification of the boundaries of safe operation together with efforts to make these boundaries visible to the actors and to give them an opportunity to learn to cope with the boundaries.

Researchers have also applied human error theory and prevention methods in health care delivery (Berwick, 1998a, 1998b; Bogner, 1994; Cook et al., 1998; Edmondson, 1996; Feldman & Roblin, 1997; Kohn et al., 1999; Leape, 1997; Taylor-Adams, Vincent, & Stanhope, 1999). Knowledge of the role of latent conditions and systemic causes is important for understanding adverse events, yet hindsight bias tends to encourage blinkered vision and foster short-sightedness. How well these concepts are understood and how widely they are believed may be critical dimensions of cultural knowledge in a health care organization (van Vuuren, 2000). This is also reflected in the categories that Pidgeon (1991) uses to identify the main elements of safety culture: norms and rules for dealing with risk, safety attitudes, and the capacity to reflect on safety practices.

Vulnerability and Failures or Resilience and Avoidance?

In addition to human factors research, studies of disasters and accidents in other industries have developed a rich literature on safety and human error in complex environments (Carroll, 1998; Perrow, 1999a, 1999b; Reason, 1990, 1998; Rochlin, 1999; Turner & Pidgeon, 1997). As noted in the introduction, three conceptual approaches have been developed to explain disaster prediction and avoidance: (1) Turner's (1978, p. 1) "man-made disasters" or disaster incubation theory (Turner & Pidgeon, 1997), (2) Normal Accident Theory (Clarke & Perrow, 1996; Perrow, 1999a), and (3) High Reliability Theory (LaPorte & Consolini, 1991; Roberts, 1990, 1993; Rochlin, 1999; Weick & Roberts, 1993). There are several reasons for considering this work here. The information failures and vulnerabilities described in the Winnipeg and Bristol reports echo many of the points made by Turner in his study of accident inquiry reports, and he contributes a theoretical model to explain such events. Recommendations in both reports stress the need to develop skills in teamwork and the ability to intervene in, and recover from, errors (Kennedy, 2001, p. 276; Sinclair, 2000, p. 497). These skills are reflected in the concepts of resilience and reliability, which have been considered in great detail by those studying reliability-seeking organizations.

"Man-Made Disasters"

Turner's remarkable *Man-Made Disasters* (Turner & Pidgeon, 1997)[1] was ahead of its time in presenting a sociotechnical model of system vulnerability and was not fully appreciated for a number of years (Short &

Rosa, 1998). In this work, Turner analyzed 84 British Government inquiry reports on disasters and accidents published between 1965 and 1975. The disasters included an unexpected array of situations with a wide range of factors and outcomes: mining and boiler explosions, marine wrecks, building collapses and fires, and a level-crossing collision. It is interesting to note that two of the accident inquiries dealt with the use of contaminated infusion fluids in a hospital and a smallpox outbreak in London.

As regards the themes of this chapter, Turner emphasized the significance of individual and organizational cultural beliefs and the social distribution of knowledge related to safety, hazards, and the adequacy of precautions. One of Turner's key observations is that disasters result from a failure of foresight and an absence of some form of shared knowledge and information among the groups and individuals involved. Sense making can be complicated by a "variable disjunction of information," that is, "a complex situation in which a number of parties handling a problem are unable to obtain precisely the same information about the problem, so that many differing interpretations of the situation exist" (Turner & Pidgeon, 1997, p. 40). To draw the parallel with Bristol and Winnipeg, neither organization used effective mechanisms to bring the appropriate individuals together to review concerns about surgical outcomes. In Bristol, there were data available but "all the data were seen in isolation" without agreed-upon standards (Kennedy, 2001, p. 236). The data "lent themselves to a variety of interpretations, not all of which pointed to poor performance ... and data were rarely considered by all members of the team together" (Kennedy, 2001, p. 240).

In considering Turner's "variable disjunction of information," Weick (1998, p. 74) has pointed out that the tendency of people to satisfice ("make do with what information they have") and to simplify interpretations (so as to be able to construct coherence from the variable and patchy information they have) creates collective blind spots that can impede perception of potential problems. Lea, Uttley, and Vasconcelos (1998) analyzed similar problems of information and interpretation that occurred in the Hillsborough Stadium disaster, using Checkland's (1999) Soft Systems Methodology to map the conflicting views of those involved in the ambiguous problem situation.

Turner's model proposes multiple stages of disaster development that can unfold over long periods of time. As Hart et al. (2001, p. 185) have pointed out, "the process nature of crises should be stressed ... they are not discrete events, but rather high-intensity nodes in ongoing streams of social interaction." The model suggests that disasters involve an element of great surprise for the majority of individuals involved or affected because they hold certain inaccurate beliefs in the initial stage—(1) that adequate safety precautions are in place, (2) that no untoward events are occurring, and (3) that the appropriate individuals are fully aware of any information that would indicate otherwise. Turner emphasized that disasters can have a prolonged incubation period during which events that are at odds with existing beliefs begin to accumulate in the environment, creating chains of unrecognized errors. During the "predisclosure" incubation period in stage two, the events may be ambiguous, unknown, or misunderstood, resulting in vague or ill-structured problem situations that are replete with information difficulties. Post-disclosure, after a transfer of information caused by a precipitating adverse event (stage three), the situation appears to be quite different and, with the benefit of hindsight, presents itself as a well-structured, recognizable problem (stages four to six). Hindsight bias can pose major problems during the efforts to piece together events after the fact, for example, during an inquiry (Henriksen & Kaplan, 2003). The ambiguity of situations facing individuals in the incubation stage is retrospectively minimized and the interpretation of events may be unwittingly (or deliberately) incomplete and/or politically driven (Brown, 2000, 2004; Gephart, 1984, 1992), as participants jockey to have their respective versions of the events accepted. The risk of hindsight bias was well recognized and articulated in the Bristol report, which discusses the problem explicitly (Kennedy, 2001, p. 36). In an ideal case, the transformation from the problematic pre-disclosure state to the well-structured post-disclosure state would be accomplished with the transfer of appropriate warning information. However, Rijpma (2003) has recently pointed out that, according to Perrow (1981), this ideal transfer is not likely to occur because the ambiguous and mixed signals are interpreted and labeled as warning information *only* with the benefit of hindsight.

Although the disasters he studied were ostensibly very different, Turner identified common features and similarities that form the basis of the man-made disasters model (Turner & Pidgeon, 1997, pp. 46–60):

1. Rigidities in perception and pervasive beliefs in organizational settings, including cultural and institutional factors that bias members' knowledge and ignorance

2. A decoy problem that distracts attention from the actual causal conditions brewing beneath the surface

3. Organizational exclusivity, which causes the organization to ignore outsiders' warnings

4. Information difficulties

   - Relevant information may be buried in a mass of irrelevant information

   - Recipients may fail to attend to information because it is only presented at the moment of crisis

   - Recipients may adopt a "passive" mode of administrative response to an issue

- Recipients may fail to put information together creatively

5. Involvement of "strangers," especially on complex sites

6. Failure to comply with existing regulations

7. Minimization of emergent danger

Turner's view of information difficulties is particularly interesting. The information in question is some form of danger sign, signal, or warnings that could potentially prevent a disaster. Information-handling difficulties can arise at any point in the development of a disaster—during the pre-disclosure incubation phase, during information transfer, and during post-disclosure—and arise from many different factors. Some difficulties relate to the nature of the signals and information itself, some depend on the characteristics of the people involved, some arise from the context or environment, yet others relate to steps in the process of information handling. In their review of a dozen examples of "great information disasters," Horton and Lewis (1991, pp. 1, 204) describe similar information difficulties as being the result of "dysfunctional information attitudes and behaviors."

Turner suggests that culture is a common influence shaping all information-handling difficulties, as was found to be the case in the Bristol and Winnipeg inquiries. He asserts that organizational culture affects the use and transfer of information by creating assumptions about what is given value as information, how it is to be communicated, and what can be ignored. "A way of seeing is always also a way of not seeing" is Turner's apt synopsis (Turner & Pidgeon, 1997, p. 49). Organizational failure of perception and collective blindness to issues may be "created, structured, and reinforced by the set of institutional, cultural, or sub-cultural beliefs and their associated practices" (Turner & Pidgeon, 1997, p. 47). The culturally sustained assumptions affect both the sense-making and decision-making processes, as "organizations strive to reduce noise, equivocation, information over-load and other ambiguous signals to politically secure and actionable messages" (Manning, 1998, p. 85). Wicks (2001) has taken a different theoretical path by emphasizing the role of institutional pressures rather than culture in organizational crises, but with similar results in terms of information breakdown. In his analysis of the Westray Mines explosion, he argues that the institutional antecedents can create a "mindset of invulnerability" that in turn creates inappropriate perception and management of risks (Wicks, 2001, p. 660). In a similar analysis of an Australian mine explosion, Hopkins (1999, p. 141) found a "culture of denial" that minimized warning signals, leading to a belief that such accidents could not happen in that venue. This was compounded by a "hierarchy of knowledge" that in this setting privileged personal experience over information from others (Hopkins, 1999, p. 141). The company managers also relied heavily on oral rather than written communication, so that written reports were ignored and issues that were communicated orally ran the risk of being forgotten.

Internal and external environmental conditions can change, creating a discrepancy between organizational assumptions and the environment. This highlights the need for environmental scanning to identify signs of hazards—a form of organizational early warning information system to support organizational intelligence and sense making (Choo, 2002).

> In studying the origins of disasters, therefore, it is important to pay attention, not just to the aggregate amount of information which is available before a disaster, but also to the distribution of this information, to the structures and communication networks within which it is located, and to the nature and extent of the boundaries which impede the flow of this information. Of particular interest are those boundaries which, by inhibiting the flow of this information, may permit disasters to occur. (Turner & Pidgeon, 1997, p. 91)

In sum, Turner's seminal work firmly established the importance of culture—that is, beliefs, values, and norms—in the occurrence of information failures and accidents.

Normal Accidents and High Reliability

Perrow's Normal Accident Theory suggests that accidents are an inevitable risk inherent in the tightly coupled and complex nature of technology-dependent systems such as nuclear or chemical plants (Perrow, 1999a). The complex nature of the functioning of these technologies can be opaque to the people charged with their maintenance and operation. This makes it almost impossible to intervene successfully when something goes wrong, unless redundancies are built into a system from the beginning (Perrow, 1999b). Interacting failures move through complex systems quickly when components have multiple functions and are closely tied to one another (Rijpma, 1997). The nature of the technology itself paves the way for unavoidable accidents. In addition, the tendency to blame individuals, the difficulty of comprehending the complexity of events in retrospect, and reluctance to report incidents make learning unlikely (Clarke & Perrow, 1996; Mascini, 1998; Perrow, 1999a).

Taking a different position are the proponents of reliability-seeking organizations (or High Reliability theorists), who argue that Normal Accident Theory is based on an overly structural view of organizations (Roberts, 1993). High Reliability theorists shift the emphasis from structure and technology to the culture and interactive processes of groups responsible for carrying out the work. They have studied exemplary aircraft carriers and nuclear plants that successfully balanced production with protection. Using the example of operations on the flight deck of an aircraft carrier, Weick and Roberts (1993) explored the concept of collective mental processes and how they mediate performance. Like Gherardi and Nicolini (2000), Weick and Roberts also drew on Lave and Wenger's (1991) concepts of legitimate peripheral participation and learning in communities of practice. They focused on the connections among the behaviors of individuals working together as an interdependent system to create a pattern of joint action (Weick & Roberts, 1993 p. 360):

> Our focus is at once on the individuals and the collective, since only individuals can contribute to a collective mind, but a collective mind is distinct from an individual mind because it inheres in the pattern of interrelated activities among many people.

The intelligent, purposeful, and careful combination of collective behaviors constitutes "heedful interrelating" (Weick & Roberts, 1993, pp. 361, 364). The more developed the heedful interrelating among the members of a group, the greater the capacity to deal with nonroutine events. As a corollary, if the activities of contributing, representing, or subordinating are carried out carelessly, then adverse results can occur (Weick & Roberts, 1993, p. 375). The Winnipeg and Bristol inquiries showed that the care teams had not been able to achieve this level of functioning and communication, thus impairing their ability to recover when problems arose with the cases during surgery and post-operatively (Kennedy, 2001, p. 214; Sinclair, 2000, pp. 475, 496).

Weick (2001, p. 307) has described reliability as a "dynamic non-event," because organizations must continuously manage and adapt to a changing and uncertain environment while producing a stable outcome, the avoidance of accidents. Weick, Sutcliffe, and Obstfeld (1999) highlight five key characteristics that allow reliability-seeking organizations to achieve this end. Weick (2002) suggests that the same characteristics may be usefully cultivated in hospitals. The first is *preoccupation with failure*. Reliability-seeking organizations anticipate that problems will occur and remain vigilant to the possibilities. They learn as much as possible from near misses and reward staff for reporting them. Second is *reluctance to simplify interpretations*. Because the task environment can be ambiguous and problems ill-structured, such organizations foster diverse viewpoints and interpretations, thus developing "conceptual slack" to avoid blind spots (Schulman, 1993, p. 346). Similarly, Westrum (1992) has suggested that requisite imagination is needed to anticipate problems, while Pidgeon and O'Leary (2000, p. 22) refer to developing "safety imagination." The third is *continuous sensitivity to operations*. Reliability-seeking organizations work on maintaining collective situational awareness, alert to the fact that this can be eroded by work overload and pressures to produce services. The next characteristic is *commitment to resilience*. The organizations provide continuous training so that teams can learn to contain the effects of errors and deal with surprises effectively. By contrast, in Winnipeg the nurses made repeated requests for orientation and practice runs through routines with the new surgeon, but these went unanswered (Sinclair, 2000, p. 132). The last characteristic is *underspecification of structure*. Paradoxically, although aircraft carriers have a clear military hierarchy, they combine this with decentralized decision making and problem solving and so possess the flexibility to link expertise with problems as needed. Once again by contrast, in Bristol responsibilities were delegated to a great degree, but without attendant decision-making powers. The result was that rigid organizational hierarchy was combined with unclear lines of authority and suppression of communications, all of which impaired the ability to solve problems (Kennedy, 2001, p. 201). Similar confusion about responsibilities eroded the situation in Winnipeg (Sinclair, 2000, p. 471).

Weick et al. (1999) have further elaborated the concept of reliability and collective mindfulness, emphasizing the need for ongoing readjustment in the face of unusual events. "Continuous, mindful awareness" means knowing how to keep track of and respond to those variations in results "that generate potential information about capability, vulnerability, and the environment" (Weick et al., 1999, p. 88). "If people are blocked from acting on hazards, it is not long before their 'useless' observations of those hazards are also ignored or denied, and errors cumulate unnoticed" (Weick et al., 1999, p. 90). Rochlin (1999) has pointed out the dangers of stifling or ignoring staff members' observations when what is really needed is active nurturing of early warning systems. A hospital "grapevine" may well carry such information, but the organizational cultures of the hospital may not support its effective use. In Winnipeg, the inquest noted that "managers ignored pertinent information that was brought to their attention, and at best, simply tolerated the bearers of bad news" (Sinclair, 2000, p. 485). When variations and anomalies are ignored or internalized and simply accepted, an organization creates the normalization of deviance, such as that which contributed ultimately to the failure of the Challenger launch (Vaughan, 1996). Weick and Sutcliffe (2003, p. 73) have described the situation that existed in Bristol as a "culture of entrapment," by which they mean "the process by which people get locked into lines of action, subsequently justify those lines of action, and search for confirmation that they are doing what they should be doing." The mindset that prevailed in Bristol and in Winnipeg accepted the learning curve, low numbers of patients, and the complexity of the case mix as explanations for the poor outcomes. Weick and Sutcliffe use a theory of behavioral commitment to explain why that mindset endured for so long.

Rijpma (1997, 2003) has given a critical assessment of the disagreements between Normal Accident Theory and High Reliability Theory researchers, observing that the ongoing debates between the camps have not produced definitive conclusions and may not have been as theoretically productive as might have been expected. Schulman (2002) has questioned the direct applicability of reliability theory derived from high-hazard operations to health care organizations. He points out that medical care is unlike nuclear power or air traffic control and that the reliability challenges differ in complex ways. The immediate impact of a failure is usually limited to a patient (not large numbers of people outside the organization), so that there may be fewer voices calling for change. There is a conflict of goals inherent in the context of limited resources: "If we organize to be more reliable, this must come at some cost to another value—speed, output, or possibly efficiency" (Schulman, 2002, p. 201). Given the demand for medical services, risk of failure at some level seems to have been accepted as inevitable. The Bristol report illustrates this with its description of the National Health Service as having a culture of making-do and muddling through in the hope that things might eventually improve (Kennedy, 2001, p. 4).

Conclusions

The goal of this chapter has been to show how failures in handling information can contribute to the occurrence of adverse events and failures in health care settings. The extensive inquiries into the care of pediatric cardiac surgery patients in Winnipeg and Bristol provided a catalogue of telling examples to illustrate how this happens. In explaining why such failures happen, research has pointed to the roles played by culture, human factors, and systems analysis.

Researchers have suggested that culture is a central influence on how information is or is not used for the purposes of learning and patient safety in health care organizations. As Westrum (2004, p. ii22) defines it, culture is "the patterned way that an organisation responds to its challenges, whether these are explicit (for example, a crisis) or implicit (a latent problem or opportunity)." Organizational and professional cultures can make it difficult to achieve a safe environment with appropriate reporting and use of information about risks and hazards. The values, norms, and assumptions that shape the response to information about problems include, for example, the status of particular disciplines (who holds authority) and norms of individual responsibility and blame. Hierarchical structures of health care disciplines can create silos that undermine communication and teamwork. The traditional emphasis on individual responsibility for adverse events combined with a propensity to "blame and shame" can create a context of fear. In such an environment, the personal costs of admitting mistakes are far greater than the incentives to report errors and mishaps. How an organization responds to information about problems or "bad news" is indicative of its culture; pathological cultures suppress such information and punish the messengers, whereas generative cultures actively encourage and reward such reporting (Hudson, 2003; Westrum, 1992, 2004).

Research in human factors and systems analysis has contributed important insights into the context and genesis of health care failures.

There is growing recognition of the contribution of latent systems factors to clinical mishaps. Individual health care providers are usually the last connection in the chain of events that results in an adverse outcome. They are at the "sharp end" of the system, the last and most visible connection to the patient and the most likely target for fault-finding (Reason, 1995, p. 80). The propensity to blame individuals for issues that should be investigated as failures of the system has made learning from adverse events more difficult. In such circumstances, potential information about systemic causes is often overlooked because investigations of health care failures do not focus on, and gather, the appropriate data.

Based on research into reliability-seeking organizations, recommendations to improve patient safety fall into several categories. At the system level, one common recommendation is to build system-wide capacity for learning and constant monitoring for problems by encouraging confidential reporting and analysis of near misses and incidents. At the local level, another recommendation is to build the capability of teams to be vigilant and questioning in all situations, and resilient and able to respond flexibly to contain problems if and when they occur. Toft and Reynolds's (1994, p. xi) description of "active foresight" provides an articulate synopsis of both the challenges and the goal to be achieved in improving the handling of information:

> By developing systems that feedback information on accident causation it should be possible to help prevent future recurrence of similar disasters. This information feedback requires an appropriate environment—a safety culture— which allows the formation of "active foresight" within an organization. Active foresight has two elements—foresight of conditions and practices that might lead to disaster and active implementation of remedial measures determined from that foresight. The analysis of organizations and people involved in disasters should not be focused on considerations of culpability or censure, but on acquiring information for the feedback process. The lessons of disasters arise at great cost in terms of human distress and damage to the living environment. We owe it to those who have lost their lives, been injured, or suffered loss to draw out the maximum amount of information from those lessons, and apply it to reduce future suffering.

Endnotes

1. The original 1978 book was updated with Nick Pidgeon and published as a second edition in 1997, after Turner's death.

References

Alvesson, M. (1993). *Cultural perspectives on organizations*. Cambridge, UK: Cambridge University Press.

Alvesson, M. (2002). *Understanding organizational culture*. Thousand Oaks, CA: Sage.

Argyris, C., & Schön, D. A. (1996). *Organizational learning II: Theory, method, and practice.* Reading, MA: Addison-Wesley.

Baker, G. R., & Norton, P. (2002). *Patient safety and healthcare error in the Canadian healthcare system: A systematic review and analysis of leading practices in Canada with reference to key initiatives elsewhere.* Ottawa: Health Canada. Retrieved July 20, 2004, from http://www.hc-sc.gc.ca/english/search/a-z/p.html

Baker, G. R., Norton, P. G., Flintoft, V., Blais, R., Brown, A., Cox, J., et al. (2004). The Canadian Adverse Events Study: The incidence of adverse events among hospital patients in Canada. *Canadian Medical Association Journal, 170*(11), 1678–1686.

Barach, P., & Small, S. D. (2000). Reporting and preventing medical mishaps: Lessons from non-medical near miss reporting systems. *British Medical Journal, 320*(7237), 759–763.

Battles, J. B., & Lilford, R. J. (2003). Organizing patient safety research to identify risks and hazards. *Quality & Safety in Health Care, 12*, ii2–ii7.

Berwick, D. M. (1998a). Crossing the boundary: Changing mental models in the service of improvement. *International Journal for Quality in Health Care, 10*(5), 435–441.

Berwick, D. M. (1998b, November). *Taking action to improve safety: How to increase the odds of success.* Paper presented at the Enhancing Patient Safety and Reducing Errors in Health Care Conference, Annenberg Center for Health Sciences at Eisenhower, Rancho Mirage, CA.

Billings, C. (1998). Incident reporting systems in medicine and experience with the aviation safety reporting system. In R. I. Cook, D. D. Woods, & C. Miller (Eds.), *Tale of two stories: Contrasting views of patient safety.* Report from a workshop on Assembling the Scientific Basis for Progress on Patient Safety (pp. 52–61, Appendix B). Chicago: National Patient Safety Foundation. Retrieved July 20, 2004, from http://www.npsf.org/exec/toc.html

Bloor, G., & Dawson, P. (1994). Understanding professional culture in organizational context. *Organization Studies, 15*(2), 275–295.

Bogner, M. S. (Ed.). (1994). *Human error in medicine.* Hillsdale, NJ: Erlbaum.

Bosk, C. L. (1979). *Forgive and remember: Managing medical failure.* Chicago: University of Chicago Press.

Brennan, T. A., Leape, L. L., Laird, N. M., Hébert, L., Localio, A. R., Lawthers, A. G., et al., (1991). Incidence of adverse events and negligence in hospitalized patients: Results of the Harvard Medical Practice Study I. *New England Journal of Medicine, 324*, 370–376.

Brown, A. D. (2000). Making sense of inquiry sensemaking. *Journal of Management Studies, 37*(1), 45–75.

Brown, A. D. (2004). Authoritative sensemaking in a public inquiry report. *Organization Studies, 25*(1), 95–112.

Brown, A. D., & Starkey, K. (1994). The effect of organizational culture on communication and information. *Journal of Management Studies, 31*(6), 807–828.

Carroll, J. S. (1998). Organizational learning activities in high-hazard industries: The logics underlying self-analysis. *Journal of Management Studies, 35*(6), 699–717.

Ceci, C. (2004). Nursing, knowledge and power: A case analysis. *Social Science & Medicine, 59*, 1879–1889.

Checkland, P. (1999). *Systems thinking, systems practice.* Chichester, UK: Wiley.

Choo, C. W. (1998). *The knowing organization.* Oxford, UK: Oxford University Press.

Choo, C. W. (2002). *Information management for the intelligent organization: The art of scanning the environment* (3rd ed.). Medford, NJ: Information Today.

Clarke, L., & Perrow, C. (1996). Prosaic organizational failure. *American Behavioral Scientist, 39*(8), 1040–1056.

Cohen, M. R. (2000). Why error reporting systems should be voluntary: They provide better information for reducing errors. *British Medical Journal, 320*(7237), 728–729.

Cook, R. I., Woods, D. D., & Miller, C. (1998). *Tale of two stories: Contrasting views of patient safety.* Chicago: National Patient Safety Foundation. Retrieved on July 20, 2004, from http://www.npsf.org/exec/toc.html

Davenport, E., & Hall, H. (2002). Organizational knowledge and communities of practice. *Annual Review of Information Science and Technology, 36*, 171–227.

Davenport, T. H. (1997). *Information ecology: Mastering the information and knowledge environment.* New York: Oxford University Press.

Davies, H. T. O., Nutley, S. M., & Mannion, R. (2000). Organisational culture and quality of health care. *Quality in Health Care, 9*, 111–119.

Davies, J. M., Hébert, P., & Hoffman, C. (2003). *The Canadian patient safety dictionary.* Ottawa: The Royal College of Physicians and Surgeons of Canada.

Denison, D. R. (1996). What is the difference between organizational culture and organizational climate? A native's point of view on a decade of paradigm wars. *Academy of Management Review, 21*(3), 619–654.

Douglas, M. (1992). *Risk and blame: Essays in cultural theory.* New York: Routledge.

Edmondson, A. C. (1996). Learning from mistakes is easier said than done: Group and organizational influences on the detection and correction of human error. *Journal of Applied Behavioral Science, 32*(1), 5–28.

Feldman, S. E., & Roblin, D. W. (1997). Medical accidents in hospital care: Applications of failure analysis to hospital quality appraisal. *Joint Commission Journal on Quality Improvement, 23*(11), 567–580.

Fisher, S. R., & White, M. A. (2000). Downsizing in a learning organization: Are there hidden costs? *Academy of Management Review, 25*(1), 244–251.

Flin, R., Mearns, K., O'Connor, P., & Bryden, R. (2000). Measuring safety climate: Identifying the common features. *Safety Science, 34*(1–3), 177–192.

Gaba, D. M. (2000). Structural and organizational issues in patient safety: A comparison of health care to other high-hazard industries. *California Management Review, 43*(1), 83–102.

Gephart, R. P. (1984). Making sense of organizationally based environmental disasters. *Journal of Management, 10*, 205–225.

Gephart, R. P. (1992). Sense making, communicative distortion and the logic of public inquiry legitimation. *Industrial and Environmental Crisis Quarterly, 6*, 115–135.

Gherardi, S., & Nicolini, D. (2000). The organizational learning of safety in communities of practice. *Journal of Management Inquiry, 9*(1), 7–18.

Gherardi, S., Nicolini, D., & Odella, F. (1998). What do you mean by safety? Conflicting perspectives on accident causation and safety management in a construction firm. *Journal of Contingencies and Crisis Management, 6*(4), 202–213.

Ginman, M. (1987). Information culture and business performance. *IATUL Quarterly, 2*(2), 93–106.

Hale, A. R. (2000). Culture's confusions. *Safety Science, 34*(1–3), 1–14.

Hart, P., Heyse, L., & Boin, A. (2001). New trends in crisis management practice and crisis management research: Setting the agenda. *Journal of Contingencies and Crisis Management, 9*(4), 181–188.

Helmreich, R. L., & Merritt, A. C. (1998). *Culture at work in aviation and medicine: National, organizational, and professional influences.* Brookfield, VT: Ashgate.

Henriksen, K., & Kaplan, H. (2003). Hindsight bias, outcome knowledge and adaptive learning. *Quality & Safety in Health Care, 12*, ii46–ii50.

Hofmann, D. A., & Stetzer, A. (1998). The role of safety climate and communication in accident interpretation: Implications for learning from negative events. *Academy of Management Journal, 41*(6), 644–657.

Hopkins, A. (1999). Counteracting the cultural causes of disaster. *Journal of Contingencies and Crisis Management, 7*(3), 141–149.

Horton, F. W., & Lewis, D. (Eds.). (1991). *Great information disasters.* London: Aslib.

Hudson, P. (2003). Applying the lessons of high risk industries to health care. *Quality & Safety in Health Care, 12*, i7–i12.

Institute of Medicine. (2003). *Patient safety: Achieving a new standard for care.* Washington, DC: Institute of Medicine. Retrieved on July 20, 2004, from http://books.nap.edu/catalog/10863.html

Johnson, C. W. (2003). How will we get the data and what will we do with it then? Issues in the reporting of adverse healthcare events. *Quality & Safety in Health Care, 12*, ii64–ii67.

Kaplan, H., & Barach, P. (2002). Incident reporting: Science or protoscience? Ten years later. *Quality & Safety in Health Care, 11*(2), 144–145.

Kaplan, H. S., & Fastman, B. R. (2003). Organization of event reporting data for sense making and system improvement. *Quality & Safety in Health Care, 12*, ii68–ii72.

Karson, A. S., & Bates, D. W. (1999). Screening for adverse events. *Journal of Evaluation in Clinical Practice, 5*(1), 23–32.

Kennedy, I. (2001). *Learning from Bristol. The report of the public inquiry into children's heart surgery at the Bristol Royal Infirmary 1984–1995.* London: HMSO. Retrieved December 10, 2004, from http://www.bristol-inquiry.org.uk/index.htm

Kohn, L. T., Corrigan, J. M., & Donaldson, M. S. (1999). *To err is human: Building a safer health system.* Washington, DC: Committee on Quality of Health Care in America, Institute of Medicine.

LaPorte, T. R., & Consolini, P. M. (1991). Working in practice but not in theory: Theoretical challenges of "high-reliability organizations." *Journal of Public Administration Research and Theory, 1*, 19–47.

Lave, J., & Wenger, E. (1991). *Situated learning: Legitimate peripheral participation.* Cambridge, UK: Cambridge University Press.

Lawton, R., & Parker, D. (2002). Barriers to incident reporting in a healthcare system. *Quality & Safety in Health Care, 11*(1), 15–18.

Lea, W., Uttley, P., & Vasconcelos, A. C. (1998). Mistakes, misjudgements and mischances: Using SSM to understand the Hillsborough disaster. *International Journal of Information Management, 18*(5), 345–357.

Leape, L. L. (1997). A systems analysis approach to medical error. *Journal of Evaluation in Clinical Practice, 3*(3), 213–222.

Leape, L. L. (1999). Why should we report adverse incidents? *Journal of Evaluation in Clinical Practice, 5*(1), 1–4.

Leape, L. L. (2000). Reporting of medical errors: Time for a reality check. *Quality in Health Care, 9*(3), 144–145.

Leape, L. L., & Berwick, D. M. (2000). Safe health care: Are we up to it? *British Medical Journal, 320*(7237), 725–726.

Leape, L. L., Brennan, T. A., Laird, N., Lawthers, A. G., Localio, A. R., & Barnes, B. A. (1991). The nature of adverse events in hospitalized patients: Results of the Harvard Medical Practice Study II. *New England Journal of Medicine, 324*, 377–384.

Manning, P. K. (1998). Information, socio-technical disasters and politics. *Journal of Contingencies and Crisis Management, 6*(2), 84–87.

Marchand, D. A., Kettinger, W. J., & Rollins, J. D. (2000). Information orientation: People, technology and the bottom line. *Sloan Management Review, 41*(4), 69–80.

Marchand, D. A., Kettinger, W. J., & Rollins, J. D. (2001). *Information orientation: The link to business performance.* Oxford, UK: Oxford University Press.

Martin, J. (1992). *Cultures in organizations: Three perspectives.* Oxford, UK: Oxford University Press.

Martin, J. (2002). *Organizational culture: Mapping the terrain.* Thousand Oaks, CA: Sage.

Mascini, P. (1998). Risky information: Social limits to risk management. *Journal of Contingencies and Crisis Management, 6*(1), 35–44.

Mearns, K. J., & Flin, R. (1999). Assessing the state of organizational safety: Culture or climate? *Current Psychology, 18*(1), 5–17.

National Steering Committee on Patient Safety. (2002). *Building a safer system: A national integrated strategy for improving patient safety in Canadian health care.* Ottawa, ON: The Committee. Retrieved July 20, 2004, from http://rcpsc.medical.org/publications/building_a_safer_system_e.pdf

Nieva, V. F., & Sorra, J. (2003). Safety culture assessment: A tool for improving patient safety in healthcare organizations. *Quality & Safety in Health Care, 12*, ii17–ii23.

Nolan, T. W. (2000). System changes to improve patient safety. *British Medical Journal, 320*(7237), 771–773.

Perrow, C. (1981). The President's Commission and the normal accident. In D. Sills, C. Wolf, & V. Shelanski (Eds.), *The accident at Three Mile Island: The human dimensions* (pp. 173–184). Boulder, CO: Westview Press.

Perrow, C. (1999a). *Normal accidents: Living with high-risk technologies.* Princeton, NJ: Princeton University Press.

Perrow, C. (1999b). Organizing to reduce the vulnerabilities of complexity. *Journal of Contingencies and Crisis Management, 7*(3), 150–155.

Pidgeon, N. (1991). Safety culture and risk management in organizations. *Journal of Cross-Cultural Psychology, 22*(1), 129–140.

Pidgeon, N., & O'Leary, M. (2000). Man-made disasters: Why technology and organizations (sometimes) fail. *Safety Science, 34*(1–3), 15–30.

Rasmussen, J. (1997). Risk management in a dynamic society: A modelling problem. *Safety Science, 27*(2/3), 183–213.

Reason, J. (1990). *Human error*. Cambridge, UK: Cambridge University Press.

Reason, J. (1995). Understanding adverse events: Human factors. *Quality Health Care, 4*(2): 80–89.

Reason, J. (1997). *Managing risks of organizational accidents*. Brookfield, VT: Ashgate.

Reason, J. (1998). Achieving a safe culture: Theory and practice. *Work & Stress, 12*(3), 293–306.

Richter, A., & Koch, C. (2004). Integration, differentiation, and ambiguity in safety cultures. *Safety Science, 42*(8), 703–722.

Rijpma, J. A. (1997). Complexity, tight-coupling, and reliability: Connecting normal accidents theory and high reliability theory. *Journal of Contingencies and Crisis Management, 5*(1), 15–23.

Rijpma, J. A. (2003). From deadlock to dead end: The normal accidents-high reliability debate revisited. *Journal of Contingencies and Crisis Management, 11*(1), 37–45.

Roberts, K. H. (1990). Some characteristics of one type of high reliability organization. *Organization Science, 1*(2), 160–176.

Roberts, K. H. (1993). Cultural characteristics of reliability enhancing organizations. *Journal of Managerial Issues, 5*(2), 165–181.

Rochlin, G. I. (1999). Safe operation as a social construct. *Ergonomics, 42*(11), 1549–1560.

Rosenthal, M. M., & Sutcliffe, K. M. (Eds.). (2002). *Medical error: What do we know? What do we do?* San Francisco: Jossey-Bass.

Sagan, S. D. (1993). *The limits of safety: Organizations, accidents, and nuclear weapons*. Princeton, NJ: Princeton University Press.

Schein, E. H. (1992). *Organizational culture and leadership* (2nd ed.). San Francisco: Jossey-Bass.

Schein, E. H. (1996). Three cultures of management: The key to organizational learning. *Sloan Management Review, 38*(1), 9–20.

Schulman, P. R. (1993). The negotiated order of organizational reliability. *Administration and Society, 25*, 353–372.

Schulman, P. R. (2002). Medical errors: How reliable is reliability theory? In M. M. Rosenthal & K. M. Sutcliffe (Eds.), *Medical error: What do we know? What do we do?* (pp. 200–216). San Francisco: Jossey-Bass.

Schultz, M. (1994). *On studying organizational cultures: Diagnosis and understanding*. Berlin: de Gruyter.

Sexton, J. B., Thomas, E. J., & Helmreich, R. L. (2000). Error, stress, and teamwork in medicine and aviation: Cross sectional surveys. *British Medical Journal, 320*(7237), 745–749.

Sharpe, V. A. (1998, November). *"No tribunal other than his own conscience": Historical reflections on harm and responsibility in medicine*. Paper presented at the Enhancing Patient Safety and Reducing Errors in Health Care Conference, Annenberg Center for Health Sciences at Eisenhower, Rancho Mirage, CA.

Short, J. F., & Rosa, E. A. (1998). Organizations, disasters, risk analysis and risk: Historical and contemporary contexts. *Journal of Contingencies and Crisis Management, 6*(2), 93–96.

Sinclair, C. M. (2000). *Report of the Manitoba paediatric cardiac surgery inquest: An inquiry into twelve deaths at the Winnipeg Health Sciences Centre in 1994.* Winnipeg: Provincial Court of Manitoba. Retrieved December 10, 2004, from http://www.pediatriccardiac inquest.mb.ca

Snook, S. A. (1996). *Practical drift: The friendly fire shootdown over northern Iraq.* Doctoral dissertation, Harvard University. Retrieved July 20, 2004, from ProQuest Digital Dissertations database.

Snook, S. A. (2000). *Friendly fire: The accidental shootdown of US Black Hawks over northern Iraq.* Princeton, NJ: Princeton University Press.

Sophar, G. (1991). $170,000 down the drain: The MRAIS story. In F. W. Horton & D. Lewis (Eds.), *Great information disasters* (pp. 151–159). London: Aslib.

Stanhope, N., Crowley-Murphy, M., Vincent, C., O'Connor, A. M., & Taylor-Adams, S. E. (1999). An evaluation of adverse incident reporting. *Journal of Evaluation in Clinical Practice, 5*(1), 5–12.

Starr, P. (1982). *The social transformation of American medicine: The rise of a sovereign profession and the making of a vast industry.* New York: Basic Books.

Sutcliffe, K. M. (2001). Organizational environments and organizational information processing. In F. M. Jablin & L. L. Putnam (Eds.), *The new handbook of organizational communication* (pp. 197–230). Thousand Oaks, CA: Sage.

Sutcliffe, K. (2004). Defining and classifying medical error: Lessons for learning. *Quality & Safety in Health Care, 13*(1), 8–9.

Tamuz, M., Thomas, E. J., & Franchois, K. E. (2004). Defining and classifying medical error: Lessons for patient safety reporting systems. *Quality & Safety in Health Care, 13*(1), 13–20.

Taylor-Adams, S., Vincent, C., & Stanhope, N. (1999). Applying human factors methods to the investigation and analysis of clinical adverse events. *Safety Science, 31*, 143–159.

Thomas, E. J., & Helmreich, R. L. (2002). Will airline safety models work in medicine? In M. M. Rosenthal & K. M. Sutcliffe (Eds.), *Medical error: What do we know? What do we do?* (pp. 217–234). San Francisco: Jossey-Bass.

Thomas, E. J., & Petersen, L. A. (2003). Measuring errors and adverse events in health care. *Journal of General Internal Medicine, 18*(1), 61–67.

Thomas, E. J., Sexton, J. B., & Helmreich, R. L. (2003). Discrepant attitudes about teamwork among critical care nurses and physicians. *Critical Care Medicine, 31*(3), 956–959.

Thomas, E. J., Studdert, D. M., Burstin, H. R., Orav, E. J., Zeena, T., Williams, E. J., et al. (2000). Incidence and types of adverse events and negligent care in Utah and Colorado. *Medical Care, 38*(3), 261–271.

Toft, B., & Reynolds, S. (1994). *Learning from disasters.* Oxford, UK: Butterworth-Heinemann.

Turner, B. A. (1976). Organizational and interorganizational development of disasters. *Administrative Science Quarterly, 21*(3), 378–397.

Turner, B. A. (1978). *Man-made disasters.* London: Wykeham Science Press.

Turner, B. A. (1991). The development of a safety culture. *Chemistry and Industry, 1*(7), 241–243.

Turner, B. A., & Pidgeon, N. F. (1997). *Man-made disasters* (2nd ed.). Oxford, UK: Butterworth-Heinemann.

Van Maanen, J., & Barley, S. R. (1984). Occupational communities: Culture and control in organizations. *Research in Organizational Behavior, 6,* 287–365.

van Vuuren, W. (1999). Organisational failure: Lessons from industry applied in the medical domain. *Safety Science, 33*(1–2), 13–29.

van Vuuren, W. (2000). Cultural influences on risks and risk management: Six case studies. *Safety Science, 34*(1–3), 31–45.

Vaughan, D. (1996). *The Challenger launch decision: Risky technology, culture, and deviance at NASA.* Chicago: University of Chicago Press.

Vaughan, D. (1999). The dark side of organizations: Mistake, misconduct, and disaster. *Annual Review of Sociology, 25,* 271–305.

Vicente, K. (2004). *The human factor: Revolutionizing the way people live with technology.* Toronto, ON: Knopf.

Vincent, C., Neale, G., & Woloshynowych, M. (2001). Adverse events in British hospitals: Preliminary retrospective record review. *British Medical Journal, 322,* 517–519.

Vincent, C., Stanhope, N., & Crowley-Murphy, M. (1999). Reasons for not reporting adverse incidents: An empirical study. *Journal of Evaluation in Clinical Practice, 5*(1), 13–21.

Wald, H., & Shojania, K. G. (2001). Incident reporting. In R. M. Wachter (Ed.), *Making health care safer: A critical analysis of patient safety practices* (Evidence Report/Technology Assessment AHRQ Publication 01–E058). Rockville, MD: Agency for Healthcare Research and Quality. Retrieved July 20, 2004, from http://www.ahrq.gov/clinic/ptsafety

Walshe, K., & Rundall, T. G. (2001). Evidence-based management: From theory to practice in health care. *Milbank Quarterly, 79*(3), 429–457.

Walshe, K., & Shortell, S. M. (2004). When things go wrong: How health care organizations deal with major failures. *Health Affairs, 23*(3), 103–111.

Weick, K. E. (1995). *Sensemaking in organizations.* Thousand Oaks, CA: Sage.

Weick, K. E. (1998). Foresights of failure: An appreciation of Barry Turner. *Journal of Contingencies and Crisis Management, 6*(2), 72–75.

Weick, K. E. (2001). *Making sense of the organization.* Oxford: Blackwell.

Weick, K. E. (2002). The reduction of medical errors through mindful interdependence. In M. M. Rosenthal & K. M. Sutcliffe (Eds.), *Medical error: What do we know? What do we do?* (pp. 177–199). San Francisco: Jossey-Bass.

Weick, K. E., & Roberts, K. H. (1993). Collective mind in organizations: Heedful interrelating on flight decks. *Administrative Science Quarterly, 38,* 357–381.

Weick, K. E., & Sutcliffe, K. M. (2003). Hospitals as cultures of entrapment: A re-analysis of the Bristol Royal Infirmary. *California Management Review, 45*(2), 73–84.

Weick, K. E., Sutcliffe, K. M., & Obstfeld, D. (1999). Organizing for high reliability: Processes of collective mindfulness. *Research in Organizational Behavior, 21,* 81–123.

Weingart, S. N., Ship, A. N., & Aronson, M. D. (2000). Confidential clinician-reported surveillance of adverse events among medical inpatients. *Journal of General Internal Medicine, 15*(7), 470–477.

West, E. (2000). Organisational sources of safety and danger: Sociological contributions to the study of adverse events. *Quality in Health Care, 9,* 120–126.

West, E., Barron, D. N., Dowsett, J., & Newton, J. N. (1999). Hierarchies and cliques in the social networks of health care professionals: Implications for the design of dissemination strategies. *Social Science & Medicine, 48,* 633–646.

Westrum, R. (1987). Management strategies and information failure. In J. A. Wise & A. Debons (Eds.), *Information systems: Failure analysis* (Vol. F32, pp. 109–127). Berlin: Springer-Verlag.

Westrum, R. (1992). Cultures with requisite imagination. In J. A. Wise, V. D. Hopkin, & P. Stager (Eds.), *Verification and validation of complex systems: Human factors issues* (Vol. 110, pp. 401–416). Berlin: Springer-Verlag.

Westrum, R. (2004). A typology of organisational cultures. *Quality and Safety in Health Care, 13*(Suppl. II), ii22–ii27.

Wicks, D. (2001). Institutionalized mindsets of invulnerability: Differentiated institutional fields and the antecedents of organizational crisis. *Organization Studies, 22*(4), 659–692.

Wild, D., & Bradley, E. H. (2005). The gap between nurses and residents in a community hospital's error-reporting system. *Joint Commission Journal on Quality and Patient Safety, 31*(1), 13–20.

CHAPTER 10

Workplace Studies and Technological Change

Angela Cora Garcia
Mark E. Dawes
Mary Lou Kohne
Felicia M. Miller
Stephan F. Groschwitz
University of Cincinnati

Introduction

> Massive, multimillion dollar debacles are not just isolated
> occurrences in IT. They are the norm. (Hugos, 2003, p. 54)

Michael Hugos, Chief Information Officer of Network Services
Company, laments in a *Computerworld* article that the typical failure
rate for IT projects is 70 percent, estimating that some "30 percent fail
outright, and another 40 percent drag on for years, propped up by huge
cash infusions until they are finally shut down" (Hugos, 2003, p. 54).
Headlines in daily newspapers and results of *Google* news searches
reveal a continual stream of failed IT projects, many of them dependent
upon high technology. One recent and salient example is the blackout of
2003, when 50 million households and businesses in the U.S. and
Canada lost power for up to seven days depending on their geographic
location (U.S.-Canada Power System Outage Taskforce, 2004; cited in
Fonseca, 2004). According to the Task Force report, faulty software and
control room procedures led to an alarm system failure and blackout,
which ultimately spread to affect a broader power grid: "The worst
power failure in U.S. history could have been avoided in part by better
business contingency planning and IT management" (Fonseca, 2004, p.
35). Apparently "inadequate situational awareness" at *First Energy* in
Akron, Ohio, contributed to the causes of the blackout (U.S.-Canada
Power System Outage Taskforce, 2004; cited in Fonseca, 2004, p. 35).

In the business world, one of the most important strategic investments that companies can make is in technology. Such purchases are
generally made with the goal of improving work practices, reducing

staff, or solving specific business problems. But the technology may not have been designed in the most effective way possible, or even if it has, wrong choices may have been made as to how to integrate it with the workforce that it was designed to support. Rather than merely accepting high rates of failures as inevitable, what can technology managers and planners do differently to anticipate and prevent failures? Insight into how people actually use technology to accomplish everyday work should help answer these questions. New research methods may help to deepen our understanding of the appropriate role of technology in the workplace. They can shed light on how problems can be prevented and provide valuable insights into what factors are of significance for those who design and implement new systems.

A branch of social science research is emerging called Workplace Studies (henceforth, "WPS"), which we believe has the potential to substantially improve the choices that companies make about the purchase, design, and implementation of technology. In this chapter, we briefly define WPS research, explain its history and methods, and show how specific WPS research findings can inform a business's decision-making process for solving typical technology-related problems. The two issues we address are the facilitation of collaboration among employees, whether co-present or geographically dispersed, and the underutilization of employees, which can lead to hidden financial losses or to excessive employee turnover. We then discuss several related research approaches to the study of technology in the workplace (requirements engineering, computer-supported cooperative work [CSCW], and social informatics), and discuss how they are similar to and different from WPS. We conclude with a discussion of how WPS can contribute to system design and how organizations can apply WPS research findings and methods to their own business problems.

What Is Workplace Studies?

WPS is a subfield of sociology that places the *social organization of work* at the center of its research agenda. WPS is interested in how "technologies, ranging from complex systems through to mundane tools, feature in the practical accomplishment of organizational activities" (Heath & Luff, 2000, p. 8).[1] At issue here is how new technologies, including computers, networks, software, and digital communications, are penetrating and fundamentally changing the organizational and interactional character of work in contemporary society. WPS researchers believe that direct examination of the social process of work is essential for understanding the problems that result in mismatches between technology and users.

WPS researchers study how people use "high-tech" tools such as expert systems (Whalen & Vinkhuyzen, 2000), control centers (Heath & Luff, 2000), graphic design software (Heath & Luff, 2000), and videophones (Heath & Luff, 2000) and "low-tech" tools such as paper (Gaver,

1991; Sellen & Harper, 2003) and whiteboards (Xiao, Lasome, Moss, Mackenzie, & Faraj, 2001; in conjunction with spatial arrangement and other artifacts (Kawatoko, 1999) and procedures for filing and coding documents (Suchman, 2000). This research, which involves direct observation, analysis of videotapes of the work process, analysis of documents, and interviews with workers (Luff, Hindmarsh, & Heath, 2000; Heath & Luff, 2000), contributes to our understanding of how users exploit "technology to inform the production and coordination of their actions and activities" (Luff et al., 2000, p. xiii). WPS research enhanced our understanding of workplace dynamics in such varied organizations as the International Monetary Fund (IMF) (Harper, 2000a; Sellen & Harper, 2003), the London Underground (Heath & Luff, 2000), the Reuters news room (Heath & Luff, 2000), and software development projects at Xerox PARC (Suchman, Blomberg, Orr, & Trigg, 1999). It has also been instructive for software development in such applications as air traffic control (Harper & Hughes, 1993), institutional projects such as bridge building and architecture (Suchman, Trigg, & Blomberg, 2002), and systemic health care delivery in Helsinki University Central Hospital (Engeström, 2000).

WPS researchers reject the approach of examining the technology in isolation or only in conjunction with a single worker or "tester"; rather, they study human interactions with technology in their natural workplace settings. WPS argues that work is inherently a social process and shows how it is done collaboratively. The focus is on the process of doing the work required in the organization with the help of technology (Luff et al., 2000). This approach situates work at the center of the analytic agenda and illuminates the critical social processes that surround and embed it.

In addition to studying the doing of work, WPS research also addresses the technical design process. Designers tend to assume that the introduction of complex systems can improve efficiency; however, these goals are often not achieved (Heath & Luff, 2000, p. 8). Conventional approaches to technical design tend to underestimate the importance of the "socially organized competencies and reasoning on which persons rely in using technologies as part of their daily work" (Heath & Luff, 2000, p. 8). Consequently, system designers often miss important aspects of *how* the work is done that are crucial for fitting and integrating technology into the work processes for which it is designed. Heath and Luff contend that these disconnects are root causes for the high failure rates in expert systems software. In short, social features of work should be the crux of the challenge for designers.

The Origins of Workplace Studies

WPS researchers view the development of workplace studies as a response to system failures in HCI (human–computer interaction) systems, CSCW, group decision support systems (GDSS), and artificial intelligence (AI) systems, areas that had been expected to produce improvements in organizational efficiency and productivity (Luff et al., 2000). According to Heath and Luff (2000, pp. 8–9) WPS was first inspired by broad critiques of the design of AI systems and the cognitive science used to develop them by researchers such as Dreyfus (1972), Coulter (1979), and Searle (1980). These researchers questioned the notion that computers could simulate human intelligence.

However, Suchman's (1987) book, *Plans and Situated Actions*, may have been the catalyst for the development of the WPS approach. According to Heath and Luff (2000; see also Arminen, 2001; Heath, Luff, & Svensson, 2003), Suchman (1987) provided a new perspective on human–machine interaction and communication by suggesting that the processes of human intelligence that designers were attempting to reproduce were likely not as formally ordered and logical as previously thought. Using a comparison of western navigational practices that reflected a focus on plans, logical order, and execution with those of Micronesian seamen who were more likely to navigate by ad hoc means, Suchman argued that successful accomplishment of tasks and activities could not proceed without accounting for the *situational* character of action. While humans clearly approach tasks with goals in mind, they are invariably confronted with situational conditions and events that require improvisational responses. For example, Whalen, Whalen, and Henderson (2002) found that although there are specific goals and tasks associated with the job of sales representatives working in a call center, improvisation was an integral part of the work.

Thomas and Kellogg (1989; cited in Bannon, 2000) broadened the critique of conventional cognitive psychological approaches to HCI research, finding several "gaps" between the laboratory setting and actual work situations. They identified "ecological gaps" that resulted from research done in laboratory environments, "user gaps" based on individual differences and motivations, "task gaps" where the laboratory is unable to faithfully reproduce the actual work situation, "problem formulation gaps," "artifact gaps," "extensionality gaps," and "work context gaps" (Thomas & Kellogg, 1989; cited in Bannon, 2000). Luff et al. (2000) show that "this has led to an increased interest in field study methods as distinct from laboratory studies, and to the rise of a new field—namely, CSCW, where explicit attention is paid to the sociality of the workplace" (Bannon, 2000, p. 233).

An early collection of WPS research edited by Graham Button (1993) illustrates that this perspective provides a coherent approach to studying workplace interaction, especially where the use of technology (broadly defined) is involved. Although some of the specific technologies studied are understandably out of date at this point (e.g., the use of paper slips for air traffic controllers to keep track of planes [Harper & Hughes, 1993], the workings of early computerized ordering and billing systems [Button & Harper, 1993], and the effectiveness—or otherwise—of systems providing automatic interpretation of electrocardiograms

[Hartland, 1993]), this book clearly demonstrates the power and consistency of the WPS approach. For example, Button and Harper's (1993) chapter on computerized ordering and billing systems declines to study the systems in isolation or only in conjunction with the employees directly involved in its operation; instead, they examine how ordering and billing are integrated into both the manufacturing process and the determination of prices for the goods sold. They also consider how customer expectations for immediate delivery factor into the overall process. The success or failure of an ordering system is dependent on its fit with these other aspects of the business.

WPS researchers point to the two primary paradigmatic assumptions made by system designers that continue to result in the problems identified by WPS research. First, some designs are developed with an ideal end state in mind, which the designers assume workers will adopt, but which they do not because of inconsistencies *with the way the work is done*. This, in turn, leads to the second problem, namely, designs that do not acknowledge the intricacy and nuances of the mundane practices and methods used in the workplace. At the crux of both are the social processes upon which the accomplishment of the work is based; these are glossed over by designers and missing in the end product, rendering it of questionable value to "users" (Luff et al., 2000).

Theoretical Perspectives and Methods

WPS is a sociotechnical research perspective that "takes the social and situational seriously, and which drives attention toward the way people use technologies to accomplish and coordinate their day-to-day practical activities" (Luff et al., 2000, p. 12). Technology lies at the center of its research agenda. Heath and Luff (2000, p. 6) contrast this with sociology's more conventional theoretical interests in the "ways in which communication and information systems influence and are influenced by such aspects as the division of labor, work-force skilling and deskilling, occupational structure and associated features such as power, job opportunity, and unionization."

Although there is considerable research on the use of technology in the workplace, much of it incorporating qualitative methods such as ethnography, detailed interviews, observation, or diary logs, we feel it is worth distinguishing between this broad category of research and the narrower subset of work that falls more closely under the penumbra of the ethnomethodological paradigm. For example, a study of communication across boundaries (Hinds & Kiesler, 1999) uses surveys, detailed interviews, and diary logs to collect information about the use of communication technology (telephone, e-mail, and voice mail) across several levels of management and employees. Although this type of study allows researchers to test hypotheses about the types of technology used by different categories of employees, it does not provide direct observations of people using the technology and so cannot provide the type of information gained by studies involving direct observation of the work being done.

Theoretical perspectives used by WPS researchers include *distributed cognition*, which recognizes the collaborative nature of knowledge development; *symbolic interactionism*, which recognizes that shared definitions and understandings form the basis of collaboration and action in the workplace; *course of action studies*, which explicate the use of technologies within the courses of action in which they are embedded; *activity theory*, which bridges the detailed nuances of microsociological research with macro-level organizational change; and most commonly, *ethnomethodology* and *conversation analysis*, which provide a method and analytical rationale for studying situated human action directly in its moment-by-moment emergence (Heath & Luff, 2000).

The investigative and data analysis methods draw upon a variety of perspectives (see Arminen, 2001; Heath & Luff, 2000; and Luff et al., 2000 for summaries of these). Two of the main perspectives in use are ethnomethodology and conversation analysis. Ethnomethodology (Garfinkel, 1967) is a qualitative sociological perspective that focuses attention on how members of a culture, community, or organization accomplish the social organization of that setting. It encourages analyzing the procedures used to do work in order to understand how participants in that setting collaboratively construct the work setting and perform the goals of that setting. Conversation analysis (e.g., Heritage, 1984) is a branch of ethnomethodology that focuses on understanding how interaction is organized and how people use speech to accomplish tasks.

Ethnomethodology

Ethnomethodology and conversation analysis draw on Harold Garfinkel's (1967) theories of social action. Garfinkel (1967, p. 11) defined ethnomethodology as "the investigation of the rational properties of indexical expressions and other practical actions as contingent ongoing accomplishments of organized artful practices of everyday life." Indexical expressions have the property of being interpreted sensibly only where and when they are heard (Garfinkel, 2002). For example, exactly where the words "here" and "there" refer to can be understood only in the context of a particular interaction or text in which the context of their use identifies their referents. Garfinkel (2002) points out that this property of indexicality applies to the performance and interpretation of all social action. Human action can only be performed and understood in the context of the stream of behavior in which it occurs. Consequently, ethnomethodology focuses on studying action, interaction, and communication as locally and situationally constructed activity (Garfinkel, 2002).

Ethnomethodology's principal observational technique is derived from Karl Mannheim's (1952) "documentary method of interpretation." According to Garfinkel (1967, p. 78):

> The method consists of treating an actual appearance as "the document of," as "pointing to," as "standing on behalf of," a presupposed underlying pattern. Not only is the underlying pattern derived from its individual documentary evidences, but the individual documentary evidences, in their turn, are interpreted on the basis of "what is known" about the underlying pattern. Each is used to elaborate the other.

Garfinkel (1967) describes the documentary method of interpretation as a method individuals use to engage in and interpret human action. The documentary method of interpretation looks at interaction as patterned accounts drawn from the common experience of the participants.

The documentary method of interpretation can serve the research process in two ways. First, it provides a useful way to frame social interaction. Individuals' utterances and other actions can be perceived as "documentary evidence" in interaction and studied as "documents" in research. They represent and rely on "underlying patterns" of common knowledge and experience. Second, many devices in the participant's environment can also be perceived as "documents" in the traditional sense and as "documents" in the ethnomethodological sense. There have been several ethnomethodological ethnographic studies of how work is done in organizations that illustrate the "documentary method of interpretation." For example, Meehan (1986) conducted an ethnographic, ethnomethodological study of "police handling" of juveniles in two adjoining suburban police departments. The aim of this study was to explore the relationship between police record keeping and the work of policing juveniles. As Meehan (1986) penetrated record-keeping practices, he found formal, informal, and personal records, and an unwritten file that he described as the "running record." The "running record" included a "particularly rich stock of local work-relevant knowledge that [was] essential and often utilized as a source of information" (Meehan, 1986, p. 74). It was composed of the "mental dossier[s]" of fellow police officers in the "police subculture." Regarded as "factual" and "authoritative," this running record was shared among fellow officers. Meehan observed that, although the running record was different in form from formal written records, it nonetheless functioned as an "important source in police decision making" (Meehan, 1986, p. 74).

Meehan (1986) found that police officers shared information with other agencies, but some was deliberately withheld. Some informal and personal information about cases was also recorded using methods deliberately designed to render it useless to outsiders, should the officer's records be called into formal proceedings. He also found occasions when police officers manipulated the selection and recording of charges to manage the advantages and disadvantages of "contractual" relationships with other outside government agencies, to reduce embarrassment for local citizens, to protect themselves, or to increase the likelihood of a conviction. Meehan found rational explanations for these activities and characterized them as necessary to enable police officers to accomplish the professional administration of a public safety service, manage personal career issues, and maintain effective and sensitive relationships with local citizens.

Zimmerman's (1969) study of a social welfare agency found rational explanations for training programs that caused social welfare employees to disregard their preexisting tendencies to support and help citizens in favor of professional skepticism and distrust. The conventional interpretation of this attitude centers on the impersonality of bureaucracies. However, Zimmerman found that the primary work activity of welfare workers was not providing aid, but rather determining who was qualified to receive aid by examining and "proving" the legitimacy of their documents and other evidence. This perspective enabled the welfare workers to increase their professional effectiveness at executing their roles as intake agents in a public welfare agency. Acting as if one distrusted clients was necessary to do the work of separating legitimate clients from dishonest ones. Both Meehan's and Zimmerman's studies illustrate how participants use the documentary method of interpretation—treating documents, actions, and utterances as indicating underlying patterns, and using these patterns to interpret and shape their actions and their understanding of their co-interactants.

Garfinkel's concept of "reflexivity" refers to the idea that actions (and hence, meaning) are reflexively produced by participants in conjunction with the interactional and social context within which they occur. For example, participants in a classroom both *create* the social setting by their actions and *shape* their behavior to fit the setting. Students who refrain from speaking until called upon by the instructor are displaying an orientation to the social context within which they are interacting. However, it is also the case that their actions (in speaking or not speaking at certain times) are what makes it a class. The same individuals in the same room, talking informally with each other in a conversational pattern would produce an interaction that is not "classroom interaction." The participants' actions (e.g., a student is silent until asked to speak by a teacher) are *documents* that others can interpret as indicating an *underlying pattern*: They are behaving in this way because they are students participating in a class. Just as participants in a setting can use and orient to these "documents" in order to construct and interpret action, so can the observer or analyst. Thus, ethnomethodology focuses researchers' attention on the fact that human action can be readily analyzed and interpreted by observing it directly in its naturally occurring context.

We believe that the ethnomethodological notion of "lived work" has become an essential element of the most effective WPS research. This concept embodies a way of conceptualizing, investigating, analyzing, and explaining work that reveals aspects that might otherwise be missed by conventional technology research and design processes. These unaccounted-for actions often emerge in postmortem reviews of system failures as the unanticipated consequences of design processes that do not or cannot capture the mission-critical detail and nuance of the "lived experience of work."

Garfinkel, Lynch, and Livingston (1981) have demonstrated the concept of "lived work" via an analysis of the discovery of a pulsar by astronomers. They sought to shift the sociological agenda away from its conventional concerns by placing social *processes* instead of social *outcomes* at its center. In essence, their task was to identify the characteristics of work that make it recognizable and accomplishable as astronomy (or as any other type of work, for example, railway engineering, air traffic controlling, or more effective integrated group activity via software such as Lotus Notes) (Garfinkel et al., 1981, p. 133). Garfinkel et al.'s analysis of the discovery process in astronomy involved studying and comparing a variety of different sources of information. They examined audiotape recordings that were made as the pulsar was discovered, and examined the notes, documents, and test results produced by the scientists during and after the discovery process, as well as the scientists' published reports of their discovery. A review of these materials highlighted the differences between them. This enabled Garfinkel et al. to compare post hoc accounts with the "lived accounting" of the discovery made possible by the tape recordings and thus to uncover aspects of the work that had previously gone unnoticed.

Garfinkel et al. (1981, p. 134) concluded that the "lived work" of discovering the pulsar has three principle "relevancies:" (1) an unaffected and unbiased quality that captures the action as a "first-time-through" event; (2) an orderliness embodied in the work's "local historicity;" and (3) a "quiddity" or essential nature that differentiates it from other methods of capturing the work as an observable and accountable orderly activity. Research tools such as audio- and videotape, which capture work in its rawest or "first-time-through" form, provide insight that is least affected by the filters that are a natural outcome of subjects' and researchers' evaluations. Recognizing that work has a "local historicity" acknowledges that work's "lived orderliness" or natural processual order is composed of a sequence of "situated" actions that occur *in real time.* "Shop talk" and "shop work," which are natural forms used by the workers themselves, most accurately identify and document the action taking place. The "quiddity" or essential nature of work reflects its real-time character; its in situ constitution as characterized by given situations and contexts; its local production, which is reliant on the people and artifacts present at the time; and its natural accountability which is reflected, for example, in the telling or observing as it happens in a tape or video recording.

Focusing on capturing a task's essential "lived work" can reveal aspects of work that are unaccounted for by conventional research methods such as focus groups, participant observation, self-reports, interviews, ethnographic accounts, or subject-produced documents (Garfinkel et al., 1981). It is the WPS focus on capturing the "lived work" in projects such as the investigation of the London Underground (Heath & Luff, 2000), the IMF (Harper, 2000a; Sellen & Harper, 2003), the Reuters news room (Heath & Luff, 2000), and software development projects at Xerox PARC (Suchman et al., 1999) that enhances the research product by more comprehensively capturing the taken-for-granted, but inescapably present components of work.

Conversation Analysis

Conversation analysis is a qualitative approach to the analysis of talk in interaction that has its origins in ethnomethodology (Prevignano & Thibault, 2003). The conversation-analytic approach (e.g., Heritage, 1984; ten Have, 1999) is well suited to studying the process of interaction. The goal of conversation analysis is to discover the commonsense understandings and procedures that people use to shape their conduct in particular interactional settings (Heritage, 1984, 1987; Sacks, 1984; Schegloff & Sacks, 1973; West & Zimmerman, 1982). Members' shared interactional competencies not only enable them to produce their own actions but also to interpret the actions of others. Because participants display their orientation to the procedures they use in the utterances they produce (see also Heritage & Atkinson, 1984; Schegloff & Sacks, 1973), analysts are able to discover conversational procedures by analyzing the talk itself (Schegloff & Sacks, 1973). The conversation is assumed to be a context with regard to which participants shape their utterances and interpret the utterances of others (Goodwin & Duranti, 1992; Heritage, 1987). Thus the sequential context—the immediately prior utterances and previous utterances in the conversation—as well as the interactional context and the physical and temporal contexts are all assumed to be potentially relevant to the participants as they structure their talk (Heritage & Atkinson, 1984). Bowers, Pycock, and O'Brien's (1996) study of interaction in a Collaborative Virtual Environment can serve as an example of the use of conversation analysis in WPS research.

WPS researchers use a combination of methods to collect and analyze data, but do not rely on survey or interview data alone. Although surveys can capture people's ideas and feelings about their work practices, they do not capture the social process of doing the work, much of which is done without our conscious awareness or attention to the details of how we're doing it. Instead, WPS researchers prefer to use naturally occurring data sources (real people working in real jobs) rather than laboratory or "test case" studies, because they find that the doing of the work is influenced by the physical, social, and organizational contexts that exist in the actual workplace. Given that the effects of these factors cannot always be predicted, the best data come from real work settings. Thus the main methods of data collection used by WPS researchers include direct observation, videotapes and audiotapes, computer records of transactions, documents or artifacts in use in the setting, and interviews with workers about their experiences doing the work. WPS researchers thus "set out to convey a sense of the real-world, real-time nature of the work as it actually takes place, rather than some idealized version of events" (Rouncefield, Hughes, Rodden, & Villers, 1994; as cited in Crabtree, Nichols, O'Brien, Rouncefield, & Twidale, 2000, p. 669).

Ethnographic methods are clearly recognized as a mainstay of WPS research (e.g., Plowman, Rogers, & Ramage, 1995), but what is not often appreciated is the centrality of videotaped and audiotaped data for capturing the actual process of the work. There is a very different level of detail and precision in those studies that use videotaped data to capture the "lived work"—the apparently unconscious aspects of actually doing tasks that workers are frequently unable to remember or recount in interviews because the tasks have become second nature or taken for granted (see Garfinkel et al., 1981). For example, how many of us can actually recount the detailed procedures for riding a bicycle? Could you do so in a fashion that is usable by system designers for writing software to automate the process of bike riding? Videotaped data can provide a detailed record of the actions involved (both physical and social) in performing the work. The work of Zuboff (2002) can serve as an example of why interviews alone may not always provide adequate data for understanding how work is done in a particular setting. Specifically, Zuboff (2002) described a situation where process engineers in a paper mill attempted to computerize aspects of the production process. They interviewed an employee about the tasks required to operate a paper machine. The employee remembered pulling one lever; however, when the program did not work, they observed the worker performing the task and discovered that the employee actually pulled *two levers* simultaneously (Zuboff, 2002, pp. 104–105). This is not a situation where employees resisted the design process, but a situation where the "lived work" of the task had become second nature and lost its distinctiveness in the retelling.

More specifically, it is the use of videotaped data that can provide the basis for the use of conversation-analytic techniques in conjunction with an ethnographic analysis of the recorded interaction to produce a sequential analysis of the entire interaction—the participants' speech, physical actions, and actions with/of the technology being used in the workplace. We believe that the detail and precision of this approach, combined with ethnomethodological attention to lived work and situated action, accounts for the distinctive "edge" WPS provides over traditional ethnographic and interview methods. For example, in an analysis of videotaped conversations, Whittaker, Frohlich, and Daly-Jones (1994) were able to show how informal interactions and visual contacts interrelated to provide opportunities for collaboration.

Workplace studies are time-consuming and labor-intensive. Although we recognize that many organizations do not have the time to conduct such extensive research in their own settings, we believe that businesses can still benefit from understanding the best methods for studying work places. With this knowledge, it may be possible to incorporate some aspects of the WPS approach into a firm's in-house research to enhance the process. Even if an organization is not able to use all of these research methods for every project, understanding the WPS approach to research will demonstrate why the information gleaned by WPS research is generally much richer and more accurate than the types of data that organizations usually collect internally. The WPS approach offers new and better ways to achieve important knowledge about the socio-technical factors that are critical to system success or failure.

Sources of Technology Failure

One reason for the emergence of WPS is:

> a series of well-publicized technological failures [that have] ... led to a growing interest amongst computer scientists and engineers in finding new and more reliable methods for the identification of requirements for complex systems. (Heath & Luff, 2000, p. 8)

In this section we describe several instances where lack of success in introducing new technology was due in part to a failure to consider the nature of particular work practices before designing and implementing a new system (see also Hughes, King, Rodden, & Andersen, 1994).

A common failing of many projects involving new technology is a "disregard for the ways in which people organise their work, coupled with a disdain for the ordinary resources on which they rely" (Heath & Luff, 2000, p. 3). For example, Heath and Luff describe a case from 1992 when control room failures created a crisis at the world's largest ambulance service (see also Heath et al., 2003). The London Ambulance Service installed an automated system "to replace the outmoded and inefficient practice of documenting the details of emergency calls on paper slips" (Heath & Luff, 2000, p. 3). Almost immediately, problems arose:

> The full introduction of the computer system effectively did away with the radio and telephone calls to stations, with the computer dispatching crews to answer calls. But, within hours, during the morning rush, it became obvious to crews and control room staff that calls were going missing in the system; ambulances were arriving late or doubling up on calls. Distraught emergency callers were also held in a queuing system which failed to put them through for up to 30 minutes.

(MacKinnon & Goodwin, 1992, cited in Heath & Luff, 2000, p. 1)

After only 10 days of operation the new system crashed and was pronounced dead "shortly after arrival." Inquiries into the reasons for its failure provided a very important insight: "Work practices do not necessarily change to make systems work" (Heath & Luff, 2000, p. 8).

Heath and Luff (2000, p. 3) cite the report of the board that was appointed to analyze the causes of the failure of the system. This report (Page, Williams, & Boyd, 1993) concluded that the design assumptions underlying the new ambulance-dispatching system were that it would "automatically" change the way employees performed their work. However,

> Management were misguided or naive in believing that computer systems in themselves could bring about [such] changes in human practices. Experience in many different environments proves that computer systems cannot influence change in this way. They can only assist in the process and any attempt to force change through the introduction of a system with the characteristics of an operational "strait-jacket" would be potentially doomed to failure. (Page et al., 1993, p. 40; cited in Heath & Luff, 2000, p. 3)

Heath and Luff (2000) attempt to generalize beyond the case of the London Ambulance Service, arguing that there is a general tendency, when designing new systems, to assume that work practices will adapt to fit the new system and that this assumption is often wrong. If a technological system is not well designed to fit in with the tasks required in the workplace, and the social process of doing the work, the system may fail or at best fulfill only part of its potential. In the case of the London Ambulance Service failure, the assumption was made that automation would inevitably improve the functioning of the service. Little regard was given to the "affordances" (Gaver, 1991; Sellen & Harper, 2003) of slips of paper that had been routinely and efficiently used by workers as they dispatched ambulances to save lives (Heath & Luff, 2000, p. 4).

From this and other studies, we gain one of the principal insights of WPS research: "high tech" is not always better. For some work tasks and settings, low-tech solutions such as pencil and paper are more effective. Sellen and Harper (2003) described how the use of paper slips in air traffic control worked well to keep track of airplanes and provided more flexibility and usable information to control tower workers than proposed new technology. Harper's (2000a) analysis of technology needs at the IMF revealed that a shared database, although much more "high-tech" than the procedures then in use, would probably not be useful to employees. Data were used primarily by IMF officers as sources for the reports they wrote; the reports were then shared with others inside or outside the organization. Harper discovered that the raw data would not have been useful to others; rather, they needed the reports created out of the data. Another example is Xiao et al.'s (2001) study of the use of a whiteboard by a hospital for scheduling surgeries. Although it would certainly be possible to create an online scheduling system for these tasks, Xiao et al. showed how the characteristics of the whiteboard facilitated the tasks involved in scheduling (in particular, the coordination of doctors, nurses, operating rooms, support staff, and patients) and made this information readily available to all concerned in an easily accessible, flexible format.

According to Heath and Luff (2000), Suchman (1987) suggests that system designers typically use conventional "goal oriented, plan-based models of human conduct ... [which] diminish the importance of the immediate context of action" (Heath & Luff, 2000, p. 10). Such models overlook how plans and rules depend on the circumstances at hand and ignore how individuals and teams jump in to prevent mishaps as contingencies arise. Real-time decisions are made to override preset routines so as to prevent major disasters. Until system designers learn to understand how people work both as individuals and collaboratively, they will continue to wonder why their technologies are not always as effective as planned.

Not all instances of technological failure are as spectacular and potentially life-threatening as the ones described here, but even the more mundane failures (which are much more common) can be extremely costly to companies in terms of resources invested. For example, Button and Harper (1996) describe a foam manufacturing firm's new invoicing system that had to be scrapped because it did not effectively support the work contingencies and practices in the plant. The designers of the system did not realize that orders were often manufactured before invoices were written (to allow same-day shipping of the product) or that production workers often spoke directly to customers and made revisions in the order before it was written, so as to fulfill the order as expeditiously as possible.

Not all systems fail outright, but if they do not work as planned, or result in less than optimal performance by workers, unnecessary losses to the organization may result. For example, Button and Harper's (1996) analysis of a computer system to help police officers write and file crime reports showed that it did not fit in with the way police officers did their work and had the unintended consequence of making it appear both that the reports were consistently filed late and that they were revised more often than was considered desirable. Grudin (1988) describes the failure of electronic calendar groupware, which unexpectedly disrupted employees' work practices (see also Rogers [1994] on this topic). A computerized workflow application installed at a printing firm was unsuccessful because it did not allow the workers sufficient flexibility to rearrange the sequence of jobs to maximize their productivity (Bowers, Button, & Sharrock, 1995; cited in Ehrlich, 1999). A groupware application "intended to improve the efficiency of assigning work to telephone company workers who are called in when there is a problem with a phone line" (Ehrlich, 1999, describing research reported by Sachs, 1995) failed to function successfully as an automated way of allocating work because it eliminated the direct contact between employees through which information about the problem to be repaired had been routinely shared. Olson and Olson (2000, p. 153) found that "many of the attempts to use distance technology either have failed outright or have resulted in Herculean efforts to adjust behavior to the characteristics of the communication media."

A WPS Approach to the Solution of Typical Business Problems

The WPS approach to how technology, individuals, and processes mutually create a work environment and work practices addresses fundamental issues that affect organizational functioning. WPS research has shown that organizations need to consider social process carefully, in order to optimize employee communication—and ultimately collaboration. Some of these solutions may involve high-tech systems, whereas others may find that low-tech solutions are most effective. In order to demonstrate how WPS research findings can be useful for businesses, we will review the WPS literature that is relevant to two common business problems: (1) the facilitation of collaboration among employees, whether co-present or geographically dispersed, and (2) the underutilization of employees, which can lead to hidden financial losses or to excessive employee turnover. We focus on these problems because we believe that they may be the most important factors in determining the effectiveness of technology used in organizations, and also because they are well covered by existing WPS research.

How Can Collaboration Be Facilitated?

The type of technology needed to facilitate collaboration will vary with the type of work being done. For example, workers editing documents together will need to see the documents simultaneously, to navigate through them, and to edit or mark them easily. They may need to compare several documents at the same time (Sellen & Harper, 2003). Architects working on the plans for a building have these needs plus other requirements, such as viewing three-dimensional models (Heath & Luff, 2000). Technology has created collaborative opportunities that businesses must selectively and deliberately deploy. These technologies can be beneficial both for workers who are co-present in a single location and for those who are geographically dispersed.

Co-Present Employees

Typically, employees at a single work location have relatively easy access to each other during the day; they can simply walk down a hall to talk to someone or attend a meeting. For some work settings, the employees are "housed" in the same room and do all or almost all of their work in that space in the company of others. Olson and Olson (2000) found that co-presence by itself caused a large increase in the productivity of workers (in this case, designers). Having direct and continual visual and aural access to each other facilitated both the work they did together and the work they did individually, allowing maximal information sharing and greater efficiency in bringing people into the problem-solving process as needed. The collaboration technologies these workers used were relatively simple, including flip charts and whiteboards with printing capabilities. Co-presence provided the work teams with fluidity of participation: "They could move from one subgroup to another, or to a meeting of the whole, by merely overhearing others' conversations, seeing what someone was working on, and being aware of how long they had worked on it with or without progress" (Olson & Olson, 2000, p. 146). Other advantages of co-presence included the ability to easily see the reactions of others to work or ideas: "If a team member wants to observe his manager's reaction to a point someone made, he can just glance quickly in her direction. A team member can refer to someone's list of ideas on a taped-up flip chart sheet by making a gesture or glance in its direction that everyone can immediately interpret" (Olson & Olson, 2000, p. 146).

In settings where workers are in close proximity, technology must make information visible and accessible for multiple users. For example, Heath and Luff (2000, p. 88) report that the Automated Train Supervision system (ATS) in the London Underground has computer monitors at each workstation to enable controllers to enter key information. The placement of the monitors in the control room allows other workers to be aware of information entered by the controllers that directly affects the work that they themselves do. In addition, the workers' proximity to each other makes information from face-to-face and telephone conversations and loudspeaker announcements available to everyone who needs it. All of the workers in the control room can overhear each other and display and make use of this overhearing in their collaborative work. The technology worked effectively because the workers and equipment were situated in relationship to each other such that they could utilize their co-presence in performing their jobs.

The London Underground uses a relatively high-tech system in which technology performs much of the work of operating the trains and also aids operators in doing the work that continues to rely on humans. But even in a technology-dependent setting such as this, the workers ultimately had to rely on commonsensical everyday procedures of hearing, seeing, and action-coordinating. The technology was dependent, for its successful operation, on the interactional skills of the employees and on the construction of the workplace to facilitate aural and visual access of the employees to each other. The lesson is that businesses planning to install expensive technology systems should consider how the technology will be used and what degree and types of access to each other the employees will continue to need, not only to use the equipment, but also to perform the tasks of their jobs with the equipment.

WPS research shows that technology may function better if it supports the social processes undergirding the work as opposed to attempting to replace those social processes. This principle is best demonstrated in Heath and Luff's (2000) study of the Reuters newsroom in which journalists assigned to particular subject areas are expected to "identify relevant stories for their particular customers and to tailor the news with regard to the practical interests of the members of the respective financial institutions working in particular areas" (Heath & Luff, 2000, p. 63). Technology assists in this effort as news stories are delivered to each journalist via a computer monitor. The journalists can then edit the stories at their keyboard and transmit them to the appropriate destination. However, the task of determining what information is useful to whom is left to the discretion of each individual journalist. The journalists must determine if any information that comes across their desks would be helpful to another desk and, if so, find the best way to communicate that information.

The journalists shared a large room furnished with rows of computer terminals. They sat only a few feet from each other, being close enough to lean over and read their neighbors' computer screens. Heath and Luff (2000) found that the journalists took advantage of this proximity to communicate with each other about articles of potential interest to one or more journalists in the room. The journalists at Reuters worked at a very fast pace, and sometimes had to "turn around" stories within as little as a minute of receiving them (Heath & Luff, 2000, p. 64). Because they were all working at such an intense pace, the journalists rarely explicitly interrupted each other to share information on a story. Rather, when they had a story they thought might be of use to someone else, they would "throw out" a comment about the article to no one in particular. If other journalists did not feel the story being "advertised" was of use to them, they did not have to stop working in order to respond. Alternatively, they could respond and elicit more information about the story or ask to have a copy sent to them. In this study, Heath and Luff (2000) make the case that neither talk nor technology can stand alone, but are in fact deeply interdependent and interrelated.

Thus decisions about which technology to purchase should be combined with a detailed understanding of how work is done in a particular setting. In this example, a technology that required journalists to wear headphones or that placed auditory or visual barriers between them might have prevented them from doing their work in an effective fashion. The deployment of the technology has to be considered in the context of how the work is done—where will the computers be placed relative to each other? What types of offices will be provided? Even though private offices are often considered a "perk" or benefit to an employee, for some types of jobs, such as at Reuters, physical co-presence in close proximity facilitates the doing of the work. It should be emphasized that the degree of proximity required will vary with the job.

Although it would be technically possible to design a computerized version of a system for sharing ideas about stories and soliciting other editors' interest in those stories, it probably could not have been done with the speed with which it was done by the Reuters employees. At the very least, the journalist would have had to stop typing or reading in order to post a message to another editor. And those messages, once sent, would intrude upon the work of the receiving editor. The beauty of the "throw-away" oral comments the journalists used to "advertise" stories was that others could choose to ignore them if they were not useful to them or if they were not at leisure to attend to them. Technology can be of great assistance to workers (as was the computer system that fed the Reuters journalists stories and allowed them to edit and transmit them to others), but the work would not have been done as effectively as it was if the workers had been located in separate private offices. Understanding in detail how work is done in a business can lead to better choices about technology and how to implement and locate it. From both the London Ambulance Service and Reuters examples, we see the importance of understanding the social processes that underlie the work of employees in close proximity to one another. Whether the systems are high-tech or low-tech, they should acknowledge the work context. Moreover, proximity does not ensure technological success. In fact, technological solutions for closely situated workers have to meet the additional challenge of supporting tasks that rely on well-coordinated, time-sensitive, and somewhat ad hoc interactions. As we will see, solutions for geographically dispersed employees pose a different set of challenges for designers.

Geographically Dispersed Employees

When employees are geographically dispersed, whether in the next office or across the globe, the challenge of facilitating collaboration is even greater than when employees are physically co-present while carrying out tasks. Bellotti and Bly (1996) found that workers who were in the same building frequently engaged in informal conversations or casual "walk-bys," which informed them of the nature and stage of the work being done by others on the design team and made them available for impromptu discussions or input into the work process. Their proximity and mobility within the local context facilitated both the accomplishment of their own work and their contributions to their colleagues' work. Those employees located in another office of the firm in a different town, however, were not able to participate in these informal exchanges or to have opportunities for visual access. The authors concluded that "while local mobility enhances local collaboration, it penalizes long distance collaboration severely"

(Bellotti & Bly, 1996, p. 209). The potential benefits of successfully integrating technology with the social process of doing work can thus be even more beneficial for geographically dispersed workers. Unfortunately, existing technologies designed to facilitate collaboration among distant employees tend to require workers to change their work practices in order to accommodate the characteristics and limitations of the technology, instead of having the technology facilitate the work (Olson & Olson, 2000, pp. 152–153):

> Our laboratory data show that, even for people who know each other and have worked together before, a simple audio connection for conversation and a shared editor for real-time work is insufficient to produce the same quality of work as that done face to face. Those with video connections produced output that was indistinguishable from that produced by people who were face to face. The process of their work changed, however, to require more clarification and more management overhead (discussions about how they will conduct the work, not actually doing the work; Olson, Olson, & Meader, 1995; Olson, Olson, Storrøsten, & Carter, 1993; see also Isaacs & Tang, 1994; Tang & Isaacs, 1993).

There are many examples of technology effectively assisting workers in collaboration over a distance. For example, Luff and Heath (2002) showed how the familiar technology of two-way radio is used successfully by controllers and train station personnel to report, communicate about, and resolve problems with the train system as they occur. Employees used their commonsense knowledge of how interaction works to coordinate their talk over radios, and learned how to communicate within the limitations of the technology. Despite the limitations, the two-way radios provided sufficient flexibility for employees to communicate successfully in their respective work settings.

However, technology introduced to work settings does not always mesh effectively with employees' communicative strategies and resources. Luff, Heath, Kuzuoka, Hindmarsh, Yamazaki, and Oyama (2003) describe why this is so in their study of the development of a video-mediated environment. They conducted a series of experiments with GestureMan, a developing "system designed to enable individuals in remote domains to interact with and through objects and artefacts" (Luff et al., 2003, p. 53). GestureMan is a device that a person can control from another location. One person controls GestureMan's movements and actions (e.g., it can point at objects, or shine a laser beam toward objects to identify them). GestureMan has embedded cameras that enable the participants to see what is seen from the perspective of GestureMan. In their experiments, the researchers put GestureMan in a room with furniture and asked two participants in separate remote locations to identify specific pieces of furniture and to move them to different locations.

The participants each had a video view of the room. Luff et al. (2003) found that, in spite of the technological sophistication of GestureMan, it was quite difficult for the participants to successfully complete the task:

> Despite the opportunities afforded by the system, the materials reveal the difficulties participants face in accomplishing object-focused actions, and the ways in which they attempt to establish and sustain common frames of reference; the system fractures the environments of action and inadvertently undermines the participants' ability to produce, interpret, and coordinate their actions in collaboration with each other. (Luff et al., 2003, p. 53)

In short, Luff et al. (2003, p. 53) discovered that the technology could not "ensure that participants have compatible views of each other's domains or ... enable a common perspective to be established within the course of interaction." The same problems experienced by the research subjects using GestureMan in attempting to collaboratively locate physical objects in a room might be experienced by employees trying to work collaboratively on documents via a videophone system. Bowers et al. (1996) studied a different Collaborative Virtual Environment and found that participants experienced difficulty with conversational turn-taking (although they noted that participants often seemed able to find ways around these challenges).

Heath and Luff (2000) reported on research and design efforts to facilitate the accomplishment of collaborative work by spatially separated coworkers. They called the infrastructures that were designed to enable audio and video access between dispersed individuals "media spaces." Some of the employees studied were located as much as 600 miles apart, while others were in different offices in the same building.

The videophone system consisted of a combination of monitors, cameras, and microphones. The goal of early videophones and videoconferencing suites was to duplicate the experience of being in the same room with the person with whom one was collaborating. In addition, these systems were designed to increase informal, chance encounters between workers, which were seen as opportunities for creating and transmitting ideas, and to give workers easy access to information about who was doing what where (Heath & Luff, 2000). In order to achieve these goals, early systems attempted to provide remote participants with visual and aural access to "everything" in the physical space of the persons with whom they were collaborating: a view of the person with whom they were working, the office in which they were, and the papers, documents, or other artifacts with which they were working. However, when these systems were tested with real workers in real offices, Heath and Luff (2000) found that a video camera view of something is not the same as actually being in the same room with one's collaborators. Here are some of the challenges that they found:

1. In face-to-face interaction, turning to look at another person in the room can serve to attract his or her attention and alert him or her that you are ready to begin interacting with them; this did not happen when the look was being transmitted over a video camera. Heath and Luff (2000) found that even smiles and waves were often ineffective in attracting the attention of remote participants.

2. In face-to-face conversation, speakers use gaze direction to monitor ongoing speech and indicate comprehension and active listening. However, with the video system it was difficult for the participants to make eye contact. Because of this, participants' speech was marred by many delays, hesitations, and repetitions when they communicated via videophone.

3. The videophone did not adequately simulate face-to-face interaction in that both verbal and nonverbal forms of communication were difficult. Although the employees could see each other, they found it difficult to create a shared visual field from the distorted image the video provided. This minimized the effectiveness of the system for collaborative work. As Sellen and Harper (2003) and others have shown, workers collaborating on a document must achieve a shared orientation to that document in order to work together on it.[2]

Heath and Luff (2000, pp. 215–216) concluded that existing technologies fail to meet the needs of geographically dispersed workers in two important ways: 1) the design focus on face-to-face communication has overlooked the use of documents and artifacts in collaborative work, and 2) subtle gestures, facial expressions, and utterances cannot be appropriately relayed via current technology. This suggests that future systems need to shift the focus from attempting to *replicate* face-to-face interaction to *representing* collaborative work. Such a shift requires developing tools that facilitate simultaneous access to documents and artifacts along with the social interaction that is required to evaluate and change them. In addition, more sophisticated technologies are needed to make intelligible the verbal and nonverbal messages that are communicated across work settings.

How Can Underutilization of Employees Be Minimized?

There are many factors that may lead businesses to underutilize employees. In this section, we first consider an example in which the choice and design of technology contributed to the underutilization of employees and then examine an example in which the underutilization of employees was related to the assignment of tasks between different categories of employees and the physical arrangement of employees relative to each other. Thus the first example deals with "high-technology" (an expert system), and the second deals with "low-technology" (the use of paper documents, pens, Post-it-Notes, and filing and coding systems).

Expert Systems and Employees' Capabilities

Businesses often fail to tap all the skills employees have. In fact, the assumption is often made that low-paid employees have "no" skills. Because of high turnover in low-paid positions, the training of employees is often seen as a waste of money. Companies may decide to use technology to automate some or all aspects of a particular job, thus attempting to eliminate the human factor from the performance of the tasks in that work setting. For example, an expert system (where the knowledge required to do the job is located in the knowledge base rather than in the employee) may be implemented to solve these business problems. However, WPS researchers (for example, Bobrow & Whalen, 2002; Whalen & Vinkhuyzen, 2000) have identified several types of mistakes that can be made when designing and implementing expert systems:

1. Ignoring the social process involved in accessing and transmitting information.

2. Ignoring the power of informal training that employees can give each other if they are located in such a way that they can share information and overhear one another.

3. Focusing only on the employee who will use the expert system and not examining its use in conjunction with the client who will be served by that system.

4. Designing a system that is not flexible enough to be used during ordinary oral interaction.

5. Underestimating the degree of knowledge about the subject matter of the expert system that is needed to use it most effectively.

In this section, we describe the results of a WPS analysis of an expert system that was implemented by a copy machine company (Whalen & Vinkhuyzen, 2000). This system was designed to be used by its Customer Service and Support Representatives (CSSRs) as they fielded calls from clients whose copy machines had broken down or malfunctioned. The company faced two challenges. First, the increasing use of software in copy machines made it necessary to distinguish reliably between problems that were caused by software and those caused by hardware. Different technicians handled software and hardware problems; thus, if the wrong technician were sent out, this could be a costly mistake. Second, a high rate of employee turnover made extensive training of CSSRs seem an unattractive proposition. The natural solution seemed to be the design of an expert system that would guide untrained CSSRs in figuring out whether a service call was warranted or whether the problem could be fixed by the client with instructions provided by the system.

Using what were at the time cutting-edge computer technologies, the company developed and implemented Case-Point, a system that was designed to identify the likely causes of a machine malfunction by matching the caller's description of the problem with its database of typical machine problems. The system was designed to identify those problems that could be easily fixed by the client after being given information over the telephone and to distinguish such problems from those (the majority) that would require a service call. Although the system was somewhat successful (it decreased the rate of unnecessary service technician visits), it also generated an array of new problems.

First, the assumption that Case-Point would be an expert system that could be used by CSSRs who had not received at least some training in the repair and working of copy machines proved to be overly optimistic. Those few employees who either already possessed (or were able to obtain on their own) some basic knowledge of the functioning of copy machines and their typical problems performed much better than the untrained CSSRs.

Second, CSSRs reported being unable to use the system as it was intended because the questions it required them to ask of the clients would have left extremely unfavorable impressions. For example, if Case-Point required the CSSR to ask the client whether the defect (e.g., smudging or blurring) also occurred on the original that was being copied, the client might be offended because such a query implies that the client is "clueless." In addition, the questions Case-Point posed often seemed random, which could leave a caller with the impression that the CSSR did not know how to carry on a coherent conversation. Although there were logical reasons for the system to request such information, the computerized "expert system" turned out to be a "communicative idiot." Fortunately for the company, the CSSRs used their commonsense knowledge of how interactions *should* work and edited or censored some of Case-Point's input in order to avoid compromising interactions with clients.

However, Whalen and Vinkhuyzen (2000) point out that even when properly utilized, Case-Point was not as successful in identifying the exact reason for a malfunction or troubleshooting problems as it could have been. Computerized expert systems are generally designed in a logical fashion, using binary choices or flow chart reasoning to help operators locate relevant data. But the decision-making and communication processes of the users are organized along different lines—flexible, open-ended, and nonlinear. For example, Case-Point used a specific set of terms to call up relevant information. A client may say there are "blotches" on a copy. However, Case-Point does not have a category for "blotches," so the CSSR has to make a judgment call and decide whether the Case-Point category of "blurry" can stand in for "blotchy." This problem of how to gloss clients' descriptions to match Case-Point's categories was not adequately handled in the design of the system and may have reduced its effectiveness.

Whalen and Vinkhuyzen (2000) identified the reliance on artificial rather than human intelligence as a main shortcoming of Case-Point. During their research they discovered some CSSRs who had figured out on their own how to navigate around the system's limitations. These CSSRs provided guidance and troubleshooting to callers, which translated into decreased costs by further reducing the number of service technician visits. Instead of following official Case-Point procedure, these CSSRs found ways to trick Case-Point by feeding the computer words and phrases that would make it provide information that the CSSR knew would be needed for troubleshooting. Such individuals showed considerable talent in understanding technology: both Case-Point and the company's copy machines. Whalen and Vinkhuyzen (2000) discovered that most of this talent could be explained by these individuals' social contacts at the workplace. By coincidence, one CSSR had some basic background knowledge of copy machines that he used to troubleshoot calls. In doing so, he figured out for himself how to manipulate Case-Point in order to aid his efforts. This individual communicated his knowledge informally to coworkers, some of whom found this active approach to their job more rewarding than merely following the prescribed protocol. It also helped them avoid the uncomfortable interactions with callers caused by Case-Point's canned question routines.

Unfortunately, as Whalen and Vinkhuyzen (2000) showed, the desire to design an expert system that would eliminate the need for knowledgeable employees resulted in a system that was largely unfit for the task of providing friendly, coherent interaction with callers. Calls usually ended with the sending out of a service technician. Although the system decreased costs somewhat, it was not as successful at providing expert customer support as it could have been. The researchers discovered that a truly ignorant user could not effectively use the system. Case-Point did not eliminate the need for intelligent employees who could think for themselves. The research demonstrated that one crucial skill required to troubleshoot a technological problem was knowing how to manage interactions with clients. Case-Point was not always able to troubleshoot copy machine problems; the ability to utilize the system for its intended purpose depended in part upon interactional skills and mutual assistance among coworkers.

As the example of Case-Point shows, ordinary employees have skills that more often than not are underutilized by companies, especially when attempts are made to accord more of the work responsibilities to technology. In communicative work, the reliance on technology and the automation of social processes may well become problematic. Dealing with a customer support call was conceptualized as a purely technical problem requiring a technical solution. However, it is not the copy machine that calls a service computer—it is a human client who calls a human CSSR. Lack of appreciation for the implications of this fact resulted in the failure to utilize employees' capacities to train one another and aid callers in troubleshooting. How much additional money the company could have saved had the system been better designed and implemented is impossible to estimate. Yet, it is obvious that solving the problem on the first call and eliminating the additional costs of sending a service mechanic is the cheapest and best solution for both the company and the customer. More attention to the social process of doing the work could have resulted in more effective CSSR performance with Case-Point.

Job Titles and Employees' Capabilities

Another way in which businesses may underutilize employees' skills is by letting job titles, job categories, and job descriptions put blinders on the organization in terms of the nature of the work each employee can or should do. Employees are often capable of doing much more challenging and useful work than their job titles imply. Another problem arises from what we refer to as *organizational geography*. By this we mean the tendency of businesses and other organizations to group, locate, or segregate employees by job title or by status level (e.g., manual vs. intellectual labor, professional vs. non-professional status, or white collar vs. blue collar work). One of the implications we draw from WPS research findings is that these practices are often counterproductive and may lead companies to underutilize existing skills and capabilities in their work force by failing to facilitate relationships between employees or create the most productive *organizational geographies* possible.

Suchman (2000) analyzed the practical implications of how organizations and firms typically divide professional work from non-professional work. These terms stand in for common job titles that distinguish mental or subjective knowledge work from manual or objective knowledge work. Suchman's goal was to reconceptualize the way work is delegated and distributed. She argued that instead of separating workers according to arbitrary categories based on the relative amounts of objective or subjective work, research should be done to identify objectively both the intellectual and manual (routine) aspects of the work. Recommendations can then be made about how best to organize, delegate, and situate it among the available workforce.

Suchman (2000) explored this issue through a study of a law office that was sorting through thousands of pages of potential evidence to decide what is important and how to store it for later retrieval by the firm's professional litigators. Her study involved looking at the physical documents and how members of the firm approached the work of coding and sorting the documents depending on their assigned job and their position in the company. She argued that an exaggerated assumption of difference between the two types of work negatively affected the output of both groups of workers.

Suchman (2000, p. 31) begins with the observation that our culture views "subjective and objective ways of knowing" as distinctive social processes. The company that she studied distinguished *subjective* coding, which was conducted by the firm's junior lawyers, and *objective* coding, which was done by temporary litigation support workers (Suchman, 2000, p. 31). Documents were initially coded by the firm's junior lawyers in terms of relevance and urgency to a given client or subject. The non-professional, temporary litigation support staff then further coded these and other lower priority documents and "create[d] a database index to the entire document set" (Suchman, 2000, p. 32) so that they could be located and retrieved easily in the future. The focus of Suchman's analysis was to discover to what extent the coding practices of the lawyers and temporary workers actually differed. She found that both types of work contained objective and subjective elements and that the supposedly more intellectual work of the junior lawyers contained much routine and mundane work, while the supposedly mundane work of the temporary workers involved much interpretive work and decision making (Suchman, 2000, pp. 32–33). Suchman (2000, p. 33) hoped to show "the hybrid, practical, and judgmental character of both forms of work." In other words, she hoped to demonstrate that although some aspects of the work differed, important elements of "judgment and interpretation" occurred in both job categories.

The junior lawyers and temporary litigation workers were located in different offices. Suchman (2000) found that partially because of their spatial separation, the two groups of coders differed in their contextual knowledge of the legal case being documented. Ultimately, both the lawyers and the temporary litigation workers developed familiarity with the details of the case and used it to perform their coding work. She found that the temporary litigation workers possessed knowledge that could have been not only helpful to the junior litigators in their coding and analysis, but also of value to the firm's professional litigators. Because of the categorical and physical separation, this knowledge went untapped and unused.

These findings led Suchman (2000, p. 43) to conclude that the distinction between subjective and objective work "orders not so much ways of knowing or acts of reasoning as it does identities, actors, and distributions of material and symbolic rewards among them." The firm had created an artificial separation of work between its professional and non-professional staff that resulted in the underutilization of lawyers as subjective knowledge workers and temporaries as objective knowledge workers. With a different approach to the design and classification of work based on a deeper appreciation of tasks and skills, the firm could have enhanced the work of its temporary litigation support staff by engaging them in what was previously considered professional work. These same steps might also have enabled their junior lawyers to expand their contribution in support of the firm's senior litigation staff. Furthermore, the firm could probably have increased the productivity of both sets of workers by locating their work in the same space, so that they could communicate about the case and the documents and share their approaches to coding with each other.

Suchman (2000, p. 44) concluded that "simple oppositions of knowledge and routine work are more ideological than descriptive, and act rather to obscure work's actual demands than to clarify them." Organizations should look past simplistic distinctions between professionals and non-professionals, subjective knowledge workers and objective knowledge workers, or mental vs. manual workers, and instead focus on the specific requirements of the tasks involved and the skills of the workers that are present in the organization (Suchman, 2000). She also found that artificial distinctions made between the work of junior lawyers and temporary litigation support staffers resulted in the firm's failure to appreciate the different types of "work intelligences" that are developed situationally in the interactions that occur while the participants are doing their work. We believe that a deeper appreciation for these nuances might be leveraged to a firm's advantage, if the job classifications were reconsidered and redesigned and if the firm's organizational geography were reconfigured.

WPS research can effectively address many different business problems. However, the problems of facilitating collaboration, both proximally and across distances, and fully utilizing employees are particularly important. These examples demonstrate several central themes of WPS research. First, it is critical to understand the social process that underlies work activities before designing technological solutions. Second, organizations should look beyond existing tools or systems that may arbitrarily impose structures unrelated to how the work is accomplished. Finally, all solutions should accommodate the current social process rather than assume that the process will adapt to the technology. These themes are central to the WPS research perspective and serve as complements to other studies of technology at work. Although WPS is sometimes criticized for focusing on identifying problems rather than suggesting solutions, sometimes the identification of the problem itself presents an obvious and potentially workable solution. For example, in the case by Suchman discussed here, locating lawyers and their paralegal helpers in the same physical location might in and of itself solve the problem by facilitating interaction and communication.

How WPS Complements Other Approaches to the Study of Technology at Work

WPS and Requirements Engineering

Jirotka and Wallen (2000) argue that the discipline of requirements engineering includes a fairly broad array of technical analysis and specification activities. In their review of the literature, they identify: (1) analyses and documentation of work practices in support of system evolution for the benefit of users and the larger system in which the technology is deployed (Davis, 1993), (2) exacting analysis and translation of user needs as inputs for software developers (Loucopolous & Karakostas, 1995), and (3) projections of the attributes that should characterize a successful system in a future environment (Gouguen, 1994). Jirotka and Wallen (2000, p. 243) argue that the purpose of requirements capture is "to analyze both the domain—the environment in which the system will be deployed—and putative aspects of system behavior." Fulfilling this second purpose requires that an understanding of the "domain as a socially ordered environment" be developed (Jirotka & Wallen, 2000, p. 243). However, requirements engineers conventionally look at the social organization of the workplace through a filter created by their intent to transform it through technology. This produces a partial and hypothetical description of a workplace in the context of a technical solution. These requirements statements are further distorted by a system design process that seeks to produce generic products that can be used in a variety of different domains (Jirotka & Wallen, 2000).

Done properly, requirements analysis begins with both an understanding of the domain and a vision of how the technology will transform the workplace (Jirotka & Wallen, 2000, p. 243). One conclusion that could be drawn from Jirotka and Wallen is that the high rate of failure of technology systems is due in part to requirements engineering's methods of analysis. Designers write systems specifications and user requirements "with little or no access to the domain in which the system will eventually be placed" (Jirotka & Wallen, 2000, p. 245). A typical process is to develop a prototype on the basis of detailed interviews with clients (e.g., Wolf, Foltz, Schlick, & Luczak, 2002). The system developed may be tested with real workers, although often in an artificial or experimental setting rather than in a real work setting (e.g., Wolf et al., 2002). In sum, "traditional techniques used in requirements analysis cannot reveal the interactional organisation of activities" (Jirotka & Wallen, 2000, p. 246).

Gause and Weinberg (1989, p. 17) discuss the costs of ambiguity in requirements: "Billions of dollars are squandered each year building products that don't meet requirements, mostly because the requirements were never clearly understood." Further, they go on to state that "some observers have estimated that approximately one-third of large software projects are never completed. Much of the enormous loss from these aborted projects can be attributed to poor requirements definition." Their book aims to teach engineers how to correctly and accurately discover appropriate requirements for design projects.

According to Glen (2003), typical excuses given as to why requirements are not properly defined include "'Users won't tell us what they want,' or 'We don't ask good questions,' or 'What they told us they wanted turned out not to be what they really wanted'" (Glen, 2003, p. 46). He explains that clients may not know what capabilities they want or need in their technology system and stakeholders in the organization and the design team may have different priorities and needs. Beyer and Holtzblatt (1998) recognize that design in face-to-face settings is a difficult task:

> The industry does not generally provide good models for face-to-face cooperation on the same project; it's easier and more common to split projects up into parts small enough for individuals to do independently. But there's no way to leverage multiple perspectives if everyone works independently. (Beyer & Holtzblatt, 1998, p. 128)

Glen's (2003, p. 46) recommended solution to this problem is for those designing IT projects for organizations to "negotiate requirements among the many stakeholders whose positions and interests need to be acknowledged." However, this still leaves a gap in knowledge about how work is done in the organization. As Glen points out, people often do not know or cannot articulate what they do or why they do it. They often are unable to imagine what their future needs will be. The application of the WPS approach, which involves the observation and analysis of how work gets done in real time and in its natural setting, could lead to improvements in requirements analysis techniques.

An orientation to workplace studies that views work as an achievement suggests that events, actions, and objects in the real world should be treated as practical accomplishments organized by those who participate in these events. Work is accomplished through interactions with others as they coordinate a range of activities and make use of different tools and resources (Jirotka & Wallen, 2000, p. 247).

WPS researchers repeatedly find that social processes and the social organization of work are elements that, although present in all work situations, may not be adequately accounted for by systems designers (although by contrast, note the system proposed by Büscher, Gill, Morgensen, and Shapiro [2001] for integrating analytical ethnography and participatory design in a reiterative process). Consequently, social effects may be missed in the design of technological workplace solutions and this results in projects that fail to deliver expected results. Electrical and software engineers and other technology designers typically do not utilize social-scientific perspectives and investigative methods to recapture these fundamental aspects of work.

WPS and CSCW

In the late 1970s and 1980s, the desire for more efficient work processes and the availability of affordable computer technology (mainframe and microprocessing) led to a number of new approaches for designing and developing application systems. As described by Grudin (1994), these systems emerged to meet the needs of three distinct types of users: organizations, individuals, and groups. Large, custom-built organizational systems were followed by off-the-shelf solutions for individuals; both enjoyed significant commercial success. However, the development of group-oriented systems has met with mixed results. In addition to interface issues, group systems have to deal with issues related to interpersonal interaction in order to support group work effectively. Examples of group-oriented collaborative systems include desktop videoconferencing, group productivity systems such as Lotus Notes, electronic bulletin boards, e-mail, and group calendars.

In the mid-1980s, the concepts of CSCW and groupware began to emerge in the literature to describe systems supporting collaborative work done by groups in a work setting. Both offered ways to address the difficulties of designing group-oriented systems. Grudin (1994) concludes that in order to address the challenges facing individuals who develop applications for group work, developers need to have a better understanding of both the user(s) and the work environment and adjust the development process accordingly.

Müller (1997) also discusses the problem of understanding accurately the user and the work environment. An example given by Winograd and Flores (1987) illustrates the difficulties of understanding the work process. They describe how the work being done by a person sitting at a desk writing on a piece of paper could be characterized in a number of very different ways. For example, the person could be described as writing a memo, making a decision, improving organization in the firm, and so on. Which type of description is chosen depends on the perspective and methods used by the researcher.

Plowman et al.'s (1995) review of WPS research in the CSCW area focuses primarily on ethnographic research. Although almost all WPS research is ethnographic, all ethnographic research in workplace settings is not necessarily Workplace Studies (see also Hughes et al., 1994). For example, Button and Harper (1996, p. 263) argue that although in CSCW research, work-practice is taken into account in the design process, it is often done in a way that is inconsistent with the perspective of WPS. The way in which work-practice is implemented and included in the research design actually leads to a subtle but profound difference between the WPS approach and that of CSCW. Specifically, Button and Harper (1996) argue that the CSCW approach largely ignores the lived work of the setting being studied/designed for. Instead, CSCW designers rely on the use of documents, interviews, and to some extent observations of workers in order to discover the work procedures that need to be taken into account for the design of the computer system. As an example of the importance of lived work, Button and Harper (1996) describe a CSCW analysis undertaken to develop a computerized invoicing system to help a firm that manufactures foam for furniture builders. The examination of documents, interviews, and observations of the firm's functioning resulted in a specific understanding of the pattern of work that would need to be done by the computerized invoicing system being developed. However, because the data collection methods used by the business to develop the system did not include "analytic explications of the lived work of participants" (Button & Harper, 1996, pp. 266–267), the developers failed to discover how the lived work of the business resulted in different contingencies, requirements, and even orderings of actions to accomplish the work of receiving orders, producing the goods, and delivering and billing for them. This resulted in a computerized invoicing system that was not effective for that firm. What needs to be done is to use a WPS approach to understand "the interactional practices through which the participants organised their work as they experienced it" (Button & Harper, 1996, p. 267). In other words, the data are only as good as the (sociological) analysis applied to them. In short, some researchers may correctly observe what was done, without understanding why it was done. Suchman (1995, p. 58) makes a similar point: "The aim [for systems design] is a design practice in which representations of work are taken not as proxies for some independently existent organizational processes but as part of the fabric of meanings within and out of which all working practices—our own and others'—are made."

If these suggestions are taken seriously, it may be that approaches to CSCW, such as those relying on scenarios, in which the essence of a job task or work-related interaction is abstracted and used as a basis of design for a collaborative system (e.g., Dewan, 2001), are not optimally effective. A brief summary of a workplace interaction or event does not contain sufficient detail or information about the members' perspectives on their actions to adequately support the design process. Harper (2000b) also critiques the prevalent use of watered-down approaches to ethnography, which may be inadequate to meet the needs of the organization that has commissioned the study:

> [In organisational ethnography] certain steps are required to ensure that the ethnographic materials cover a sufficient spectrum of organisationally situated tasks to enable proper examination of any particular subset of those tasks. It is only then ... that the materials generated by the research get taken seriously by members of the organisation itself. (Harper, 2000b, p. 240)

Another trend in CSCW research that is compatible with a WPS approach is Participatory Design, discussed by Kensing and Blomberg (1998). Participatory Design is an approach to "designing systems (both technical and organizational) that are informed by and responsive to people's everyday work practices" (Kensing & Blomberg, 1998, p. 180). This is done by involving the users of a technology (the workers) in the process of design. It not only helps to create design that provides "a better fit between technology and the ways people (want to) perform their work" (Kensing & Blomberg, 1998, p. 168), but also is an active and conscious effort to intervene in the "politics of design." Although a variety of methods is used in the Participatory Design process, the common thrust is to incorporate the input and participation of workers who will be using the technology in the design process, for example by setting up prototypes, which they then use. The workers' experiences in working with the prototypes can then become part of the revision process (see, for example, Johansson, Fröst, Brandt, Binder, & Messeter, 2002). Ethnographic and field study methods are becoming increasingly common in Participatory Design approaches (Kensing & Blomberg, 1998). A recent development that attempts to combine ethnographic approaches with Participatory Design (the "Bricolage" approach) seems promising (Büscher et al., 2001).

CSCW and WPS researchers share a common interest in understanding and supporting collaborative work. However, although CSCW espouses the use of ethnographic and other qualitative research methods, it does not embrace the lived work that is a central theme of WPS. As a result, CSCW researchers often can miss the subtle contingencies and requirements that emerge in the work setting and this can ultimately lead to design failure. The trend toward Participatory Design in CSCW appears to be drawing the field closer to the WPS perspective. This process, which includes the worker in the design process, uses ethnographic methods to capture the interplay between user, prototype, and designer.

WPS and Social Informatics

Social informatics is "the interdisciplinary study of the design, uses, and consequences of information technologies that takes into account their interaction with institutional and cultural contexts (Kling, 1999, p. 1)."[3] Kling (1999, p. 16) argues from a social informatics perspective that design requires the appropriate use of "discovery processes" including workplace ethnography, focus groups, user participation in design teams, and Participatory Design strategies. Social informatics researchers have found that "computerized systems" (technical systems) frequently do not work because designers, managers, and administrators do not adequately account for the social processes in which they are embedded. For example, Kling (1999, p. 19) finds that inadequately considered social processes involving workplace incentives are frequently at the root of technical design failures and that barriers to more effective use include issues of "technological access" (equipment, software) and "social access" (know-how, skill, and economic resources).

Workable technical applications are usually supported by "strong socio-technical infrastructures" (Kling, 1999, p. 20). These findings have led Kling and other social informatics researchers to conclude broadly that:

> Social informatics research pertains to information technology use and social change in any sort of social setting, not just organizations. Social informatics researchers are especially interested in developing reliable knowledge about information technology and social change, based on systematic empirical research, to inform both public policy debates and professional practice. Many of us have developed concepts to help understand the design, use, configuration, and/or consequences of information and communication technologies so that they are actually workable for people. This contrasts with high spirited but largely a-priori promotions of technologies that occasionally work well for people, occasionally are valuable, are sometimes abandoned, are sometimes unusable, and thus incur predictable waste and inspire misplaced hopes. That is one important way that "social informatics matters." (Kling, 1999, p. 21)

Clearly an effective end product is the goal for designers and users. Social informatics researchers have a long tradition of searching for clues and refining methods that help produce these types of solutions (for example, see the research on communities of practice by Chaiklin & Lave, 1993; Davenport & Hall, 2002; Hara, 2000; Hara & Kling, 2002; Jubert, 1999; Kling, 1999; Orlikowski & Yates, 1994; and Wenger, 1998). In the process, Kling (1999) has redefined what it means for a technical system to "work." He draws the metaphor from a critique of two relatively newly launched e-journals, both of which make ingenious use of Internet technology (see also Hara & Kling, 2002). One journal integrates both a preliminary public and self-identified peer review process and a final juried review process (with prominent figures in the field), whereas the second uses an anonymous voting procedure to review articles. The first journal became quite successful, but the second one did not. Kling (1999) argued that the journal that more effectively integrated the preexisting social processes and incentives forming part of the "commonly known" publishing process was the one that worked. He concluded from this example that our definition of what constitutes a "working" technological system (in this case, an e-journal) cannot be defined strictly in technical terms, but must take account of the social process within which the technology exists. This highlights the importance of an existing "sociotechnical" ensemble—an environment composed of technology or technological solutions and the social processes in which they are embedded—which demands consideration as part of a new technology design program.

The components of a "sociotechnical" ensemble (Kling, 1999, p. 15) are (1) people in various roles and relationships with each other and with other system elements, (2) hardware (computer mainframes, workstations, peripherals, telecommunications equipment), (3) software (operating systems, utilities and application programs), (4) techniques (management science models, voting schemes), (5) support resources (training, support, help), and (6) information structures (content and content providers, rules, norms, regulations, and access controls). This list does not explicitly include the physical setting within which the technology is used. Davenport and Bruce (2002, p. 225) explore the relationships between knowledge management and the use of space. They point out that "office activities are embedded in complex infrastructures whose effects are difficult to unravel." They investigate the process of developing and implementing an "office of the future" designed to facilitate communication and knowledge sharing among users. They conclude that:

> Managers who initiate new office projects should not have firm expectations that a de-construction of traditional habitats will liberate the knowledge management practices of employees. A totally open environment may confuse those who work there; an environment that offers generic spaces in an attempt to bound the interactions that take place may constrain those involved. Those who are required to work in non-traditional office space, just like those who work with new technology, may require instruction and mediation. (Davenport & Bruce, 2002, p. 229)

The social informatics approach to the study of the use of technology in the workplace shares many characteristics with WPS. Both tend to use qualitative research approaches. Social informatics tends to emphasize interviews and observations, whereas WPS includes these methods but grounds its analysis in videotaped data that provides the foundation for an understanding of precisely how work is done in the work setting being studied. Consequently, in social informatics research there tends to be less emphasis on the explication of the lived work that occurs as workers interact with the technology. For example, in Orlikowski's (1992) study of groupware, which combined observational data with detailed interviews, the focus of the analysis was on the interview data. Orlikowski argued that two organizational elements affect the utilization of groupware: cognitions and structural properties. *Cognitions* are the mental models, or "technological frames," that people use to understand and orient to software (Orlikowski, 1992, p. 364). Unless people are tuned into the properties of collaborative software, their preexisting models lead them to approach the new groupware as they would existing tools such as spreadsheets, e-mail programs, and word processing software. This results in underutilization of the new groupware—employees filter out and ignore many of the special features and advantages of groupware, thus reducing its effectiveness. Furthermore, when the properties of the new software run counter to the existing cultural paradigm, it is "unlikely to facilitate collective use and value" (Orlikowski, 1992, p. 362). Employees' technological frames are influenced or changed through education about product features and functionality (Orlikowski, 1992, p. 364).

According to Orlikowski (1992), *structural properties* include reward systems, policies, procedures, firm culture, and work norms. In the organization that served as the site of her groupware study, Orlikowski found that employees were rewarded for billing hours to clients. This structural condition under which employees worked made it difficult for them to justify spending time learning the new technology (Orlikowski, 1992, p. 366). In addition, employee uncertainty about changes in work routines, data quality, confidentiality, and access control impeded the implementation of the groupware (Orlikowski, 1992, p. 366). From a cultural/normative perspective, "managers and senior consultants were also more anxious about personal liability or embarrassment" (Orlikowski, 1992, p. 366). These organizational issues were perceived as being outside the CIO's domain of responsibility and therefore were left unattended in order to avoid turf issues (Orlikowski, 1992, p. 367). Additionally, corporate norms around issues of "competitive individualism" at lower organizational levels limited employees' willingness to share expertise, thus undermining the main objective of collaborative groupware implementations (Orlikowski, 1992, p. 367).

When groupware is implemented in organizations, there may be established habits and relationships that affect the success of the implementation: "A major premise [of] underlying groupware is the coordination of activities and people across time and space. For many users, such a premise may represent a radically different understanding of technology than they experienced before" (Orlikowski, 1992, p. 368). Without appropriate changes to their mental maps or technological frames, employees will experience difficulty making adjustments to their work routines and the software implementation will suffer as a result. Further, if the concept of collaborative work embodied in groupware is incongruent with an organization's culture and incentives, its implementation is likely to be ineffective unless structural changes are made.

Orlikowski (1996) reports on an organization that experienced organizational changes after the implementation of groupware for customer service. Changes that occurred in the organization included "the nature and distribution of work, the form of collaboration and interaction, the coordination among units, and the utilization of the knowledge accumulating in the groupware repository" (Orlikowski, 1996, p. 53).

Social informatics research makes a valuable contribution to our understanding of the use of technology, yet it exhibits some differences from the WPS approach. Although observational data are obtained, these tend to be of a general nature and researchers often do not conduct detailed analysis of the sequence of events and actions in which participants engage. Furthermore, social informatics research tends not to include videotaped or tape-recorded data that could enhance our understanding of how participants actually use the technology and how it affects interactions among coworkers in the organization. Perhaps because of this, there tends not to be an understanding of the lived work in the setting—resulting in a less precise understanding of the nature of the work being done than could be obtained from WPS research based on a more ethnomethodological approach.

However, one similarity between social informatics and WPS seems to be a shared understanding of the reflexive nature of context and human action. Sawyer and Echenfelder (2002, p. 438) state that a "reciprocal relationship exists between ICT design, implementation, use, and the context in which these occur." Thus in both approaches, the social context is not treated as a "container" for workers' behaviors, but as part of the web of social life that is also shaped by the actions of individuals and the technology itself. For example, Sawyer and Eschenfelder (Sawyer & Eschenfelder, 2002, p. 453) assert that "social informatics research, ideally, investigates how the influencers and nodes in a socio-technical network shape each other." However, their review article shows that social informatics research tends to focus on how contexts shape ICT design, implementation, and use and to neglect the shaping effect of these elements on their context.

Furthermore, social informatics and WPS researchers tend to mean different things when they refer to context. In a word, social informatics takes a broad brush or macro approach to the study of context, whereas WPS is more focused on the micro-level context. Walsham and Sahay's (1999) article on the difficulty of having a geographical information system (GIS) accepted in India illustrates the social informatics approach to context. They found that there were cultural characteristics that made scientific approaches less attractive, as well as a lack of cultural emphasis on maps, which decreased the need for and interest in a GIS system. On the other hand, Heath and Luff's (2000) earlier-discussed study on the work of editors in the Reuters newsroom (in which the researchers viewed and analyzed the social and interactional context created by the editors as they worked on their articles sitting side by side at computer terminals) illustrates the WPS approach to context.

How WPS Can Contribute to Systems Design

Grinter's (1997) review of WPS research concludes that even if it is not practical to implement the full range of methods advocated by WPS researchers, systems designers can still benefit by incorporating key findings of WPS research into their design process. Her review of the literature led to the following "eight observations about how individuals collaborate and the role of technology in that collaboration" (Grinter, 1997, p. 231):[4]

1. People make assessments about data based, in part, on the status of the provider.

2. Individuals make some of their work visible to others, and also monitor each other.

3. People's perception of technology effects the ways that they use it.

4. Researchers have shown that work is dynamic and involves many channels of communication.

5. Spatial information provided by the arrangement of papers and personnel lets others know the current activities of the entire group.

6. People construct and share interpretations of the work-in-progress.

7. Work often deviates from the planned activity in order to accommodate situated action.

8. Maintaining context supports long-term collaborations. (Grinter, 1997, pp. 231–232)

It is worth noting that Grinter's observations apply not only to the work of systems designers, but could be productively used by managers and other leaders in organizations who make decisions in a variety of arenas, not just technology. For example, Dawes's (2005) ethnographic study of the decision-making process around product innovation issues in a large corporation found that, when managers were reluctant to support a new product idea, they did not share these reservations in meetings with higher-level directors, if the director supported the product. This example illustrates how status differences can affect the work process, even when no technology is involved. Dawes argues that the WPS perspective and methodological approach could usefully be expanded to include a much broader concept of technology than has been the case in the past. Specifically, organizational hierarchies and procedures could be treated as "technologies" and studied in their situated, lived-work context to learn how organizations could function more effectively.

Conclusion

In this chapter, we have shown how WPS research can lead to results and insights that have practical implications for business. Social scientists are helping business leaders reassess and improve the design of, and design processes for, many different technological applications. In so doing, they are demonstrating how and why conventional approaches to requirements engineering and the design of technological solutions for workplace problems are insufficient and consequently fail to accomplish their design goals. Designers have inadvertently, but nonetheless fatally, filtered out the inherently mundane and routine social processes of work, the lived work that employees both do and depend upon in their day-to-day job activities. WPS researchers are also reconceptualizing the way we think about technology itself. Findings regarding how employees are organized and their relationships to the work being done raise legitimate questions about how technological systems are organized, how work is distributed, and how the office environment is conceived and organized geographically as a resource for accomplishing tasks at hand.

Even though WPS has been growing in scope and influence in Europe over the last two decades, it has not had a similar impact in the U.S., with the exception of work by Whalen, Suchman, and some others. We feel that this has a great deal to do with difficulties in bridging an historic gap between the social science academy and the U.S. business community. This gap has left us with two distinct institutional languages and publishing arenas, making it difficult to communicate findings from basic research. However, as we have attempted to show, there are clear applications of these findings for decision-making processes in many types of organizations. The opportunities to enhance workplace productivity, particularly in those places where humans and technology intersect, are significant.

When do you need the full-blown workplace study and when can you get by with a quick and dirty version? Plowman et al. (1995, p. 4) argue that "the 'quick and dirty' study, which is much shorter (a few days) than the traditional study (several months or years) and uses predetermined research questions, is better accommodated within the timescales of a system design cycle" (see also Hughes et al., 1994). However, Plowman et al. also criticize WPS for too often failing to produce specific design suggestions (at least as far as is evident from published studies; they acknowledge that for proprietary reasons, many of the successful design ideas coming out of WPS research may not be made public). Perhaps the investment of time and energy to do a more extensive and analytically correct workplace study would pay off in the long run more often than is currently believed.

It is not clear from published studies whether WPS researchers have worked out a *collaborative* design process that provides *in progress* information for systems designers. Such a process has the advantage of making social research available for design review and comment as it is being produced. This type of ongoing interchange between researchers, designers, and managers may well increase the efficiency and effectiveness of the research. Ideally, this type of integrated research/design process could lead to more comprehensive results with more productive solutions. On the one hand, social scientists may voice concerns that such an approach exposes the research to risks of premature and inaccurate findings. Yet, on the other hand, it may open spaces for exploration by both parties that neither had envisioned. In the end, does it make sense for the development process to be any less collaborative than the end product?

Future research on WPS could also try to determine to what extent the results of studies are left unpublished and to assess their hidden effectiveness. However, we should also keep in mind that WPS is not a substitute for the design process—it is perhaps legitimate that the actual design ideas should come from the designers themselves—but with accurate, relevant, and precise understandings of the work processes, work practices, and social meaning of the work involved, designers will have a more solid foundation on which to base their work. An interesting avenue for further research would be to compare cases of system design development that have incorporated the findings of WPS with those that have not (see Plowman et al., 1995, p. 14). This type of comparison might help WPS researchers to know more clearly what it is designers need to know, and might also show businesses more clearly the contribution WPS is capable of making.

Acknowledgments

We would like to thank Tyler Anderson and Kathryn Martin for their contributions to the first draft of this chapter and the anonymous reviewers at *ARIST* for their helpful suggestions.

Endnotes

1. In addition to Luff et al.'s (2000) and Heath and Luff's (2000) books, both of which contain excellent review chapters, Ilkka Arminen has published a useful review essay on these two books (Arminen, 2001).

2. In an experimental study of media spaces, Barnard, May, and Salber (1996) also discovered that participants found it difficult to understand which object or direction the other was referring to; these problems were greatest when a face-to-face video image was provided. Okada and Matsushita (1999) also report difficulties with eye contact and gaze direction in videoconferences.

3. Library and Information Science (LIS) tends to rely on a social informatics approach to research. LIS is concerned with the development of information technologies that support browsing, searching, and retrieving of information in library contexts (Crabtree et al., 2000). In pursuit of this goal, the field seeks to understand user behavior through the use of surveys, observational techniques, and artifact analysis. As such, there are clear connections between LIS and WPS as described here. Crabtree et al. (2000) argue that although LIS has made modest technological gains, the field needs to do a better job of understanding the context in which information retrieval occurs (Bates, 1989; Crabtree et al., 2000; Frohmann, 1992). Given this criticism, Crabtree and his colleagues go on to describe the benefits of ethnomethodologically informed ethnography as an alternate ethnographic method for conducting LIS research.

4. The idea of differences between planned and situated action is referenced later by Grinter (1997, p. 236), who attributes it to Suchman (1987).

References

Arminen, I. (2001). Workplace studies: The practical sociology of technology in action. *Acta Sociologica, 44,* 183–189.

Bannon, L. (2000). Situating workplace studies within the human–computer interaction field. In P. Luff, J. Hindmarsh, & C. Heath (Eds.), *Workplace studies: Recovering work practice and informing system design* (pp. 230–241). Cambridge, UK: Cambridge University Press.

Barnard, P., May, J., & Salber, D. (1996). Deixis and points of view in media spaces: An empirical gesture. *Behaviour & Information Technology, 15*(1), 37–50.

Bates, M. J. (1989). The design of browsing and berry picking techniques for the online search interface. *Online Review, 13*(5), 407–424.

Bellotti, V., & Bly, S. (1996). Walking away from the desktop computer: Distributed collaboration and mobility in a product design team. *Computer Supported Cooperative Work '96,* 209–218.

Beyer, H., & Holtzblatt, K. (1998). *Contextual design: Defining customer-centered systems.* San Francisco: Morgan-Kaufmann.

Bobrow, D. G., & Whalen, J. (2002). Community knowledge sharing in practice: The Eureka story. *Reflections, 4*(2), 47–59.

Bowers, J., Button, G., & Sharrock, W. (1995). Workflow from within and without: Technology and cooperative work on the print industry shopfloor. In H. Marmolin, Y. Sundblad, & K. Schmidt (Eds.), *Proceedings of the Fourth European Conference on Computer-Supported Cooperative Work (ECSCW '95),* 51–66.

Bowers, J., Pycock, J., & O'Brien, J. (1996). Talk and embodiment in collaborative virtual environments. *Proceedings of the SIGCHI Conference on Human Factors in Computing Systems,* 58–65.

Büscher, M., Gill, S., Mogensen, P., & Shapiro, D. (2001). Landscapes of practice: Bricolage as a method for situated design. *Computer Supported Cooperative Work, 10,* 1–28.

Button, G. (Ed.). (1993). *Technology in working order: Studies of work, interaction, and technology.* London: Routledge.

Button, G., & Harper, R. H. R. (1993). Taking the organisation into accounts. In G. Button (Ed.), *Technology in working order: Studies of work, interaction, and technology* (pp. 98–110). London: Routledge.

Button, G., & Harper, R. (1996). The relevance of "work-practice" for design. *Computer Supported Cooperative Work (CSCW), 4,* 263–280.

Chaiklin, S., & Lave, J. (1993). *Understanding practice: Perspectives on activity and context.* Cambridge, UK: Cambridge University Press.

Coulter, J. (1979). *The social construction of mind: Studies in ethnomethodology and linguistic philosophy.* London: Macmillan.

Crabtree, A., Nichols, D. M., O'Brien, J., Rouncefield, M., & Twidale, M. B. (2000). Ethnomethodologically informed ethnography and information system design. *Journal of the American Society for Information Science, 51*(7), 666–682.

Davenport, E., & Bruce, I. (2002). Innovation, knowledge management and the use of space: Questioning assumptions about non-traditional office work. *Journal of Information Science, 28*(3), 225–230.

Davenport, E., & Hall, H. (2002). Organizational knowledge and communities of practice. *Annual Review of Information Science and Technology, 36,* 171–227.

Davis, A. (1993). *Software requirements: Objects, functions, and states* (2nd ed.). Englewood Cliffs, NJ: Prentice Hall.

Dawes, M. (2005). *Doing culture change: An examination of the implications of time and space for sustainable innovation processes.* Unpublished masters thesis, University of Cincinnati.

Dewan, P. (2001). An integrated approach to designing and evaluating collaborative applications and infrastructures. *Computer Supported Cooperative Work, 10,* 75–111.

Dreyfus, H. L. (1972). *What computers still can't do: A critique of artificial reason.* Cambridge, MA: MIT Press.

Engeström, Y. (2000). From individual action to collective activity and back: Developmental research as an interventionist methodology. In P. Luff, J. Hindmarsh, & C. Heath (Eds.), *Workplace studies: Recovering work practice and informing system design* (pp. 150–166). Cambridge, UK: Cambridge University Press.

Ehrlich, K. (1999). Designing groupware applications: A work-centered design approach. In M. Beaudouin-Lafon (Ed.), *Computer-supported co-operative work* (pp. 1–28). Chichester, UK: Wiley.

Fonseca, B. (2004, April 19). IT role cited in blackout. *Eweek*, 35.

Frohmann, B. (1992). The power of images. *Journal of Documentation, 48*(4), 365–386.

Garfinkel, H. (1967). *Studies in ethnomethodology.* Cambridge, UK: Polity Press.

Garfinkel, H. (2002). *Ethnomethodology's program.* New York: Rowman & Littlefield.

Garfinkel, H., Lynch, M., & Livingston, E. (1981). The work of a discovering science construed with materials from the optically discovered pulsar. *Philosophy of the Social Sciences, 11,* 31–58.

Gause, D., & Weinberg, G. (1989). *Exploring requirements: Quality before design.* New York: Dorset House.

Gaver, W. W. (1991). Technology affordances. *Proceedings of the SIGCHI Conference on Human Factors in Computing Systems*, 79–84.

Glen, P. (2003). Stop gathering IT requirements. *Computerworld, 37*(40), 46.

Goodwin, C., & Duranti, A. (1992). Rethinking context: An introduction. In A. Duranti & C. Goodwin (Eds.), *Rethinking context: Language as an interactive phenomenon* (pp. 1–43). Cambridge, UK: Cambridge University Press.

Gouguen, J. (1994). Requirements engineering as the reconciliation of social and technical issues. In M. Jirotka & J. Gouguen (Eds.), *Requirements engineering: Social and technical issues* (pp. 105–200). London: Academic Press.

Grinter, R. E. (1997). From workplace to development: What have we learned so far and where do we go. *Proceedings of the International ACM SIGGROUP Conference on Supporting Group Work, GROUP'97*, 231–240.

Grudin, J. (1988). Why CSCW applications fail: Problems in the design and evaluation of organisational interfaces. *Proceedings of the 1988 ACM Conference on Computer-Supported Cooperative Work*, 85–93.

Grudin, J. (1994). Groupware and social dynamics: Eight challenges for developers. *Communications of the ACM, 37*(1), 93–105.

Hara, N. (2000). *Social construction of knowledge in professional communities of practice: Tales in courtrooms.* Unpublished doctoral dissertation, Indiana University, Bloomington.

Hara, N., & Kling, R. (2002). Communities of practice with and without information technology. *Proceedings of the 65th Annual Meeting of the American Society for Information Science and Technology*, 338–349.

Harper, R. H. R. (2000a). Analysing work practice and the potential role of new technology at the International Monetary Fund: Some remarks on the role of ethnomethodology. In P. Luff, J. Hindmarsh, & C. Heath (Eds.), *Workplace studies: Recovering work practice and informing system design* (pp. 169–186). Cambridge, UK: Cambridge University Press.

Harper, R. H. R. (2000b). The organisation in ethnography: A discussion of ethnographic fieldwork programs in CSCW. *Computer Supported Cooperative Work, 9*, 239–264.

Harper, R. H. R., & Hughes, J. A. (1993). 'What a f-ing system! Send 'em all to the same place and then expect us to stop 'em from hitting': Making technology work in air traffic control. In G. Button (Ed.), *Technology in working order: Studies of work, interaction, and technology* (pp. 127–146). London: Routledge.

Hartland, J. (1993). The use of 'intelligent machines' for electrocardiographic interpretation. In G. Button (Ed.), *Technology in working order: Studies of work, interaction, and technology* (pp. 55–80). London: Routledge.

Heath, C., & Luff, P. (2000). *Technology in action.* Cambridge, UK: Cambridge University Press.

Heath, C., Luff, P., & Svensson, M. S. (2003). Technology and medical practice. *Sociology of Health and Illness, 25,* 75–96.

Heritage, J. (1984). *Garfinkel and ethnomethodology.* Cambridge, UK: Polity Press.

Heritage, J. (1987). Ethnomethodology. In A. Giddens & J. Turner (Eds.), *Social theory today* (pp. 224–272). Cambridge, UK: Polity Press.

Heritage, J., & Atkinson, J. M. (1984). Introduction. In J. M. Atkinson & J. Heritage (Eds.), *Structures of social action: Studies in conversation analysis* (pp. 1–15). Cambridge, UK: Cambridge University Press.

Hinds, P., & Kiesler, S. (1999). Communication across boundaries: Work, structure, and use of communication technologies in a large organization. In G. De Sanctis & J. Fulk (Eds.), *Shaping organization form: Communication, connection, and community* (pp. 211–246). Thousand Oaks, CA: Sage.

Hughes, J., King, V., Rodden, T., & Andersen, H. (1994). Moving out from the control room: Ethnography in system design. *Proceedings of the 1994 ACM Conference on Computer Supported Cooperative Work*, 429–439.

Hugos, M. (2003). Working hard: Making the same mistakes over and over. *Computerworld, 37*(13), 54.

Isaacs, E. A., & Tang, J. C. (1994). What video can and cannot do for collaboration: A case study. *Multimedia Systems, 2,* 63–73.

Jirotka, M., & Wallen, L. (2000). Analysing the workplace and user requirements: Challenges for the development of methods for requirements engineering. In P. Luff, J. Hindmarsh, & C. Heath (Eds.), *Workplace studies: Recovering work practice and informing system design* (pp. 242–251). Cambridge, UK: Cambridge University Press.

Johansson, M., Fröst, P., Brandt, E., Binder, T., & Messeter, J. (2002). Partner engaged design: New challenges for workplace design. In T. Binder, J. Gregory, & I. Wagner (Eds.), *PDC 02 Proceedings of the Participatory Design Conference* (162–172). Palo Alto, CA: CPSR.

Jubert, A. (1999). Developing an infrastructure for communities of practice. *Proceedings of the 19th International Online Meeting*, 165–168.

Kawatoko, Y. (1999). Space, time and documents in a refrigerated warehouse. *Human Studies, 22*, 315–337.

Kensing, F., & Blomberg, J. (1998). Participatory design: Issues and concerns. *Computer Supported Cooperative Work, 7*, 167–185.

Kling, R. (1999). What is social informatics and why does it matter? *D-Lib Magazine, 5(1)*, 1–32.

Loucopoulos, P., & Karakostas, V. (1995). *System requirements engineering*. London: McGraw Hill.

Luff, P., & Heath, C. (2002). Broadcast talk: Initiating calls through computer mediated technology. *Research on Language and Social Interaction, 35*(3), 337–366.

Luff, P., Heath, C., Kuzuoka, H., Hindmarsh, J., Yamazaki, K., & Oyama, S. (2003). Fractured ecologies: Creating environments for collaboration. *Human–Computer Interaction, 18*, 51–84.

Luff, P., Hindmarsh, J., & Heath, C. (Eds.). (2000). *Workplace studies: Recovering work practice and informing system design*. Cambridge, UK: Cambridge University Press.

MacKinnon, I., & Goodwin, S. (1992, October 29). Ambulance chief quits after patients die in computer crash. *Independent*, p. 1

Mannheim, K. (1952). On the interpretation of Weltanschauung. In *Essays on the sociology of knowledge* (pp. 33–83). London: Routledge & Kegan Paul.

Meehan, A. J. (1986). Record-keeping practices in the policing of juveniles. *Urban Life, 15*, 70–102.

Müller, R. (1997). Coordination in organizations. In S. Kirn & G. O'Hare (Eds.), *Cooperative knowledge processing: The key technology for intelligent organizations* (pp. 26–42). London: Springer.

Okada, K., & Matsushita, Y. (1999). MAJIC videoconferencing system. In Y. Matsushita (Ed.), *Designing communication and collaboration support systems* (pp. 17–32). Amsterdam: Gordon and Breach.

Olson, G. M., & Olson, J. S. (2000). Distance matters. *Human–Computer Interaction, 15*, 139–178.

Olson, J. S., Olson, G. M., & Meader, D. K. (1995). What mix of video and audio is useful for remote real-time work? *Proceedings of the SIGCHI Conference on Human Factors in Computing Systems*, 362–368.

Olson, J. S., Olson, G. M., Storrøsten, M., & Carter, M. (1993). Group work close up: A comparison of the group design process with and without a simple group editor. *ACM Transactions on Information Systems, 11*, 321–348.

Orlikowski, W. J. (1992). Learning from notes: Organizational issues in groupware implementation. *Proceedings of the 1992 ACM Conference on Computer-Supported Cooperative Work*, 362–369.

Orlikowski, W. J. (1996). Evolving with notes: Organizational change around groupware technology. In C. U. Ciborra (Ed.), *Groupware and teamwork: Invisible aid or technical hindrance?* (pp. 23–60). Chichester, UK: Wiley.

Orlikowski, W. J., & Yates, J. (1994). Genre repertoire: The structuring of communicative practices in organizations. *Administrative Sciences Quarterly, 39*, 541–574.

Page, D., Williams, P., & Boyd, D. (1993). *Report of the inquiry into the London Ambulance Service*. London: South West Thames Regional Health Authority.

Plowman, L., Rogers, Y., & Ramage, M. (1995). What are workplace studies for? *Proceedings of the Fourth European Conference on Computer-Supported Cooperative Work (ECSCW '95)*, 309–324.

Prevignano, C. L., & Thibault, P. J. (Eds.). (2003). *Discussing conversation analysis: The work of Emanuel A. Schegloff*. Amsterdam: John Benjamins.

Rogers, Y. (1994). Integrating CSCW in evolving organizations. *Proceedings of the 1994 ACM Conference on Computer Supported Cooperative Work*, 67–78.

Rouncefield, M., Hughes, J. A., Rodden, T., & Villers, S. (1994). Working with "constant interruption": CSCW and the small office. *Proceedings of the 1994 ACM Conference on Computer Supported Cooperative Work*, 275–286.

Sachs, P. (1995). Transforming work: Collaboration, learning and design. *Communications of the ACM, 38*(9), 36–44.

Sacks, H. (1984). Notes on methodology. In J. M. Atkinson & J. Heritage (Eds.), *Structures of social action: Studies in conversation analysis* (pp. 21–27). Cambridge, UK: Cambridge University Press.

Sawyer, S., & Eschenfelder, K. R. (2002). Social informatics: Perspectives, examples, and trends. *Annual Review of Information Science and Technology, 36*, 427–466.

Schegloff, E. A., & Sacks, H. (1973). Opening up closings. *Semiotica, 7*, 289–327.

Searle, J. R. (1980). Minds, brains, and programs. *Behavioural and Brain Sciences, 3*(3), 427–457.

Sellen, A., & Harper, R. H. R. (2003). *The myth of the paperless office*. Cambridge, MA: MIT Press.

Suchman, L. (1987). *Plans and situated actions: The problem of human–machine communication*. Cambridge, UK: Cambridge University Press.

Suchman, L. (1995). Making work visible. *Communications of the ACM, 38*(9), 56–64.

Suchman, L. (2000). Making a case: 'Knowledge' and 'routine' work in document production. In P. Luff, J. Hindmarsh, & C. Heath (Eds.), *Workplace studies: Recovering work practice and informing system design* (pp. 29–45). Cambridge, UK: Cambridge University Press.

Suchman, L., Blomberg, J., Orr, J. E., & Trigg, R. (1999). Reconstructing technologies as social practice. *American Behavioral Scientist, 43*(3), 392–408.

Suchman, L., Trigg R., & Blomberg, J. (2002). Working artefacts: Ethnomethods of the prototype. *British Journal of Sociology, 53*(2), 163–179.

Tang, J. C., & Isaacs, E. A. (1993). Why do users like video? *Computer Supported Cooperative Work*, 163–196.

ten Have, P. (1999). *Doing conversation analysis: A practical guide*. London: Sage.

Thomas, J., & Kellogg, W. A. (1989). Minimizing ecological gaps in interface design. *IEEE Software, 6*(1), 78–86.

U.S.-Canada Power System Outage Task Force. (2004). *Final report on the August 14, 2003 blackout in the U.S. and Canada: Causes and recommendations*. Retrieved January 17, 2005, from http://www.nrcan-rncan.gc.ca/media/docs/final/BlackoutFinal.pdf

Walsham, G., & Sahay, S. (1999). GIS for district-level administration in India: Problems and opportunities. *MIS Quarterly, 23*(1), 39–65.

Wenger, E. (1998). *Communities of practice: Learning, meaning, and identity.* Cambridge, UK: Cambridge University Press.

West, C., & Zimmerman, D. H. (1982). Conversation analysis. In K. R. Scherer & P. Ekman (Eds.), *Handbook of methods in nonverbal behavior research* (pp. 506–541). Cambridge, UK: Cambridge University Press.

Whalen, J., & Vinkhuyzen, E. (2000). Expert systems in (inter)action: Diagnosing document machine problems over the phone. In P. Luff, J. Hindmarsh, & C. Heath (Eds.), *Workplace studies: Recovering work practice and informing system design.* Cambridge, UK: Cambridge University Press.

Whalen, J., Whalen, M., & Henderson, K. (2002). Improvisational choreography in teleservice work. *British Journal of Sociology, 53*(2), 239–259.

Whittaker, S., Frohlich, D., & Daly-Jones, O. (1994). Informal workplace communication: What is it like and how might we support it? *Proceedings of the SIGCHI Conference on Human Factors in Computing Systems*, 131–137.

Winograd, T., & Flores, F. (1987). *Understanding computers and cognition: A new foundation for design.* Reading, PA: Addison-Wesley.

Wolf, M., Foltz, C., Schlick, C., & Luczak, H. (2002). Development and evaluation of a groupware system to support chemical design processes. *International Journal of Human–Computer Interaction, 14*(2), 181–198.

Xiao, Y., Lasome, C., Moss, J., Mackenzie, C. F., & Faraj, S. (2001). Cognitive properties of a whiteboard: A case study in a trauma centre. *Proceedings of the Seventh European Conference on Computer-Supported Cooperative Work*, 16–20.

Zimmerman, D. H. (1969). Record keeping and the intake process in a public welfare agency. In S. Wheeler (Ed.), *On record: Files and dossiers in American life* (pp. 319–345). Beverly Hills, CA: Sage.

Zuboff, S. (2002). In the age of the smart machine. In A. Wharton (Ed.), *Working in America: Continuity, conflict and change* (2nd ed.). Boston: McGraw Hill.

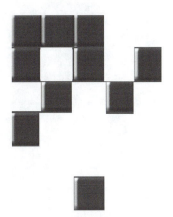

Theoretical Perspectives

CHAPTER 11

Information History

Alistair Black
Leeds Metropolitan University

Introduction

This chapter attempts to offer a definition of "information history," although the pervasiveness of information makes this a difficult task. By means of reference to a healthy sample of recent work and an economical selection of seminal studies, the chapter offers a tentative agenda for a new field. The history of information can be broken down into a number of component parts. These include the history of print and written culture, including the history of libraries and librarianship, and the history of more recent information disciplines and practices, namely information management, information systems, and information science. Contiguous areas like the history of the information society, surveillance, and the information infrastructure are also considered. Beyond these fairly predictable areas, however, an effort is made to advertise work in which social historians have highlighted the existence of past informal information networks. The chapter's conclusion includes brief comments on the importance of historical perspectives to education for information work.

Information History

The pervasiveness and intangibility of information renders elusive any normative definition of "information history." The task of defining terms prefaced by the word "information" is always difficult as it inevitably raises questions about the nature and properties of information in the context of human relations and endeavor (as distinct from information in the context of human biology, such as electro-chemical signaling in the brain and genetic transmission). A fairly common assumption, it might be argued, is that:

> Anything contains information and information may relate to anything ... Information may be contained in the smell of a new perfume ...; in Landseer's "Stag at Bay"; in a letter to a sweetheart; a laboratory slide; a computer tape. (Ritchie, 1982, p. 96)

More often than not, attempts to settle the semantics of information produce confusion instead of the desired clarity. The passage just quoted, for example, immediately invites speculation on the enduring dialectic between "information as process" and "information as thing" (Buckland, 1991). Historical and etymological analysis of the word *information* can throw light on this tension, as well as provide, of course, a first example in this chapter of information history in action. The classical meaning of *information* revolved around information as an activity or happening (information as verb), our actions being "informed" by the metaphysical or, from the Enlightenment onwards, by the hidden powers of reason (Peters, 1988). Bailey's *A New Universal Etymological English Dictionary*, published in 1775, defines information as "the act of informing or actuating." Derived from the Latin word *informare* (to instruct), it has a long history of being used in the sense of the receiving or giving of new knowledge about something. However, in recent centuries, and emphatically in recent decades, what has come to the fore is the notion of information as an item (information as noun)—a change of meaning wholly in keeping, of course, with the increased commercialization and commodification of information. The reification of information, the conversion of it from a processual concept into a "thing," implies that almost any aspect of human culture, material or otherwise, can be considered to have an information dynamic. Even in discourses where the "process" definition of information receives support, there can be a tendency to emphasize—unhelpfully for our purposes, it might be suggested—the perceived universality of information, as in the case of the following statement:

> The concept of information science is based on the assumption that all organisms are information systems .... The information system is an environment of persons, machines, and procedures that augment [sic] human biological potential to acquire, process, and act upon data. (Debons, Horne, & Croneweth, 1988, pp. 8–9)

All this leaves those attempting to theorize and formulate a discrete subject named "information history" with a difficult problem. If information itself defies precise definition, what chance is there that its definition might be historicized? There is a danger that information history, like information, can be conceptualized in such vague and heterogeneous ways that it is rendered unwieldy and thus unsuitable for the award of disciplinary status. A subject field that potentially accommodates the history of communications, of the book, of copyright, of printing, of libraries and librarianship, of publishing, of organizational bureaucracy and infrastructures, of computing, of newspapers and periodicals, of the media, of clerical work, of intelligence, of market research, of propaganda, of professionals and their expert record systems, of manual and automated information technologies and systems, of information science, of accountancy, of knowledge, of scientific dissemination, of information warfare, of administration, of records and archives management, of software development, of Web site creation and management, or of information management would be difficult to defend as a cohesive paradigm and might even attract accusations of naivety and scornful laughter.

More positively, however, the same cannot perhaps be said of attempts to place an informational spin on *particular* past events or topics, whether this may be, to offer some highly selective examples, the role of information management in business history; the contributing of record-keeping expertise to the planning of D-Day; telephone exchanges, television stations and other communication facilities as prime targets in wars and revolutions; or the contribution of mechanical, but sophisticated, information technology to the Holocaust, as revealed by Edwin Black (2001). The key issue, of course, is whether such diverse topics of historical interest can ever be shaped by, and included in, a single cognate field, even if they deserve attention in historical studies greater than the footnote status they normally achieve.

In truth, this question would at present tend to stimulate a negative answer. However, if, as the information society enthusiasts would have it, information becomes the defining feature of our civilization and the information disciplines continue to gain in academic recognition, a more positive response may not be too far off. Certainly, an affirmative answer becomes a more realistic possibility when one considers that historical fields have, to a degree, owed their development to the maturation of the "umbrella" disciplines under which they are located. Broadly, the development of the history discipline over the past century has tracked the development of "parent" domains of knowledge. The great flowering of history in the 20th century into a variety of subsets was in many ways linked to the emergence of the social sciences—economics, anthropology, sociology, psychology, geography, political science, and their supporting statistical methods—giving rise to economic history, social history, and so on. Equally, as the pure sciences became more sophisticated and, above all, accessible, the history of science and technology became less the monopoly of an exclusive group of specialists focusing on the internal development of their subjects and more the domain of historians in pursuit of the "social" causes and effects of scientific and technological discovery and innovation (Marwick, 1981, pp. 99–133 and p. 221; Soffer, 1994; Taylor, 1997).

Prompted by the digital "revolution," information has become a fashionable topic for historical investigation. Unsurprisingly, there has been curiosity as to how information was "created, diffused and manipulated in the past, and with what effects" (Slack, 2004, p. 33). But this does not make information history a recognized history subfield. Unlike other sub-sets of history, it commands no immediately identifiable canon of literature or methodological discourse. However, like most history subfields, it *can* point to an interdisciplinarity and an eclecticism that, although not conducive to a tidy disciplinary definition, suggests it to be an interesting, intriguing, and rewarding "area" (as opposed to recognized subject) of study and research. The large variety of potential topics and perspectives that can *potentially* be incorporated under the umbrella of information history makes for an exciting and fresh scholarly domain. Evidence of the width and vibrancy that can be achieved is clearly visible in Chandler and Cortada's (2000) collection of edited essays on how information has shaped the U.S. since colonial times.

Notwithstanding the pervasiveness of information and its consequential detectable and unremitting presence throughout the historical record, it is nonetheless possible, I suggest, to map out at least the contours of an information history field with identifiable boundaries (of course, to use the term "information history," as opposed to the looser term "history of information," engagingly discussed by Stevens in 1986, has a more definitive ring to it, perhaps bringing nearer an ultimate realization of the field). Even if the various tribes that constitute the information professions may not be accustomed to visiting each other's territories, let alone the small historical enclaves that nestle within them, much of the subject field outlined in this chapter is terrain recognizable across the various information academies. Familiar landscapes are the long-established history of libraries and the history of print culture, including book history, the history of reading, and the history of publishing. Also familiar, although because of their recent emergence perhaps slightly less so than the tradition historical landscapes of print culture, are the history of information science, information management, and information systems, including computer systems.

Thereafter, the journey moves into what for many in the information professions is *terra incognita*: areas that are known to exist but about which there is *relatively* little knowledge imparted in either professional preparation or continuing professional development. These are the largely unknown lands bounded by the social sciences—in particular the history and sociology of the information society—including, if one is taking a holistic approach, important associated areas such as the history of surveillance, globalization, mass media, information warfare, modernity (and post-modernity), and industrialism (and post-industrialism), each of which displays a potent information dynamic (Kumar, 1995; Robins & Webster, 1999, pp. 89–110). This is not to deny, of course, the contributions, disclosed later in this chapter, of those in the information field who are aware of the "long view" of the information society (e.g., Feather, 2004; Gorman, 2001; Rice, 1991) or those who have understood the historical dimensions of the social foundations of information, as in the landmark work of Jesse Shera, Margaret Egan, and Steve Fuller on social epistemology (Furner, 2004; Zandonade, 2004; see also Fallis in the present volume).

My mapping of the subject field culminates with the presentation of a selection of examples from social and economic history which emphasizes the past development and workings of informal information networks.

Whereas, in the information world, the territory of "information society" history may seem mysterious to most—that is to say, distant but at least known to exist and occasionally visited by missionaries teaching the existence of "historic information societies"—that of social history and its information aspects remains largely undiscovered. However, in reality this domain may not be too far removed from the information science concept of "social intelligence" (Cronin & Davenport, 1993), in which case the features of the landscape, once disclosed, are hopefully less likely in the future to generate surprise.

Although maps can be drawn for individual use, they naturally gain in value the more widely they are disseminated. Thus, the chapter's conclusion includes some brief thoughts on how information history can contribute to education for information professionalism. Dissemination and promotion of the subject to this community is arguably a prerequisite not only of the development of information history but also of its acceptance as a legitimate field in the general history discipline.

To summarize, this chapter explores issues of discipline definition and legitimacy by segmenting information history into its various components:

- The history of print and written culture, including relatively long-established areas such as the histories of libraries and librarianship, book history, publishing history, and the history of reading.

- The history of more recent information disciplines and practice, that is to say, the history of information management, information systems, and information science.

- The history of contiguous areas, such as the history of the information society and information infrastructure, necessarily enveloping communication history (including telecommunications history) and the history of information policy.

- The history of information as social history, with emphasis on the importance of informal information networks.

Given the current lack of recognition and definition of information history, to provide a thorough literature review of a field that in effect does not yet exist is, unsurprisingly, problematic. The extensive width of the proposed field, containing the many constituent elements alluded to here, further deters any effort to offer a comprehensive bibliographic audit. Some readers may be disappointed, for example, that little or no direct reference is made here to the history of cybernetics, statistics, or social signification—or, indeed, knowledge, often used as a synonym for information (e.g., Bennett & Mandelbrote, 1998; Frängsmyr, Heilborn, & Rider, 1990; Mahoney, 1990). However, this chapter is in any case intended to be less a full review of academic work across information history's potential domains than an attempt, by means of reference to both a sample of recent work and a fairly selective sample of seminal studies, to set a tentative agenda, to write a draft manifesto, for a new field.

In conceptualizing this new field, this chapter, by mobilizing sociological and social-historical perspectives, goes beyond the discussion of the history of information science by Buckland and Liu in their 1995 *ARIST* chapter, which mostly addressed the history of practice and professional knowledge in the information disciplines, including librarianship. Those seeking to access an exhaustive and authoritative review of sources in these specific areas will find considerable value in Buckland and Liu's (1995) review, as well as in contributions in recent years by Rayward (1998, 2004b) and Buckland (1999). Like all these reviews, this chapter mostly discusses scholarship, issues, developments, and activities in American and British contexts, with occasional reference to work in other countries.

History of Print and Written Culture: Libraries, Books, Reading, and Publishing

The work of historians would not be possible without access to libraries and the service rendered by those who work in them. Yet, considering the use they make of libraries, it is surprising how, with the occasional exception, library history has, as Bartlett (1966, p. 13) noted four decades ago at an early gathering of library historians in the U.S., "been for the most part forgotten by historians." Little has changed since this observation was made. In the absence of "professional" historians (those whose main job it is to research and teach history), the field of library history has mostly been left to the library world itself. Library history research has been conducted mostly by librarians, students of librarianship, and library educators. Inevitably, historians from these groups have less time than "full-time" historians to devote to the study of contextual historical knowledge and to methodological debates in history, with the result that library history is always vulnerable to the criticism that it lacks rigor (Shiflett, 1984, p. 386). However, many "library historians," it must be emphasized, have received extensive education and training in history and historical methodology. One should certainly not underestimate, therefore, the admirable historical scholarship and competency present in the field. Similar arguments apply to those working in the field of information science history.

The absence of the professional historian in the field has meant that library history has evolved without the *full* influence of knowledge of "other" history. However, let there be no doubt that in recent decades library history has moved a long way from its antiquarian origins. Its scholars have begun to appreciate the importance of importing into their research theories drawn from other disciplines, thereby endeavoring to match methodological standards existing in "mainstream" history fields. Much more than in the past, library historians now are prepared to explore the contexts in which libraries operated, as in the case of an account of the history of Cambridge University Library in which the author declares an attempt "to pay close attention to the place of the Library not only within the University, but also in the country at large and in the international scholarly community" (McKitterick, 1986, p. 2).

Bodies influential in the field include the American Library Association's Round Table on Library History, the Library History Section of the International Federation of Library Associations and Institutions (IFLA), and, in the U.K., the Library and Information History Group of the Chartered Institute of Library and Information Professionals (CILIP). Each of these bodies has been busy in recent years, running conferences and seminars, and initiating scholarly projects. Active library history groups function in many other countries, including Germany (The Wolfenbüttel Round Table on Library History, the History of the Book and the History of Media, located at the Herzog August Bibliothek), Denmark (The Danish Society for Library History, located at the Royal School of Library and Information Science), Finland (The Library History Research Group, University of Tamepere), and Norway (The Norwegian Society for Book and Library History). Sweden has no official group dedicated to the subject, but interest is generated by the existence of a museum of librarianship in Borås, established by the Library Museum Society and directed by Magnus Torstensson. Activity in Argentina, where, as in Europe and the U.S., a "new library history" has developed, is described by Parada (2004).

As one would expect in a field associated with librarianship, the bibliographical apparatus of the subject is extensive. The field is serviced in the U.S. by the journal *Libraries and Culture*, and in the U.K. by the journal *Library History*. The *British Library History Bibliography*, of which six volumes were issued between 1972 and 1991, and which has sadly now been discontinued, covers work in the period 1960–1988 (Keeling, 1972, 1975, 1979, 1983, 1987, 1991). In the U.S., those seeking an initial grounding in the subject would benefit from a reading of the many valuable historiographical and bibliographical contributions to Wertheimer and Davis's (2000) *Library History Research in America*. The current state of library history in Germany has been discussed by Vodosek (2001). Researchers at Princeton University have created an astonishing database of 10,000 libraries, of all sorts, in existence in the U.S. before 1876 (The Davies Project at Princeton University, 2004); international coverage of the history of libraries and librarianship is afforded through the wide-ranging encyclopedia by Wiegand and Davis (1994).

A recent and prestigious activity undertaken by the Library and Information History Group in the U.K. has been its directing of *The Cambridge History of Libraries in Britain and Ireland*, to be published in three volumes with contributions from over a hundred scholars (Hoare, in press). This ambitious project follows in the footsteps of the impressive four-volume history of libraries in France edited by Poulain (1992).

Encouragingly, a number of general histories of libraries, crossing time periods and geographical boundaries, is available, including texts by Tolzmann, Hessel, and Peiss (2001), Staikos (2000), Battles (2003), and Lerner (2001), each telling the story of libraries since the invention of writing; Harris's (1999) seminal account of the history of libraries in the western world is now in its fourth edition. Despite the fact that the explosion in library provision is a phenomenon of the modern society of the past three centuries, coverage of premodern and ancient libraries remains vibrant (Casson, 2001)—not least in regard to the enduring subject of the Alexandrian library (MacLeod, 2002), work that has built on scholarship by Parsons (1952) a half century earlier.

Studies of the "modern" library scene have been numerous. The theme of the public library has naturally been prominent. Black (1996, 2000), Hewitt (2000), and Snape (1995) have each attempted to provide a fresh approach or perspective to British public library history. A history of American public libraries is offered by Martin (1998). Also in an American context, Jones (1997) has provided an account of the history of Carnegie libraries, and Dain (2000) has written an equally accessible popular history of the New York Public Library. Outside the Anglo-American context, contributions have been forthcoming from Mäkinen (2001) with respect to Finnish libraries and from Stieg (1992) with respect to a particular aspect of the German context: public libraries under the Nazis. But library historians of the "modern" period have also interested themselves in a wide variety of library types other than public libraries, including national, academic, special (for example, industrial, corporate, research, government), subscription (including Mechanics' Institute), circulating, works/plant (for recreational purposes), radical, artisan, ecclesiastical, and private/personal libraries. Library historians have concerned themselves with an array of themes common to many of these institutions: for example, Raven (2004), Knuth (2003), and Samuel (2004) have addressed the enduring theme of library and collection destruction. Thematic approaches have formed the basis of several international conferences, addressing such topics as library provision of the 1960s and 1970s in the context of radicalism, utopianism, and social protest (Torstensson, 2002); libraries in the context of the Cold War (Anghelescu & Poulain, 2001); libraries and literature (Vodosek & Jefcoate, 1999); libraries and philanthropy (Arnold, in press; Davis, 1996); and libraries and librarianship viewed through the lens of gender (Kerslake & Moody, 2000).

One of the problems facing library history is the institutional focus that the term implies. In the past, much work has centered on the history of libraries as institutions; as Stam's (2001) extensive encyclopedia and Brooks and Haworth's (1993) history of the Portico Library in Manchester have shown, this approach remains popular and important. More recently, however, increasing attention has been paid to the history of librarianship (e.g., Garrett, 1999), including a renaissance in the study of leading librarians of the past, thereby building on a tradition expertly constructed, in the U.K. for example, by biographers such as Miller (1967) and Munford (1963). Many biographical works have been characterized by high scholarship, such as Wiegand's (1996) study of Melvil Dewey and the collection of essays on the professional lives of leading 19th-century librarians in Boston by Davis, Carpenter, Wiegand, and Aikine (2002). Fourteen essays investigating various pioneers of library and information science have been edited by Rayward (2004a). Such is the continuing importance of biography that new editions have appeared of the Dictionary of American Library Biography (Davis, 2003; Wiegand, 1990), a work first edited by Wyner in 1978. Librarians themselves continue to display a penchant for autobiography (e.g., Harrison, 2000).

Critical biographical work incorporating gender analysis has been produced by Kerslake (2001). Garrison's (2003) exceptional history of early American librarianship has rightly achieved a second edition. The past image and role of the librarian continue to generate interest, work often being complemented by imaginative use of sources from past popular culture, such as films, comic books, children's literature, and advertising (Arant & Benefiel, 2002; Nagl, 1999). Drawing on "official" professional sources, by contrast, Samek (2001) has carefully explored the heated debate in American librarianship in the 1960s and 1970s between, on one hand, those defending neutrality in selection and policy initiatives and, on the other, those seeking greater social responsibility in librarianship through a prioritization of lower socioeconomic groups. A similar tension is explored by Black and Muddiman (1997) in their history of community librarianship in late-20th-century Britain.

Some of the most interesting work on the history of libraries has addressed issues of architecture and design in their social context (Markus, 1993, pp. 172–183). Grime's (1998) detailed catalog of Carnegie libraries in Ireland can be contrasted in terms of approach with Van Slyke's (1995) "social causes of design" analysis of Carnegie libraries in America. Demonstrating how history can throw light on current issues and practices, Garrett (2004) has intriguingly identified a legacy of the Baroque in virtual representations of library space.

It is also worth noting that research has been undertaken on institutions that conform in part to the definition of a library. Examples of work on such quasi-library institutions include the history of patent collections (Hewish, 2000) and the history of archives (Cantwell, 1991; Schwartz & Cook, 2002).

Options for the future direction of library history have been hotly debated (Black, 1998, 2001a; Davis & Aho, 2001; Mäkinen, 2004). Fears that the subject will slide into academic oblivion have not been realized. However, scholars have been busy speculating on its future. Davis and Aho (2001) have suggested four models relating to library history's possible trajectory: retain the current model; change to an information science model; change to a history model; or change to a history of the book model. If library history were to become more closely aligned with the last of these options, book history, and the linked subjects of publishing and reading history—a synergy explored by Rose (1994)—it would be allying itself with areas where scholarship in recent years has been of the highest standard. A major development in book history has been the publication of two initial volumes (five more are projected) of the *Cambridge History of the Book in Britain* (Barnard & McKenzie, 2002; Hellinga & Trapp, 1999). A thriving area, book history has been well served over the decades by the journal *The Library*, published by the Bibliographical Society. Another leading journal in the field is *Book History*, co-edited by Jonathan Rose, founder of the Society for the History of Authorship, Reading, and Publishing (SHARP), which runs a large annual international conference. The vitality of book history is demonstrated in the continuing publication of the extensive *ABHB: Annual Bibliography of the History of the Printed Book and Libraries*, sponsored by the Rare Books and Manuscripts Section of IFLA. The subject span of book history has been extremely wide—geographically, temporally, and culturally. Research has recently covered subjects ranging from the 8th-century Lindisfarne Gospels (Brown, 2003) and 15th-century incunabula (Jensen, 2003), to periodicals of the British Empire (Vann & Van Arsdel, 1996) and the early history of printing (Moran, 2003). The related area of the history of publishing and the book trade remains highly active (e.g., Epstein, 2001; Feather, 1987, 2003, pp. 1–26; Myers, Harris, & Mandelbrote, 2001; Tebbel, 1987).

The history of reading has been addressed, among many others, by Fischer (2003) and Manguel (1996), as well as in a collection of essays edited by Raven, Small, and Tadmor (1996). Rose (2002) has emphasized the importance of researching the interpretations readers give to texts, rather than simply focusing on what was read. The associated issue of literacy has been explored by Clanchy (1993), whose study revealed a significant development in literacy in medieval England, which can be characterized as an early information society; and by Vincent (1993), who examines the history of working-class reading and writing. The entire story of writing is ambitiously told by Robinson (1995); Sacks (2003), in examining the origins and development of the alphabet, suggests it to be humankind's most far-reaching invention.

The History of the Information Disciplines: Documentation, Information Science, Information Management, and Information Systems

Buckland and Liu's (1995, p. 387) assessment that the "amount of published material on the history of IS [information science] remains small" is happily not something that will be repeated in this review. Three major works of collected essays have been published in recent years. By reprinting major papers on the subject in recent issues of information society journals, Hahn and Buckland's (1998) collection of studies in the history of information science laid down an important marker for the field. This momentum was maintained by the organization, in 1998, of a large international conference, sponsored by the American Society for Information Science and Technology (ASIST) and the Chemical Heritage Foundation (CHF), on the history and heritage of science information systems, the extensive proceedings of which were published the following year (Bowden, Hahn, & Williams, 1999). The conference was repeated in 2002, again giving rise to a valuable set of published proceedings (Bowden & Rayward, 2004). Overviews of the subject area are offered in these three volumes by Rayward (1998, 2004b) and by Buckland (1999). In short, information science history—not to be confused with "historical information science," being information science applied to problems connected with the creation, preservation, storage, retrieval, and so forth of historical information (Higgs, 1998; McCrank, 2002; Rosenzweig, 2003)—is an emergent field with a strong and growing academic base.

Those seeking a succinct introduction to the history of information science will find Hjørland (2000) to be of value; whereas those interested in the history of scientific communication generally are advised to commence with authoritative texts by Richards (1994) and Ditmas (1948), as well as a more recent detailed overview by Vickery (2000), which lends good weight to the classical, medieval, and early modern periods. The best source on the origins of information science in the U.S. remains Farkas-Conn's (1990) detailed account of the American Documentation Institute (ADI), established in 1937 and forerunner of the American Society for Information Science and Technology. In the British context, Muddiman (in press) has taken a similar institutional approach in researching the early history of Aslib (Association of Special Libraries and Information Bureaux, established 1924), partly by means of reference to the association's extensive but unlisted archive.

As befits the ethos and practice of information science, efforts have been made to provide wide access to information science history sources via the digital environment. Web sites have been constructed by Williams (2004) on pioneers of information science in North America and by Williams and LaMotte (2004) on bibliographic sources for the history of information science in the 20th century, also in North America. In addition, Williams and Bowden (2000) offer a chronology of chemical information science from the late 18th century on. Elsewhere, information science in Germany has been documented by Hapke (2004). In Britain, efforts to build an information science community have been disappointing, although positive strides may follow from the amalgamation in 2002 of the Institute of Information Scientists and the Library Association to form the Chartered Institute of Library and Information Professionals, which inspired the subsequent change of name, the following year, of the Library History Group to the Library and Information History Group. The lack of activity in Britain compares poorly, of course, with activity in ASIST, which has welcomed information science history panels at its conferences since 1991.

Information management (IM) may be a recent addition to university programs for education and training in information work, but its formal, "scientific" practice can be traced back to the 19th century and perhaps beyond; indeed, one historian has highlighted the complexity of the internal organization and procedures operating in the 18th-century merchant's counting house (Price, 1987). New techniques and technologies of information management—improvements in the methods and "machinery" of collecting and communicating information in organizations and in the way that data and documents were controlled—emerged in response to the complex tasks and operational requirements that confronted the burgeoning state bureaucracies and large-scale business enterprises of the late 19th century. Such were the administrative problems arising from increased state intervention in social and economic life and from the need of businesses to plan production, monitor markets, audit activity, and speed throughput that a range of information technologies and techniques was introduced in a relatively short period of time at the end of the 19th century and beginning of the 20th century. A variety of innovative methods appeared to improve organizations' information and communication infrastructures: statistical analysis, graphic representation, the internal memo, the staff magazine, the management meeting, schemes for classifying documents, the procedural manual, and written protocols. These, in turn, were supported by a series of "device" innovations: stencil duplicators, typewriters, telephones, accounting machines, filing cabinets, and card indexes—supplemented in the decades that followed by the punch-card machine, microfilm, and the photostat. These developments occurred in the context of the growth and increased centralization of office departments, where (increasingly semi-skilled) workers were grouped according to function and subjected to Taylorist supervision; and where, for greater speed, methods of work and forms of documentation were standardized to pave the way for investment in office machinery.

Methodological issues connected with studying the history of such business communication has been the subject of expert commentary by leading researchers in the field (Locker, 1996). In the U.S., groundbreaking research on the early history of IM in large corporations has been conducted by Chandler (1977), Yates (1989), and Beniger (1986). In Britain, the work of Campbell-Kelly (1992, 1994, 1998) with respect to data processing in railway and insurance companies has been equally impressive. The same can be said of Agar's (2003b) history of information management and systems, especially the mechanization of bureaucracy, in British government since early in the 19th century. Orbell (1991) has examined the late-19th- and early-20th-century revolution in the ability of organizations to copy, record, organize, analyze, and communicate information. Rhodes and Streeter (1999) focus on a particular aspect of this revolution and its origins: mechanical copying.

Black and Brunt (2001) have examined the history of IM in the murky world of British secret intelligence. Founded in 1909, the British Security Service, which became known as MI5, was given the task of countering sabotage, espionage, and subversion at home and across the British Empire. The torrent of information that flowed into the organization before and during the First World War led it to construct a relatively efficient manual information management system. The hub of MI5's information activity was its registry, where documents were arranged by hundreds of clerks in subject and personal files, backed up by detailed indexing and cross-referencing in a card catalogue. In the inter-war period, however, this system was allowed to fall into disrepair; so much so that, at the start of the Second World War, MI5's information management system virtually collapsed due to data flooding the system. This crisis forced a reform of information systems in the organization, including the introduction of punched-card machines, microfilming, improved indexing, and the replacement of unwieldy subject files by individual files. Research on indexing techniques in British intelligence has been continued by Brunt (2001) in his analysis of knowledge organization at Bletchley Park during the Second World War.

The history of information systems incorporates, most obviously and immediately, the history of computing. Computing history is a wide-ranging subject that, at one end of its spectrum, examines the evolution of computer hardware (Morris, 1990; Reid, 1985), and at the other, centers on the past application of computer technology in society. It is at the applications end of the spectrum, especially in the context of the organization, that computing history can be seen to merge into the history of (soft) information systems (Boyns & Wale, 1996).

An accessible history of computing, one that links the eras before and after the advent of the electronic computer, has been offered by Campbell-Kelly and Aspray (1996). Work in the area has ranged from accounts pitched at the industrial and sector level (Campbell-Kelly, 1989, 1995) to systems applications within organizations, such as the appearance in the early 1950s, in the confectionery and catering giant J. Lyons and Co., of Britain's first business computing system (Bird, 1994; Camier, Aris, Hermon, & Land, 1997). Early corporate computing in the U.S. has been covered by Haigh (2001a; 2001b); other American perspectives have been discussed in the past by Collen (1995) with regard to the evolution of medical informatics and, earlier yet, by Copeland and McKenney (1988) with regard to the history of airline reservation systems. Beyond the business environment, some researchers have begun to research, with keen sociological insight, the relatively short history of virtual communities and community informatics (Agar, Green, & Harvey, 2002; Schuler, 2001).

An interesting ratification for conceptualizing a discursive information history field is the special issues of the *IEEE Annals of the History of Computing* devoted to the history of computer applications in libraries (Rayward, 2002).

The History of the Information Society: Surveillance, Infrastructure, Communications, and Policy

Enthusiasm for the idea that we have entered a fundamentally new age, termed the "information society," has been tempered by skeptical appraisal of its novelty and epochal legitimacy. The contemporary approach to the questioning of the information society—the debunking of predictions and the questioning of visions—has been strident (Preston, 2001; Webster, 2002). Such analyses of current society and recent change argue that over the past 20 years the predicted large-scale beneficial effects of information technology have been exaggerated. For example, the belief that printed documentation in administration would become a thing of the past, leading to the paperless office and increased leisure time for the post-industrial workforce, has proved to be groundless (Sellen & Harper, 2002).

In questioning the legitimacy of the information society concept, historical perspectives have proved as valuable as contemporary analysis. Recent social and technological change is seen by Dearnley and Feather (2001, p. 131) "as no more than the latest part in a continuum that has stretched over many centuries." All societies have, to a degree and in various ways, been information societies and, as Duff (2000, p. 171) argues, it can be difficult to prove that modern societies are more information-based than others. It is possible, therefore, to conceptualize the existence of "historic information societies," the landscapes and landmarks of which Mattelart (2003) has expertly mapped. May (2002, pp. 19–47) has stressed the importance of locating the "information age" in history, a methodology adopted, among others, by R. E. Day (2001) in outlining two 20th-century information ages predating the "virtual age": the age of European documentation (before the Second World War) and the age of information theory and cybernetics (immediately after the War). Historicizing the information society is a prerequisite to understanding the impact of recent and current technological revolutions, a motive that underpins Blok and Downey's (2003) collection of edited essays on labor in information revolutions, past and present, and Bud-Frierman's (1994) notion of "information acumen" in business history.

History has witnessed a number of sudden shifts in the evolution of human communication, beginning with the development of speech and the invention of writing. The invention of printing in the 15th century represented, in the words of Eisenstein (1983, p. 275), the next great "cultural change of phase." Fueled by the new print culture, the Enlightenment generated new institutions, techniques, and formats— the encyclopedia, the scientific academy, the scholarly journal, the salon, mathematically accurate maps, statistical analysis, and libraries and rational clubs and societies of all kinds—designed to further knowledge and enhance the storage and communication of information (Burke, 2000; Clarke, 2001; Headrick, 2000; Hoare, 1998). In regard to the 19th century, a strong case can be made for the appearance of what I have referred to elsewhere as the "Victorian information society," under-pinned by the arrival of the telegraph and telephone and the rapid devel-opment of the postal service, mechanized printing, the publishing industry, and "public sphere" memory institutions like libraries, muse-ums, and art galleries (Black, 2001b). A Victorian perspective is simi-larly pursued by Richards (1993) who explains how British colonial rule led to the accumulation in various institutions—such as the British Museum—of vast amounts of information from far-flung corners of the Empire, thereby constituting an historic information society. An imper-ial dimension to information is also constructed by Bayly (1996) in an exhaustive account of intelligence gathering by administrative elites in India in the late 18th and early 19th centuries, specifically the network of Indian spies, news writers, and knowledge secretaries deployed by British officials to secure military, political, and social intelligence.

Armed with a knowledge of these historic information societies, prac-tices, and trends, the information historian is well placed to confront the predictions of information society millenarians. A mentality that seeks out continuities inevitably comes to focus on the large-scale cultural and economic shifts of the past. The continuities of industrialism, capitalism, and modernity vastly outweigh any change in human thought and rela-tions wrought by digital technologies. The same might be said of the sur-veillance capabilities and tendencies displayed by modern states over recent centuries. In the era of modernity, the major agency of surveil-lance has been the nation-state (Giddens, 1987). Higgs (2004) tells the story of how and why over the past 500 years the central state in England and Wales has increasingly involved itself in the collection and manipulation of information on the private citizen, observing that mod-ern states are essentially "information states." A similar analysis has been offered by Lyon (1994, 2000), who detects an uninterrupted line of development from the mechanical surveillance of the 19th-century state to the electronic eye of the digital state of today; Lloyd (2003) has chron-icled and discussed how the same period in the evolution of surveillance has witnessed the birth and increasing sophistication of the passport as a means of tracking citizens. Slack (2004) examines the information role of government in 17th-century Britain, with particular emphasis on the rise of "political arithmetic"—the accounting of the wealth, strength, and trade of the nation—as the necessary complement to the rapid growth of the military-fiscal state. Surveillance in its most overt form is addressed by Herman (1996, pp. 1–35), who introduces his analysis of military intelligence services in the modern world with an account of their evolution to date.

The information role of the state in times of war is emphasized in work on subjects as varied as the history of propaganda as a tool of warfare (Taylor, 1995), the communication system underpinning Britain's aircraft detection system in the Battle of Britain (Checkland & Howell, 1998), and indexing techniques at Bletchley Park, again in the Second World War (Brunt, 2004). The subject of the role of intelligence in war has increasingly commanded the interest of historians, although Keegan's (2003, pp. 5, 293) view that intelligence "does not point out unerringly the path to victory" and is "secondary to the age-old business of fighting it out," represents a sobering thought for those who would place information at the center of everything. This said, the emergent concept of "information warfare" will undoubtedly generate further historical studies of the way war and information have coalesced.

In the setting of the professions—especially in the context of professional knowledge as a source of social power—information as the core of surveillance might be explored as the lynchpin of the modern expert's categorization and recording of his or her clients or of society at large. This perspective has been investigated by Driver (1993) with respect to what he refers to as the administrative landscapes of the Victorian Poor Law. Information extracted by the state and its experts from the monitoring of "deviant" groups—such as criminals, delinquents, paupers, prostitutes, vagrants, drunkards, and the insane—amounted to a "science of moral statistics" (Driver, 1993, p. 10); a science enshrined in the work of pioneering 19th-century British social investigators such as Charles Booth and Seebhom Rowntree (Englander & O'Day, 1995). More generally, the scientific planning of society has been assisted by the grandest of surveillance mechanisms, the census, the history of which in the U.S. is told by Anderson (1988, 2001).

The history of surveillance has been theorized by Calhoun (1992) at the level of social relationships. Increasingly, these have become indirect in nature, for premodern "direct," face-to-face, *Gemeinschaft* relationships tend to be overshadowed in industrial and urban societies by mediated relationships, whether via the telegraph or the Internet. Surveillance—and privacy—become an issue in social relationships that are indirect, especially if mediated by digital information technology.

From incunabula to the Internet, it is possible to conceptualize not only the development of past information and surveillance societies but also the evolution of the "information infrastructure" (Borgman, 2000; Lebow, 1995). Tracing the history of the information infrastructure entails the study of both communication history and the history of information policy, in addition to the history of library and information systems already outlined. The institutions, structures, and technologies that have shaped the information infrastructure are worthy of considered historical investigation, if only because, as Bawden and Robinson (2000, p. 56) correctly observe, the examination of earlier communications revolutions may provide lessons for understanding the one currently underway.

The history of the information infrastructure incorporates the history of communication, a subject that commands a scholarly pedigree and comprises many parts (Solymar, 1999). Communication technologies have been at the heart of the shaping and networking of the modern world (Mattelart, 2000), as in the case, for example, of the rise of nationalism, the fundamental basis of which Deutsch (1966) took to be a complementary "social communication" akin to shared culture. Communication history has popular appeal because it often addresses "everyday" technologies, whether these be the cell/mobile phone (Agar, 2003a), the fountain pen (Dragoni & Fichera, 1998), the pencil (Petroski, 1990), the telephone (Graham & Marvin, 2001, pp. 50–51; Sterling, 1995; Young, 1991), or the primitive technology of the speaking tube, as employed in early-19th-century factories and highlighted in Charles Babbage's (1835, p. 9) observations on proto-industrial capitalism. The bewildering array of communications media developed by human societies, from the drums of our ancient ancestors to digital telecommunications, is laid out in descriptive studies by Meadow (2002), as well as by Gardner and Shortelle (1997) in their accessible encyclopedia of communications technology. More analytical contributions have come from Levinson (1997) and, in particular, Flichy (1995), who wins credit for addressing, among other media, the sociology of early photography, just one aspect of the emergence of the "optical culture" of the 19th century, which is also the focus of Crary (1990). A history of telecommunications in the 20th century is presented by Jensen (2000).

No sooner do technologies arrive than they speedily become the subject of historical inquiry concerning their origins and evolution. Even the history of the Internet can now be studied, and has duly become the focus of serious historical scholarship (Abbate, 1999; Castells, 2001, pp. 9–35; Caygill, 1999; Moschovitis, Poole, Schuyler, & Senft, 1999). Furthermore, the Internet has spawned new interest in the history of communication generally and in the history of the telegraph in particular. Standage (1998) has posited the electric telegraph as the direct ancestor of the modern digital superhighway, describing it as the "Victorian internet;" while Morus (2000) has researched the telegraph as a metaphor for Victorian views on bodily and social discipline and on the nervous and social systems. Histories of the telegraph are more than justified, given that this technology presented the first real-time communication over large distances. Other antecedents of the Internet and the World Wide Web were the focus of a recent conference at the Mundaneum in Mons, Belgium, which celebrated the vision and legacy of Paul Otlet (Rayward, 1975; Van den Heuval, Rayward, & Uyttenhove, 2003).

In the 20th century also, precomputer and noncomputer "information society" developments can be identified: the evolution of the mass media, popular newspapers, radio, film, and television. Such information and communication technologies, it might be argued, have made as much, if not more, of an impact on people's lives as the "digital turn." As an antidote to deterministic and purely technological analyses, thereby mirroring the approach of theorists working in the field of social informatics (Sawyer & Rosenbaum, 2000), historical treatments of communication technologies ideally need to be located in social contexts and causes (Smith & Marx, 1994). This has been the trend in media history in recent decades, as evidenced, for example, in the pages of the international journal *Media History*, in Samuel's (2001) history of post-war television advertising in America, and in Briggs's (1961–1995) magisterial history of broadcasting in Britain. Contextualization has also been evident in various histories of "communication institutions," such as Perry's (1992) study of the Victorian Post Office and postal system, which he researches in the context of the 19th-century revolution in government and the emergence of the modern corporate state, and in Fryer and Akerman's (2000) treatment of the same subject matter, although with broader reference to the economy and society. In fact, the history of mail systems and services appears to be a growing area, as seen in the organization of a recent conference on mail services in Europe over the past three centuries (http://www.laposte.fr/chp/Fichiershtm/Colloque.htm).

The history of museums—one of the main memory institutions of modernity—is an area where significant efforts have been made to produce analytical studies rooted in social and intellectual causation. As a subject, the history of museums has moved a considerable distance from its original anchorage in scholarly description (Miller, 1973; Wittlin, 1949). Studies have paid close attention to the social and epistemological dimensions and causes of museum provision: How they reflected the assumptions and values of past societies and shaped their knowledge (Bennett, 1995; Hooper-Greenhill, 1992; Preziosi & Farago, 2004).

Throughout most of human history the information infrastructure has evolved haphazardly. However, in the era of modernity it has been increasingly planned and regulated. The term "information policy," which encapsulates the shift to regulation and planning, is relatively recent in origin. Yet, as well as addressing the recent history of information policy (P. Day, 2001), scholars have attempted to map its development before it became formally recognized as a legitimate function of government and the professions (Duff, 2004; Willmore, 2002). It is also evident that just as current information policy is highly discursive—interfacing, like transport policy, for example, with initiatives and concerns across the policy spectrum—so also in the past, information has played a key role in many areas of governmental and expert planning. The past pervasive importance of information in policy formulation promises, therefore, to be a fruitful area for future research—for example, in the area of economic planning, where "information" is a key indexing term in guiding researchers through government archives on the subject (Alford, Lowe, & Rollings, 1992).

Information History as Social History: Informal Information Networks

Creating the information infrastructures in past information societies involved the purposeful development of technologies, systems, and institutions. Whether they were cultural spaces for the construction of the "enlightened citizen," such as libraries, museums, and scientific clubs, or technological modes of information transmission, such as the telegraph or telephone, the networks of communication that defined emergent, modern information societies from the 18th century onward were, as Mattelart (1996, p. 85) points out, "envisaged as creators of a new universal bond." However, focusing on the information infrastructure in its *formal* sense, in the sense of constructing communication networks reliant on planned capital investment—as in the case of the information superhighway, Otlet's universal bibliographic index, or a system of libraries—tends to obscure the naturally occurring *informal* communication networks and means of information exchange that have always existed, but were accelerated and expanded by modernity, industrialization, and urbanization. The importance of informal networks (primary sources that are difficult to obtain) has become increasingly recognized by historians (e.g., Bayly, 1996), in addition to the social dimensions of formal networks that have interested them for longer.

Much of the information we demand and consume—certainly a substantial part of that which is immediately practical or personal—is not obtained from librarians, databases, or the Internet. Use of other people has always been one of the most frequently employed and successful methods of obtaining information (Fisher, Marcoux, Miller, Sánchez, & Cunningham, 2004; Shenton & Dixon, 2003). Such means of information transmission and reception are mostly intimate and local, occurring between kin, neighbors, friends, and fellow workers in the public house, factory, political society, or cobbler's shop.

The massive growth of the "impersonal" city in the 19th century might appear, at first glance, to have worked against such *Gemeinschaft* information networks. However, although urbanization encouraged the growth of anonymity, it also led to a huge escalation in informal information exchange among citizens. Improvements in the means by which people could move about the city—better transport, paved sidewalks, detailed maps—opened up the city and created what Joyce (2003) has called a "republic of the streets": a notion of the city as a liberal space, rather than as a repressive phenomenon of modernity, capitalism, and the techno-administrative state.

The informal informational activity that accompanied and strengthened the "republic of the streets" was complemented by an explosion in ephemeral advertising reflecting the growth, from the second half of the 19th century on, of a commodity culture (Richards, 1990). Recalling life in early-20th-century Salford, near Manchester, Roberts (1971, p. 134) noted that:

> A culture of the streets existed from which the young especially profited: one soaked in information of every kind from posters and advertisements pasted on gable end and massive hoarding. But above all young intelligence learned from a regular scrutiny of newsagents' windows.

The knowledge obtained from informal information networks might be termed "social knowledge," akin in some respect to the concept of "social intelligence" articulated by information theorists (Cronin & Davenport, 1993). Social knowledge is formed, firstly, by the experience we gain from our environment. Secondly, it is formed through informal information sources, whether this be a hotel doorman advising guests on the safest areas of a city or the information dynamic of the traditional role of women in knitting together working-class neighborhoods. Social knowledge may include the type of information exchange one would categorize as rumor or gossip, which may or may not be accurate (Clarke, 1989; Tebbutt, 1995).

Historically, some of the most valuable social knowledge has been of the financial kind. Examples of this include the information sought and exchanged by women in pre-industrial and early industrial economies in relation to the credit and stock markets (Spicksley, in press); freemasonry as a business information network in the 19th century (Burt, 2003); and the role played by 18th-century Paris notaries who, before the emergence of formal financial institutions, acted as intermediaries between borrowers and lenders, using their social knowledge to find reliable borrowers for lenders and advising the latter on the creditworthiness of the former—a practice that encourages us to push back the full transition to a capitalist economy, certainly in France, from the 19th to the 18th century (Hoffman, Postel-Vinay, & Rosenthal, 1999). Social knowledge has been important in certain other worldly, life-enhancing ways. There exists a long tradition in the medical treatment of illness—for centuries before the arrival of self-diagnosis via the Internet—of healers having to respond to, or "mediate," the medical knowledge that patients gained from popular sources like almanacs or simply through "word of mouth" and "gossip" (de Blécourt & Usborne, 2004).

Conclusion

The subtext of this chapter has been formed by the question: "What is information history?" Posing this question involves an appropriate academic and professional curiosity. As I have argued elsewhere (Black, 2004), for any body of knowledge or field of expert practice to claim disciplinary status, it must be able to summon up a history. This occurs in the case of areas like medicine, law, or sociology. It also occurs in the information disciplines, although some (e.g., librarianship and information science) can command a richer and longer historiography than others (e.g., information management).

History is important. Historical perspectives contribute significantly to our understanding of contemporary information issues. Even those in the information community who claim no substantial association with information history—whether academics, researchers, or practitioners—can often be found offering brief historical introductions to their published discourses beyond the ubiquitous and compulsory review of the literature. The conscious promotion of an information history field may encourage further engagement by commentators on contemporary issues in the historical routes and contexts of their subjects. But for that to happen, the historical component of education for information work needs to be strengthened, especially in light of the fact that historical subjects have virtually disappeared from library and information science curricula, crowded out by the emergence of a virulent vocationalism, which prioritizes knowledge and skills that can be shown to be directly and immediately pertinent for the workplace. A renaissance in information history education will only happen if a case can be made for the value of historicism to professional practice. Briefly, historical knowledge can contribute to professional identity and enhance the self-criticism and self-reflection that any discipline or practice requires. By rooting out past principles, historical awareness can support philosophical and ethical deliberations. History also enhances critical thinking, invites interdisciplinarity, and encourages societal awareness and a heightened knowledge of context.

Despite the value of historicism for the information disciplines, the validity of information history as a term that carries meaning not only for these disciplines but also for historians remains questionable. If information does become the defining feature of our culture, as information society protagonists predict, then an information history field will be more likely to emerge, even if working against this will be the possible future poverty of primary sources, due to their low survival rate in an electronic communications environment. We have no way of telling for certain if information history will become a reality, but history tells us that history itself has a lively capacity for generating subsets within itself, particularly in response to the appearance of new areas of knowledge.

References

Abbate, J. (1999). *Inventing the Internet*. Cambridge, MA: MIT Press.

Agar, J. (2003a). *Constant touch: A global history of the mobile phone*. Cambridge, UK: Icon.

Agar, J. (2003b). *The government machine: A revolutionary history of the computer.* Cambridge, MA: MIT Press.

Agar, J., Green, S., & Harvey, P. (2002). Cotton to computers: From industrial to information revolutions. In S. Woolgar (Ed.), *Virtual society? Technology, cyberbole, reality* (pp. 24–285). Oxford, UK: Oxford University Press.

Alford, B. W. E., Lowe, R., & Rollings, N. (1992). *Economic planning: A guide to documentation in the Public Record Office.* London: HMSO.

Anderson, M. J. (1988). *The American census: A social history.* New Haven, CT: Yale University Press.

Anderson, M. J. (2001). *Who counts? The politics of census-taking in contemporary America.* New York: Russell Sage Foundation.

Anghelescu, H. G. B., & Poulain, M. (Eds.). (2001). *Books, libraries, reading and publishing in the Cold War.* Washington DC: Library of Congress Center for the Book and University of Texas Press.

Arant, W., & Benefiel, C. R. (Eds.). (2002). The image and role of the librarian [Special issue]. *Reference Librarian, 78.*

Arnold, W. (Ed.). (in press). *Libraries and philanthropy.* Wiesbaden, Germany: Harrassowitz Verlag.

Babbage, C. (1835). *On the economy of machinery and manufactures.* London: Charles Knight.

Bailey, N. (1775). *A new universal etymological English dictionary.* London: William Cavell.

Barnard, J., & McKenzie, D. F. (2002). *The Cambridge history of the book in Britain. Volume IV: 1557–1695.* Cambridge, UK: Cambridge University Press.

Bartlett, R. A. (1966). The state of the library history art. In J. D. Marshall (Ed.), *Approaches to library history: Proceedings of the 2nd Library History Seminar* (pp. 13–23). Tallahassee, FL: Journal of Library History.

Battles, M. (2003). *Library: An unquiet history.* London: Random House.

Bawden, D., & Robinson, L. (2000). A distant mirror? The Internet and the printing press. *Aslib Proceedings, 52*(2), 51–57.

Bayly, C. A. (1996). *Empire and information: Intelligence gathering and social communication in India, 1780–1870.* Cambridge, UK: Cambridge University Press.

Beniger, R. (1986). *The control revolution: Technological and economic origins of the information society.* Cambridge, MA: Harvard University Press.

Bennett, J., & Mandelbrote, S. (1998). *The garden, the ark, the tower and the temple: Biblical metaphors of knowledge in early modern Europe.* Oxford, UK: Museum of the History of Science, in association with the Bodleian Library.

Bennett, T. (1995). *The birth of the museum: History, theory, politics.* London: Routledge.

Bird, P. (1994). *LEO: The first business computer.* London: Hasler.

Black, A. (1996). *A new history of the English public library: Social and intellectual contexts 1850–1914.* London: Leicester University Press.

Black, A. (1998, May). Information and modernity: The history of information and the eclipse of library history. *Library History, 14,* 39–45.

Black, A. (2000). *The public library in Britain.* London: The British Library.

Black, A. (2001a). A response to "Whither library history?" *Library History, 17,* 37–39.

Black, A. (2001b). The Victorian information society: Surveillance, bureaucracy and public librarianship in nineteenth-century Britain. *The Information Society, 17*(1), 63–80.

Black, A. (2004). Every discipline needs a history: Information management and the early information society in Britain. In W. B. Rayward (Ed.), *Aware and responsible* (pp. 29–47). Lanham, MD: Scarecrow Press.

Black, A., & Brunt, R. (2001). Information management in MI5 before 1945: A research note. *Intelligence and National Security, 16*(2), 158–165.

Black, A., & Muddiman, D. (1997). *Understanding community librarianship: The public library in post-modern Britain.* Aldershot, UK: Avebury.

Black, E. (2001). *IBM and the Holocaust: The strategic alliance between Nazi Germany and America's most powerful corporation.* London: Little, Brown and Co.

Blok, A., & Downey, G. (Eds.). (2003). Uncovering labour in information revolutions, 1750–2000 [Special issue]. *International Review of Social History, 48*, Supplement 11.

Borgman, C. L. (2000). *From Gutenberg to the global information infrastructure.* Cambridge, MA: MIT Press.

Bowden, M. E., Hahn, T. B., & Williams, R. V. (Eds.). (1999). *Proceedings of the 1998 Conference on the History and Heritage of Information Science Systems,* Medford, NJ: Information Today. Retrieved December 1, 2004, from www.chemheritage.org/explore/ ASIS_documentsASIS98_main.htm

Bowden, M. E., & Rayward, W. B. (Eds.). (2004). *Proceedings of the Second Conference on the History and Heritage of Information Science Systems.* Medford, NJ: Information Today.

Boyns, T., & Wale, J. (1996). The development of management information systems in the British coal industry 1880–1947. *Business History, 38*(2), 55–80.

Briggs, A. (1961–1995). *History of broadcasting in the United Kingdom.* Oxford, UK: Oxford University Press.

Brooks, A., & Haworth, B. (1993). *Boomtown Manchester 1800–1850: The Portico connection.* Manchester, UK: The Portico Library.

Brown, M. P. (2003). *The Lindisfarne Gospels: Society, spirituality and the scribe.* London: British Library.

Brunt, R. (2001, October). Indexing the intelligence. *The Indexer,* 187–190.

Brunt, R. (2004). Indexes at the Government Code and Cipher School, Bletchley Park 1940–45. In M. E. Bowden & W. B. Rayward (Eds.), *Proceedings of the Second Conference on the History and Heritage of Information Science Systems* (pp. 291–299). Medford, NJ: Information Today.

Buckland, M. K. (1991). Information as thing. *Journal of the American Society for Information Science, 42*(5): 351–360.

Buckland, M. K. (1999). Overview of the history of science information systems. In M. E. Bowden, T. B. Hahn, & R. V. Williams (Eds.), *Proceedings of the 1998 Conference on the History and Heritage of Information Science Systems* (pp. 3–7). Medford NJ, Information Today.

Buckland, M. K., & Liu, Z. (1995). History of information science. *Annual Review of Information Science and Technology, 30*, 385–416.

Bud-Frierman, L. (Ed.). (1994). *Information acumen: The understanding and use of knowledge in modern business.* London: Routledge.

Burke, P. (2000). *A social history of knowledge from Gutenberg to Diderot*. Cambridge, UK: Polity Press.

Burt, R. (2003). Freemasonry and business networking during the Victorian period. *Economic History Review, 56*(4), 657–688.

Calhoun, C. (1992). The infrastructure of modernity: Indirect social relationships, information technology, and social integration. In H. Haferkamp & N. J. Smelser (Eds.), *Social change and modernity* (pp. 205–236). Berkeley: University of California Press.

Camier, D., Aris, J., Hermon, P., & Land, F. (1997). *LEO: The incredible story of the world's first business computer*. London: McGraw Hill.

Campbell-Kelly, M. (1989). *ICL: A business and technical history*. Oxford, UK: Clarendon Press.

Campbell-Kelly, M. (1992). Large-scale data-processing in the Prudential 1850–1930. *Accounting, Business and Financial History, 2*, 117–139.

Campbell-Kelly, M. (1994). The railway clearing house and Victorian data-processing. In L. Bud-Frierman (Ed.), *Information acumen: The understanding and use of knowledge in modern business* (pp. 51–74). London: Routledge.

Campbell-Kelly, M. (1995). Development and structure of the international software industry, 1950–1990. *Business and Economic History, 24*(2), 73–111.

Campbell-Kelly, M. (1998). Information in the business enterprise. In E. Higgs (Ed.), *History and electronic artefacts* (pp. 59–67). Oxford, UK: Clarendon Press.

Campbell-Kelly, M., & Aspray, W. (1996). *Computer: A history of the information machine*. New York: Basic Books.

Cantwell, J. D. (1991). *The Public Record Office 1838–1958*. London: HMSO.

Casson, L. (2001). *Libraries in the ancient world*. New Haven, CT: Yale University Press.

Castells, M. (2001). *The Internet galaxy: Reflections on the Internet*. Oxford, UK: Oxford University Press.

Caygill, H. (1999). Meno and the Internet: Between memory and the archive. *History of the Human Sciences, 12*(2), 1–11.

Chandler, A. D. (1977). *The visible hand: The managerial revolution in American business*. Cambridge, MA: Harvard University Press.

Chandler, A. D., & Cortada, J. W. (2000). *A nation transformed by information: How information shaped the United States from colonial times to the present*. Oxford, UK: Oxford University Press.

Checkland, P., & Howell, S. (1998). *Information, systems and information systems: Making sense of the field*. Chichester, UK: Wiley and Sons.

Clanchy, M. T. (1993). *From memory to written record: England 1066–1307* (2nd ed.). Oxford, UK: Blackwell.

Clarke, A. (1989). Whores and gossips: Sexual reputations in London 1770–1825. In A. Angerman, G. Binnema, A. Keunen, V. Poels, & J. Zirkzee (Eds.), *Current issues in women's history* (pp. 231–248). London: Routledge.

Clarke, P. (2001). *British clubs and societies 1580–1800: The origins of an associational world*. Oxford, UK: Oxford University Press.

Collen, M. (1995). *A history of medical informatics in the United States: 1950–1990*. Bethesda, MD: American Medical Informatics Association.

Copeland, D., & McKenney, J. (1988). Airline reservation systems: Lessons from history. *MIS Quarterly, 12*(3), 353–370.

Crary, J. (1990). *Techniques of the observer: On vision and modernity in the nineteenth century.* Cambridge, MA: MIT Press.

Cronin, B., & Davenport, E. (1993). Social intelligence. *Annual Review of Information Science and Technology, 28*, 3–44.

Dain, P. (2000). *The New York Public Library: A universe of knowledge.* New York: New York Public Library.

The Davies Project at Princeton University. (2004). *American libraries before 1876.* Retrieved December 3, 2004, from www.princeton.edu/~davpro/databases

Davis, D. G., Jr. (Ed.). (1996). *Libraries and philanthropy: Proceedings of Library History Seminar IX.* Austin, TX: Graduate School of Library and Information Science, University of Texas at Austin.

Davis, D. G., Jr. (2003). *Dictionary of American library biography: Second supplement.* Westport, CT: Libraries Unlimited.

Davis, D. G., Jr., & Aho, J. A. (2001). Whither library history? A critical essay on Black's model for the future of library history, with some additional options. *Library History, 17*, 21–37.

Davis, D. G., Jr., Carpenter, K. E., Wiegand, W. A., & Aikine, J. (2002). *Winsor, Dewey and Putnam: The Boston experience* (Occasional Paper 212). Champaign, IL: Graduate School of Library and Information Science, University of Illinois.

Day, P. (2001). Participating in the information society: Community development and social inclusion. In L. Keeble & B. Loader (Eds.), *Community informatics: Shaping computer-mediated social relations* (pp. 305–323). London: Routledge.

Day, R. E. (2001). *The modern invention of information: Discourse, history, and power.* Carbondale: Southern Illinois University Press.

de Blécourt, W., & Usborne, C. (Eds.). (2004). *Cultural approaches to the history of medicine: Mediating medicine in early modern and modern Europe.* Basingstoke, UK: Palgrave Macmillan.

Dearnley, J., & Feather, J. (2001). *The wired world: An introduction to the theory and practice of the information society.* London: Library Association Publishing.

Debons, A., Horne, E., & Croneweth, S. (1988). *Information science: An integrated view.* Boston: G.K. Hall.

Deutsch, K. W. (1966). *Nationalism and social communication* (2nd ed.). Cambridge, MA: MIT Press.

Ditmas, E. M. R. (1948). Co-ordination of information: A survey of schemes put forward in the last fifty years. *Journal of Documentation, 3*(4), 209–221.

Dragoni, G., & Fichera, G. (Eds.). (1998). *Fountain pens: History and design.* Woodbridge, Suffolk: The Antique Collectors Club.

Driver, F. (1993). *Power and pauperism: The workhouse system 1834–1884.* Cambridge, UK: Cambridge University Press.

Duff, A. (2000). *Information society studies.* London: Routledge.

Duff, A. (2004). The past, present, and future of information policy. *Information, Communication and Society, 7*(1), 69–87.

Eisenstein, E. (1983). *The printing revolution in early modern Europe.* Cambridge, UK: Cambridge University Press.

Englander, D., & O'Day, R. (1995). *Retrieved riches: Social investigation in Britain 1840–1914*. Aldershot, UK: Scholar Press.

Epstein, J. (2001). *Book business: Publishing past, present and future*. New York: W.W. Norton.

Farkas-Conn, I. (1990). *From documentation to information science: The beginnings and early development of the American Documentation Institute – American Society for Information Science*. Westport, CT: Greenwood Press.

Feather, J. (1987). *A history of British publishing*. London: Croom Helm.

Feather, J. (2003). *Communicating knowledge: Publishing in the 21st century*. Munich, Germany: K. G. Saur.

Feather, J. (2004). *The information society: A study of continuity and change* (4th ed.). London: Facet Publishing.

Fischer, S. R. (2003). *A history of reading*. London: Reaktion Books.

Fisher, K. E., Marcoux, E., Miller, L. S., Sánchez, A., & Cunningham, E. R. (2004). Information behaviour of migrant Hispanic farm workers and their families in the Pacific Northwest. *Information Research: An Internet Electronic Journal, 10*(1). Retrieved December 3, 2004, from http://informationr.net/ir/10-1/paper199.html

Flichy, P. (1995). *Dynamics of modern communication: The shaping and impact of new communication technologies*. London: Sage.

Frängsmyr, T., Heilborn, J. L., & Rider, R. E. (1990). *The quantifying spirit in the Enlightenment*. Berkeley: University of California Press.

Fryer, G., & Akerman, C. (2000). *The reform of the Post Office in the Victorian era and its impact on social and economic activity*. London: Royal Philatelic Society.

Furner, J. (2004). "A brilliant mind": Margaret Egan and social epistemology. *Library Trends, 52*(4), 792–809.

Gardner, R., & Shortelle, D. (1997). *From talking drums to the Internet: An encyclopedia of communications technology*. Santa Barbara, CA: ABC-CLIO.

Garrett, J. (1999). Redefining order in the German library, 1775–1825. *Eighteenth-Century Studies, 33*(1), 103–123.

Garrett, J. (2004). The legacy of the Baroque in virtual representations of library space. *Library Quarterly, 74*(1), 42–62.

Garrison, D. (2003). *Apostles of culture: The public librarians and American society, 1876–1920*. Madison: University of Wisconsin Press.

Giddens, A. (1987). *The nation-state and violence*. Cambridge, UK: Polity Press.

Gorman, M. (2001). Human values in a technological age. *Information Technology & Libraries, 20*(1), 4–11.

Graham, S., & Marvin, S. (2001). *Splintering urbanism: Networked infrastructures, technological mobilities and the urban condition*. London: Routledge.

Grimes, B. (1998). *Irish Carnegie libraries: A catalogue and architectural history*. Dublin, Ireland: Irish Academic Press.

Hahn, T. B., & Buckland, M. (1998). *Historical studies in information science*. Medford, NJ: Information Today.

Haigh, T. (2001a). The chromium-plated tabulator: Institutionalizing an electronic revolution 1954–58. *IEEE Annals of the History of Computing, 23*(4), 75–104.

Haigh, T. (2001b). Inventing information systems: The systems men and the computer, 1950–1968. *Business History Review, 75,* 15–61.

Hapke, T. (2004). *History of scientific information, communication, and documentation in Germany.* Retrieved December 3, 2004, from www.tu-harburg.de/b/hapke/infohist.htm# germany

Harris, M. H. (1999). *History of libraries in the western world* (4th ed.). Lanham, MD: Scarecrow Press.

Harrison, K. C. (2000). A librarian's odyssey: Episodes of autobiography. Eastbourne, UK: Author.

Headrick, D. R. (2000). *When information came of age: Technologies of knowledge in the Age of Reason and Revolution 1700–1850.* Oxford, UK: Oxford University Press.

Hellinga, L., & Trapp, J. B. (1999). *The Cambridge history of the book in Britain. Volume III: 1400–1557.* Cambridge, UK: Cambridge University Press.

Herman, M. (1996). *Intelligence and power in peace and war.* Cambridge, UK: Cambridge University Press.

Hewish, J. (2000). *Rooms near Chancery Lane: The Patent Office under the Commissioners 1852–1883.* London: The British Library.

Hewitt, M. (2000). Confronting the modern city: The Manchester Free Public Library, 1850–80. *Urban History, 27*(1), 62–88.

Higgs, E. (Ed.). (1998). *History and electronic artefacts.* Oxford, UK: Clarendon Press.

Higgs, E. (2004). *The information state in England.* Basingstoke, UK: Palgrave.

Hjørland, B. (2000). Documents, memory institutions and information science. *Journal of Documentation, 56*(1), 27–41.

Hoare, P. (1998). The development of the European information society. *Library Review, 47*(7/8), 377–382.

Hoare, P. (Ed.). (in press). *Cambridge history of libraries in Britain and Ireland.* Cambridge, UK: Cambridge University Press.

Hoffman, P. T., Postel-Vinay, G., & Rosenthal, J. (1999). Information and economic history: How the credit market in old regime Paris forces us to rethink the transition to capitalism. *American Economic History Review, 104*(1), 69–94.

Hooper-Greenhill, E. (1992). *Museums and the shaping of knowledge.* London: Routledge.

Jensen, P. (2000). *From the wireless to the Web: The evolution of telecommunications 1901–2001.* Sydney, Australia: University of New South Wales.

Jensen, K. (2003). *Incunabula and their readers: Printing, selling and the use of books in the fifteenth century.* London: The British Library.

Jones, T. (1997). *Carnegie libraries across America: A public legacy.* New York: Wiley.

Joyce, P. (2003). *The rule of freedom: Liberalism and the modern city.* London: Verso.

Keegan, J. (2003). *Intelligence in war: Knowledge of the enemy from Napoleon to Al-Quaeda.* London: Hutchinson.

Keeling, D. F. (Ed.). (1972). *British library history: Bibliography 1962–1968.* London: The Library Association.

Keeling, D. F. (Ed.). (1975). *British library history: Bibliography 1969–1972.* London: The Library Association.

Keeling, D. F. (Ed.). (1979). *British library history: Bibliography 1973–1976.* London: The Library Association.

Keeling, D. F. (Ed.). (1983). *British library history: Bibliography 1977–1980*. London: The Library Association.

Keeling, D. F. (Ed.). (1987). *British library history: Bibliography 1981–1984*. London: The Library Association.

Keeling, D. F. (Ed.). (1991). *British library history: Bibliography 1985–1988*. London: The Library Association.

Kerslake, E. (2001). No more the hero: Lionel McColvin, women library workers, and impacts of othering. *Library History, 17*(3), 181–188.

Kerslake, E., & Moody, N. (2000). *Gendering library history*. Liverpool, UK: Liverpool John Moores University Press.

Knuth, R. (2003). *Libricide: The regime sponsored destruction of books and libraries in the twentieth century*. Westport, CT: Praeger.

Kumar, K. (1995). *From post-industrial to post-modern society*. Oxford, UK: Blackwell.

Lebow, I. (1995). *Information highways and byways: From the telegraph to the 21st century*. New York: IEEE Press.

Lerner, F. (2001). *The story of libraries: From the invention of writing to the computer age*. New York: Continuum.

Levinson, P. (1997). *The soft edge: A natural history and future of the information revolution*. London: Routledge.

Lloyd, M. (2003). *The passport: The history of man's most travelled document*. Stroud, UK: Sutton Publishing.

Locker, K. O. (1996). Studying the history of business communication. *Business Communication Quarterly, 59*(2), 109–127.

Lyon, D. (1994). *The electronic eye: The rise of surveillance society*. Cambridge, UK: Polity Press.

Lyon, D. (2000). *The surveillance society*. Cambridge, UK: Polity Press.

MacLeod, R. (Ed.). (2002). *The Library of Alexandria: Centre of learning in the ancient world*. London: IB Tauris Publishers.

Mahoney, M. S. (1990). Cybernetics and information technology. In R. C. Olby, G. N. Cantor, & J. R. R. Christie (Eds.), *Companion to the history of modern science* (pp. 537–553). London: Routledge and Kegan Paul.

Mäkinen, I. (2001). *Finnish public libraries in the twentieth century*. Tampere, Finland: University of Tampere.

Mäkinen, I. (2004). Information history, library history, or history by and large? Remarks on recent discussions of library and information history. In W. Boyd Rayward (Ed.), *Aware and responsible* (pp. 103–114). Lanham, MD: Scarecrow Press.

Manguel, A. (1996). *A history of reading*. London: Penguin.

Markus, T. A. (1993). *Buildings and power: Freedom and control in the origin of modern building types*. London: Routledge.

Martin, L. A. (1998). *Enrichment: A history of the public library in the United States in the twentieth century*. Lanham, MD: Scarecrow Press.

Marwick, A. (1981). *The nature of history*. London: Macmillan Press.

Mattelart, A. (1996). *The invention of communication*. Minneapolis: University of Minnesota Press.

Mattelart, A. (2000). *Networking the world 1794–2000*. Minneapolis: University of Minnesota Press.

Mattelart, A. (2003). *The information society: An introduction*. London: Sage.

May, C. (2002). *The information society: A sceptical view*. Oxford, UK: Polity Press.

McCrank, L. J. (2002). *Historical information science: An emerging unidiscipline*. Medford, NJ: Information Today.

McKitterick, D. (1986). *Cambridge University Library: A history*. Cambridge, UK: Cambridge University Press.

Meadow, C. T. (2002). *Making connections: Communication through the ages*. Lanham, MD: Scarecrow Press.

Miller, E. (1967). *Prince of librarians: The life and times of Antonio Panizzi of the British Museum*. London: Andre Deutsch.

Miller, E. (1973). *That noble cabinet: A history of the British Museum*. London: Andrew Deutsch.

Moran, J. (2003). *Wynkyn de Worde: Father of Fleet Street*. London: The British Library.

Morris, P. R. (1990). *A history of the world semiconductor industry*. London: Peter Peregrinus.

Morus, I. R. (2000). The nervous system of Britain: Space time and the electric telegraph in the Victorian age. *British Journal for the History of Science, 33*(4), 455–475.

Moschovitis, C. J. P., Poole, H., Schuyler, T., Senft, T. M. (1999). *History of the Internet: A chronology 1943 to the present*. Santa Barbara, CA: ABC-CLIO.

Muddiman, D. (in press). A new history of Aslib. *Journal of Documentation*.

Munford, W. (1963). *Edward Edwards 1812–1886: Portrait of a librarian*. London: The Library Association.

Myers, R., Harris, M., & Mandelbrote, G. (2001). *Under the hammer: Book auctions since the seventeenth century*. New Castle, DE: Oak Knoll Press.

Nagl, M. (1999). Stille, Ordnung, Katastrophen. Bibliotheken im Film—Bibliotheken aus mannlichem Blick? In P. Vodosek & G. Jefcoate (Eds.), *Bibliotheken in der literarischen Darstellung* [Libraries in literary representation] (pp. 115–126). Wiesbaden, Germany: Harrassowitz Verlag.

Orbell, J. (1991). The development of office technology. In. A. Turton (Ed.), *Managing business archives* (pp. 60–83). Oxford, UK: British Archives Council.

Parada, A. E. (2004). The new history of books and libraries in Argentina: Background, history and periods. In *Library research in Argentina: New approaches* (pp. 73–80). Buenos Aires: Instituto de Investigaciones Bibliotecológicas, Facultad de Filosofiá y Letras, Universidad de Buenos Aires. Produced in association with the IFLA Division 7: Education and Research for the IFLA World Library and Information Congress, Buenos Aires, 22–27 August 2004.

Parsons, E. A. (1952). *The Alexandrian Library: Glory of the Hellenic world. Its rise, antiquities and destructions*. London: Cleaver-Hume Press.

Perry, C. R. (1992). *The Victorian Post Office: The growth of a bureaucracy*. London: Royal Society.

Peters, J. D. (1988). Information: Towards a critical history. *Journal of Communication Inquiry, 12*(2), 9–23.

Petroski, H. (1990). *The pencil: A history of design and circumstances*. New York: Knopf.

Poulain, M. (Ed.). (1992). *Histoire des bibliothèques françaises* [History of French libraries]. Paris: Promidis—Editions du Cercle de la Librarie.

Preston, P. (2001). *Re-shaping communications*. London: Sage.

Preziosi, D., & Farago, C. (2004). *Grasping the world: The idea of the museum*. Aldershot, UK: Ashgate.

Price, J. M. (1987). Directions for the conduct of a merchant's counting house. In R. P. T. Davenport-Hines & J. Liebenau (Eds.), *Business in the age of reason* (pp. 134–150). London: Frank Cass.

Raven, J. (Ed.). (2004). *Lost libraries: The destruction of great book collections since antiquity*. Basingstoke, UK: Palgrave Macmillan.

Raven, J., Small, H., & Tadmor, N. (1996). *The practice and representation of reading in England*. Cambridge, UK: Cambridge University Press.

Rayward, W. B. (1975). *The universe of information: The work of Paul Otlet for documentation and international organisation*. Moscow: All-Union Institute for Scientific and Technological Information, for the International Federation for Documentation.

Rayward, W. B. (1998). The history and historiography of information science: Some reflections. In T. B. Hahn & M. K. Buckland (Eds.), *Historical studies in information science* (pp. 7–21). Medford, NJ: Information Today.

Rayward, W. B. (Ed.). (2002). Computer applications in libraries [Special issue]. *IEEE Annals of the History of Computing, 24*(2/3).

Rayward, W. B. (Ed.). (2004a). Pioneers in library and information science [Special issue]. *Library Trends 52*(4).

Rayward, W. B. (2004b). Scientific and technological information systems in their many contexts: The imperatives, clarifications, and inevitability of historical study. In M. E. Bowden & W. B. Rayward (Eds.), *Proceedings of the Second Conference on the History and Heritage of Scientific and Technological Information Systems* (pp. 1–11). Medford NJ: Information Today.

Reid, T. R. (1985). *Microchip: The story of a revolution and the men who made it*. London: Pan Books.

Rhodes, B., & Streeter, W. W. (1999). *Before photocopying: The art and history of mechanical copying 1780–1938*. New Castle, DE: Oak Knoll Press.

Rice, J. (1991). The evolution of early visions: An historical perspective of today's information technology. *Reference Librarian, 33*, 111–124.

Richards, P. S. (1994). *Scientific information in wartime: The Allied German rivalry, 1939–1945*. Westport, CT: Greenwood Press.

Richards, T. (1990). *The commodity culture of Victorian England: Advertising and spectacle 1851–1914*. Stanford, CA: Stanford University Press.

Richards, T. (1993). *The imperial archive: Knowledge and the fantasy of empire*. London: Verso.

Ritchie, S. (1982). *Modern library practice*. Buckden, UK: ELM Publications.

Roberts, R. (1971). *The classic slum: Salford life in the first quarter of the twentieth century*. Manchester, UK: Manchester University Press.

Robins, K., & Webster, F. (1999). *Times of technoculture*. London: Routledge.

Robinson, A. (1995). *The story of writing*. London: Thames and Hudson.

Rose, J. (1994). How to do things with book history. *Victorian Studies, 37*, 461–471.

Rose, J. (2002). *The intellectual life of the British working classes*. New Haven, CT: Yale Nota Bene.

Rosenzweig, R. (2003). Scarcity or abundance? Preserving the past in a digital era. *American Historical Review, 108*(3), 734–762.

Sacks, D. (2003). *The alphabet*. London: Hutchinson.

Samek, T. (2001). *Intellectual freedom and social responsibility in American librarianship 1967–1974*. Jefferson, NC: McFarland.

Samuel, L. R. (2001). *Brought to you by: Postwar television advertising and the American dream*. Austin: University of Texas Press.

Samuel, J. (Producer). (2004). *Save and burn* [Film]. (Available from Julian Samuel, at: jjsamuel@vif.com).

Sawyer, S., & Rosenbaum, H. (2000). Social informatics in the information sciences: Current activities and emerging directions. *Informing Science, 3*(2), 89–95.

Schuler, D. (2001). Cultivating society's civic intelligence: Patterns for a new "world brain." In L. Keeble & B. Loader (Eds.), *Community informatics: Shaping computer-mediated social relations* (pp. 284–304). London: Routledge.

Schwartz, J. M., & Cook, T. (2002). Archives, records and power [Special issue]. *Archival Science: International Journal on Recorded Information, 2*(1/2 and 3/4).

Sellen, A. J., & Harper, R. (2002). *The myth of the paperless office*. Cambridge, MA: MIT Press.

Shenton, A. K., & Dixon, P. (2003). Youngsters' use of other people as an information-seeking method. *Journal of Librarianship and Information Science, 35*(4), 219–233.

Shiflett, O. L. (1984). Clio's claim: The role of historical research in library and information science. *Library Trends 32*, 385–406.

Slack, P. (2004). Government and information in seventeenth-century England. *Past and Present. 184*(1) 33–68.

Smith, M., & Marx, L. (1994). *Does technology drive history? The dilemma of technological determinism*. Cambridge, MA: MIT Press.

Snape, R. (1995). *Leisure and the rise of the public library*. London: Library Association.

Soffer, R. N. (1994). *Discipline and power: The university, history, and the making of an intellectual elite, 1870–1930*. Stanford, CA: Stanford University Press.

Solymar, L. (1999). *Getting the message: A history of communications*. Oxford, UK: Oxford University Press.

Spicksley, J. (in press). A dynamic model of social relations: Celibacy, credit and the identity of the spinster. In H. R. French & J. Barry (Eds.), *Identity and agency in English society 1500–1800*. Basingstoke, UK: Palgrave.

Staikos, K. S. (2000). *The great libraries: From antiquity to the Renaissance (3000 BC to AD 1600)*. New Castle, DE: Oak Knoll Press.

Stam, D. H. (2001). *International dictionary of library histories*. Chicago: Fitzroy Dearborn.

Standage, T. (1998). *The Victorian Internet: The remarkable story of the telegraph and the nineteenth century's on-line pioneers*. London: Weidenfeld and Nicolson.

Sterling, B. (1995). The hacker crackdown: Evolution of the US telephone network. In N. Heap, R. Thomas, G. Einon, & H. MacKay (Eds.), *Information technology and society: A reader* (pp. 33–40). London: Sage and Open University Press.

Stevens, N. D. (1986). The history of information. *Advances in Librarianship, 14*, 1–48.

Stieg, M. F. (1992). *Public libraries in Nazi Germany*. Tuscaloosa: University of Alabama Press.

Taylor, M. (1997, Spring). The beginnings of modern British social history. *History Workshop Journal*, 155–176.

Taylor, P. M. (1995). *Munitions of the mind: A history of propaganda from the ancient world to the present day*. Manchester, UK: Manchester University Press.

Tebbel, J. (1987). *Between covers: The rise and transformation of book publishing in America*. Oxford, UK: Oxford University Press.

Tebbut, M. (1995). *Women's talk: A social history of 'gossip' in working-class neighbourhoods, 1880–1960*. Aldershot, UK: Scholar Press.

Tolzmann, D. H., Hessel, A., & Peiss, R. (2001). *The memory of mankind: The story of libraries since the dawn of history*. New Castle, DE: Oak Knoll Press.

Torstensson, M. (Ed.). (2002). Libraries in times of utopian thought and social protest: The libraries of the late 1960s and 1970s [Special issue]. *Journal of Swedish Library Research, 14*(1).

Van den Heuval, C., Rayward, B., & Uyttenhove, P. (2003). L'architecture du savoir: Une recherche sur le Mundaneum et les précurseurs europééns de l'Internet. *Transnational Associations: The Review of the Union of International Associations 1–2*, 16–28.

Van Slyke, A. A. (1995). *Free to all: Carnegie libraries and American culture 1890–1920*. Chicago: University of Chicago Press.

Vann, J. D., & Van Arsdel, R. T. (1996). *Periodicals of Queen Victoria's empire*. London, Mansell.

Vickery, B. C. (2000). *Scientific communication in history*. Lanham, MD: Scarecrow Press.

Vincent, D. (1993). *Literacy and popular culture: England 1750–1914*. Cambridge, UK: Cambridge University Press.

Vodosek, P. (2001). Library history in Germany: A progress report. *Library History, 17*(2), 119–126.

Vodosek, P., & Jefcoate, G. (Eds.). (1999). *Bibliotheken in der literarischen Darstellung* [Libraries in literary representation]. Wiesbaden, Germany: Harrassowitz Verlag.

Webster, F. (2002). *Theories of the information society* (2nd ed.). London: Routledge.

Wertheimer, A. B., & Davis, D. G., Jr. (Eds.). (2000). *Library history research in America: Essays commemorating the fiftieth anniversary of the Library History Round Table*. Washington DC: Library of Congress Center for the Book.

Wiegand, W. A. (Ed.). (1990). *Supplement to the dictionary of American library biography*. Englewood, CO: Libraries Unlimited.

Wiegand, W. A. (1996). *Irrepressible reformer: A biography of Melvil Dewey*. Chicago: American Library Association.

Wiegand, W. A., & Davis, D. G., Jr. (Eds.). (1994). *The encyclopedia of library history*. New York: Garland.

Williams, R. V. (2004). *Pioneers of information science in North America*. Retrieved December 3, 2004, from www.libsci.sc.udu/bob/ISP/ISP.htm

Williams, R. V., & Bowden, M. E. (2000). *Chronology of chemical information science*. Retrieved December 3, 2004, from www.chemheritage.org/explore/timeline/CHCHRON.HTM

Williams, R. V., & LaMotte, V. (2004). *Bibliography of the history of information science in North America, 1900–2000*. Retrieved December 3, 2004, from www.libsci.sc.edu/bob/istchron/Isbiblio4.PDF

Willmore, L. (2002). Government policies towards information and communication technologies: A historical perspective. *Journal of Information Science, 28(2)*, 89–96.

Wittlin, A. S. (1949). *The museum: Its history and its tasks in education*. London: Routledge and Kegan Paul.

Wyner, B. S. (Ed.). (1978). *Dictionary of American library biography*. Littleton, CO: Libraries Unlimited.

Yates, J. (1989). *Control through communication: The rise of system in American management*. Baltimore, MD: Johns Hopkins University Press.

Young, P. (1991). *Person to person: The international impact of the telephone*. Cambridge, UK: Granta Editions.

Zandonade, T. (2004). Social epistemology from Jesse Shera to Steve Fuller. *Library Trends, 52(4)*, Spring, 810–832.

CHAPTER 12

Social Epistemology and Information Science

Don Fallis
University of Arizona

Philosophy and Information Science

Recently, information science has been turning to philosophy for insight. For example, several monographs (e.g., Budd, 2001; Dick, 2002a; Fuller, 2002) have applied work in philosophy to issues in information science. "Philosophy of Information" was the topic of a recent issue of *Library Trends* (Herold, 2004). A few years earlier, the question of whether information science needs a philosophy was the focus of an extended debate in *Library Quarterly* (see Radford & Budd, 1997; Zwadlo, 1997, 1998).

Philosophy is traditionally divided into three main areas. *Metaphysics* looks at what existence is and at what kinds of things exist in the world. *Ethics* looks at what the good is and at how people ought to behave. *Epistemology* looks at what knowledge is and at how people come to know things about the world. All three of these areas have relevance for information science.[1]

Some of the applications of philosophy to information science address important metaphysical questions, such as *what is information?* (e.g., Capurro & Hjørland, 2003). Most of the applications, however, have been in the area of *information ethics*, which addresses issues such as intellectual freedom, censorship, equitable access to information, privacy, and intellectual property (e.g., Doctor, 1992; Froehlich, 1992; Lievrouw & Farb, 2003; Smith, 1997). Several journals, such as *Ethics and Information Technology* and the *Journal of Information Ethics*, are completely devoted to information ethics. A recent issue of *Library Trends* (Wengert, 2001) explored the topic of "Ethical Issues of Information Technology."

The main thesis of this chapter, however, is that epistemology (and in particular social epistemology) is actually the area of philosophy most central to information science. This chapter provides support for the thesis primarily by laying out the many applications of social epistemology to information science.[2] The chapter begins with a discussion of the case, made by Jesse Shera and other information scientists, for the importance of epistemology to information science. This is followed by a brief introduction to epistemology in general and social epistemology in particular. Several important objections to applying social epistemology to information science are then addressed. The chapter concludes with a look at some applications that illustrate the important connections between social epistemology and information ethics.

Epistemology and Information Science

Libraries and other information services (museums, archives, search engines, bookstores, database providers, publishing companies, etc.) help people acquire knowledge. They do this by collecting, organizing, and providing access to recorded information (books, journals, Web sites, images, etc.).[3] In order to help people to acquire knowledge, we have to understand what knowledge is and how people acquire it. As a result, epistemology is arguably of central importance to information science. In fact, according to Heilprin (1968, p. 35), "lack of knowledge of epistemology is possibly the greatest barrier to improving library and information science."

Furthermore, information science provides many important applications of epistemology. Most of our knowledge about the world is acquired through communication with other members of society rather than through direct observation. Much of this socially acquired knowledge is transmitted to us via recorded information. For example, most of what we know about what is going on in other parts of the world comes from books, newspapers, television, and the Internet. As a result, information issues are some of the most pressing, concrete concerns within the scope of epistemology (cf. Fallis, 2004a; Goldman, 1999, pp. 161–182).

Several authors (Budd, 1995; Dick, 1999; Fallis, 2000; Froehlich, 1989; Hjørland, 2002; Radford, 1998) have discussed the importance of epistemology to information science, but Margaret Egan and Jesse Shera were the first to make this point. In a landmark article in *Library Quarterly* (Egan & Shera, 1952), they claimed that, in order for libraries and other information services to effectively facilitate the acquisition of knowledge, we need to know more about how knowledge is acquired and used.[4] In other words, they felt that there was a gap in our knowledge about knowledge. According to Egan and Shera (1952, p. 132), "a new discipline must be created that will provide a framework for the effective investigation of the whole complex problem of the intellectual processes of society." The name that they gave to this new discipline was *social epistemology*. Social epistemology encompasses a number of research areas within information science and related disciplines (e.g., bibliometrics, economics of information, information retrieval, sociology of knowledge).

Although the term was coined by information scientists, social epistemology has subsequently been developed (independently) by a number of philosophers (e.g., Fuller, 1988; Goldman, 1987, 1999; Longino, 1990;

Schmitt, 1994b).[5] Social epistemology is simply the branch of epistemology that focuses on the role that social factors and social institutions play in knowledge acquisition. Such factors and institutions are clearly important when people acquire knowledge from other people (e.g., via recorded information). A more detailed introduction to epistemology in general and social epistemology in particular will be provided in the sections that follow.

Jesse Shera and Social Epistemology

The idea of social epistemology was first developed in an article co-written with Egan, but Shera's is the name that is most widely associated with social epistemology and its application to information science.[6] Shera (1961, 1968, 1970) went on to develop the idea of social epistemology in a number of other publications. His most comprehensive discussion of social epistemology is in his *Sociological Foundations of Librarianship* (Shera, 1970).

Jonathan Furner (2002) provides a comprehensive analysis of Shera's position on social epistemology (see also Budd, 2002a; Dick, 2002b; Zandonade, 2004), tracing the influences of epistemologists, sociologists of knowledge, cognitive psychologists, and documentalists on Shera's thinking. I focus here on what I take to be the most critical aspects of Shera's position. Some of its more idiosyncratic aspects (e.g., his views on whether knowledge has to be *true* and on what it means for *society as a whole* to have knowledge) will be addressed in subsequent sections.

First, Shera (1970, pp. 108–109) intended that social epistemology provide a *theoretical foundation* for information science.[7] That is, he wanted something that would do for our understanding of how knowledge is acquired and used what quantum theory and relativity theory do for our understanding of how the physical world works. This is some indication of the importance Shera assigned to social epistemology for information science.

Luciano Floridi (2002) has recently argued, however, that social epistemology by itself will not provide the requisite theoretical foundation (see also Floridi, 2004). He points out that several important aspects of information science are not epistemological in nature (e.g., the metaphysical questions and ethical questions mentioned earlier). Thus, Floridi suggests that a more broadly based *philosophy of information* provides a better theoretical foundation for information science. Even if this proposal is correct, social epistemology nevertheless remains central to information science, as it provides a theoretical framework for addressing a large number of the activities performed by information services (namely, those focused on facilitating the acquisition of knowledge).

Second, although social epistemology is a theoretical discipline, Shera (1970, p. 88) believed that it must have practical implications (cf. Dick, 1999, p. 305).[8] He held that social epistemology should be pursued in order to make information services more effective at "facilitating intellectual access to ... knowledge" (Shera, 1970, p. 84). According to Egan and Shera (1952, p. 132), this new discipline should "not only result in understanding and appreciation but also make possible future national planning and implementation."

Shera (1970, pp. 90–92) himself focused primarily on one particular application of social epistemology to information science: namely, classification. In order to acquire knowledge, people have to be able to find the information they need. Information services use classification systems to organize information resources so that people will be able to find this information. Shera believed that a better understanding of how knowledge is acquired and used would allow us to improve classification systems.[9]

Of course, information services perform a number of activities to facilitate the acquisition of knowledge (e.g., reference work, collection management, outreach). In fact, social epistemology is connected to almost all areas of information science. As a result, a wide variety of applications of social epistemology has been discussed in the information science literature.

Finally, because he hoped that it would make recommendations for what information services *ought* to do to be more effective, Shera clearly thought of social epistemology as a normative project (cf. Furner, 2002, p. 7). It does not restrict itself to describing how people acquire knowledge from information services. By contrast, the *sociology of knowledge* is an example of a descriptive project, for it aims simply to describe the causal role that social factors play in the creation of knowledge (cf. Bloor, 1976).

Connections Between Social Epistemology and Information Science

In line with Shera's position, this chapter focuses on how social epistemology can help information services to facilitate knowledge acquisition on the part of the users of the services. It should be noted, however, that there are two further important connections between social epistemology and information science.

First, information science techniques can sometimes be used to do research in social epistemology. For example, bibliometric techniques have been used to study the structure and dynamics of scientific knowledge (e.g., Marshakova-Shaikevich, 1993). These bibliometric techniques might in turn enhance information professionals' efforts to aid users of information services in the acquisition of knowledge.

Second, there are epistemological questions about information science research itself. In his discussion of epistemology and information science, Budd (1995), for example, is primarily concerned with identifying the appropriate methodological approach for information scientists. In other words, he deals with questions concerning the knowledge acquired by information scientists rather than with the knowledge acquired by users of information services.[10] In this vein, researchers can ask, as has Dick (1999, p. 307), how much of what "LIS claims to know on the basis of its modes of professional practice and research traditions can indeed be justified?" More specifically, researchers can investigate whether information scientists really know that particular practices will facilitate knowledge acquisition on the part of users of information services (see Henige, 1987).

Epistemology

As Shera (1970, p. 87) pointed out, several areas of study (e.g., psychology, cognitive science, and even sociology) deal with the question of how people acquire knowledge. Epistemology, however, is uniquely concerned with the question of what knowledge is. As a result, epistemology has great potential for helping to clarify the epistemic objectives of information services. In this section, I provide a brief introduction to epistemology and the main questions it addresses.

Epistemologists traditionally try to understand what knowledge is by identifying *necessary* and *sufficient* conditions for something to count as knowledge (cf. Feldman, 2003; Steup, 2001).[11] This is what scientists generally do when they are trying to understand some natural phenomenon (e.g., what is a mammal?, what is copper?). For example, having 29 protons in the nucleus is a necessary and sufficient condition for a substance to be copper. This is a necessary condition because anything that is copper must have an atomic number of 29. It is a sufficient condition because anything that has an atomic number of 29 is copper.

Various conditions have been proposed (cf. Furner, 2002, p. 8; Goldman, 2002, pp. 183–188) but almost all epistemologists agree that knowledge is some form of *justified true belief* (Feldman, 2003, pp. 15–16; Steup, 2001). In other words, they hold that (a) being a *belief*, (b) being *true*, and (c) being *justified* are all necessary conditions for something to be knowledge.

These three conditions on knowledge arguably capture (at least part of) the epistemic objectives of information services. When someone goes to a library in order to acquire knowledge, it is reasonable to assume that he or she would like to acquire justified true beliefs. For example, suppose that a student is doing a report on the Eiffel Tower and wants to be able to say how high it is. When she goes to the library to do her research, she wants to end up with a *belief* about how high it is (e.g., that it is 300 meters high). Otherwise, she will have nothing to put in her report about the height of the Eiffel Tower. In addition, she wants this belief to be *true* (Meola, 2000). Otherwise, her teacher may very well take off points from the grade given for the report. Finally, she probably also wants to be *justified* in having this belief; she wants to be able to say, for example, that she obtained this information from an authoritative source, such as an encyclopedia. If she simply made a guess at the height of the Eiffel Tower (and were unable to cite a source for the fact), her teacher might again deduct points (even if the guess happened to be correct).

Finally, it should be noted that epistemologists focus primarily on what it means to *know that* something is the case (e.g., that the Eiffel Tower is 300 meters high) rather than on what it means to *know how* to do something (e.g., how to build a house). By contrast, information services are concerned with both sorts of knowledge. However, in providing access to recorded information, information services typically facilitate the acquisition of *procedural* knowledge by facilitating the acquisition of *propositional* knowledge. For example, someone might read a book, which consists of a large number of propositions, in order to learn how to build a house. Thus, it is reasonable for us to follow the epistemologists and focus (at least initially) on propositional knowledge.

Major Projects in Epistemology

Although most epistemologists agree that knowledge is some form of justified true belief, this does not mean that the question of what constitutes knowledge has been settled. Several important projects in contemporary epistemology address this topic. Here I provide a brief survey of these projects; later in this chapter I discuss their relevance for information science.

The Nature of Justification

First, let us consider the project of determining the precise conditions under which a belief is *justified* (Feldman, 2003, pp. 39–107; Steup, 2001). There are several competing accounts of justification—e.g., reliabilism, evidentialism, foundationalism, coherentism—which can be divided into two main camps: namely, *externalist* accounts of justification and *internalist* accounts.

Reliabilism is a prototypical example of an externalist account of justification. According to a reliabilist account of justification, a belief is justified if it was produced by a reliable process (i.e., a process that tends to produce true beliefs). For example, a student is justified in believing that the Eiffel Tower is 300 meters high if she obtained this information from a reliable source (i.e., a source that mostly contains accurate information). This is an externalist account of justification because what makes the belief justified is *external* to what is going on in the student's mind. In particular, it does not matter whether she has any reason to believe that the source is reliable.

Evidentialism is a prototypical example of an internalist account of justification. According to an evidentialist account of justification, a belief is justified if a person has good grounds (i.e., evidence) for this belief. For example, a student is justified in believing that the Eiffel Tower is 300 meters high if she obtained this information from a source that she has reason to believe is reliable. This is an internalist account of justification because it does not matter whether the source actually is reliable. It just matters that her beliefs about the world fit together (i.e., that they support each other).

In addition to disagreement about what *type* of justification is required, there is also disagreement about how *much* justification is required in order for a belief to count as knowledge. For example, how good does your evidence have to be? Or, for a reliabilist, how reliable does the process that produced the belief have to be?

Descartes (1641/1996, p. 14), in particular, set an extremely high standard for justification. He claimed that a belief only counts as knowledge if it is "certain and indubitable." However, almost all of our beliefs are open to some small degree of doubt (e.g., our senses deceive us, people do not always tell the truth).[12] As a result, if this is the correct account of knowledge, then almost none of our beliefs counts as knowledge. Our belief in our own existence may be the only thing that is not open to doubt (Descartes, 1641/1996, p. 17).

Most epistemologists, however, do not set such a high standard for justification. In other words, the degree of justification that is required for a belief to count as knowledge does not have to completely eliminate the possibility of the belief being false. For example, the fact that the *Encyclopædia Britannica* says so is probably good enough evidence that the Eiffel Tower is 300 meters high even though it is certainly possible that the encyclopedia is incorrect. Goldman (1999, pp. 23–25) even thinks that there is a "weak" sense of knowledge in which a true belief counts as knowledge even if it is not justified at all.

The Gettier Problem

Another epistemological project seeks to find a *fourth* necessary condition for something to be knowledge (Feldman, 2003, pp. 25–38; Steup, 2001). Gettier (1963) has convincingly argued that a justified true belief does not always count as knowledge. For example, suppose (1) that you believe that Matt owns a Saturn, (2) that he really does own a Saturn (i.e., that your belief is true), and (3) that you have good grounds for this belief because you have seen Matt driving a Saturn to work each day. However, suppose that the Saturn that you have seen him driving each day is one that he has borrowed from a friend, while his own Saturn sits safely locked in his garage. Even though your belief is true and justified, your justification is not appropriately connected to the truth of the proposition that you believe. Thus, it seems that you do not really *know* that Matt owns a Saturn.

Because the belief, truth, and justification conditions are not jointly sufficient for something to constitute knowledge, some additional condition must be satisfied. After almost half of a century, epistemologists continue to debate what this additional "Gettier condition" might be.

Replying to the Skeptics

A third project concerns establishing that knowledge is possible at all (see Feldman, 2003, pp. 108–129). It seems clear that we have all sorts of knowledge about the world (e.g., that the Eiffel Tower is 300 meters high). But some epistemologists wonder whether we even know that there is an Eiffel Tower, much less that it is 300 meters high. After all, as Descartes (1641/1996, p. 15) pointed out, it is possible that a "malicious demon of the utmost power and cunning has employed all his energies in order to deceive [us]." A more modern version of this sort of worry has an evil scientist removing our brains from our bodies, placing them in vats, and hooking them up to a computer (see Feldman, 2003, p. 94), which then stimulates our brains so that we falsely believe that we are still moving around in the world. This is essentially the scenario depicted in the movie *The Matrix*.

These sorts of scenarios do not seem very likely, but they do establish that (almost) all of our beliefs *could be* false. Skeptics agree with Descartes that a belief is not knowledge if there is any possibility that we might be wrong. Epistemologists are still debating how best to reply to such a thoroughgoing skeptic.

Socializing Epistemology

Finally, there is the project of *socializing* epistemology. As Shera (1970, p. 85) points out, "epistemologists have studied the origins, growth, and development of knowledge ... with reference to the *individual*" (cf. Goldman, 1999, p. 4). In other words, epistemologists have traditionally tended to think about how an individual working alone acquires knowledge. The prototypical case is that of Descartes (1641/1996, p. 12) locking himself away and working "quite alone" to acquire knowledge about the world. However, several epistemologists (e.g., Fuller, 1988; Goldman, 1987, 1999; Longino, 1990; Schmitt, 1994b) have recently tried to develop a more socially based epistemological framework, which looks at how people working together acquire knowledge. In the next section, I discuss this particular project in epistemology and its importance to information science.

Social Epistemology

Like all epistemologists, social epistemologists are concerned with what knowledge is and how people acquire it. They focus on the case where knowledge acquisition takes place within a social context. Because it is concerned with such socially acquired knowledge (in particular, knowledge acquired from other people via recorded information), information science needs to look to social epistemology in particular, not just epistemology in general. As Shera (1970, p. 86) put it, information science must be concerned with the "production, flow, integration, and consumption of all forms of communicated thought throughout the entire *social* fabric."

Different Senses of Social

There are several different ways in which epistemology might be considered social. Each of these different senses of the concept *social* has relevance for information science. Shera (1970) touched on all of these ways in which epistemology might be socialized. First, epistemology might look at how *social processes* (e.g., social practices and institutions) can lead to knowledge acquisition. That is, instead of focusing on how we can acquire knowledge by working alone, we might look at how we can acquire knowledge by enlisting the aid of other people. Goldman (1999) and Kitcher (1993, pp. 303–389), for example, do social epistemology in this sense of "social."

This first sense of "social" is especially relevant for information science. As Wilson (1983) put it, information science focuses on "secondhand knowledge." In other words, information services facilitate knowledge acquisition by transferring information from one person to another.[13] Effective transfer usually require cooperation among a number of people.

Second, epistemology might look at how the members of a social group can acquire knowledge. Social practices and institutions often have an effect on how much knowledge large numbers of people acquire. The public library, for example, would like all the members of a community to be able to acquire knowledge. Because its resources are usually limited, however, an information service has to consider how knowledge ought to be distributed among the members of a social group. On this topic, Goldman (1999, p. 94) has suggested that we should maximize the average amount of knowledge acquisition in a social group. However, it is not clear that this is necessarily the appropriate objective. Maximizing the knowledge acquisition of the "least advantaged" members of society (Rawls, 1971), for example, is more in line with a concern for equitable access to information (cf. Fallis, 2004c).

Third, epistemology might look at how social groups themselves can acquire knowledge. Even when many people are acquiring knowledge, each of them may be acquiring knowledge as individuals. However, some epistemologists (e.g., Mathiesen, 2003; Schmitt, 1994a) have suggested that groups themselves might possess knowledge (cf. Van House, 2004, pp. 40–41). For example, we often attribute knowledge to juries, governments, and other organizations. In fact, *society as a whole* could be the agent that has knowledge. This seems to be what Shera (1970, p. 83) had in mind when he claimed that "the society collectively knows all the contents of all the encyclopaedias, the reference books, the proceedings of learned societies, *et cetera*, that have ever been published" (cf. Furner, 2002, p. 9).

Finally, epistemology might look at how social factors can affect knowledge acquisition. Knowledge acquisition always occurs within a social context. As a result, the knowledge that people acquire is always influenced by social and cultural factors, such as income, social status, ethnicity, and gender.[14] For example, Kitcher (1994, p. 125) points out that when women first became primatologists in the 1970s, studies of primates yielded all sorts of new results because the female researchers did not make the same unconscious assumptions that male primatologists did when observing the world. *Sociologists of knowledge* (e.g., Bloor, 1976; Shapin 1994) study the impact of such social factors on knowledge acquisition (cf. Van House, 2004, pp. 9–11).

It used to be a popular view (known as the "arationality hypothesis") that we need not consider social factors when explaining how people end up with *true* beliefs (cf. Bloor, 1976, pp. 5–10; Goldman, 2001). According to this view, social factors come into play only when we explain how people end up with *false* beliefs (i.e., why people make mistakes). Bloor (1976, p. 5) has convincingly argued against this view and in favor of the "symmetry principle," which states that social factors can help to explain the formation of true as well as false beliefs.

In any event, this final sense of "social" is very important for information science. Social institutions, such as schools and libraries, need to be aware of how social and cultural factors affect people's ability to acquire knowledge. Burbules and Callister (1997), for example, discuss the impact of social factors on people's ability to acquire information from the Internet. In fact, language itself is a social factor that has an appreciable impact on the acquisition of propositional knowledge (cf. Goldman, 1999, p. 21). Thus, social factors play a role in almost any sort of knowledge-gathering activity.

Evaluating Social Practices and Institutions

As noted earlier, each of these senses of social has relevance for information science. However, because information services facilitate knowledge acquisition by transferring information from one person to another, the first sense of "social" has special relevance for information science. Accordingly, this chapter focuses on identifying social practices and institutions that facilitate knowledge acquisition.[15]

Advances in communication and information technology have greatly expanded the ways in which people communicate with each other. People can now communicate with *more people*, over *longer distances*, and *faster* than ever before (cf. Goldman, 1999, pp. 161–162; Shera, 1961, p. 768). Thus, it is especially important for information professionals to understand how social processes lead to knowledge acquisition.[16]

Even before epistemologists began to socialize epistemology, there was some work in epistemology on social processes. In particular, the *epistemology of testimony* studies the conditions under which you can trust that what another person says is true (e.g., Coady, 1992; Goldman,

1999, pp. 103–130, 2002, pp. 139–163; Hume, 1748/1977, pp. 72–90; Lipton, 1998). Information science is especially interested in the specific case where testimony is communicated via recorded information (cf. Fallis, 2004b, p. 466). In other words, information science concerns itself with the conditions under which we can trust that what an information resource says is true. For example, when can we trust what a Web site says about the height of the Eiffel Tower or about the treatment of a particular medical condition?[17]

More recently, Goldman (1999) has studied how social processes in a number of different areas (e.g., science, law, education, and politics) can lead to knowledge acquisition. In particular, he has examined areas of great concern to information science, such as communication technology, scholarly publishing, and freedom of speech (cf. Fallis, 2000; Thagard, 2001).

Classical and Revolutionary Social Epistemology

In addition to the various different senses of "social" in social epistemology, there is another important way in which work in social epistemology can be divided. According to Floridi (2002, p. 40), one can distinguish between a *classical* approach to social epistemology and a *revolutionary* approach (cf. Goldman, 2001; Resnik, 1996, pp. 567–578; Van House, 2004, p. 8).

The classical (or *objectivist*) approach is continuous with traditional epistemology. In particular, it adopts the standard account of knowledge as a form of justified true belief. This approach simply socializes (in one or more of the various senses of "social") traditional epistemology. Kitcher (1994) and Goldman (1999), for example, take this approach to social epistemology.

By contrast, the revolutionary (or *non-objectivist*) approach abandons traditional epistemology and starts over from scratch. In particular, it rejects the standard account of knowledge as a form of justified true belief, preferring instead to view knowledge as a form of "institutional belief." What counts as knowledge is a result of negotiation; in other words, knowledge is "socially constructed." This approach to social epistemology often takes its inspiration from the sociology of knowledge. For example, according to Bloor (1976, pp. 2–3), knowledge is "whatever men take to be knowledge ... what is collectively endorsed." In addition to Bloor, Fuller (1988), and Harding (1992) have taken this approach to social epistemology.

Shera (1970, p. 96) also clearly took a revolutionary approach to social epistemology, as is exemplified by his claim that whatever has gone through "a filtering process ... through the ethical system or the intellectual system, or the system of scholarship ... of the individual who receives it" constitutes knowledge. Given Shera's influence on the field, a revolutionary approach is more common than a classical approach when social epistemology is applied to information science (e.g., Budd,

1995; Radford, 1998). This chapter, however, takes the position that a classical approach provides a more useful theoretical framework for information science. It focuses on the various ways in which classical social epistemology can be (and already has been) fruitfully applied to information science.[18]

Objections to Applying Social Epistemology to Information Science

Information scientists have discussed the importance of applying social epistemology to information science and several have raised important objections to such an undertaking. First, some information scientists have claimed that helping people to acquire knowledge is not the main objective of information services. They suggest that something else (e.g., increasing exposure to recorded information) is the true objective of information services.[19]

Second, although most epistemologists think that knowledge is some form of justified true belief, many information scientists disagree. In particular, they have (either explicitly or implicitly) criticized the truth condition and the belief condition. It should be noted that this objection (as well as the following two objections) are directed specifically at *classical* social epistemology.

Third, even though they have yet to reach a consensus on these conditions, epistemologists are looking for a *single* set of necessary and sufficient conditions for something to count as knowledge. However, it seems clear that people using information services have a variety of epistemic interests and values. For example, people want to acquire knowledge that is *significant*, they want to acquire knowledge *quickly*, and they sometimes want to acquire *wisdom* and *understanding*.

Finally, much work in contemporary epistemology is fairly esoteric. For example, real people (e.g., library patrons) are rarely worried about being deceived by malicious demons or evil scientists. The Gettier problem also seems rather divorced from the concerns of people who use information services; the standard examples of justified true beliefs that do not count as knowledge seem to be simply bizarre cases of bad luck.[20]

In the following sections, I discuss these objections to applying epistemology to information science. Considering how we might respond to these objections will suggest how epistemology can be successfully applied to information science.

The Main Objective of Information Services

There is a strong presumption that the *raison d'être* of information services is to allow people to acquire knowledge. This is certainly the presumption behind Shera's claim that social epistemology can provide a theoretical foundation for information science. In fact, this presumption is so obvious and commonplace that it formed the basis of a recent *New Yorker* cartoon: A woman walks into a bookstore and asks "Where's your section of books that tell you simple things you already know?" It is not clear why this cartoon is funny if the acquisition of knowledge is not the main objective of information services.[21]

And yet, knowledge acquisition is rarely mentioned explicitly as an objective of information services (cf. Baker & Lancaster, 1991; Evans, Borko, & Ferguson, 1972). In fact, some scholars (e.g., Hamburg, Ramist, & Bommer, 1972, p. 111) have even claimed that knowledge acquisition is not the main objective of information services. In this section, I defend the view that knowledge acquisition actually is the main objective of information services and, thus, that social epistemology is critical for information science (cf. Fallis, 2001, pp. 176–178). I do so by responding to three important objections that might be raised against this view.[22]

The Ultimate Goals of Information Services

One potential objection to the thesis that knowledge acquisition forms the main objective of information services is that knowledge acquisition is clearly not always the *ultimate goal* of information services. For example, public libraries are ultimately interested in promoting the public good. In fact, Shera (1961, p. 770) claims that "the aim of librarianship ... is to maximize the *social utility* of graphic records." Also, publishing companies, database providers, and bookstores are ultimately interested in making a profit (cf. Mattlage, 1999, p. 313).

Be that as it may, facilitating knowledge acquisition is the means by which information services go about achieving these ultimate goals. For example, public libraries do not promote the public good by just any means (e.g., by feeding the homeless or putting out fires). They do so specifically by providing patrons with access to information that will allow them to acquire knowledge (cf. Shera, 1961, 1970).[23] Similarly, database providers make a profit specifically by providing customers with access to information that will allow them to acquire knowledge. Thus, knowledge acquisition constitutes the immediate aim of information services.

Admittedly, there are a few other means by which information services achieve their ultimate goals. For example, public libraries also try to promote the public good by providing entertainment to patrons (e.g., by lending popular novels and movies). However, facilitating knowledge acquisition is still the *definitive* means by which information services achieve their ultimate goals.[24] In other words, helping people to acquire knowledge is arguably the *sine qua non* of information services. A library that did not help patrons to acquire knowledge would not be much of a library.

In addition, there are several good reasons for information services to focus on knowledge acquisition rather than, say, social utility. For instance, information services have a reasonable degree of control over their success at facilitating knowledge acquisition. By contrast, numerous factors beyond the control of information services have an impact on social utility (cf. Hamburg et al., 1972, p. 110). A library can (by providing access to the necessary information) help a patron learn how to build a house but it has very little control over whether the patron actually builds one (cf. Baker & Lancaster, 1991, p. 14).[25] Furthermore, knowledge acquisition promotes social utility in so many different ways that it is very difficult to measure the impact of information services on social utility. As a result, the objective of knowledge acquisition can serve as a useful proxy for the ultimate goals of information services (cf. Kirkwood, 1997, pp. 24–25).

Even so, it should be noted that although knowledge acquisition is the main objective of information services, this does not mean that information services should try to facilitate knowledge acquisition *at all costs*. In particular, there are ethical constraints on how information services can achieve this epistemic objective (cf. American Library Association, 1995). For example, a school librarian should not make copies of materials in violation of the author's intellectual property rights, even if doing so would help students to acquire more knowledge.[26] Of course, the fact that there are ethical constraints on our pursuit of an objective does not show that it is not our objective. Thus, knowledge acquisition can still be the main objective of information services even if information services clearly need to pursue this objective in an ethical manner.

Finally, the ultimate goals of information services (especially the profit motive) can sometimes conflict with the objective of knowledge acquisition. For example, an academic journal publisher might be able to make more money by selling to a few people at a very high price than by selling to many people at a more reasonable price (cf. Mattlage, 1999, p. 314). Even this scenario fails to demonstrate that the acquisition of knowledge is not the main objective of these information services. As noted earlier, publishing companies and database providers are in the business of making a profit *by facilitating knowledge acquisition*.[27] An academic journal publisher that did not help its customers acquire the knowledge that they needed would not stay in business for long.

The Standard Objectives of Information Services

A second potential objection is that knowledge acquisition is not one of the commonly discussed objectives of information services. Knowledge may be a means to the ultimate goals of information services but it is rarely explicitly taken as an objective.[28] For example, many of the standard objectives of information services focus on the properties of collections. We want a collection to be *comprehensive*, we want it to include materials for which there is a high *demand*, and we want these materials to be easily *accessible* (e.g., Baker & Lancaster, 1991; Evans et al., 1972). Other standard objectives focus on the properties of retrieval systems (e.g., Blair, 2003, p. 7; Harter & Hert, 1997). For example, we want retrieval systems to have high *recall* (i.e., many relevant documents are retrieved) and to have high *precision* (i.e., most retrieved documents are relevant).

However, even if an information service possesses all of these proper- ties, it would be worthless if no one actually used it to acquire knowl- edge. These are certainly valuable properties for collections and retrieval systems to have, but they are not *intrinsically* valuable. They are valuable precisely because they are instrumental to the acquisition of knowledge by the service's clients: people are more likely to acquire knowledge if they have access to materials on a wide range of topics that are of interest to them. [29] In order to achieve our objectives, it is impor- tant not to confuse our means with our ends (cf. Kirkwood, 1997, p. 22).

Exposure to Recorded Information

A third potential objection is that something other than knowledge acquisition constitutes the main objective of information services. For example, according to Hamburg et al. (1972, p. 111), information ser- vices aim to increase the "exposure of individuals to documents of recorded human experience" (cf. Baker & Lancaster, 1991, pp. 14–16). In other words, they contend that the objective is something short of knowl- edge acquisition.

Hamburg et al. agree that knowledge acquisition is a means to the ultimate goals of information services but they argue that exposure to recorded information is the means by which information services go about facilitating knowledge acquisition. They also agree that the stan- dard objectives of information services are not intrinsically valuable. However, they hold that the value of these standard objectives lies in the fact that they serve as a means to exposing a service's clients to recorded information.

Exposure to recorded information is certainly a critical means by which information services facilitate knowledge acquisition. Even so, there are good reasons for information services to focus on knowledge acquisition rather than something such as exposure to recorded infor- mation. Just like social utility, knowledge is intrinsically valuable (cf. Goldman, 1999, p. 75). However, as is the case with the standard objec- tives of information services, exposure to recorded information is not intrinsically valuable; indeed, it is worthless if it does not lead to knowl- edge acquisition.

Information services can increase people's exposure to recorded information by, for example, ignoring the quality of information sources. To take a concrete case, one of the all-time best-selling publi- cations of the University of Arizona Press is *I Married Wyatt Earp* (Earp, 1976). This book was advertised as the "recollections of Josephine Sarah Marcus Earp collected and edited by Glenn Boyer," but appears to have been a hoax (see Sharlet, 1999). It has thus misled many readers (including historical researchers). Exposing many people to such false or misleading information is clearly not a good way for an information service to facilitate knowledge acquisition or to achieve its ultimate goals.[30]

Hamburg et al. might try to get around this specific problem by suggesting that increasing exposure to *quality* information is the main objective of information services. However, exposure to quality information is not intrinsically valuable either. For example, a library patron might not acquire the knowledge that she needs about house building even if she is exposed to quality information on the topic. For example, the patron might not have the reading skills to comprehend the library's books on this topic or simply might not have a quiet enough place to concentrate on reading them.

Nevertheless, Hamburg et al. (1972, pp. 110–111) are correct to point out two important advantages to focusing on exposure to recorded information rather than knowledge acquisition. First, information services have more control over exposure to recorded information than they have over knowledge acquisition. For example, families and schools arguably have a much larger impact than information services on whether people acquire knowledge. Even so, information services clearly have a role to play in mitigating factors (such as lack of reading skills) that interfere with knowledge acquisition. With literacy programs in public libraries, for example, information services can and do go beyond merely exposing people to recorded information.

Second, it is admittedly much easier to measure success at exposing people to recorded information than to measure success at facilitating knowledge acquisition. However, in order to make good decisions about policies and practices, information services need to identify (and keep in mind) their "true" objective rather than some objective that happens to be easier to measure.[31] As Rawls (1971, p. 91) observed, "it is irrational to advance one end rather than another simply because it can be more accurately estimated." We do not want to be like the proverbial drunk who looks under a lamppost for his keys, even though he lost them elsewhere, simply because the light is better under the lamppost.[32]

Objections to Knowledge as Justified True Belief

In the previous section, I argued that knowledge acquisition is the main objective of information services. Because epistemology is concerned with what knowledge is, one might expect that work in epistemology would help to clarify this objective. However, many information scientists (e.g., Hannabuss, 2001; Radford, 1998, p. 631; Shera, 1970, p. 97; Wilson, 1983), as well as many non-objectivist social epistemologists, have objected to the claim that knowledge constitutes some form of justified true belief.

The main objections have to do with the truth condition.[33] For example, according to Shera (1970, p. 97), knowledge "has nothing to do really with truth or falsehood. Knowledge may be false knowledge, or it may be true knowledge. It is still knowledge, it is knowable and known." Furthermore, some information scientists have also objected to the belief condition. In this section, I respond to these objections to applying epistemology to information science.

The Truth Condition

According to the truth condition in the standard account of knowledge, you can know that Matt owns a Saturn only if it is true that Matt owns a Saturn. But what does it means to say that such a proposition is true? A number of different theories of truth have been proposed (cf. Goldman, 1999, pp. 41–68). However, the basic idea is that a proposition is true if it captures the way that the world actually is. Thus, truth is an objective property independent of what anyone happens to believe. For example, it is true that Matt owns a Saturn if Matt *really does* own a Saturn. Similarly, it is true that the Eiffel Tower is 300 meters high if the Eiffel Tower *really is* 300 meters high.

The objections to the truth condition can be divided roughly into three categories: metaphysical, epistemological, and ethical. I consider these three types of objection in turn.

Metaphysical Objections

Proponents of metaphysical objections to the truth condition argue there is no reason to think that there is such a thing as "objective truth." Accordingly, they often point to cases where there appears to be no truth of the matter. For example, consider the following question that might be found on an intelligence test: "There are five birds sitting on a fence. You shoot one of these birds. How many birds are left on the fence?" There seem to be (at least) two correct answers to this question. Some people (and cultures) are inclined to say that there will be four birds left because five minus one is four. However, other people (and cultures) are inclined to say that there will be no birds left because the four other birds will be scared off by the shot.

These sorts of considerations have led many information scientists to conclude that all knowledge is *socially constructed* (see Meola, 2000, p. 174). In other words, they hold that what counts as knowledge is determined by the beliefs of the people in a particular culture. As Shera (1970, p. 97) puts it, "what is true in one society, in one culture, may be completely false in another." [34]

I would argue to the contrary, however, that there only appears to be no truth of the matter because such questions are ambiguous. There are two correct answers because two different questions are being asked (namely, "What is five minus one? Four." and "What happens when you shoot at birds? They fly away."). Once we are clear about which question is being asked, there is only one correct answer. Social and cultural factors certainly play an important role in determining which question is being asked (i.e., in determining how to interpret the question), but do not come into play in determining the correct answer to that specific question (cf. Kitcher, 1994, pp. 26–128). [35]

The metaphysical objection to the truth condition might also involve Bloor's symmetry principle (i.e., the claim that social factors can explain the formation of true as well as false beliefs). As Goldman (2002, p. 192) has pointed out, if social factors can explain the formation of both true and false beliefs, then it is easy to conclude that truth itself plays no role. [36] However, even though one may concede that social factors always play an important role in belief formation, this does not necessarily mean that they give us the whole story. As Kitcher (1994, pp. 120–121) has observed, some methods of belief formation (e.g., performing scientific experiments) are more likely to lead to true beliefs than others (e.g., reading tea leaves).

Epistemological Objections

According to the second type of objection to the truth condition, even if there is such a thing as objective truth, there is no reason to think that we can ever find out (with certainty) what it is. For example, it would seem to be rather difficult (if not impossible) for information services to ensure that their users acquire true beliefs. As Dick (1999, p. 315) has put it, "libraries themselves are, of course, unable to sift truth from falsehood." Similarly, according to Wolkoff (1996, p. 87), "it is difficult for librarians to judge objectively the accuracy of all materials, and they should not undertake the role of arbiters of truth."

However, even if an epistemic objective of information services is that their users acquire true beliefs, this does not mean that the information professionals have to be able to tell these users what things are true (and what things are not). For example, the librarian does not have to determine for the patron what the shape of the earth really is or what the height of the Eiffel Tower really is. In fact, this would probably be a rather inefficient means of facilitating the acquisition of knowledge. Nevertheless, there are many techniques that librarians can (and often do) use to make it more likely that patrons will acquire true beliefs. For example, a reference librarian might check two sources rather than just one (e.g., Connell & Tipple, 1999, p. 363). By doing so, the librarian is more likely to give patrons correct answers to their queries (cf. Hume, 1748/1977, pp. 72–90). Also, when selecting materials on a particular topic, librarians are more likely to pick accurate materials if they consult with experts in that area (cf. Goldman, 2002). In addition, according to Mill (1859/1978, pp. 15–52), giving people access to a diversity of viewpoints makes it more likely that people will acquire true beliefs. [37] This is because people are better able to determine what the truth is if they are exposed to all sides of an issue. In a similar vein, Harding (1992, pp. 578–585) argues that we need to adopt diverse perspectives (or "standpoints") in order to expose hidden assumptions and biases (cf.

Trosow, 2001). Thus, librarians do not have to be arbiters of truth in order to facilitate the acquisition of knowledge.

Ethical Objections

A third type of objection holds that, even if there is such a thing as objective truth and we can (at least sometimes) find out what it is, there is reason to think that it is not a goal that we should be pursuing. In particular, it has been claimed that appeals to objective truth are often used as a means of repression. For example, Harding (1992, p. 568) claims that the supposedly unbiased pursuit of truth in science can serve "male supremacy, class exploitation, racism, and imperialism" (cf. Foucault, 1979, p. 305).

In the context of information science specifically, taking truth as our goal might seem to conflict with certain ethical values to which information professionals are deeply committed. Pursuing truth might seem to require restricting access to information that is deemed to be objectively false. Thus, the pursuit of truth potentially involves information services in restricting access to information and possibly even censorship.

Mill (1859/1978), however, showed that it is perfectly possible to argue convincingly for the value of unrestricted access to information without giving up truth as a goal. For example, he argued that because we are fallible and cannot be absolutely sure which pieces of information are objectively false, restricting access to information that is deemed to be false is actually not a good means of pursuing truth.

Furthermore, setting the acquisition of true beliefs as an epistemic objective does not say anything about the best way to attain them. Even if there is only one correct answer to a question, there may be many different ways to find that answer. In fact, different people in different contexts may be better off using different techniques. In addition, saying that the acquisition of true beliefs is our epistemic objective does not say anything about which true beliefs we should try to acquire. According to Goldman (1999, p. 89), we really acquire knowledge only by getting answers to "questions of interest." Thus, libraries should try to help people to find answers to the questions that they actually have. For example, it is very important in reference work to determine exactly what question is being asked.

The Truth Condition Reconsidered

I have tried to respond to the most common objections to the truth condition. There is, however, one further important reason why information services should help people to acquire *true* beliefs. Namely, true beliefs are what people are typically after when they gather information (cf. Dretzke, 1981, p. 46; Goldman, 1999, pp. 30–35). For example, the student doing a report on the Eiffel Tower wants to know how high the Eiffel Tower *really* is. [38] Furthermore, this goal is not unique to our particular culture. As Goldman (1999, pp. 32–33) has observed, people in many other cultures seek true beliefs. [39]

As a result of this, there are many instances where information services clearly try to help people to acquire true beliefs. For example, reference services are evaluated with respect to the degree to which they provide *accurate* information to their users (e.g., Hernon & McClure, 1986; Meola, 2000). Also, many information scientists are concerned with the accuracy of information on the Internet (e.g., Connell & Tipple, 1999; Fallis & Frické, 2002). In fact, a number of information scientists (e.g., Doyle, 2001, pp. 62–63; Fallis, 2000, p. 314; Meola, 2000) explicitly acknowledge that people are typically seeking true beliefs when they use information services.

In any event, even if we do decide to eliminate the truth condition, it is important not to retreat completely into *epistemological relativism* (cf. Harding, 1992, p. 573). As Dick (1999, p. 315) has pointed out, "the rejection of truth and reason weaken the value of and need for the institution [of the library] itself" (cf. Wilson, 1999, p. 164). At the very least, information services need to identify some property of beliefs that serves a similar function (i.e., that ensures the *quality* of the information provided). Even if we do decide to eliminate the truth condition, information services will still want to engage in practices very similar to those designed to get at the truth. For example, if we want users of information services to believe "what is collectively endorsed" or what has gone through "a filtering process," we will still want to check multiple sources and consult with experts in answering their queries.[40]

The Belief Condition

Although the primary objections to applying work in epistemology to information science concern the truth condition, some information scientists have objected to the belief condition as well. As noted earlier, epistemologists typically use the term "knowledge" to refer only to a special type of belief. It is not something that exists out there in the world independently of human minds. In other words, they concur with Shera (1970, p. 93) that "there can be no knowledge without a knower ... we cannot talk about knowledge as existing somewhere out in space or in some kind of vacuum." However, people sometimes talk about pieces of knowledge that are not beliefs in the heads of human beings. For example, information scientists often say that the books and journals in libraries contain knowledge (cf. Van House, 2004, p. 4). In fact, according to Popper (1972, p. 115), "almost every book ... contains objective knowledge, true or false, useful or useless; and whether anybody ever reads it and really grasps its contents is almost accidental."[41]

But despite the importance of Popper's "epistemology without a knowing subject" to information science, information services are ultimately concerned that people acquire knowledge. If no one ever reads the books and journals in libraries, then there is no point in collecting them. Preservation of information sources, for instance, is valuable only because it ensures that people will have access to these sources in the future and will be able to acquire knowledge from them. Thus, an account of knowledge that includes the belief condition is needed if we want to capture the main objective of information services.[42]

Of course, even if we are ultimately concerned that people acquire knowledge, an understanding of the structure of knowledge is often a critical means of achieving this goal. After all, the effective management of knowledge *qua* document in libraries is one of the best ways to get knowledge *qua* belief into the heads of human beings.

Varieties of Knowledge

In the previous section I argued that information services can and do facilitate the acquisition of knowledge, taking knowledge as a form of justified true belief. But this account of knowledge does not tell us everything we need to know about people's epistemic interests and values when they seek to acquire knowledge. As a number of information scientists (e.g., Neill, 1982b; Shera, 1970, pp. 102–106) have pointed out, there are several different *varieties* of knowledge that people might want to acquire. For example, information services might need to provide "knowledge that can give meaning to life" or provide "information to answer factual questions" (Neill, 1982b, p. 73). Also, different people (e.g., students, academic researchers, doctors, medical patients, moviegoers) demand different degrees of reliability from information resources. As Levi (1962, p. 56) has put it, different individuals who "are attempting to replace doubt by true belief ... [may] ... exercise different 'degrees of caution' in doing so." In fact, although very few epistemologists agree with Descartes (1641/1996) that absolute certainty is required for knowledge, there are some people (e.g., mathematicians) who do seek absolute certainty (cf. Fallis, 2002).[43]

Because somewhat different practices might be required to help people to acquire these different varieties of knowledge, it is important for information services to determine what these different epistemic interests and values are. Fortunately, there is work in epistemology that can help here. I begin by showing how it can be used to identify a number of epistemically valuable properties of beliefs. I then discuss some of the suggestions that have been made as to how information services can facilitate the acquisition of beliefs with these properties. I also show how work in epistemology can be used to identify some additional epistemic objectives that information services commonly have. I then consider how information services might deal with the conflicts that can arise between these epistemic objectives.

Epistemically Valuable Beliefs

Some epistemologists (e.g., DeRose, 1992; Smith, 2002), known as *contextualists*, claim that the standards for justification actually differ from context to context. Thus, whether something counts as knowledge depends on the context.[44] For example, consider the proposition that a particular drug is a safe and effective treatment for a particular disease. We would probably say that a student who believes this proposition on the basis of reading it in an encyclopedia knows that it is true. However, we would probably not say that a doctor who believes this proposition on the same grounds knows that it is true. In general, the standards for justification are typically higher if the costs of being wrong are higher in a particular context (cf. Smith, 2002, pp. 71–72). For example, if the proposition turns out to be false, will a patient die or will a student just receive a bad grade?

Contextualists usually focus on the fact that different *degrees* of justification are required in different contexts. However, it may also be the case that different *types* of justification (e.g., internalist justification versus externalist justification) are required in different contexts. In some contexts, for example, it may be important that a person have good reasons for her belief. In other contexts, it may be more important simply that the belief has been produced by a reliable process.

It is important *to epistemology* to determine whether the evidentialist or the reliabilist is right about whether something counts as knowledge, but it may not be important *to information science* to identify a single set of necessary and sufficient conditions for knowledge. For example, providing users with reliable sources and ensuring that they end up with good evidence both seem to be reasonable epistemic objectives of information services.[45] Thus, information science may simply look to the various conflicting accounts of justification to identify different epistemically valuable properties (cf. Dick, 1999, p. 318).

In fact, there are epistemically valuable properties of beliefs beyond merely different degrees and types of justification.[46] In many contexts, we often want to acquire knowledge that is significant, cognitively rich, explanatorily powerful, and systematically organized (cf. Paterson, 1979). Indeed, in some contexts, we want to acquire knowledge that provides understanding and wisdom (cf. Bruce, 2000; Rodman Public Library, 2003). In any event, it is clear that we often want more than just knowledge (i.e., more than justified true beliefs that meet the Gettier condition).

In addition, it is important that we acquire epistemically valuable beliefs on topics that are of interest to us (cf. Goldman, 1999, pp. 88–89). A person could acquire many justified true beliefs, for example, simply by memorizing the telephone directory. However, because these beliefs are not likely to be of much interest to this person, it would probably not be very useful to her to have all these beliefs. Among other things, this explains why we want to retrieve documents that are relevant to our queries and, thus, why we want retrieval systems to have high recall and high precision.

Furthermore, a belief might be epistemically valuable even if it does not count as knowledge.[47] For example, Goldman (1999, pp. 24–25) has pointed out that true beliefs are epistemically valuable even if they are not justified. Similarly, as Kvanvig (1998, p. 444) has suggested, justified true beliefs are epistemically valuable even if the Gettier condition is not satisfied. In fact, a belief might be epistemically valuable even if it is not true. For example, it may sometimes be valuable to believe something that is "collectively endorsed" even if is not true (cf. Bloor, 1976; Hongladarom, 2002). In the previous section, I argued that people typically seek the truth, but I do not mean to claim that they always seek the truth and nothing but the truth.[48]

Finally, in addition to contextualism, there is other work in epistemology that can help us to identify epistemically valuable properties of beliefs. For example, work has been done in the philosophy of science to clarify the "epistemic utilities" of scientists (e.g., Levi, 1962; Maher, 1993). In addition, Paterson (1979) has tried to identify the epistemic values and interests that arise in education. Because epistemic objectives in education are often very similar to the epistemic objectives of information services, this work is likely to be especially relevant to information science. Goldman (1999) has examined the epistemic values and interests that people have in a number of different areas, including law and politics as well as science and education.

Facilitating the Acquisition of Epistemically Valuable Beliefs

Once an information service determines what kinds of beliefs are epistemically valuable to its users, it can try to engage in practices that produce many epistemically valuable beliefs. In other words, it can try to ensure that its users are as well informed as possible (cf. Frické, 1998). The main way in which an information service can facilitate the acquisition of epistemically valuable beliefs is by providing access to materials on topics that are of interest to its users.[49] An information service does this both by acquiring such materials and by providing access to materials it does not own through interlibrary loan (ILL) and the Internet (cf. Goldman, 1999, pp. 165–173; Thagard, 2001).[50]

In addition, the information service must ensure that its users can find these materials in the mass of recorded information to which they now have access. In this vein, several scholars (e.g., Budd, 2002b, pp. 94–96; Fallis, 2000, p. 306; Marco & Navarro, 1993; Shera, 1970, pp. 97–107; Smiraglia, 2001, pp. 198–199) have proposed applications of epistemology to work on classification. For example, Andersen (2002) has suggested that we put materials into categories based on the types of claims made and evidence offered (e.g., to distinguish articles that present new knowledge from articles that review existing knowledge). Also, Popper's epistemological project of determining the *structure* of knowledge has been applied to classification (cf. Neill, 1982a, p. 37).

In addition to organizing its materials, there are other ways that an information service can help its users to find materials that will lead to epistemically valuable beliefs. Goldman (1999, pp. 169–170), for example, discusses the epistemic benefits of hypertext, which, within an electronic environment, provides people with immediate links to further information on a topic. Several authors (e.g., Budd, 2002b, p. 96; Meola, 2000) have considered how work in epistemology can inform work in "information mediation services." For example, an understanding of how social factors can affect knowledge acquisition is often necessary for conducting a successful reference interview.

Finally, an information service can facilitate the acquisition of epistemically valuable beliefs only if people actually use the information service. An information service can, for example, increase its patronage by advertising its services or by making itself an attractive and comfortable place for people to visit. As McDowell (2002, p. 56) has noted, "a library that maintains what patrons perceive as a pleasant or agreeable environment could increase its patronage and increase knowledge." For example, libraries might install couches and coffee bars (cf. Sannwald, 1998). Because all of these activities facilitate knowledge acquisition, they fall within the scope of social epistemology.

Further Epistemic Objectives

Simply knowing what properties are epistemically valuable, however, does not tell us all we need to know about our epistemic objectives. Goldman (1987, p. 128) refers to a practice that produces many epistemically valuable beliefs as being *powerful*.[51] However, he points out that we also want practices to be speedy, fecund, and reliable.

In addition to wanting to acquire many epistemically valuable beliefs, we typically want to acquire them as quickly as possible. As Frické (1998, p. 387) has observed, "late arrivals, among documents, tend to lose their ability to meet an information need." This epistemic objective of speed is, of course, reminiscent of Ranganathan's admonition to "save the time of the reader" (Baker & Lancaster, 1991, p. 15). One notable way for an information service to ensure that its users can acquire knowledge as quickly as possible is to provide access to information electronically. For example, Thagard (2001, p. 478) noted in his epistemic evaluation of the Internet that "the Internet and the World Wide Web have enormous advantages with respect to the speed of producing answers."

We also typically want as many people as possible to acquire epistemically valuable beliefs.[52] One way for an information service to increase its *fecundity* is to provide access to a wide range of materials that will meet the information needs of a wide range of people. In addition, an information service can ensure that as many people as possible acquire knowledge by engaging in outreach activities. For example, bookmobiles, telephone reference services, and access to materials through the Internet can provide access to information for people who would not otherwise have it (cf. Thagard, 2001, p. 477).

Finally, we do not usually want to encourage the acquisition of beliefs that lack these epistemically valuable properties to be acquired. For example, we do not want many false beliefs to be acquired. As noted earlier, one way to ensure that people are not misled is to provide them with access to highly reliable materials.[53] However, in order to provide people with all the information they need, information services often have to provide access to information from a variety of sources of varying quality (cf. Van House, 2002). For example, the Internet contains much inaccurate (as well as accurate) information and it is difficult to distinguish between the two (cf. Goldman, 1999, pp. 168–169; Impicciatore, Pandolfini, Casella, & Bonati, 1997). Information services have addressed this issue by supplying users with guidelines for evaluating information (e.g., Alexander & Tate, 1999). For example, the authority of the author and the currency of the information are often taken to be good indicators of quality. Also, it has long been suggested that the citation of an information resource by a large number of other sources is a reliable indicator of quality (cf. Davenport & Cronin, 2000; Garfield, 1972).[54]

An information service can also increase its reliability by organizing information in particular ways (cf. Fallis, 2000, p. 312). For example, Atkinson (1996, pp. 254–257) has suggested that libraries should provide access to as many materials as possible and they should also make the more reliable materials easier to access. Indeed, simply making it easier for people to find more information on a particular topic can help them to evaluate the information that they already have. For example, Holocaust denial literature could be collocated with refutations of that literature (cf. Wolkoff, 1996, p. 93). Similarly, document retrieval systems could link retractions of scientific articles to the original articles (cf. Pfeifer & Snodgrass, 1990, p. 1422).

Goldman (1987, p. 129) has said that we want to engage in practices that, in addition to having power, speed, fecundity, and reliability, are *efficient*. In other words, we prefer practices that produce many epistemically valuable beliefs at a low cost. This is certainly true. However, contra Goldman, cost-effectiveness itself is not an epistemic objective. Financial considerations (like ethical considerations) are non-epistemic constraints on our epistemic objectives.

Even so, in order to achieve our epistemic objectives, it is critically important to pay attention to the cost-effectiveness of the various options. Given that we usually have a limited budget with which to work, we will typically want to do a comparative, rather than an absolute, evaluation of the practices available (cf. Goldman, 1999, p. 92). That is, we do not need to know exactly how powerful/speedy/fecund/reliable any given practice is; rather, we need to know how the various practices that we can afford to engage in compare in terms of power/speed/fecundity/reliability.

Finally, it should be noted that Goldman characterizes these epistemic objectives (e.g., power, speed, fecundity, and reliability) in terms of *true* beliefs. For example, a practice is powerful if it leads to many true beliefs on topics that are of interest to us. However, we can characterize these epistemic objectives in terms of any kind of epistemically valuable beliefs. For example, we might say that a practice is speedy if it quickly produces *justified* beliefs on topics that interest us. Also, we might say that a practice is reliable if it tends not to produce beliefs that are not "collectively endorsed." Thus, even if we disagree with Goldman's account of knowledge as mere true belief, it is still fruitful to consider the power, speed, fecundity, and reliability of the practices in which we engage.

Difficulties with Multiple Epistemic Objectives

As we use social epistemology to study people's epistemic interests and values, it becomes clear that there are several possible epistemic objectives. First, there are many different kinds of epistemically valuable beliefs that people want to acquire (true beliefs, justified beliefs, systematically organized beliefs, etc.). Second, even with regard to just one kind of epistemically valuable belief, there are still many possible epistemic objectives (power, speed, fecundity, reliability, etc.). Unfortunately, the existence of several possible epistemic objectives creates additional difficulties for information services.

First, different users of the same information service may have different epistemic objectives. For example, a doctor treating a patient will typically require a greater degree of reliability than a student doing an assignment. As a result, it is unlikely that an information service will be able to aim for a degree of reliability that accords exactly with the epistemic objectives of each individual making use of it. Thus, an information service can aim only for that degree of reliability that best serves the epistemic objectives of its users in aggregate (cf. Atkinson, 1996, pp. 214–242).

Second, if the standards for justification differ in different contexts, we might worry that authors would sometimes adopt a different (possibly lower) standard than readers. As a result, it can sometimes be difficult for people to acquire knowledge (via recorded information) from what other people already know about the world. Smith (2002) has suggested that information services might deal with this problem by promoting a limited set of publicly recognized standards for justification.

Finally, and most importantly, if an individual or an organization has more than one objective, these objectives can sometimes come into conflict. In other words, increasing the achievement of one objective may decrease the achievement of another. Thus, if an information service has more than one epistemic objective, these epistemic objectives can sometimes come into conflict.[55] In order to make decisions with epistemic consequences, an information service will have to determine how to resolve conflicts between these epistemic objectives.

In order to illustrate this difficulty, I will give an example from collection management involving conflicting epistemic objectives. For the sake of simplicity, I will assume that the epistemically valuable beliefs with which we are concerned in this case are simply true beliefs.

The Case of the Old Encyclopedia

Suppose that a librarian has to decide whether to remove (i.e., weed) an old edition of an encyclopedia from a collection. Because patrons may be led into error by out-of-date information, there are potential epistemic benefits to removing this edition.[56] However, retention of the item may also have potential epistemic benefits. For example, it is possible that this edition contains valuable information not found in more recent editions of the encyclopedia.[57]

Interestingly enough, precisely this sort of scenario was depicted in an episode of the television show *Picket Fences*. One of the kids, Zachary, does a report for school and everyone is horrified by the racist content (viz., that blacks are less intelligent than whites). It turns out that Zachary was innocently reporting "facts" that he had found in an old edition of the *Encyclopædia Britannica* in the school library. Of course, it is not only elementary school students who can acquire false beliefs from information resources. For example, Pendergrast (1988, p. 84), a librarian at a college library, wrote that "our students may unfortunately be naïve enough and ill-informed enough so that if they find a book in the library, they might automatically assume that the views expressed in it are accurate."[58]

In deciding whether to retain or remove the old edition of the encyclopedia (containing information not found in the new edition) from the collection, two epistemic objectives come into conflict.[59] First, the librarian wants patrons to acquire true beliefs on topics that are of interest to them. In other words, she is concerned with power. Retaining the old edition seems to be the best way to achieve this objective because it gives patrons access to more information. Second, the librarian does not want patrons to acquire false beliefs. In other words, she is also concerned with reliability. Removing the old edition seems to be the best way to achieve this objective.

The mere fact that an information service seeks to facilitate the acquisition of knowledge does not dictate a particular course of action in this case. Epistemologists agree that, all other things being equal, having more true beliefs is epistemically better and that, all other things being equal, having fewer false beliefs is epistemically better. However, it is not clear what the librarian should do when these two objectives come into conflict.[60]

In order to make a decision that involves conflicting objectives, we have to determine the relative importance of each of our objectives (Kirkwood, 1997, p. 53). In other words, because we cannot accomplish everything we want to, we have to determine how much of one objective we are willing to trade off for more of the other. The appropriate decision in the old encyclopedia example depends on the relative importance to the library of power and reliability.[61] For example, if it is much more important that patrons avoid error (e.g., if they are engaged in safety-critical activities such as medicine), then removing the old edition will probably be the appropriate decision. However, if it is much more important that patrons acquire more true beliefs on a broad range of topics, then retaining the old edition will probably be the appropriate decision. Furthermore, the appropriate tradeoff between power and reliability might also be different for different parts of the collection. For example, we might strive for a higher degree of reliability in the reference collection.

Finally, it should be noted that the problem of conflicting objectives is not unique to organizations that have epistemic objectives. In fact, such conflicts are common in nonprofit organizations (cf. Kirkwood, 1997, p. 19). Indeed, even if we were to take increasing exposure to recorded information to be the main objective of information services, we would still face the problem of conflicting objectives. Because there are many different types of exposure, there is potential for conflict. For example, are "browsing exposures to ten different documents" better or worse than "in-depth exposure to one document" (Hamburg et al., 1972, p. 113)?

Other Cases of Conflicting Epistemic Objectives

In addition to conflicts between power and reliability, other conflicts between epistemic objectives can complicate matters for information services. For example, outreach activities can certainly increase the fecundity of an information service. However, given limited resources, a library may have to decide between increasing the number of patrons it serves and improving its service to its current patrons. In other words, there is a potential conflict here between fecundity and power.

The epistemic objective of speed often comes into conflict with other epistemic objectives. For example, it typically takes time for the reference librarian (or the patron herself) to check whether an information resource is reliable. Thus, some speed may be sacrificed in order to achieve greater reliability, and vice versa.

In many cases an information service has to determine the relative importance of speed and power. For example, a library can increase its power by spending more money on ILL requests, which provide patrons with access to an extremely wide range of information resources. However, a library can increase its speed by spending more money on materials for its permanent collection, which patrons can access much faster.[62] Similarly, because compact shelving allows a library to hold more materials, it will increase the amount of knowledge that patrons can acquire on topics of interest to them. However, because compact shelving tends to make it more difficult to access these materials, it will take patrons more time to acquire this knowledge.

Predicting Epistemic Consequences

So far, I have focused on the importance of clarifying our epistemic objectives (and how to resolve conflicts among them). However, this is certainly not all that information services need to do. In addition to knowing what our ends are, we have to find means to those ends. As a result, information services have to be able to predict the epistemic consequences of the various strategies that they might adopt.

The effectiveness of a strategy depends to a large degree on the intellectual profile of the information service's users. As a result, in order to achieve its epistemic objectives, an information service needs to be sensitive to its clients' skills and abilities. For example, if its users do not have good critical thinking skills, then an information service probably needs to restrict itself to authoritative information resources in order to be reliable.[63] However, if its users have good critical thinking skills, then an information service can be both more powerful and reliable simply by providing access to diverse viewpoints (cf. Mill, 1859/1978, pp. 15–52).

As noted earlier, just being clear on what our epistemic objectives are can often suggest effective strategies. However, it may be necessary to gather empirical data about the information-seeking behavior of the users of information services. For example, empirical studies have found that people (including professional researchers) tend to stop searching once they find an information resource that addresses their questions, even if much better information resources are available (e.g., Mann, 1993, pp. 91–101). Empirical studies have also looked at what sorts of information resources people consider to be authoritative (e.g., Rieh, 2002; Van House, 2004, p. 41). Knowing about the principle of least effort and about people's credibility judgments will certainly have an impact on how information services attempt to facilitate knowledge acquisition. In fact, such empirical data may come from a variety of social scientific disciplines (e.g., psychology, sociology, economics, cognitive science).

Information Ethics and Social Epistemology

In this chapter, I have tried to establish that social epistemology is the area of philosophy that is most central to information science. This is not to say, however, that other areas of philosophy are unimportant. Ethics, in particular, is also critical to information science. It is important that information services and information professionals behave in an ethical manner (cf. American Library Association, 1995). However, it should be emphasized that, in the context of information science, ethics and epistemology are not completely separate issues (cf. Fallis, 2004a, 2004b, pp. 481–482, 2004c; McDowell, 2002).

Ethical considerations typically act as a constraint on behavior. For example, an athlete has a moral obligation not to take performance-enhancing drugs even though doing so would increase his chances of winning a race. Similarly, ethical considerations act as a constraint on the ability of information services to provide access to information. For example, as noted earlier, protecting people's intellectual property rights requires restricting access to certain kinds of information resources. Protecting people's privacy rights also requires restricting access to personal information.[64] As a result ethical considerations can constrain our ability to pursue the main goal of information services.

Although ethical considerations sometimes interfere with knowledge acquisition, it is also important to note that they frequently facilitate knowledge acquisition. In other words, ethical behavior on the part of information services and information professionals is a social practice that often helps people to acquire knowledge.[65] This connection between ethics and epistemology suggests further applications of social epistemology to information science.

Intellectual Freedom and Censorship

One of the most important principles in information ethics is that information services should promote intellectual freedom and resist censorship (cf. American Library Association, 1995). Mill (1859/1978, pp. 15–52) gave one of the most famous arguments for these principles (cf. Doyle, 2001). And, interestingly, Mill argued for unrestricted access to information on epistemic grounds (i.e., on the grounds that it helps people to acquire knowledge).

Mill argued that, if we restrict access to information, we will sometimes end up restricting access to the truth (because we are fallible human beings). Thus, we will be interfering with people's ability to acquire true beliefs. Mill further maintained that, even if we were able to restrict access only to information that is false, it would still be epistemically better not to do so. His reasoning was that people cannot be justified in their beliefs unless they are exposed to all sides of an issue (and know how to respond to those who disagree with them). In fact, people may be able to find the truth in the first place only if they are exposed to all sides of an issue. Thus, people need to have access to a wide variety of viewpoints in order to acquire knowledge. As a result, Lee (1998, p. 24) has suggested that libraries should try "to include as many different versions of 'Truth' as possible, in the hopes that somewhere in the mass of material, some truth may be found by the discerning reader."

Goldman (1999), however, argued that unrestricted access to information has epistemic costs as well as epistemic benefits. He responded to Mill by noting that "if hearers are disposed to believe certain categories of statements, even when the evidence or authority behind them is dubious, it may sometimes be preferable not to disseminate those statements" (Goldman, 1999, p. 212).

The dissemination of certain pieces of information may lead to more false beliefs than true beliefs but Goldman's response to Mill ignores several important points. First, it assumes that the only way to ensure reliability is to restrict people's access to unreliable information sources. However, instead of restricting access to sources, we might work on improving critical thinking skills (e.g., by providing guidelines for evaluating information). Second, it assumes that reliability is our only epistemic objective. As has been noted, providing access only to authoritative sources (e.g., respected reference works) may make an information service very reliable but it will also make the service much less powerful. Third, it assumes that we are concerned only with the immediate epistemic consequences of our practices. But even if they are initially deceived by unreliable information sources, people are arguably more likely to acquire true beliefs about a topic in the long run if their access to information is not restricted (cf. Fallis, 2004a, p. 111).

Goldman (1999, p. 212) has also responded to Mill's position by noting that, even with unrestricted access to information, people will not necessarily be exposed to all sides of an issue. In fact, it is easy for many sides of an issue to simply disappear into the mass of information that is now available through the Internet and other information sources. As Lee (1998, p. 25) put it, "the mass of data has grown to overshadow truth, so that we often lose sight of what is important." However, restricting access to that which is bad is not the only way (and probably not the best way) to make it more likely that people will find that which is good. For example, as noted earlier, information sources that have different viewpoints on the same topic should be collocated.

Privacy

Another important principle in information ethics is that information services should protect people's privacy rights (cf. American Library Association, 1995). In particular, information services should maintain the confidentiality of records about their users. As noted earlier, protecting people's privacy rights does require restricting access to some information; thus, it can interfere with knowledge acquisition. Strickland, Baldwin, and Justsen (2005) discuss the shifting balance between domestic security and civil liberties, for example whether Federal Bureau of Investigation (FBI) agents should be able to acquire knowledge about the information sources consulted by suspected terrorists. However, protecting people's privacy rights also has important epistemic benefits.[66]

A failure to protect people's privacy rights can have a "chilling effect" on their use of an information service such as a library (cf. Garoogian, 1991, p. 229; Johnson, 2000, p. 512). If people think that their actions may be observed, they often do not act freely and may not seek information that they want or need. For example, a person might decide not to consult a particular information source if she is concerned that the information service providing this souce may report her activities to the FBI. In fact, she might decide not to visit the information service at all. But if she does not visit the information service or does not consult the information source, she cannot acquire knowledge from them. Thus, if information services are to achieve their main objective, it is important that they protect people's privacy rights.

It should be noted that keeping circulation records confidential is not the only way in which information services can protect privacy rights. For example, self-checkout machines and privacy screens on computers also help to keep a library patron's information-seeking behavior from being observed by other human beings. Thus, installing such mechanisms can also potentially increase the power of an information service. In addition, self-checkout machines can potentially increase the speed of the information service.

The "chilling effect" of privacy violations is an example of what McDowell (2002) calls *communication-worthiness* effects. That is, if an information service engages in unethical behavior, people may decide not to use it as a source of information. However, unethical behavior can also interfere with knowledge acquisition by means of what McDowell calls *inferential* effects. That is, if an information service engages in unethical behavior, people may decide that it is not a trustworthy source of information.[67] For example, a librarian who is perceived to be uncivil toward patrons may not be trusted to give accurate answers to reference questions. Shapin (1994) and Burbules (2001) also discuss such connections between morality and credibility.

Intellectual Property

Another important principle in information ethics is that information services should protect people's intellectual property rights (cf. American Library Association, 1995). A few authors (e.g., Fallis, 2004a, pp. 105–109; Goldman, 1999, pp. 181–182; Warner, 1993, p. 307) have discussed the connections between intellectual property and social epistemology. For example, because protecting people's intellectual property rights restricts access to information, it can interfere with knowledge acquisition. However, protecting people's intellectual property rights can also have epistemic benefits. In fact, intellectual property rights are usually justified on epistemic grounds. For example, according to the United States Constitution (Article 1, Section 8), creators are given intellectual property rights specifically in order to "promote the progress of science and useful arts."

A failure to protect intellectual property rights can reduce the motivation to produce new intellectual property. As a result, people will have access to less information. Thus, although the protection of people's intellectual property rights does require restricting access to information, it arguably brings about access to more information (and more knowledge) in the long run.

It should also be noted that protecting intellectual property rights does not simply mean respecting an author's *legal* rights (e.g., under copyright law). For example, researchers may be less likely to disseminate information (e.g., the locations of rare plants and animals) if they fear that it will be misused (cf. Van House, 2002, p. 105). Researchers may also be less motivated to publish their work if they are worried that it will be plagiarized (cf. Resnik, 1996, p. 579).

Of course, protecting intellectual property rights does not necessarily lead to good epistemic consequences. If intellectual property laws give people too much control over their intellectual property, bad epistemic consequences will result. For example, an extension of the length of copyright does not seem to significantly increase the motivation to create new intellectual property, but it does significantly restrict people's access to information (Langvardt & Langvardt, 2004, pp. 228–230).[68] Intellectual property laws should be strict enough to motivate people to create new intellectual property, but they should be no stricter. Otherwise, they will needlessly interfere with people's ability to access information and acquire knowledge.

Recently Vaidhyanathan (2002) has pointed out another way in which intellectual property laws can potentially interfere with people's ability to acquire knowledge. In order to criticize your viewpoint, I often have to repeat some of what you say. In other words, I have to use some of your intellectual property. As a result, intellectual property laws can potentially be used to keep me from criticizing your viewpoint. In fact, the Church of Scientology has effectively used the threat of litigation on these grounds to silence some of its critics (cf. Vaidhyanathan, 2002). Thus, intellectual property laws that are too strict can potentially decrease the number of viewpoints to which people have access.[69] And, as Mill has argued, such censorship can interfere with people's ability to acquire knowledge.

Other Ethical Considerations

There are also other ways in which information professionals can facilitate knowledge acquisition by promoting ethical policies and ethical behavior. First, in cases where existing laws interfere with the main objective of information services, information professionals can work to have those laws changed. For example, librarians were instrumental in having laws passed to protect the confidentiality of patron records (cf. Johnson, 2000). This is also what Vaidhyanathan recommends in the case of overly strict intellectual property laws. Second, information professionals can encourage the users of information services to engage in ethical behavior in their use of information resources (cf. Association of College and Research Libraries, 2000). After all, violations of intellectual property rights on the part of users as well as information professionals can reduce the motivation to produce new intellectual property.

Conclusion

For the reasons that Floridi (2002) gives, social epistemology may not provide a *complete* theoretical foundation for information science. Even so, helping people to acquire knowledge is the *main* objective of libraries and other information services. In other words, most activities that take place in information services are intended to facilitate knowledge acquisition. As a result, social epistemology does provide a unified theoretical framework that may allow us to improve the policies and practices of information services (cf. Shera, 1970, pp. 108–109).

Because epistemology is concerned with what knowledge is and how people acquire it, work in epistemology can help to clarify this central objective of information services. It can also suggest ways in which information services might achieve this objective. In other words, it can help information services to identify policies and practices that facilitate knowledge acquisition. Information scientists (and a few philosophers) have applied work in social epistemology to several issues in information science (e.g., collection management, organization of information, evaluation of information, reference work). However, much work remains to be done. For example, given the existence of several different possible epistemic objectives, information services need to clarify exactly what theirs are. Also, many other activities (and potential activities) in information services can be evaluated in light of these epistemic objectives.[70]

Acknowledgments

I would like to thank Alvin Goldman, Kay Mathiesen, and my students in IRLS 617 "Social Epistemology and Information Science" at the University of Arizona for many helpful discussions on this topic. I would also like to thank Blaise Cronin, Tony Doyle, Carrie Fang, Alan Mattlage, and three *ARIST* reviewers for helpful suggestions on this chapter.

Endnotes

1. Philosophy and information science are connected in other important ways, as well. For example, Wittgenstein's philosophy of language has been applied to information retrieval (see Blair, 2003) and Foucault's method of discourse analysis has been used to understand theoretical discourses in information science (see Frohmann, 1994).

2. A full defense of this thesis would require comparing these applications with all of the applications of other areas of philosophy to information science. Such an undertaking, however, is beyond the scope of this chapter.

3. Recorded information itself is sometimes referred to by information scientists as "knowledge." In this chapter, however, "knowledge" will refer exclusively to a mental state of human beings (cf. the section on "Belief Condition").

4. In attempting to contrast social epistemology with the sociology of knowledge, Shera (1970, pp. 107–108) made it sound as if social epistemology is not about the acquisition of knowledge, but only about its use. However, as Furner (2002, pp. 14–15) has pointed out, this is not correct.

5. Interestingly, philosophers have paid little attention to the work of Egan and Shera. Fuller (1996) is one of the few who even cites their work.

6. Furner (2004) makes a very good case that Egan deserves much more credit for the development of this idea than she has received so far.

7. Shera (1970, p. 88) famously suggested that library and information science needs a theoretical foundation, writing that "librarianship must be more than a bundle of tricks, taught in a trade school, for finding a particular book, on a particular shelf, for a particular patron, with a particular need." Interestingly, Zwadlo (1997) has recently argued the contrary position—that information science would be better off without any theoretical foundation at all.

8. As Floridi (2004) has pointed out, foundational work does not always have practical implications. In many disciplines, foundational work is often pursued simply to gain a better understanding of the concepts involved (e.g., mathematics).

9. The effectiveness of information systems may vary across cultures, times, and individuals (cf. Shera, 1970, pp. 89–92).

10. "Epistemic" means simply that something has to do with knowledge acquisition. For example, we have an *epistemic objective* if we want a person (or a group of people) to acquire knowledge.

11. Note that some philosophers (e.g., Williamson, 2002) do not think it is possible to give necessary and sufficient conditions for something to count as knowledge.

12. Beliefs based on empirical evidence are certainly open to doubt and thus do not count as knowledge for Descartes. This is why, *pace* Budd (2001, p. 33), Descartes is usually considered to be a rationalist rather than an empiricist.

13. Schools and libraries are social institutions that have the (more or less) explicit objective of facilitating knowledge acquisition. However, *epidemiological* and *contagion* models of belief transmission (also known as *memetics*) have also been used to study the unintentional diffusion of beliefs and knowledge throughout a society (cf. Dawkins, 1976, pp. 189–201; Goffman, 1966; Goldman, 2002, pp. 164–181).

14. Feminist epistemologists (e.g., Harding, 1992; Longino, 1990) focus specifically on the influence that gender has on knowledge acquisition (cf. Van House, 2004, pp. 30–33).

15. Of course, the other senses of "social" (e.g., social and cultural factors) can certainly play a role in determining which social practices and institutions are going to be effective.

16. These social processes include the communication and information technology itself. Note that these processes may often be automated (as is the case of information retrieval systems).

17. An answer to this question would help information professionals teach their clients how to evaluate information sources more effectively (cf. Fallis & Frické, 2002).

18. Van House (2004) essentially provides a survey of how revolutionary social epistemology can be applied to information science. As a result, her *ARIST* chapter provides a useful complement to this one. Much of the research she cites focuses on *scientific*

knowledge, but as Van House (2004, p. 4) notes, this research is potentially applicable to knowledge in general.

19. In addition to such proposals, it has also been suggested that information services should actually promote people's *ability* to acquire knowledge rather than promoting knowledge acquisition itself (e.g., Garnham, 1999). But this project still requires an understanding of what knowledge is and how it is acquired and used. Also, the capability of acquiring knowledge is valuable only because knowledge itself is valuable (cf. Goldman, 1999, p. 351).

20. As Kvanvig (1998, p. 444) has observed, "once one notices that eliminating the fortuity involved in failing to know what one justifiedly and truly believes makes no difference at all for one's present or future, it is hard to see how eliminating it could serve any important human need, interest, or purpose."

21. Admittedly, people may not explicitly ask for knowledge (much less justified true belief) when they use an information service. But the fact that they do not explicitly ask for it does not constitute proof that it is not what they are really after. In fact, they do not ask for knowledge because it goes without saying that it is what they are after. Of course, as I shall argue in the section on "Varieties of Knowledge," the justified true belief account of knowledge probably does not capture *all* of what users of information services are after.

22. It would require considerable empirical data to establish conclusively this claim about information services. I will simply try to eliminate some important counter-arguments to it.

23. Shera (1970, p. 83) tended to focus on the fact that "the individual mind deteriorates when it is deprived of knowledge or information." However, there are many other ways in which knowledge acquisition promotes social utility (cf. Shera, 1970, pp. 102–103).

24. In fact, provision of entertainment can also facilitate knowledge acquisition indirectly by getting people to use the library in the first place.

25. Social utility is arguably increased merely because the patron knows how to build a house. However, Shera presumably had more in mind by "social utility" than just the intrinsic value of knowledge.

26. Although ethical considerations sometimes interfere with knowledge acquisition, they often promote it (see the section on "Information Ethics and Social Epistemology").

27. Goldman (1999, p. 260) makes a similar point with respect to the motivations of scientists. He notes that a "profit motive ... need not conflict with a veritistic motivation; in fact, discovering truths may be an essential means to profitable technological application."

28. The objective of knowledge acquisition does show up explicitly in the mission statements of some information services (e.g., Holmes Community College, 2004; Rodman Public Library, 2003).

29. In addition, these properties are a means to the ultimate goals of information services because they are a means to knowledge acquisition.

30. If people did not come back to information services that provided them with false or misleading information, information services would have to provide information that met certain quality standards in order to succeed in increasing exposure in the long run. Unfortunately, empirical studies indicate that "public library patrons do not differentiate better books from the poorer books" (Baker & Lancaster, 1991, p. 103).

31. Of course, because it is easier to measure, the objective of exposure to recorded information might often serve as a useful proxy for the objective of knowledge acquisition.

32. The story of the lost keys apparently derives from a story about the Sufi master, Mulla Nasrudin (Shah, 1966, p. 24).

33. Goldman (1999, pp. 9–40) discusses a number of different objections to the truth condition.

34. There is a mundane sense in which this is indisputable. For example, it is polite to eat with one's fingers in Morocco, but not in England. But Shera clearly has something more radical in mind.

35. Saying that there is objective truth does not necessarily mean that there is an objective truth of the matter about *everything*. For example, the truth of propositions in certain areas, such as aesthetics, may be completely determined by social and cultural factors (cf. Goldman, 1999, p. 26).

36. Harding (1992, pp. 585–587), for example, argues that there is no such thing as objective truth on the grounds that this concept plays no role in science. She claims that the truth plays no role because scientific theories cannot be definitively confirmed or falsified.

37. For further discussion of Mill's arguments, see the section on "Intellectual Freedom and Censorship" and Doyle (2001).

38. Admittedly, some students may just want an answer from a source that they can cite, but presumably this is not what the teacher had in mind.

39. True belief is not necessarily the *main* epistemic goal in all cultures (cf. Hannabuss, 2001; Hongladarom, 2002; Maffie, 2000, pp. 248–249).

40. In fact, as Goldman (1999, p. 71) has noted, collective endorsement can often be "diagnostic of truth."

41. This realm of objective knowledge is what Popper refers to as World 3. By contrast, the world of physical objects is World 1 and the world of mental states (e.g., in the minds of human beings) is World 2 (cf. Neill, 1982a, pp. 33–35).

42. Knowledge does not have to be completely in the head. As Clark (2003) has pointed out, mental processes often involve objects in the world (e.g., paper and pencil) other than our bodies. Thus, what matters is whether the knowledge is easily retrievable (e.g., from our notes, from our computers, or from the library). After all, we have to retrieve knowledge from memory.

43. Thus, information services that assist mathematicians in carrying out their research may even want to facilitate the acquisition of absolutely certain knowledge.

44. Contextualism originally developed as a way of replying to the skeptics (i.e., as a way of showing that knowledge is possible). The contextualist concedes to the skeptics that there are contexts in which the standards for justification are very high (e.g., when we are discussing the possible existence of malicious demons). However, the contextualist also claims that there are also contexts where the standards for justification are not nearly so high (e.g., when we are discussing the height of the Eiffel Tower). In other words, there are contexts where we can meet the standards and so have knowledge.

45. In order to ensure that its users end up with good evidence, an information service should have information resources that give the evidence for the claims that they make (so that users will have good reasons for believing these claims). In addition, the information service can make sure that its users know why reliable sources are reliable.

46. Goldman (2002, pp. 51–70) thinks that truth is the only thing that is epistemically valuable. Justification is still valuable, but it is only valuable as a means to acquiring true beliefs.

47. Even so, for the sake of simplicity, I will often continue to talk about the acquisition of *knowledge* rather than the acquisition of *epistemically valuable beliefs* in this chapter.

48. The motivation for information scientists to object to the truth condition may be reduced once it is pointed out that there are other kinds of epistemically valuable beliefs.

49. Because the economics of publishing (especially, scholarly publishing) have made it expensive for information services to provide people with access to information, several authors (e.g., Atkinson, 1996, p. 253; Ortega y Gasset, 1961, p. 153) have suggested that information services move into the publishing business.

50. The Web is not the only part of the Internet that facilitates knowledge acquisition. For example, as Goldman points out, e-mail provides a convenient means of directly contacting experts on a particular topic.

51. Power (i.e., many epistemically valuable beliefs acquired) is somewhat analogous to recall (i.e., many relevant documents retrieved).

52. As noted earlier, in addition to being concerned that many people acquire knowledge, we also want to foster equitable distribution of knowledge (cf. Fallis, 2004c).

53. Reliability (i.e., most beliefs acquired are epistemically valuable) is somewhat analogous to precision (i.e., most retrieved documents are relevant).

54. There is now empirical evidence that this is indeed the case (cf. Frické & Fallis, 2004; Lee, Schotland, Bacchetti, & Bero, 2002). Thus, information services need to provide their users with easy access to such metadata.

55. As has been noted, the *non-epistemic* objectives of an information service can also conflict with its epistemic objectives.

56. Another potential epistemic benefit of removing the old edition is that it would free up shelf space that could be used for more up-to-date and reliable materials.

57. As Mill (1859/1978, pp. 44–46) has observed, even predominantly false doctrines may contain a grain of truth not available elsewhere. In addition, even if the old edition of the encyclopedia contained only inaccurate information about the world, it might be a useful resource for historians studying what people believed about the world at the time it was published.

58. These days, their credulity also extends to Web pages found on the Internet.

59. For the sake of simplicity, I am assuming that retaining the old edition of the encyclopedia and removing it are the only two courses of action open to the librarian. However, he or she could also teach patrons to be better evaluators of information (cf. Fallis, 2004b). In that case, the old edition could be retained without so great a risk of patrons being misled by out-of-date information.

60. Epistemology does not seem to dictate any particular tradeoff between power and reliability (cf. Fallis, 2004a, pp. 110–111; Levi, 1962, p. 57). As a result, an information service has to decide for itself which relationship between power and reliability is in the best interests of its users.

61. The appropriate decision also depends on how many true beliefs and how many false beliefs are likely to result from retaining the old edition.

62. ILL might also allow a library to provide access to materials that are of interest to a greater number of patrons, thus generating a conflict between speed and fecundity.

63. As has been noted, the information service can also try to improve its users' skills and abilities (e.g., by teaching them better critical thinking skills).

64. As Goldman (1999, p. 173) has put it, "epistemology focuses on the means to knowledge *enhancement*, whereas privacy studies focus on the means to knowledge *curtailment*."

65. In a similar vein, Resnik (1996) argues that ethical behavior on the part of scientists can further their scientific objectives.

66. Authors who discuss the value of privacy rarely mention its epistemic value (e.g., Moore, 2003).

67. People may not be justified in drawing this inference, but as long as they do, such unethical behavior can interfere with knowledge acquisition.

68. Similarly, by making it illegal to circumvent copy protection measures, the Digital Millennium Copyright Act has the effect of unduly restricting the fair use of copyrighted material.

69. In fact, intellectual property laws were originally enacted in order for the government to "suppress anything that upset royal sensibilities or ran contrary to their interests" (Mann, 1998, p. 73).

70. This will certainly require more empirical data about the information-seeking behavior of the people who use these services.

References

Alexander, J. E., & Tate, M. A. (1999). *Web wisdom*. Mahwah, NJ: Erlbaum.

American Library Association. (1995). Code of ethics. Retrieved April 12, 2004, from http://www.ala.org/ala/oif/statementspols/codeofethics/codeethics.htm

Andersen, J. (2002). The role of subject literature in scholarly communication: An interpretation based on social epistemology. *Journal of Documentation, 58*, 463–481.

Association of College and Research Libraries. (2000). Information literacy competency standards for higher education. Retrieved July 4, 2004, from http://www.ala.org/ala/acrl/acrlstandards/informationliteracycompetency.htm

Atkinson, R. (1996). Library functions, scholarly communication, and the foundation of the digital library: Laying claim to the control zone. *Library Quarterly, 66*, 239–265.

Baker, S. L., & Lancaster, F. W. (1991). *The measurement and evaluation of library services*. Arlington, VA: Information Resources Press.

Blair, D. C. (2003). Information retrieval and the philosophy of language. *Annual Review of Information Science and Technology, 37*, 3–50.

Bloor, D. (1976). *Knowledge and social imagery*. London: Routledge and Kegan Paul.

Bruce, B. (2000). Credibility of the Web: Why we need dialectical reading. *Journal of Philosophy of Education, 34*, 97–109.

Budd, J. M. (1995). An epistemological foundation for library and information science. *Library Quarterly, 65*, 295–318.

Budd, J. M. (2001). *Knowledge and knowing in library and information science: A philosophical framework*. Lanham, MD: Scarecrow.

Budd, J. M. (2002a). Jesse Shera, sociologist of knowledge? *Library Quarterly, 72*, 423–440.

Budd, J. M. (2002b). Jesse Shera, social epistemology and praxis. *Social Epistemology, 16*, 93–98.

Burbules, N. C. (2001). Paradoxes of the Web: The ethical dimensions of credibility. *Library Trends, 49*, 441–453.

Burbules, N. C., & Callister, T. A. (1997). Who lives here? Access to and credibility within cyberspace. In C. Bigum & C. Lankshear (Eds.), *Digital rhetorics, 3* (pp. 95–108). Canberra City: Commonwealth of Australia.

Capurro, R., & Hjørland, B. (2003). The concept of information. *Annual Review of Information Science and Technology, 37,* 343–411.

Clark, A. (2003). *Natural-born cyborgs: Minds, technologies, and the future of human intelligence.* New York: Oxford University Press.

Coady, C. A. J. (1992). *Testimony.* New York: Oxford University Press.

Connell, T. H., & Tipple, J. E. (1999). Testing the accuracy of information on the World Wide Web using the AltaVista search engine. *Reference & User Services Quarterly, 38,* 360–368.

Davenport, E., & Cronin, B. (2000). The citation network as a prototype for representing trust in virtual environments. In B. Cronin & H. B. Atkins (Eds.), *The web of knowledge: A Festschrift in honor of Eugene Garfield* (pp. 513–530). Medford, NJ: Information Today.

Dawkins, R. (1976). *The selfish gene.* Oxford, UK: Oxford University Press.

DeRose, K. (1992). Contextualism and knowledge attribution. *Philosophy and Phenomenological Research, 52,* 913–929.

Descartes, R. (1641/1996). *Meditations on first philosophy.* Cambridge, UK: Cambridge University Press.

Dick, A. L. (1999). Epistemological positions and library and information science. *Library Quarterly, 69,* 305–323.

Dick, A. L. (2002a). *Philosophy, politics and economics of information.* Pretoria, South Africa: Unisa.

Dick, A. L. (2002b). Social epistemology, information science and ideology. *Social Epistemology, 16,* 23–35.

Doctor, R. D. (1992). Social equity and information technologies: Moving toward information democracy. *Annual Review of Information Science and Technology, 27,* 43–96.

Doyle, T. (2001). A utilitarian case for intellectual freedom in libraries. *Library Quarterly, 71,* 44–71.

Dretske, F. I. (1981). *Knowledge & the flow of information.* Cambridge, MA: MIT Press.

Earp, J. S. M. (1976). *I married Wyatt Earp.* Tucson: University of Arizona Press.

Egan, M., & Shera, J. (1952). Foundations of a theory of bibliography. *Library Quarterly, 44,* 125–137.

Evans, E., Borko, H., & Ferguson, P. (1972). Review of criteria used to measure library effectiveness. *Bulletin of the Medical Library Association, 60,* 102–110.

Fallis, D. (2000). Veritistic social epistemology and information science. *Social Epistemology, 14,* 305–316.

Fallis, D. (2001). Social epistemology and LIS: How to clarify our epistemic objectives. *Proceedings of the Twenty-Ninth Annual Conference of the Canadian Association for Information Science,* 175–183.

Fallis, D. (2002). What do mathematicians want? Probabilistic proofs and the epistemic goals of mathematicians. *Logique et Analyse, 45,* 373–388.

Fallis, D. (2004a). Epistemic value theory and information ethics. *Minds and Machines, 14,* 101–117.

Fallis, D. (2004b). On verifying the accuracy of information: Philosophical perspectives. *Library Trends, 52*, 463–487.

Fallis, D. (2004c). Social epistemology and the digital divide. In J. Weckert & Y. Al-Saggaf (Eds.), *Conferences in research and practice in information technology, 37* (pp. 79–84). Sydney: Australian Computer Society.

Fallis, D., & Frické, M. (2002). Indicators of accuracy of consumer health information on the Internet. *Journal of the American Medical Informatics Association, 9*, 73–79.

Feldman, R. (2003). *Epistemology*. Upper Saddle River, NJ: Prentice Hall.

Floridi, L. (2002). On defining library and information science as applied philosophy of information. *Social Epistemology, 16*, 37–49.

Floridi, L. (2004). LIS as applied philosophy of information: A reappraisal. *Library Trends, 52*, 658–665.

Foucault, M. (1979). *Discipline and punish*. New York: Vintage Books.

Frické, M. (1998). Jean Tague-Sutcliffe on measuring information. *Information Processing & Management, 34*, 385–394.

Frické, M., & Fallis, D. (2004). Indicators of accuracy for answers to ready-reference questions on the Internet. *Journal of the American Society for Information Science and Technology, 55*, 238–245.

Froehlich, T. J. (1989). The foundations of information science in social epistemology. *Proceedings of the Twenty-Second Annual Hawaii International Conference on System Sciences, 4*, 306–314.

Froehlich, T. J. (1992). Ethical considerations of information professionals. *Annual Review of Information Science and Technology, 27*, 291–324.

Frohmann, B. (1994). Discourse analysis as a research method in library and information science. *Library & Information Science Research, 16*, 119–138.

Fuller, S. (1988). *Social epistemology*. Bloomington: Indiana University Press.

Fuller, S. (1996). Recent work in social epistemology. *American Philosophical Quarterly, 33*, 149–166.

Fuller, S. (2002). *Knowledge management foundations*. Boston: Butterworth-Heinemann.

Furner, J. (2002). Shera's social epistemology recast as psychological bibliology. *Social Epistemology, 16*, 5–22.

Furner, J. (2004). "A brilliant mind": Margaret Egan and social epistemology. *Library Trends, 52*, 792–809.

Garfield, E. (1972). Citation analysis as a tool in journal evaluation. *Science, 178*, 471–479.

Garnham, N. (1999). Amartya Sen's "capabilities" approach to the evaluation of welfare: Its application to communications. In A. Calabrese & J. C. Burgelman (Eds.), *Communication, citizenship and social policy* (pp. 113–124). Lanham, MD: Rowman & Littlefield.

Garoogian, R. (1991). Librarian/patron confidentiality: An ethical challenge. *Library Trends, 40*, 216–233.

Gettier, E. (1963). Is justified true belief knowledge? *Analysis, 23*, 121–123.

Goffman, W. (1996, October). Mathematical approach to the spread of scientific ideas. *Nature, 212*, 449–452.

Goldman, A. I. (1987). Foundations of social epistemics. *Synthese, 73*, 109–144.

Goldman, A. I. (1999). *Knowledge in a social world*. New York: Oxford University Press.

Goldman, A. I. (2001). Social epistemology. *Stanford encyclopedia of philosophy*. Retrieved December 1, 2003, from http://plato.stanford.edu/entries/epistemology-social

Goldman, A. I. (2002). *Pathways to knowledge*. New York: Oxford University Press.

Hamburg, M., Ramist, L. E., & Bommer, M. R. W. (1972). Library objectives and performance measures and their use in decision making. *Library Quarterly, 42*, 107–128.

Hannabuss, S. (2001). A wider view of knowledge. *Library Management, 22*, 357–363.

Harding, S. (1992). After the neutrality ideal: Science, politics, and "strong objectivity." *Social Research, 59*, 567–587.

Harter, S. P., & Hert, C. A. (1997). Evaluation of information retrieval systems: Approaches, issues and methods. *Annual Review of Information Science and Technology, 32*, 3–94.

Heilprin, L. B. (1968). Response. In E. B. Montgomery (Ed.), *The foundations of access to knowledge* (pp. 26–35). Syracuse, NY: Syracuse University Press.

Henige, D. (1987). Epistemological dead end and ergonomic disaster? The North American Collections Inventory Project. *Journal of Academic Librarianship, 13*, 209–213.

Hernon, P., & McClure, C. R. (1986, April 15). Unobtrusive reference testing: The 55 percent rule. *Library Journal*, 37–41.

Herold, K. (Ed.). (2004). Philosophy of information [Special issue]. *Library Trends, 52*(3).

Hjørland, B. (2002). Epistemology and the socio-cognitive perspective in information science. *Journal of the American Society for Information Science and Technology, 53*, 257–270.

Holmes Community College. (2004). *Library system mission statement*. Retrieved November 10, 2004, from http://www.holmes.cc.ms.us/library/abouthcclibraries.html

Hongladarom, S. (2002). Cross-cultural epistemic practices. *Social Epistemology, 16*, 83–92.

Hume, D. (1748/1977). *An enquiry concerning human understanding*. Indianapolis, IN: Hackett.

Impicciatore, P., Pandolfini, C., Casella, N., & Bonati, M. (1997). Reliability of health information for the public on the World Wide Web: Systematic survey of advice on managing fever in children at home. *British Medical Journal, 314*, 1875–1879.

Johnson, S. D. (2000). Rethinking privacy in the public library. *International Information and Library Review, 32*, 509–517.

Kirkwood, C. W. (1997). *Strategic decision making*. Belmont, California: Duxbury.

Kitcher, P. (1993). *The advancement of science: Science without legend, objectivity without illusions*. New York: Oxford University Press.

Kitcher, P. (1994). Contrasting conceptions of social epistemology. In F. Schmitt (Ed.), *Socializing epistemology* (pp. 111–134). Lanham, MD: Rowman & Littlefield.

Kvanvig, J. L. (1998). Why should inquiring minds want to know? *Meno* problems and epistemological axiology. *Monist, 81*, 426–451.

Langvardt, A. W., & Langvardt, K. T. (2004). Unwise or unconstitutional? The Copyright Term Extension Act, the Eldred decision, and the freezing of the public domain for private benefit. *Minnesota Intellectual Property Review, 5*, 193–292.

Lee, E. (1998). *Libraries in the age of mediocrity*. Jefferson, NC: McFarland.

Lee, K. P., Schotland, M., Bacchetti, P., & Bero, L. A. (2002). Association of journal quality indicators with methodological quality of clinical research articles. *Journal of the American Medical Association, 287*, 2805–2808.

Levi, I. (1962). On the seriousness of mistakes. *Philosophy of Science, 29*, 47–65.

Lievrouw, L. A., & Farb, S. E. (2003). Information and equity. *Annual Review of Information Science and Technology, 37*, 499–540.

Lipton, P. (1998). The epistemology of testimony. *Studies in the History and Philosophy of Science, 29*, 1–31.

Longino, H. E. (1990). *Science as social knowledge*. Princeton, NJ: Princeton University Press.

Maffie, J. (2000). Alternative epistemologies and the value of truth. *Social Epistemology, 14*, 247–257.

Maher, P. (1993). *Betting on theories*. New York: Cambridge University Press.

Mann, C. C. (1998, September). Who will own your next good idea? *Atlantic Monthly, 282*, 57–82.

Mann, T. (1993). *Library research models: A guide to classification, cataloging, and computers*. New York: Oxford University Press.

Marco, F. J. G., & Navarro, M. A. E. (1993). On some contributions of the cognitive sciences and epistemology to a theory of classification. *Knowledge Organization, 20*, 126–132.

Marshakova-Shaikevich, I. V. (1993). Bibliometrics as a research technique in epistemology and philosophy of science. *International Forum on Information and Documentation, 18*, 3–9.

Mathiesen, K. (2003, March). What do *we* know? Collective knowledge and collective knowers. Paper presented at the 5th St. Louis Philosophy of Social Science Roundtable.

Mattlage, A. (1999). Networked scholarly publication. *Journal of Academic Librarianship, 25*, 313–321.

McDowell, A. (2002). Trust and information: The role of trust in the social epistemology of information science. *Social Epistemology, 16*, 51–63.

Meola, M. (2000). Review of *Knowledge in a social world* by Alvin I. Goldman. *College & Research Libraries, 61*, 173–174.

Mill, J. S. (1859/1978). *On liberty*. Indianapolis, IN: Hackett.

Moore, A. D. (2003). Privacy: Its meaning and value. *American Philosophical Quarterly, 40*, 215–227.

Neill, S. D. (1982a). Brookes, Popper, and objective knowledge. *Journal of Information Science, 4*, 33–39.

Neill, S. D. (1982b). Knowledge or information: A crisis of purpose in libraries. *Canadian Library Journal, 39*, 69–73.

Ortega y Gasset, J. (1961). The mission of the librarian. *Antioch Review, 2*, 133–154.

Paterson, R. W. K. (1979). Towards an axiology of knowledge. *Journal of Philosophy of Education, 13*, 91–100.

Pendergrast, M. (1988). In praise of labeling; or, when shalt thou break commandments? *Library Journal, 113*, 83–85.

Pfeifer, M. P., & Snodgrass, G. L. (1990). The continued use of retracted, invalid scientific literature. *Journal of the American Medical Association, 263*, 1420–1423.

Popper, K. (1972). *Objective knowledge*. Oxford, UK: Oxford University Press.

Radford, G. P. (1998). Flaubert, Foucault, and the Bibliothèque Fantastique: Toward a postmodern epistemology for library science. *Library Trends, 46*, 616–634.

Radford, G. P., & Budd, J. M. (1997). We do need a philosophy of library and information science—We're not confused enough: A response to Zwadlo. *Library Quarterly, 67,* 315–321.

Rawls, J. (1971). *A theory of justice.* Cambridge, MA: Harvard University Press.

Resnik, D. (1996). Social epistemology and the ethics of research. *Studies in History and Philosophy of Science, 27,* 565–586.

Rieh, S. Y. (2002). Judgment of information quality and cognitive authority in the Web. *Journal of the American Society for Information Science and Technology, 53,* 145–161.

Rodman Public Library. (2003). *Mission statement.* Retrieved November 10, 2004, from http://www.rodmanlibrary.com/rpl/mission.htm

Sannwald, W. (1998). Espresso and ambience: What public libraries can learn from book-stores. *Library Administration and Management, 12,* 200–211.

Schmitt, F. (1994a). The justification of group beliefs. In F. Schmitt (Ed.), *Socializing epistemology* (pp. 257–287). Lanham, MD: Rowman & Littlefield.

Schmitt, F. (Ed.). (1994b). *Socializing epistemology.* Lanham, MD: Rowman & Littlefield.

Shah, I. (1966). *The exploits of the incomparable Mulla Nasrudin.* New York: Simon and Schuster.

Shapin, S. (1994). *A social history of truth: Civility and science in seventeenth-century England.* Chicago: University of Chicago Press.

Sharlet, J. (1999, June 11). Author's methods lead to showdown over much-admired book on old west. *Chronicle of Higher Education, 45,* A19.

Shera, J. (1961). Social epistemology, general semantics and librarianship. *Wilson Library Bulletin, 35,* 767–770.

Shera, J. (1968). An epistemological foundation for library science. In E. B. Montgomery (Ed.), *The foundations of access to knowledge* (pp. 7–25). Syracuse, NY: Syracuse University Press.

Shera, J. (1970). *Sociological foundations of librarianship.* New York: Asia Publishing House.

Smiraglia, R. P. (2001). Works as signs, symbols, and canons: The epistemology of the work. *Knowledge Organization, 28,* 192–202.

Smith, C. (2002). Social epistemology, contextualism and the division of labor. *Social Epistemology, 16,* 65–81.

Smith, M. M. (1997). Information ethics. *Annual Review of Information Science and Technology, 32,* 339–366.

Steup, M. (2001). The analysis of knowledge. *Stanford encyclopedia of philosophy.* Retrieved December 1, 2003, from http://plato.stanford.edu/entries/knowledge-analysis

Strickland, L. S., Baldwin, D. A., & Justsen, M. (2005). Domestic security surveillance and civil liberties. *Annual Review of Information Science and Technology, 39,* 433–513.

Thagard, P. (2001). Internet epistemology: Contributions of new information technologies to scientific research. In K. Crowley, C. D. Schunn, & T. Okada (Eds.), *Designing for science* (pp. 465–485). Mawah, NJ: Erlbaum.

Trosow, S. E. (2001). Standpoint epistemology as an alternative methodology for library and information science. *Library Quarterly, 71,* 360–382.

Vaidhyanathan, S. (2002, August 2). Copyright as cudgel. *The Chronicle of Higher Education, 48,* B7.

Van House, N. A. (2002). Digital libraries and practices of trust: Networked biodiversity information. *Social Epistemology, 16*, 99–114.

Van House, N. A. (2004). Science and technology studies and information science. *Annual Review of Information Science and Technology, 38*, 3–86.

Warner, J. (1993). Writing and literary work in copyright: A binational and historical analysis. *Journal of the American Society for Information Science, 44*, 307–321.

Wengert, R. G. (Ed.). (2001). Ethical issues of information technology [Special issue]. *Library Trends, 49*(3).

Williamson, T. (2002). *Knowledge and its limits*. Oxford, UK: Oxford University Press.

Wilson, F. (1999). Flogging a dead horse: The implications of epistemological relativism within information systems methodological practice. *European Journal of Information Systems, 8*, 161–169.

Wilson, P. (1983). *Second-hand knowledge: An inquiry into cognitive authority*. Westport, CT: Greenwood Press.

Wolkoff, K. N. (1996). The problem of holocaust denial literature in libraries. *Library Trends, 45*, 87–96.

Zandonade, T. (2004). Social epistemology from Jesse Shera to Steve Fuller. *Library Trends, 52*, 810–832.

Zwadlo, J. (1997). We don't need a philosophy of library and information science—We're confused enough already. *Library Quarterly, 67*, 103–121.

Zwadlo, J. (1998). Comment. *Library Quarterly, 68*, 114–117.

CHAPTER 13

Formal Concept Analysis in Information Science

Uta Priss
Napier University

Introduction

Formal Concept Analysis (FCA) is a method for data analysis, knowledge representation, and information management that is largely unknown among information scientists in the U.S. FCA was invented by Rudolf Wille in the early 1980s (Wille, 1982). For the first 10 years, FCA was developed mainly by a small group of researchers and Wille's students in Germany. Because of the mathematical nature of most of the early publications, knowledge of FCA remained restricted to a group of "insiders." Through funded research projects, FCA was implemented in several larger-scale applications, most notably a knowledge exploration system for civil engineering in cooperation with the Ministry for Civil Engineering of North-Rhine Westfalia (cf. Eschenfelder, Kollewe, Skorsky, & Wille, 2000). But these applications were not publicized widely beyond Germany.

During the last 10 years, however, FCA has grown into an international research community with applications in many disciplines, such as linguistics, software engineering, psychology, artificial intelligence (AI), and information retrieval. This development is due to a variety of factors (cf. Stumme, 2002). A few influential papers stirred interest in FCA in several fields. For example, Freeman and White's (1993) paper on social network analysis sparked interest in the use of FCA software among sociologists. In software engineering, several FCA papers (such as Fischer, 1998, and Eisenbarth, Koschke, & Simon, 2001) won Best Paper Awards at conferences (Snelting, in press) because FCA happened to facilitate a type of analysis that was previously not available in that field. As Stumme (2002) explains, FCA shifted emphasis to applications in computer science partly due to a merger with the Conceptual Graphs community (Sowa, 1984). An overview of the relationship between Conceptual Graphs and FCA is provided by Mineau, Stumme, and Wille (1999).

Some of the structures of FCA appear to be fundamental to information representation and were independently discovered by different researchers. For example, Godin, Gecsei, and Pichet's (1989) use of concept lattices (which they call "Galois lattices") in information retrieval is based on an independent discovery by Barbut and Monjardet (1970). Apart from creating a following in France and francophone Canada, Godin's work also influenced Carpineto and Romano's (1993) work in Italy. Over time, these groups have grown into a joint international FCA community, stretching as far as Peter Eklund's group in Australia. FCA groups can also be found in Germany, France, and Eastern Europe. Although FCA is probably not in the mainstream in the U.K., it has been used in a prestigious multimillion-pound, multiuniversity collaboration on Advanced Knowledge Technologies (Kalfoglou, Dasmahapatra, & Chen-Burger, 2004).

So far, however, FCA is relatively unknown to information scientists in the U.S. To our knowledge, only one paper on FCA has appeared in the *Journal of the American Society for Information Science* (Carpineto & Romano, 2000). Several FCA papers were presented at conferences of the American Society for Information Science and Technology (ASIST), but did not receive much attention. Several FCA papers have also been published by the Association for Computing Machinery (ACM), but more in the field of software engineering (for example, Godin & Mili, 1993; Mili, Ah-Ki, Godin, & Mcheick, 1997). There was some interest in the use of mathematical lattices for information retrieval by Soergel (1967) and others in the 1950s and 1960s (cf. Priss, 2000). But their use of lattices is different from FCA and did not result in implemented systems.

This *ARIST* chapter, therefore, attempts to fill a gap by both introducing information scientists to FCA and providing an overview of current research. We decided not to include many areas of FCA in this paper because they seemed less relevant for information science, such as Wolff's (2004) research on Temporal Concept Analysis; Ferre's research on Logical Concept Analysis (Ferre & King, 2004); work on performance of algorithms by Kuznetsov and Obiedkov (2002) and others; research in linguistics (cf. Priss, in press); research in software engineering (e.g., Snelting, in press); educational research (e.g., Hara, 2002); applications in machine learning (Kuznetsov, 2004), knowledge discovery, and data mining (Valtchev, Missaoui, & Godin, 2004); and older research on applications of FCA in psychology (Spangenberg & Wolff, 1993).

FCA Introductions, Bibliographies, and Software

With the exception of Wolff's (1994) and Wille's (1997c) introductions to FCA, introductory material in English has not been readily available. The mathematical foundations of FCA are described by Ganter and Wille (1999b), but only a very small portion of that book discusses topics of interest to information scientists. With respect to applications, the best overview will probably be the "state of the art" volume of the 2003

International Conference on Formal Concept Analysis (ICFCA), which is yet to be published but is available in parts on the Web. This volume will contain overviews of FCA applications in information retrieval (Carpineto & Romano, in press), software analysis (Snelting, in press), linguistics (Priss, in press), and many other domains. Carpineto and Romano's (2004a) recent book provides an overview of FCA applications in information retrieval.

An online bibliography of FCA can be found at www.fcahome.org.uk. This bibliography does not attempt to list papers, which would be an impossible task, but instead contains links to bibliographies maintained by FCA research groups and to conference proceedings. Many FCA publications of recent years appear in the proceedings of two series of international conferences: the International Conference on Conceptual Structures (ICCS), which has included FCA papers since 1995, and the International Conference on FCA, which was started in 2003. Many of the older publications, however, are more difficult to obtain. For example, the conference proceedings of conferences held in Germany before 2000 are mostly out of print.

An overview of FCA software is provided by Tilley (2004). Until recently, there was no open-source FCA software available. Free software could only be obtained if one knew someone who was developing such software. Furthermore, the documentation and interfaces were usually written in German or were comprehensible only to mathematicians. The availability of several Java-based open-source tools since 2003, such as ConExp (sourceforge.net/projects/conexp) and ToscanaJ (Becker & Hereth Correia, in press), is probably another factor contributing to the recent growth of interest in FCA. These tools are cross-platform compatible, easy to install, and fairly easy to use.

Current FCA software is still far from realizing the full potential of applications as detailed in theoretical research papers. Eklund, Groh, Stumme, and Wille (2000) describe a possible implementation of some of the more advanced FCA aspects, but to our knowledge none has been implemented. Because of the complexity of the underlying lattice data structures and the visualizations, FCA software usually cannot be developed in the kinds of short-term projects that are normally funded by national research agencies. With a few exceptions, such as the open-source software mentioned earlier and Carpineto and Romano's (2004b) Credo engine (accessible at credo.fub.it), FCA software often exists virtually rather than actually.

There are at least two companies that apply and develop FCA software. The Germany-based company Navicon (www.navicon.de) was founded 10 years ago by some of Wille's former students whose goal was to employ FCA software for information management tasks. The company is doing well but currently is more focused on database technologies than on FCA (F. Vogt, personal communication, February, 2003). A second commercial venture is an Australian company that is marketing an e-mail analysis tool based on FCA (www.mail-sleuth.com).

Basic Notions of Formal Concept Analysis

The Duality of Extension and Intension

The basic notions of FCA are described in this section but without their mathematical details, which can be found in Ganter and Wille (1999b). A central notion of FCA is a duality called a "Galois connection." This duality can often be observed between two types of items that relate to each other in an application, such as objects and attributes or documents and terms. A Galois connection implies that if one makes the sets of one type larger, they correspond to smaller sets of the other type and vice versa. Consider, for example, documents and terms in information retrieval: Enlarging a set of terms will reduce the set of documents containing all of these terms, whereas a smaller set of terms will correspond to a larger set of documents.

In FCA, the elements of one type are called "formal objects" and the elements of the other type are called "formal attributes." The adjective "formal" is used to emphasize that these are formal notions. "Formal objects" need not be "objects" in any kind of commonsense meaning of "object." But the terms "object" and "attribute" are helpful because, in many applications, it may be useful to select object-like items as formal objects and to select their features or characteristics as formal attributes. In an information retrieval application, documents could be considered object-like and terms considered attribute-like. Other examples of sets of formal objects and formal attributes are tokens and types, values and data types, data-driven facts and theories, words and meanings, and so on.

The sets of formal objects and formal attributes together with their relation to each other form a "formal context," which can be represented by a cross table (see Figure 13.1). The elements on the left side are formal objects; the elements at the top are formal attributes; and the relation between them is represented by the crosses. In this example, the formal objects are animals that are famous in certain parts of the world: the cartoon characters Garfield and Snoopy, Bill Clinton's cat Socks, Greyfriar's Bobby (a Scottish dog faithful to his master even after the latter's death), and Harriet the tortoise (claimed to be the planet's oldest animal, brought to England by Darwin and now living in the Australia Zoo). The attributes describe whether these animals are cartoon characters or "real" animals and whether they are dogs, cats, mammals, or tortoises. This is, of course, a small-scale example, but it is sufficient to explain the basic features of FCA.

A further interesting feature of Galois connections is that a certain "closure" of the relation is implied. Starting with any set of formal objects, one can identify all the formal attributes they have in common. Using the example from Figure 13.1, the only common attribute of Harriet and Bobby is that they are both real. But they are not the only real animals, because Socks is also real. If one starts with Harriet and Bobby, derives all their shared attributes (i.e., "real"), and then derives

	cartoon	real	tortoise	dog	cat	mammal
Garfield	X				X	X
Snoopy	X			X		X
Socks		X			X	X
Greyfriar's Bobby		X		X		X
Harriet		X	X			

Figure 13.1   A formal context of "famous animals."

all other animals that are also real, one obtains the object set "Harriet, Bobby, Socks" and the attribute set "real." At this point the relation is "closed" because one can neither enlarge the attribute nor the object set. If one starts with Bobby and Socks, one obtains "mammal" and "real" and then no further objects or attributes because no other formal objects have both these formal attributes. Pairing a set of formal objects with a set of formal attributes that is closed in this manner is called a "formal concept." Thus "Harriet, Bobby, Socks" and "real" form a formal concept, and "Bobby, Socks" and "mammal, real" form a different formal concept. The set of formal objects of a formal concept is called its "extension"; the set of formal attributes its "intension." For a given formal context, the formal concepts, their extensions and intensions are uniquely defined and fixed.

Concept Lattices

An important advantage of FCA is that the Galois connections and the sets of formal concepts can be visualized. Figure 13.2 shows a so-called line diagram of a concept lattice corresponding to the formal context in Figure 13.1. A concept lattice consists of the set of concepts of a formal context and the subconcept–superconcept relation between the concepts (cf. Ganter & Wille [1999b] for the mathematical details). The nodes in Figure 13.2 represent formal concepts. Formal objects are noted slightly below and formal attributes slightly above the nodes that they label. Continuing with the example from the previous section, the node on the right side, which is labeled with the formal attribute "real," shall be referred to as Concept A.

To retrieve the extension of a formal concept, one needs to trace all paths that lead down from the node to collect the formal objects. In this example, the formal objects of Concept A are Socks, Bobby, and Harriet. To retrieve the intension of a formal concept, one needs to trace all paths that lead up in order to collect all the formal attributes. In this example, there is a node above Concept A, but that node has no formal attributes attached. Thus Concept A represents the formal concept with the extension "Harriet, Bobby, Socks" and the intension as the single-element set

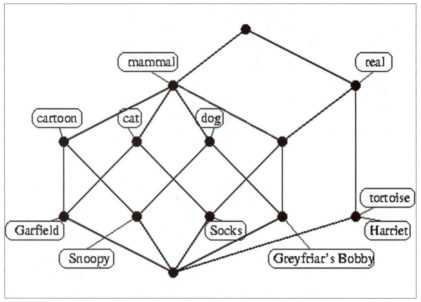

Figure 13.2 A concept lattice for the formal context in Figure 13.1.

"real." The other formal concept mentioned in the last section was the one with extension "Socks, Bobby" and intension "real, mammal." This concept is connected to Concept A by an edge (the line going down from Concept A to the left) and not labeled by any object or attribute in the line diagram in Figure 13.2. This concept shall be referred to as Concept B.

Figure 13.3 summarizes the relationship between Concept A and Concept B. Concept B is a subconcept of Concept A because the extension of Concept B is a subset of the extension of Concept A and the intension of Concept B is a superset of the intension of Concept A. All edges in the line diagram of a concept lattice represent this subconcept–superconcept relation.

The top and bottom concepts in a concept lattice are special. The top concept has all formal objects in its extension. Its intension is often empty, but does not need to be so. In the example in Figure 13.2, the top concept could have a formal attribute "animal." The bottom concept has all formal attributes in its intension. If any of the formal attributes mutually exclude each other (such as "dog" and "cat"), then the extension of the bottom concept must be empty (because no formal object can be a dog and cat at the same time). The top concept can be thought of as representing the "universal" concept and the bottom concept, the "null" or "contradictory" concept of a formal context.

The subconcept–superconcept relation is transitive, which means that a concept is a subconcept of any concept that can be reached by

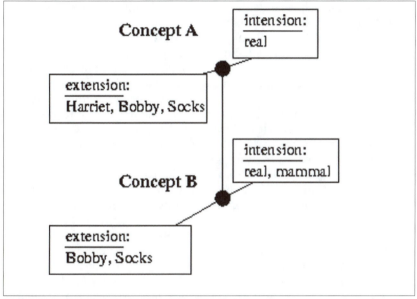

Figure 13.3 A subconcept-superconcept relation.

traveling upwards from it. If a formal concept has a formal attribute, then its attributes are inherited by all its subconcepts. This corresponds to the notion of "inheritance" used in the class libraries of object-oriented modeling. For this reason, FCA is suitable for modeling and analyzing object-oriented class libraries; this has been described, for example, by Godin and Mili (1993) and by Huchard and Leblanc (2000).

Conceptual Scaling

A single concept lattice of large sets of objects and attributes can become fairly complex. Visualizations of concept lattices are only of interest if they are not too "messy" to be comprehensible to a human user. The only information that a messy visualization provides is that the underlying lattice is complex. Lindig and Snelting (1997) treat this as a positive case: by showing that the concept lattice of dependencies between different pieces of software code is extremely messy, they provide an argument for not attempting to re-engineer such code. Otherwise, concept lattices should either be human-readable or the information should be displayed in a different format. For example, in some applications, such as Carpineto and Romano's (2004b) Credo engine, concept lattices are mainly used as internal structures that are not displayed as a whole to the user. In Credo, only parts of the lattices are displayed in a manner similar to file/folder displays, where a second level of the hierarchy is indented and can be expanded or collapsed interactively by users. Other applications use fish-eye displays, as described by Carpineto and Romano (1995) and Furnas (1986).

Another approach for reducing the complexity of concept lattices is that of dividing lattices into different components based on groupings of related formal attributes (similar to the notion of "facets" in library science classification theory). These groups are then separately visualized as lattices. For some such groups, it is possible to predict the structure of the formal attributes without even considering the formal objects. For example, the structure of data derived from a survey often represents rank orders ("agree strongly," "agree," "neutral," "disagree," "disagree strongly"). The lattice for such a rank order can be drawn without considering the actual results of the survey. Ganter and Wille (1989) call such lattices "conceptual scales." After drawing a conceptual scale, formal objects can be mapped to their positions on the scale. In this manner, a single scale can be reused for different formal contexts. In the survey example, the same scale could be used for different surveys to compare their results. According to Ganter and Wille (1989), typical conceptual scales are "nominal scales" (cf. Figure 5), "ordinal scales," and "interval scales" (cf. Figure 4). The names of these scales are adopted from traditional statistics. In addition to facilitating reuse, conceptual scales can also be utilized for analyzing dependencies between attributes (Ganter & Wille, 1989).

An interval scale as shown in Figure 13.4 is suitable for formal attributes that have a range of possible values. In this example, the formal objects are trees that are claimed to be the world's oldest trees according to information presented on Web sites. The formal attributes describe the claimed ages. For example, some sites claim that the Japanese Jomonsugi tree is 2,000 year old, whereas others claim that it is 7,000 years old. Therefore, this tree is assigned the formal attributes "at least 2,000" and "at most 7,000." The age range for the North American Methusala is much smaller. It is usually claimed to be about 4,700 years old. Because 4,000 is larger than 2,000, this tree is also assigned the formal attributes "at least 2,000" and "at least 3,000" and so on, but its more precise values are "at least 4,000" and "at most 5,000." The further down a formal object is on an interval scale, the more precise its range of values.

Figure 13.5 shows a second scale for the same set of formal objects as in Figure 13.4. In this case, the formal attributes are the botanical classifications of the trees. All four trees are conifers, but each belongs to a different botanical family. As mentioned earlier, this kind of scale, which is really just a partition of the data into separate classes, is called a "nominal scale." If several scales are applied to the same set of objects, they can be combined in so-called nested line diagrams (Ganter & Wille, 1999b). The nested line diagram in Figure 13.6 shows the scales from Figures 13.5 and 13.4 combined (although in a somewhat simplified

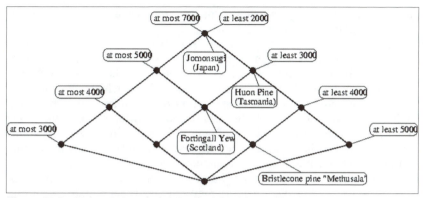

Figure 13.4   An interval scale for the world's oldest trees.

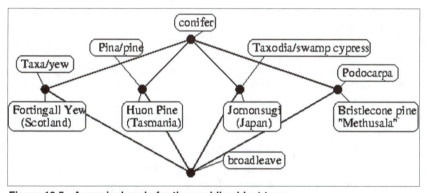

Figure 13.5   A nominal scale for the world's oldest trees.

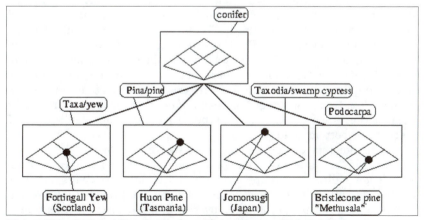

Figure 13.6   A nested line diagram combining Figures 13.4 and 13.5.

manner). The interval scale is used as the inner scale, whereas the nominal scale is used as the outer scale.

The purpose of Toscana software (Kollewe, Skorsky, Vogt, & Wille, 1994) is to facilitate interactive exploration of such nested line diagrams. In Toscana, a user can select any scales that have been prepared and combine them. The user can choose the order in which the scales are combined, specifically, which are the outer or the inner scales. It is possible to zoom in and out of scales and to choose different parameters for the display of the formal objects. The initial setup of a "Toscana system" for a new application is quite labor-intensive because usually only some of the formal attributes can be depicted with standard scales. For qualitative data, new scales may need to be designed for every application. Toscana software has been used in many applications to date, including some large-scale applications that required the manual design of many scales. Examples are the Toscana system for the civil engineering building codes of the North-Rhine Westfalian Ministry (Eschenfelder et al., 2000) mentioned before and the Toscana system for a library catalog (Rock & Wille, 2000).

Toscana systems do not represent the only possible manner for combining scales. A different approach is suggested in the Faceted Information Retrieval (FaIR) system (Priss, 2000), where scales are not nested but presented side by side. The relationship between the scales is explored by selecting elements in one scale that will cause the corresponding elements in other scales to be highlighted.

Implications

It is beyond the scope of this chapter to discuss all FCA techniques that have been used or described for particular applications. But an important area that must be mentioned is the use of "implications." The example in Figure 13.2 shows some dependencies among formal attributes. For example, the formal attribute "mammal" is attached to a node that is above the nodes for "dog" and "cat." That means that the formal attributes "dog" or "cat" imply the formal attribute "mammal." This is, of course, a biological truth. But there are more attribute implications in this example, some of which may hold only for the example but not in general, such as "cartoon" implies "mammal" and "tortoise" implies "real." Implications can also involve combinations of attributes. In Figure 13.2, "cat" and "dog" together imply all other formal attributes because they are contradictory and the paths down from "cat" and "dog" meet in the bottom node.

Implications have been studied since Ganter and Wille (1986). They can be used for a stepwise, computer-guided construction of conceptual knowledge called "attribute exploration" (Ganter & Wille, 1999b). Support for attribute exploration is, for example, provided by ConImp software (Burmeister, 2000) and by the ConExp program mentioned before. Starting with a preliminary formal context, the software prompts a user with a series of questions about the relationships among attributes. In the example from Figure 13.2, the software could ask whether all cartoon animals are mammals. The user must then either agree that this is always true or must provide a counterexample of a cartoon animal that is not a mammal. This counterexample is then added to the formal context. The process stops when all possibly missing implications have been checked. Attribute exploration was studied further by Stumme (1996a, 1996b) and developed into "concept exploration" (Stumme, 1997), which can be used to explore sub-lattices of larger data sets. But, to the best of our knowledge, concept exploration is another example of advanced FCA technology that so far only exists virtually and not as publicly available software.

The Relationship of FCA Notions to Similar Notions in Other Fields

One can argue that "formal concepts" as defined in FCA describe a natural feature of information representation that is as fundamental to hierarchies and object/attribute structures as set theory or relational algebra is to relational databases. The reason for this claim is that the basic structures of FCA have been rediscovered over and over by different researchers and in different settings. Gerard Salton's (1968) document/term lattices are essentially concept lattices in the sense of FCA. Interestingly, however, his discussion of lattices was removed from the second edition of his book because at that time no one could envisage applications for such lattices in information retrieval. Two years later, Barbut and Monjardet (1970) described what they called "Galois Lattices"; these are also equivalent to concept lattices. Shreider from the "Russian School of Taxonomy" (cf. Gusakova & Kuznetsov, 2002) again independently discovered what are, in essence, concept lattices. Feature structure lattices, as used in linguistic componential analysis, are likewise very similar to concept lattices (cf. Dyvik's [1998] work). Last, but not least, Barwise and Seligman's (1997) "classifications" are also concept lattices. The main difference between all of these independent discoveries and FCA is that Wille (1982) had the vision to develop concept lattices into a complex theory with many applications, whereas others noted the structural elegance of both the duality and the lattices, but did not know how to exploit these in applications.

Formal concepts in FCA can be seen as the mathematical formalization of what has been called the "classical theory of concepts" in psychology and philosophy, which states that a concept is formally definable via its features. This theory has been refuted by Wittgenstein (1953), Rosch (1973), and others, but as Medin (1989, p. 1476) states, "despite the overwhelming evidence against the classical view, there is something about it that is intuitively compelling." Even though, from a psychological standpoint, the classical view does not accurately represent human cognition, it nevertheless dominates the design of computerized information systems because it is much easier to implement and manage in an electronic environment. The classical view implicitly underlies many knowledge representation formalisms used in AI and in traditional information retrieval and library systems. Even if non-classical approaches, such as cluster analysis or neural networks, are implemented, the resulting concepts are still sometimes represented in the classical manner.

To avoid confusion with non-classical theories or non-mathematical versions of the classical theory, concepts in FCA are called "formal concepts." Formal concepts are primarily mathematical entities. The criticism against the classical theory of concepts is not relevant as long as FCA is used in a formal domain (such as in software engineering). But if FCA is to be used in domains that are primarily concerned with human cognition, such as psychology or linguistics, the same amount of careful modeling and caution is required for FCA as is required for statistical methods. Novice users often need to be reminded that although many formal concepts in an application may correspond to their own intuitive notions, not all formal concepts need to do so. Linguists might argue that formal concepts are quite different from cognitive processes relating to natural languages. This is why current FCA applications in linguistics focus more on formal structures found in lexica and dictionaries than on cognitive linguistic phenomena. FCA is thus not to be understood as a formal analysis of (human) concepts, but instead as a mathematical method using "formal concepts."

It should be remarked that although FCA has adapted the notions of "extension" and "intension" from philosophy, it uses them in a slightly different way. For example, Frege's (1892) "Sinn" (sense) and "Bedeutung" (reference) do not correspond to extension and intension as defined by FCA. According to Frege's example, "morning star" and "evening star" both have the same referent (Venus), but a different sense ("seen in the morning" vs. "seen in the evening"). In FCA, an extension occurring with respect to one formal context can have only one corresponding intension, not two different intensions. To resolve this problem, FCA would model Frege's example either by defining the intension of Venus as the set of "morning star" and "evening star" or by stating that morning star and evening star do not belong to the same formal context. According to FCA, Frege's notions might be better translated as "denotation" for "Bedeutung" and "connotation" for "Sinn" (cf. Priss [1998] for a more detailed discussion). Philosophers might disagree with these definitions but, in general, mathematical formalizations as achieved by FCA only approximate non-formal notions held in non-mathematical disciplines. The advantage of formalizations, however, is that notions are defined with absolute precision within the formal realm and therefore may be implementable in software.

FCA in Information Retrieval

There was some interest in the use of lattices for information retrieval by Salton (1968) and others with respect to document/term lattices and lattices of Boolean query combinations (cf. Priss [2000] for a summary of these early attempts). But none of these resulted in practical implementations, and for a long time the dominant mathematical model of information retrieval was the vector space model, which excluded a lattice approach. Interest in lattices was again spurred by Godin et al. (1989), who developed an information retrieval system based on document/term lattices. Godin's system was text-based without graphical representations of the lattices, but that was due to the hardware limitations of that time: He did discuss fish-eye (Furnas, 1986) and other techniques suitable for visualizations. Godin's group compared information retrieval based on concept lattices to Boolean queries and to navigation in hierarchical classifications (Godin, Missaoui, & April, 1993) and concluded that Boolean queries and lattice navigation performed similarly and that both were better than the use of hierarchical classification. Later, their interest shifted to retrieval of software components, which differs from general information retrieval because the search space is delimited by the formal nature of programming languages. Mili et al. (1997) discovered that the use of faceted classification is not advisable for software component retrieval because the cost of developing such classifications outweighs any benefits. Furthermore, controlled vocabularies (as used in a faceted classification) may be too restrictive with respect to programming languages. Although these results are interesting, they should probably be treated with caution because, at least to our knowledge, no one outside of Godin's group has tried to replicate any of these experiments.

Carpineto and Romano's (1993) research was initially influenced by Godin's work but since then has been independently advanced to a high level. Their Credo engine (Carpineto & Romano, 2004b) facilitates a lattice-based, metasearch of Google results. An overview of their work and FCA applications in information retrieval in general can be found in Carpineto and Romano (2004b). In that paper, Carpineto and Romano argue that FCA can serve three purposes in information retrieval. First, FCA can support query refinement. Because a document/term lattice structures a search space into clusters of related documents, lattices can be used to make suggestions for query enlargement in cases where too few documents are retrieved and for query refinement in cases where too many documents are retrieved. Second, lattices can support the integration of querying and navigation (or browsing). An initial query identifies a start node in a document/term lattice. Users can then navigate to related nodes. Further queries are used to "prune" a document/term lattice so as to help users focus their search (Carpineto & Romano, 1996a). Third, a thesaurus hierarchy can be integrated with a concept lattice—an idea that has been independently discussed by various researchers (e.g., Carpineto & Romano, 1996b; Priss, 1997; Skorsky, 1997) but probably still leaves room for further research.

Apart from Credo, a second FCA application that has attained professional quality is Mail-Sleuth (Eklund, Ducrou, & Brawn, 2004). This software is marketed by an Australian company and consists of a plugin for MS Outlook e-mail software, which can be used to mine large e-mail archives. This software was developed using earlier research on retrieval of information from semi-structured texts (Cole & Eklund, 2001; Cole & Stumme, 2000).

In general, FCA software appears to show promise for applications in information retrieval, albeit with some limitations. Like latent semantic analysis (LSA) (Dumais, 2004), FCA is not suited to direct manipulation of very large data sources. It is difficult to give precise upper limits because these depend on the application. It also matters whether both the object and attribute sets are of similar size. FCA has been applied to thousands of documents in a small library (Rock & Wille, 2000). However, it would presumably not be possible to apply FCA (or LSA for that matter) directly to the complete Google database because of the computational limitations involving large matrices. On the other hand, either method can be applied as a secondary tool to reorganize a set of documents resulting from a Google query (as demonstrated by the Credo engine). Because both FCA and LSA employ matrices (although LSA matrices are "many-valued contexts" in FCA terminology), it would be interesting to compare both methods more closely, something that, to our knowledge, has not yet been done. Further research should also be conducted with regard to the usability of FCA software. FCA technology claims to be human-centered due to its philosophical basis, but only a few practical usability studies exist (such as Eklund et al., 2004). It is to be hoped that Credo will be extensively tested for usability.

FCA as a Tool for Knowledge Representation and Knowledge Discovery

FCA provides a contrast to some of the traditional, statistical means of data analysis and knowledge representation because of its focus on human-centered approaches. In his first paper on FCA, Wille (1982) explains that he was influenced by H. von Hentig's (1972) concerns about the status of sciences in the modern world. Von Hentig's idea was "to 'restructure' theoretical developments in order to integrate, rationalize origins, connections, interpretations, and applications" (Wille, 1982, p. 447). FCA began as an attempt at restructuring mathematical lattice theory in a manner that both facilitates communication about mathematical theory to a wider non-mathematical audience and facilitates exploitation of mathematical theory for a wide range of applications. The concept lattices of FCA serve as a means for communication, exploration, and discussion, which is consistent with both Habermas's

Theory of Communicative Action and Peirce's pragmatism (cf. Wille, 1997b).

Because they facilitate discussion and exploration of conceptual structures, concept lattices can be characterized as a means of external cognition in the sense of Scaife and Rogers (1996). The use of diagrams for reasoning has been formally investigated by Dau (2004). He observes that mathematicians often include diagrams in their descriptions of mathematical facts, but that normally such diagrams themselves are not permissible as arguments. By formally distinguishing between a mathematical structure and its diagrammatic representation, Dau provides a framework in which diagrams can be used for formal reasoning. Thus, in addition to an intuitive notion of the importance of visualizations, such as concept lattices, Dau can even formally evaluate their usefulness within a formal framework itself.

FCA has been examined with respect to principles of knowledge representation. Wille (1997b) identifies 10 functions of knowledge processing (exploring, searching, recognizing, identifying, analyzing, investigating, deciding, improving, restructuring, and memorizing) and investigates how these are supported by FCA. Stumme (2002) analyses FCA with respect to Davis, Shrobe, and Szolovits's (1993) five principles of knowledge representation: Knowledge representations as a medium of human expression, a set of ontological commitments, a surrogate, a fragmentary theory of intelligent reasoning, and a medium for pragmatically efficient computation.

Conceptual Knowledge Discovery (Hereth, Stumme, Wille, & Wille, 2000; Stumme, Wille, & Wille, 1998) is mainly supported by Toscana systems (Kollewe et al., 1994), as has been described. In contrast to statistical software, which attempts to provide probable answers to narrow questions, Toscana systems facilitate browsing and interactive exploration of implicit and explicit structures. Because the preparation of data for input into a Toscana system is labor-intensive and requires substantial knowledge of FCA, Toscana systems are usually compiled by an FCA expert in cooperation with a domain expert. Wille (2001) argues that this is an advantage because the processes involved in creating a conceptual representation (in the sense of FCA) encourages the discovery of implicit information and facilitates the conversion of information into knowledge. Nevertheless, the effort required for setting up Toscana systems may be a reason why their use is not more widespread. It should be emphasized, however, that only the preparation of a Toscana system requires expertise. End-users can use such a system after reading a brief introduction. A side effect of the careful setup of a Toscana system is that it can be less error prone than some statistical methods because careful conceptual modeling prevents data misrepresentation.

We believe that there are further research opportunities for FCA in the area of knowledge discovery that have not yet been exploited. To date, two workshops have been held on the topic of "concept lattices and knowledge discovery in databases." But many of the papers at these workshops focused on algorithmic issues (such as Kuznetsov & Obiedkov, 2002) or abstract issues (such as Wille, 2001). These issues are important, but it would be interesting to see more realistic applications. Furthermore, it might be useful to compare "relational scaling," as described by Prediger and Stumme (1999), to methods employed in business intelligence and data warehousing, especially as they appear to pursue similar goals.

Applications of FCA in Logic and Artificial Intelligence

Since about 1996, attempts have been made to combine FCA with other formalisms, such as Sowa's (1984) Conceptual Graphs. Wille (1997a) describes a translation of Conceptual Graphs into formal contexts and concept lattices. Mineau et al. (1999) investigate the commonalities between both theories at a more general level. Conceptual Graphs (Sowa, 1984) are a formalism for knowledge representation that is similar to semantic networks, entity relationship diagrams, and the Semantic Web standard Resource Description Framework (RDF). Sowa developed his Conceptual Graphs on the basis of Peirce's Existential Graphs—a graphical, symbolic notation for reasoning that incorporates aspects of context and modal logic. In contrast to the hierarchical relations that are expressed in concept lattices, Conceptual Graphs can be used to represent semantic relations, such as part/whole, and linguistic case relations, such as Agent (or subject of a sentence), Patient (or direct object of a sentence), and so on. Natural language sentences can be translated more or less directly into Conceptual Graphs. A connection with FCA is established via types: Each concept of a Conceptual Graph contains information about its type. The types form a hierarchy that can be modeled with FCA. According to Wille (1997a), a combination of FCA and Conceptual Graphs can facilitate a formalization of elementary logic and thus presents a powerful formalism for the representation and analysis of human reasoning and argumentation.

Apart from Conceptual Graphs, connections have been established between FCA and Description Logics (Prediger & Stumme, 1999). In contrast to Conceptual Graphs and FCA, which primarily focus on representations, Description Logics investigate the expressivity and computability of logical representations. For example, Description Logics can check whether a concept subsumes another concept (with respect to a knowledge base) or whether an instance belongs to a concept.

A combination of FCA and Conceptual Graphs is not just another ontology formalism because of its philosophical foundation. Wille (2000b) perceives logic (and human reasoning and argumentation) as a Kantian triad of concepts, judgments, and conclusions. The goal is to use FCA to achieve a mathematization of these three philosophical concepts in a framework of "Contextual Logic" (Prediger, 1998). Whereas FCA is used to mathematize concepts, Conceptual Graphs are used to mathematize judgments. A combination of Conceptual Graphs, FCA, and

Description Logics can then be used to mathematize conclusions. Wille sees this as a continuation of Boole's logic of signs and classes (Wille, 2000a, 2004). But in contrast to Boole (1854/1958), who envisioned a "Universal" set or class, FCA focuses on formal contexts, which are finite in most applications (Ganter & Wille, 1999a) and so avoid some of the confusion caused by the assumption of universality.

A challenge for this kind of mathematization of logic is the treatment of negation (cf. Dau, 2000; Kwuida, Tepavcevic, & Seselja, 2004; Wille, 2000a). For example, what is the meaning of a negated concept, such as "not a piano," or of a negated attribute, such as "not green"? Intuitively, one might in both cases construct some hypothetical superconcept and then negate with respect to that concept. "Not a piano" might be other musical instruments that are not pianos. "Not green" might be other colors. Without such a superconcept, negation would be too ambiguous. For example, one would probably not accept "information science" in the extension of a concept called "not a piano."

With respect to concept lattices, it can occur that negations of formal concepts are not materialized as formal concepts themselves. This is due to the fact that a formal context does not normally explicitly specify negation. If an object does not have an attribute assigned, it can mean that the attribute is irrelevant; the relationship is unknown; or the object does sometimes, usually, or never have the attribute (Burmeister & Holzer, 2000). Therefore, adding negation to a concept lattice requires considering so-called protoconcepts or semi-concepts (Wille 2000a), which are mathematical structures similar to formal concepts, but whose mathematical properties are more complex and more difficult to describe. The study of the structures arising from such proto- or semi-concepts is ongoing (e.g., Hereth Correia & Klinger, 2004). Other aspects of Contextual Logic pertain to implicit knowledge, incorporation of background knowledge (Hereth Correia & Klinger, 2004), and the incorporation of existential quantifiers (Wille, 2002).

It remains to be seen how much of the framework of Contextual Logic will eventually lead to implementations in computational ontology software or whether the focus of this research will remain on mathematical structures and philosophical implications. Some suggestions for implementations were made by Groh and Eklund (1999) and Eklund et al. (2000), but as far as we know, those suggestions have not yet been instantiated. In any case, this research provides an interesting contrast to the commercially driven developments occurring within the Semantic Web community.

Conclusion

An *ARIST* chapter can provide only a glimpse of FCA's bases and potential. FCA provides a large set of methods, of which a fairly small, but powerful, subset has so far been implemented in software that is readily available. Applications demonstrate the utility of such software.

It is hoped that this overview will stir more interest in the topic among information scientists.

References

Barbut M., & Monjardet B. (1970). *Ordre et classification: Algèbre et combinatoire* [Order and classification: Algebra and combinatories].Tome II, Paris: Hachette.

Barwise, J., & Seligman, J. (1997). *Information flow. The logic of distributed systems.* Cambridge, UK: Cambridge University Press.

Becker, P., & Hereth Correia, J. (in press). The ToscanaJ Suite for implementing conceptual information systems. In G. Stumme (Ed.), *Formal concept analysis: State of the art, Proceedings of the First International Conference on Formal Concept Analysis.* Berlin: Springer.

Boole, G. (1854/1958). *An investigation of the laws of thought, on which are founded the mathematical theories of logic and probabilities.* New York: Dover.

Burmeister, P. (2000). ConImp: Ein Programm zur formalen Begriffsanalyse [A program for formal term analysis]. In G. Stumme & R. Wille (Eds.), *Begriffliche Wissensverarbeitung. Methoden und Anwendungen* [Conceptual knowledge processing: Methods and applications] (pp. 25–56). Berlin: Springer.

Burmeister, P., & Holzer, R. (2000). On the treatment of incomplete knowledge in formal concept analysis. In B. Ganter & G. Mineau (Eds.), *Conceptual structures: Logical, linguistic and computational issues* (Lecture Notes in Artificial Intelligence No.1867, pp. 385–398). Berlin: Springer.

Carpineto, C., & Romano, G. (1993). GALOIS: An order-theoretic approach to conceptual clustering. *Proceedings of the 10th Conference on Machine Learning,* 33–40.

Carpineto, C., & Romano, G. (1995). Ulysses: A lattice-based multiple interaction strategy retrieval interface. In B. Blumenthal, J. Gornostaev, & C. Unger (Eds.), *Proceedings of the Fifth East–West International Conference on Human–Computer Interaction* (Lecture Notes in Computer Science No. 1015, pp. 91–104). Berlin: Springer.

Carpineto, C., & Romano, G. (1996a). Information retrieval through hybrid navigation of lattice representations. *International Journal of Human–Computer Studies, 45*(5), 553–578.

Carpineto, C., & Romano, G. (1996b). A lattice conceptual clustering system and its application to browsing retrieval. *Machine Learning, 24*(2), 1–28.

Carpineto, C., & Romano, G. (2000). Order-theoretical ranking. *Journal of the American Society for Information Science, 51,* 587–601.

Carpineto, C., & Romano, G. (2004a). *Concept data analysis: Theory and applications.* New York: Wiley.

Carpineto, C., & Romano, G. (2004b). Exploiting the potential of concept lattices for information retrieval with CREDO. *Journal of Universal Computing, 10,* 985–1013.

Carpineto, C., & Romano, G. (in press). Using concept lattices for text retrieval and mining. In G. Stumme (Ed.), *Formal concept analysis: State of the art, Proceedings of the First International Conference on Formal Concept Analysis.* Berlin: Springer.

Cole, R., & Eklund, P. (2001). Browsing semi-structured Web texts using formal concept analysis. In H. Delugach & G. Stumme (Eds.), *Conceptual structures: Broadening the base* (Lecture Notes in Artificial Intelligence No. 2120, pp. 319–332). Berlin: Springer.

Cole, R., & Stumme, G. (2000). CEM: A conceptual email manager. In B. Ganter, & G. Mineau (Eds.), *Conceptual structures: Logical, linguistic and computational issues* (Lecture Notes in Artificial Intelligence No. 1867, pp. 438–452). Berlin: Springer.

Dau, F. (2000). Negations in simple concept graphs. In B. Ganter & G. Mineau (Eds.), *Conceptual structures: Logical, linguistic and computational issues* (Lecture Notes in Artificial Intelligence No. 1867, pp. 263–276). Berlin: Springer.

Dau, F. (2004). Types and tokens for logic with diagrams. In K. E. Wolff, H. Pfeiffer, & H. Delugach (Eds.), *Conceptual Structures at Work: 12th International Conference on Conceptual Structures* (pp. 62–93). Berlin: Springer.

Davis, R., Shrobe, H., & Szolovits, P. (1993). What is a knowledge representation? *AI Magazine, 14*(1), 17–33.

Dumais, S. (2004). Latent semantic analysis. *Annual Review of Information Science and Technology, 38*, 189–230.

Dyvik, H. (1998). A translational basis for semantics. In H. Johansson & S. Oksefjell (Eds.), *Corpora and crosslinguistic research: Theory, method and case studies* (pp. 51–86). Amsterdam: Rodopi.

Eisenbarth, T., Koschke, R., & Simon, D. (2001). Feature-driven program understanding using concept analysis of execution trace. *Proceedings of the Ninth International Workshop on Program Comprehension. International Conference on Software Maintenance*, 300–309.

Eklund, P., Ducrou, J., & Brawn, P. (2004). Concept lattices for information visualization: Can novices read line diagrams? In P. Eklund (Ed.), *Concept Lattices: Second International Conference on Formal Concept Analysis* (Lecture Notes in Computer Science No. 2961, pp. 14–27). Berlin: Springer.

Eklund, P., Groh, B., Stumme, G., & Wille, R. (2000). A contextual-logic extension of TOSCANA. In B. Ganter & G. Mineau (Eds.), *Conceptual structures: Logical, linguistic and computational issues*. (Lecture Notes in Artificial Intelligence No. 1867, pp. 453–467). Berlin: Springer.

Eschenfelder, D., Kollewe W., Skorsky, M., & Wille, R. (2000). Ein Erkundungssystem zum Baurecht: Methoden der Entwicklung eines TOSCANA-Systems [A system for conceptual exploration of civil engineering laws: Methods for developing a TOSCANA system]. In G. Stumme & R. Wille (Eds.), *Begriffliche Wissensverarbeitung. Methoden und Anwendungen* [Conceptual knowledge processing: Methods and applications] (pp. 254–272). Berlin: Springer.

Ferre, S., & King, R. (2004). BLID: An application of logical information systems to bioinformatics. In P. Eklund (Ed.), *Concept lattices: Second International Conference on Formal Concept Analysis* (Lecture Notes in Computer Science No. 2961, pp. 47–54). Berlin: Springer.

Fischer, B. (1998). Specification-based browsing of software component libraries. *Proceedings of the 13th IEEE International Conference on Automated Software Engineering*, 246–254.

Freeman, L. C., & White, D. R. (1993). Using Galois lattices to represent network data. In P. Marsden (Ed.), *Sociological methodology* (pp. 127–146). Cambridge, MA: Blackwell.

Frege, G. (1892). Über Sinn und Bedeutung [On sense and meaning]. *Zeitschrift für Philosophie und philosophische Kritik, 100*, 25–50.

Furnas, G. W. (1986). Generalized fisheye views. *Proceedings of the SIGCHI Conference on Human Factors in Computing Systems*, 16–23.

Ganter, B., & Wille, R. (1986). Implikationen und Abhängigkeiten zwischen Merkmalen [Implications and dependencies between attributes]. In P. O. Degens, H.-J. Hermes, & O. Opitz (Eds.), *Die Klassifikation und ihr Umfeld* [Classification and its environment] (pp. 171–185). Frankfurt, Germany: Indeks Verlag.

Ganter, B., & Wille, R. (1989). Conceptual scaling. In F. Roberts (Ed.), *Applications of combinatorics and graph theory to the biological and social sciences* (pp. 139–167). Berlin: Springer.

Ganter, B., & Wille, R. (1999a). Contextual attribute logic. In W. Tepfenhart & W. Cyre (Eds.), *Conceptual Structures: Standards and Practices. Proceedings of the 7th International Conference on Conceptual Structures* (Lecture Notes in Artificial Intelligence No. 1640, pp. 377–388). Berlin: Springer.

Ganter, B., & Wille, R. (1999b). *Formal concept analysis. Mathematical foundations.* Berlin: Springer.

Godin, R., Gecsei, J., & Pichet, C. (1989). Design of browsing interface for information retrieval. *Proceedings of the 12th Annual International ACM SIGIR Conference on Research and Development in Information Retrieval,* 32–39.

Godin, R., & Mili, H. (1993). Building and maintaining analysis-level class hierarchies using Galois lattices. *Proceedings of the Eighth Annual Conference on Object-Oriented Programming Systems, Languages, and Applications,* 394–410.

Godin, R., Missaoui, R., & April, A. (1993). Experimental comparison of navigation in a Galois lattice with conventional information retrieval methods. *International Journal of Man-Machine Studies, 38,* 747–767.

Groh, B., & Eklund, P. (1999). Algorithms for creating relational power context families from conceptual graphs. In W. Tepfenhart & W. Cyre (Eds.), *Conceptual structures: Standards and practices. Proceedings of the 7th International Conference on Conceptual Structures* (Lecture Notes in Artificial Intelligence No. 1640, pp. 389–400). Berlin: Springer.

Gusakova, S. M., & Kuznetsov, S. O. (2002). On Moscow St. Petersburg taxonomy schools. In G. Angelova, D. Corbett, & U. Priss (Eds.), *Foundations and applications of conceptual structures contributions to ICCS 2002* (pp. 92–101). Sofia: Bulgarian Academy of Sciences.

Hara, N. (2002). Analysis of computer-mediated communication using formal concept analysis as a visualizing methodology. *Journal of Educational Computing Research, 26,* 25–49.

Hereth, J., Stumme, G., Wille, R., & Wille, U. (2000). Conceptual knowledge discovery in data analysis. In B. Ganter & G. Mineau (Eds.), *Conceptual structures: Logical, linguistic and computational issues* (Lecture Notes in Artificial Intelligence No. 1867, pp. 421–437). Berlin: Springer.

Hereth Correia, J., & Klinger, J. (2004). Protoconcept graphs: The lattice of conceptual contents. In P. Eklund (Ed.), *Concept lattices: Second International Conference on Formal Concept Analysis* (Lecture Notes in Computer Science No. 2961, pp. 14–27). Berlin: Springer.

Huchard, M., & Leblanc, H. (2000). Computing interfaces in Java. *IEEE International Conference on Automated Software Engineering,* 317–320.

Kalfoglou, Y., Dasmahapatra, S., & Chen-Burger, Y. (2004). FCA in knowledge technologies: Experiences and opportunities. In P. Eklund (Ed.), *Concept lattices: Second International Conference on Formal Concept Analysis* (Lecture Notes in Computer Science No. 2961, pp. 252–260). Berlin: Springer.

Kollewe, W., Skorsky, M., Vogt, F., & Wille, R. (1994). TOSCANA: Ein Werkzeug zur begrifflichen Analyse und Erkundung von Daten [TOSCANA: A tool for the conceptual analysis and investigation of data]. In R. Wille & M. Zickwolff (Eds.), *Begriffliche Wissensverarbeitung: Grundfragen und Aufgaben* [Conceptual knowledge processing: Basic questions and tasks] (pp. 267–288). Mannheim, Germany: B.I.-Wissenschaftsverlag.

Kuznetsov, S. (2004). Machine learning and formal concept analysis. In P. Eklund (Ed.), *Concept lattices: Second International Conference on Formal Concept Analysis* (Lecture Notes in Computer Science No. 2961, pp. 287–312). Berlin: Springer.

Kuznetsov, S., & Obiedkov, S (2002). Comparing performance of algorithms for generating concept lattices. *Journal of Experimental and Theoretical Artificial Intelligence, 14,* 189–216.

Kwuida, L., Tepavcevic, A., & Seselja, B. (2004). Negation in contextual logic. In K. E. Wolff, H. Pfeiffer, & H. Delugach (Eds.), *Conceptual structures at work: 12th International Conference on Conceptual Structures* (pp. 227–241). Berlin: Springer.

Lindig, C., & Snelting, G. (1997). Assessing modular structure of legacy code based on mathematical concept analysis. *Proceedings of the 19th International Conference on Software Engineering,* 349–359.

Medin, D. L. (1989). Concepts and conceptual structure. *American Psychologist, 44,* 1469–1481.

Mili, H., Ah-Ki, E., Godin, R., & Mcheick, H. (1997). Another nail to the coffin of faceted controlled-vocabulary component classification and retrieval. *ACM SIGSOFT Software Engineering Notes, 22*(3), 89–98.

Mineau, G., Stumme, G., & Wille, R (1999). Conceptual structures represented by conceptual graphs and formal concept analysis. In W. Tepfenhart & W. Cyre (Eds.), *Conceptual structures: Standards and practices. Proceedings of the 7th International Conference on Conceptual Structures,* (Lecture Notes in Artificial Intelligence No. 1640, pp. 423–441). Berlin: Springer.

Prediger, S. (1998). *Kontextuelle Urteilslogik mit Begriffsgraphen. Ein Beitrag zur Restrukturierung der mathematischen Logik* [Contextual judgment logic with concept graphs: A contribution to the restructuring of mathematical logic]. Aachen, Germany: Shaker.

Prediger, S., & Stumme, G. (1999). Theory-driven logical scaling: Conceptual information systems meet description logics. In P. Lambrix, A. Borgida, M. Lenzerini, R. Muller, & P. Patel-Schneider (Eds.), *International Workshop on Description Logics Central European Workshop, 22.* Retrieved December 20, 2004, from http://sunsite.informatik.rwth-aachen.de/Publications/CEUR-WS

Priss, U. (1997). A graphical interface for document retrieval based on formal concept analysis. In E. Santos (Ed.), *Proceedings of the 8th Midwest Artificial Intelligence and Cognitive Science Conference* (AAAI Technical Report CF-97-01, pp. 66–70). Menlo Park, CA: American Association for Artificial Intelligence.

Priss, U. (1998). *Relational concept analysis: Semantic structures in dictionaries and lexical databases.* Aachen, Germany: Shaker.

Priss, U. (2000). Lattice-based information retrieval. *Knowledge Organization, 27,* 132–142.

Priss, U. (in press). Linguistic applications of formal concept analysis. In G. Stumme, (Ed.), *Formal concept analysis: State of the art, Proceedings of the First International Conference on Formal Concept Analysis.* Berlin: Springer.

Rock, T., & Wille, R. (2000). Ein TOSCANA-Erkundungssystem zur Literatursuche [A TOSCANA exploration system for literature searching]. In G. Stumme & R. Wille (Eds.), *Begriffliche Wissensverarbeitung. Methoden und Anwendungen* [Conceptual knowledge processing: Methods and applications] (pp. 239–253). Berlin: Springer.

Rosch, E. (1973). Natural categories. *Cognitive Psychology, 4*, 328–350.

Salton, G. (1968). *Automatic information organization and retrieval.* New York: McGraw-Hill.

Scaife, M., & Rogers, Y. (1996). External cognition: How do graphical representations work? *International Journal of Human–Computer Studies, 45*, 185–213.

Skorsky, M. (1997). *Graphische Darstellung eines Thesaurus* [Diagrammatic representation of thesauri]. Regensburg, Germany: Deutscher Dokumentartag.

Snelting, G. (in press). Concept lattices in software analysis. In G. Stumme (Ed.), *Formal concept analysis: State of the art, Proceedings of the First International Conference on Formal Concept Analysis.* Berlin: Springer.

Soergel, D. (1967). Mathematical analysis of documentation systems. *Information Storage and Retrieval, 3*, 129–173.

Sowa, J. (1984). *Conceptual structures: Information processing in mind and machine.* Reading, MA: Addison-Wesley.

Spangenberg, N., & Wolff, K. E. (1993). Datenreduktion durch die Formale Begriffsanalyse von Repertory Grids [Data reduction of repertory grids by means of formal concept analysis]. In J. W. Scheer & A. Catina (Eds.), *Einführung in die Repertory Grid Technik. Klinische Forschung und Praxis* [Introduction to repertory grid technology: Clinical research and practice] (Vol. 2, pp. 38–54). Göttingen, Germany: Hans Huber.

Stumme, G. (1996a). Attribute exploration with background implications and exceptions. In H.-H. Bock & W. Polasek (Eds.), *Data analysis and information systems: Statistical and conceptual approaches* (Studies in Classification, Data Analysis, and Knowledge Organization No. 7, pp. 457–469). Berlin: Springer.

Stumme, G. (1996b). Exploration tools in formal concept analysis. In E. Diday, Y. Lechevallier, & O. Opitz (Eds.), *Ordinal and symbolic data analysis: Proceedings of the International Conference on Ordinal and Symbolic Data Analysis* (Studies in Classification, Data Analysis, and Knowledge Organization No. 8, pp. 31–44). Berlin: Springer.

Stumme, G. (1997). Concept exploration: A tool for creating and exploring conceptual hierarchies. In D. Lukose, H. Delugach, M. Keeler, L. Searle, & J. F. Sowa (Eds.), *Conceptual structures: Fulfilling Peirce's dream. Proceedings of the International Conference on Computer Simulation* (Lecture Notes in Artificial Intelligence No. 1257, pp. 318–331). Berlin: Springer.

Stumme, G. (2002). Formal concept analysis on its way from mathematics to computer science. In U. Priss, D. Corbett, & G. Angelova (Eds.), *Conceptual structures: Integration and interfaces, 10th International Conference on Conceptual Structures,* (Lecture Notes in Computer Science No. 2393, pp. 2–19). Berlin: Springer.

Stumme, G., Wille, R., & Wille U. (1998). Conceptual knowledge discovery in databases using formal concept analysis methods. In J. M. Zytkow & M. Quafofou (Eds.), *Principles of data mining and knowledge discovery* (Lecture Notes in Artificial Intelligence No. 1510, pp. 450–458). Berlin: Springer.

Tilley, T. (2004). Tool support for FCA. In P. Eklund (Ed.), *Concept lattices: Second International Conference on Formal Concept Analysis* (Lecture Notes in Computer Science No. 2961, pp. 104–111). Berlin: Springer.

Valtchev, P., Missaoui, R., & Godin, R. (2004). Formal concept analysis for knowledge discovery and data mining: The new challenges. In P. Eklund (Ed.), *Concept lattices: Second International Conference on Formal Concept Analysis* (Lecture Notes in Computer Science No. 2961, pp. 352–371). Berlin: Springer.

von Hentig, H. (1972). *Magier oder Magister? Über die Einheit der Wissenschaft im Verständigungsprozess* [Magician or master? On the unity of science in the communication process]. Stuttgart, Germany: Klett.

Wille, R. (1982). Restructuring lattice theory: An approach based on hierarchies of concepts. In I. Rival (Ed.), *Ordered sets* (pp. 445–470). Dordrecht, Netherlands: Reidel.

Wille, R. (1997a). Conceptual graphs and formal concept analysis. In D. Lukose, H. Delugach, M. Keeler, L. Searle, & J. F. Sowa (Eds.), *Conceptual structures: Fulfilling Peirce's dream. Proceedings of the International Conference on Computer Simulation* (Lecture Notes in Artificial Intelligence No. 1257, pp. 290–303). Berlin: Springer.

Wille, R. (1997b). Conceptual landscapes of knowledge: A pragmatic paradigm for knowledge processing. In W. Gaul & H. Locarek-Junge (Eds.), *Classification in the Information Age: Proceedings of the 22nd Annual Conference of the Gesellschaft für Klassifikation* (pp. 344–356). Berlin: Springer.

Wille, R. (1997c). Introduction to formal concept analysis. In G. Negrini (Ed.), *Modelli e modellizzazione* [Models and modelling] (pp. 39–51). Rome: Consiglio Nazionale delle Ricerche, Instituto di Studi sulli Ricerca e Documentazione Scientifica.

Wille, R. (2000a). Boolean concept logic. In B. Ganter, & G. Mineau (Eds.), *Conceptual Structures: Logical, Linguistic and Computational Issues* (Lecture Notes in Artificial Intelligence No. 1867, pp. 317–331). Berlin: Springer.

Wille, R. (2000b). Contextual logic summary. In G. Stumme (Ed.), *Working with conceptual structures: Contributions to ICCS 2000* (pp. 265–276). Aachen, Germany: Shaker.

Wille, R. (2001). Why can concept lattices support knowledge discovery in databases? In E. Mephu Nguifo (Ed.), *International Workshop on Concept Lattice-Based Theory, Methods and Tools for Knowledge Discovery in Databases*, 7–20.

Wille, R. (2002). Existential concept graphs of power context families. In U. Priss, D. Corbett, & G. Angelova (Eds.), *Conceptual Structures: Integration and Interfaces, 10th International Conference on Conceptual Structures*, (Lecture Notes in Computer Science No. 2393, pp. 382–395). Berlin: Springer.

Wille, R. (2004). Preconcept algebras and generalized double Boolean algebras. In P. Eklund (Ed.), *Concept lattices: Second International Conference on Formal Concept Analysis*, (Lecture Notes in Computer Science No. 2961, pp. 1–13). Berlin: Springer.

Wittgenstein, L. (1953). *Philosophical investigations* (G. E. M. Anscombe, trans.). Oxford, UK: Basil Blackwell.

Wolff, K. E. (1994). A first course in formal concept analysis: How to understand line diagrams. In F. Faulbaum (Ed.), *SoftStat'93, Advances in statistical software* (Vol. 4, pp. 429–438). Jena, Germany: Gustav Fischer Verlag.

Wolff, K. E. (2004). Particles and waves as understood by temporal concept analysis. In K. E. Wolff, H. Pfeiffer, & H. Delugach (Eds.), *Conceptual structures at work: 12th International Conference on Conceptual Structures* (pp. 126–141). Berlin: Springer.

Index

A

Abbate, J., 70, 457
Abbott, R., 310
Abe, N., 195
Abel, M., 343
Abelson, R. P., 165
Abler, R., 64
Abu-Salem, H., 200
academic communities, collaborative
information seeking in, 330–333
academic journals, access to, *see* open
access to scholarly journals
access to information
and economic decision making, 27
selection of information and, 296
through Internet, economic potential
and, 55
access to Internet in rural locations, 65
ACM (Association for Computing
Machinery), 522
ACSys (Cooperative Research Centre for
Advanced Computational
Systems), 139
Adamczyk, P. D., 344

Adams, P., 57, 62–63, 67
Adderley, R., 244
adolescents, information behavior of,
307–308
Adorno, T. W., 29
Advanced Knowledge Technologies col-
laboration, 522
Advanced Research and Development
Activity (ARDA), 114
AEA (American Economic Association),
10
Agar, J., 453–454, 457
agents, economic, 26–29, 31
Agichtein, E., 245
Agosto, D. E., 308
Agrawal, R., 240
Ah-Ki, E., 522
Ahn, W.-K., 163
Aho, J. A., 450
AI (artificial intelligence), formal concept
analysis and, 521, 531, 536–537
Aikine, J. (2002)., 449
Aitchison, J., 181–182
Akerlof, G. A., 11, 28
Akerman, C., 458

M

S

W

X

Further Reading in
Information Science & Technology

Theories of Information Behavior

Edited by Karen E. Fisher, Sanda Erdelez, and Lynne (E. F.) McKechnie

This unique book presents authoritative overviews of more than 70 conceptual frameworks for understanding how people seek, manage, share, and use information in different contexts. Covering both established and newly proposed theories of information behavior, the book includes contributions from 85 scholars from 10 countries. Theory descriptions cover origins, propositions, methodological implications, usage, and links to related theories.

456 pp/hardbound/ISBN 1-57387-230-X
ASIST Members $39.60 • Nonmembers $49.50

Covert and Overt

Recollecting and Connecting Intelligence Service and Information Science

Edited by Robert V. Williams and Ben-Ami Lipetz

This book explores the historical relationships between covert intelligence work and information/computer science. It first examines the pivotal strides to utilize technology to gather and disseminate government/military intelligence during WWII. Next, it traces the evolution of the relationship between spymasters, computers, and systems developers through the years of the Cold War.

276 pp/hardbound/ISBN 1-57387-234-2
ASIST Members $39.60 • Nonmembers $49.50

ASIS&T Thesaurus of Information Science, Technology, and Librarianship, Third Edition

Edited by Alice Redmond-Neal
and Marjorie M. K. Hlava

The *ASIST Thesaurus* is the authoritative reference to the terminology of information science, technology, and librarianship. This updated third edition is an essential resource for indexers, researchers, scholars, students, and practitioners. An optional CD-ROM includes the complete contents of the print thesaurus along with Data Harmony's Thesaurus Master software. In addition to powerful search and display features, the CD-ROM allows users to add, change, and delete terms, and to learn the basics of thesaurus construction while exploring the vocabulary of library and information science and technology.

Book with CD-ROM: 272pp/softbound/ISBN 1-57387-244-X
ASIST members $63.95 • Nonmembers $79.95

Book only: 272pp/softbound/ISBN 1-57387-243-1
ASIST members $39.95 • Nonmembers $49.95

Understanding and Communicating Social Informatics

A Framework for Studying and Teaching the Human Contexts of Information and Communication Technologies

By Rob Kling, Howard Rosenbaum, and Steve Sawyer

Here is a sustained investigation into the human contexts of information and communication technologies (ICTs), covering both research and theory. The authors demonstrate that the design, adoption, and use of ICTs are deeply connected to people's actions as well as to the environments in which ICTs are used. They offer a pragmatic overview of social informatics, articulating its fundamental ideas for specific audiences and presenting important research findings.

240 pp/hardbound/ISBN 1-57387-228-8/$39.50

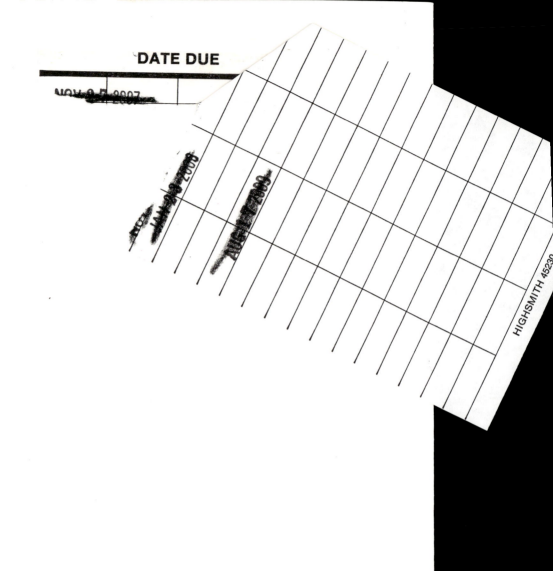